THE ATLAS OF INSECT AND PLANT VIRUSES

ULTRASTRUCTURE IN BIOLOGICAL SYSTEMS

Edited by

Albert J. Dalton

National Cancer Institute
National Institutes of Health
Bethesda, Maryland

Françoise Haguenau

Laboratoire de Médecine Expérimentale
Collège de France
Paris, France

VOLUME 1 / Tumors Induced by Viruses: Ultrastructure Studies, 1962

VOLUME 2 / Ultrastructure of the Kidney, 1967

VOLUME 3 / The Nucleus, 1968

VOLUME 4 / The Membranes, 1968

VOLUME 5 / Ultrastructure of Animal Viruses and Bacteriophages: An Atlas, 1973

VOLUME 6 / C. E. Challice and S. Virágh, eds. Ultrastructure of the Mammalian Heart, 1973

VOLUME 7 / A. Tixier-Vidal and Marilyn G. Farquhar, eds. The Anterior Pituitary, 1975

VOLUME 8 / Karl Maramorosch, ed. The Atlas of Insect and Plant Viruses, 1977

THE ATLAS OF INSECT AND PLANT VIRUSES

Including Mycoplasmaviruses and Viroids

Edited by

KARL MARAMOROSCH

Rutgers-The State University
Waksman Institute of Microbiology
Busch Campus
New Brunswick, New Jersey

ACADEMIC PRESS New York San Francisco London 1977

A Subsidiary of Harcourt Brace Jovanovich, Publishers

ACADEMIC PRESS, INC.
111 Fifth Avenue, New York, New York 10003

United Kingdom Edition published by
ACADEMIC PRESS, INC. (LONDON) LTD.
24/28 Oval Road, London NW1

Library of Congress Cataloging in Publication Data

Main entry under title:

The atlas of insect and plant viruses.

(Ultrastructure in biological systems)
Supplement to A. J. Dalton's Ultrastructure of animal viruses and bacteriophages.
Includes bibliographies.
1. Plant viruses–Atlases. 2. Insect viruses–Atlases. I. Maramorosch, Karl. II. Dalton, Albert Joseph, Date Ultrastructure of animal viruses and bacteriophages. III. Series. [DNLM: 1. Insect viruses–Atlases. 2. Plant viruses–Atlases. W1 UL755 / QW17 I59]
QR363.I57 576'.64 76-55973
ISBN 0–12–470275–9

PRINTED IN THE UNITED STATES OF AMERICA

Contents

List of Contributors xiii

Foreword xvii

Preface xix

PART I INSECT VIRUSES

Chapter 1 Baculoviruses (Baculoviridae)

MAX D. SUMMERS

I. Introduction 3
II. Baculovirus Structure 4
III. Baculovirus–Host Interactions 7
IV. Serology 9
V. Conclusion 9
References 9

Chapter 2 Entomopoxviruses (Poxviruses of Invertebrates)

EDOUARD KURSTAK AND SIMON GARZON

I. Introduction 29
II. Taxonomy 30
III. The Virion 30
IV. Inclusion Bodies 31
V. Pathology 34
VI. Morphogenesis 35
VII. Conclusion 37
References 37

Chapter 3 Densonucleosis Viruses (Parvoviridae)

E. KURSTAK, P. TIJSSEN, AND S. GARZON

I. Introduction 67
II. Classification and Nomenclature 68

III. Physicochemical Properties of Virions and Subviral Components 68
IV. Morphology 72
V. Biological Properties 73
VI. Replication and Morphogenesis 74
VII. Conclusion 75
References 75

Chapter 4 Iridoviruses (Iridoviridae)

PETER E. LEE

I. Introduction 93
II. Fine Structure of Purified Virions 93
III. Virus–Cell Relationships 94
References 95

Chapter 5 Cytoplasmic Polyhedrosis Viruses

C. C. PAYNE AND K. A. HARRAP

I. Introduction 105
II. Members of the Group 106
III. The Structure of CPV's 106
IV. Virus Replication 109
References 111

Chapter 6 Rhabdoviruses of Insects (Sigma Virus of *Drosophila*)

EDWARD S. SYLVESTER

I. Introduction 131
II. Sigma Virus of *Drosophila* 131
References 134

Chapter 7 Bee Viruses

L. BAILEY AND R. D. WOODS

I. Chronic Bee-Paralysis Virus 141
II. Chronic Bee-Paralysis Virus Associate 142
III. Sacbrood Virus 142
IV. Acute Bee-Paralysis Virus 143
V. Arkansas Bee Virus 143
VI. Bee Virus X 143
VII. Slow Bee-Paralysis Virus 144
VIII. Black Queen-Cell Virus 144
IX. Kashmir Bee Virus 144
X. *Apis* Iridescent Virus 144
References 145

PART II PLANT VIRUSES

Chapter 8 Cauliflower Mosaic Virus (DNA Virus of Higher Plants)

ROBERT J. SHEPHERD

I. Introduction 159

II. Biological Properties and Transmissibility 159
III. Physical and Chemical Properties 161
IV. Cytopathological Effects in Infected Plants 162
V. Summary 163
References 163

Chapter 9 Comoviruses

A. VAN KAMMEN AND J. E. MELLEMA

I. Introduction 167
II. Viral Genome and Genetic Properties 168
III. Composition and Morphology of the Virus Caspid 169
IV. Cytopathic Strucures in Infected Cells 170
V. Replication Process 171
References 173

Chapter 10 Rhabdoviruses of Plants

G. P. MARTELLI AND M. RUSSO

I. Introduction 181
II. Morphological and Physicochemical Properties 186
III. Serological Properties 190
IV. Interactions with the Host Cells
V. Relationships with Other Viruses 192
References 194

Chapter 11 Tobravirus (Tobacco Rattle Virus) Group

E. M. J. JASPARS

I. Introduction 215
II. Particle Structure and Composition 215
III. Genome Constitution and Replication 216
References 216

Chapter 12 Nepovirus (Tobacco Ringspot Virus) Group

R. I. B. FRANCKI AND T. HATTA

I. Virus Composition and Structure 222
II. Cytology of Virus-Infected Plants 224
References 225

Chapter 13 Tabamovirus (Tobacco Mosaic Virus) Group

EISHIRO SHIKATA

I. General Description (R/1:2/5:E/E:S/O) 237
II. Composition of Virions 237
III. Localization of the Virus Particles *in Situ* 239
References 241

Chapter 14 Tombusvirus (Tomato Bushy Stunt Virus) Group

G. P. MARTELLI, M. RUSSO, AND A. QUACQUARELLI

I. Definition of the Group 257
II. General Characteristics 258
III. Morphological and Physicochemical Properties 259
IV. Serological Properties 261
V. Ultrastructure and Interaction with the Host Cell 262
VI. Relationships with Other Viruses 264
Note Added in Proof 265
References 265

Chapter 15 Tobacco Necrosis Virus Group

B. KASSANIS

I. Introduction 281
II. Biological Properties 281
III. Physicochemical Properties 283
References 284

Chapter 16 Bromovirus (Brome Mosaic Virus) Group

J. B. BANCROFT AND R. W. HORNE

I. Introduction 287
II. Preparation and Characteristics of Bromoviruses 287
III. Translation *in Vitro* and Genetics 289
IV. Multiplication 290
V. Morphological Properties 290
References 291

Chapter 17 Cucumovirus (Cucumber Mosaic Virus)

SUE A. TOLIN

I. Introduction 303
II. Members of the Cucumovirus Group 304
III. Structure and Morphogenesis 305
References 306

Chapter 18 Potyvirus (Potato Virus Y) Group

D. S. TEAKLE AND R. D. PARES

I. Introduction 311
II. Members of the Potyvirus Group 312
III. General Properties 312
IV. Conclusions 314
References 315

Chapter 19 Watermelon Mosaic Virus Group

M. H. V. VAN REGENMORTEL

I. Introduction 323

II. Biological Properties 324
III. Morphology 324
References 324

Chapter 20 Potexvirus (Potato Virus X) Group

D. E. LESEMANN AND RENATE KOENIG

I. Structure and Particles 331
II. Physical and Chemical Properties of Particles 335
III. Structural Alterations Induced by Potexviruses in Host Cells 336
References 337

Chapter 21 Tymovirus (Turnip Yellow Mosaic Virus) Group

R. E. F. MATTHEWS

I. Introduction 347
II. Members of the Group 348
III. Components of the Virus 348
IV. Structure of the Particle 350
V. Empty Protein Shells and Minor Nucleoproteins 351
VI. Cytological Evidence Concerning Virus Replication 351
References 352

Chapter 22 Luteovirus (Barley Yellow Dwarf Virus) Group

W. F. ROCHOW AND H. W. ISRAEL

I. Introduction 363
II. Members of the Group 364
III. General Properties 364
IV. Virion Structure 365
V. Cell–Virus Interactions 365
References 366

Chapter 23 Ilarvirus (Necrotic Ringspot) Group, Subgroup B

L. S. LOESCH-FRIES, E. L. HALK,
AND R. W. FULTON

Text 371
References 372

Chapter 24 Plant Reovirus Group

EISHIRO SHIKATA

I. General Description 377
II. Structure of the Virions 378
III. Localization of the Virus Particles *in Situ* 379
References 381

Chapter 25 Pinwheel Inclusions and Plant Viruses

H. W. ISRAEL AND H. J. WILSON

I. Introduction 405
II. Morphology 406

III. Morphogenesis 407
IV. Structure and Composition 407
V. Serology 408
VI. Conclusions 408
References 409

PART III MYCOVIRUSES AND VIROIDS

Chapter 26 Mycoviruses

K. N. SAKSENA

I. Introduction 421
II. Biological Implications 421
III. Physicochemical Properties 422
IV. Ultrastructural Aspects 423
V. Concluding Remarks 424
References 424

Chapter 27 Viroids

T. O. DIENER

I. Introduction 431
II. Members of the Group 432
III. Structure 433
IV. Subcellular Location 434
V. Replication 434
References 435

PART IV MYCOPLASMA- AND SPIROPLASMAVIRUSES

Chapter 28 Mycoplasmaviruses

JACK MANILOFF, JYOTIRMOY DAS,
AND RESHA M. PUTZRATH

I. Review of the Class Mollicutes 439
II. Group 1 Mycoplasmaviruses 441
III. Group 2 Mycoplasmaviruses 443
IV. Group 3 Mycoplasmaviruses 444
References 445

Chapter 29 Spiroplasmaviruses

ROGER M. COLE

I. Discovery of Mycoplasmaviruses 451
II. Spiroplasmas: The Hosts of Spiroplasmaviruses 452
III. The Spiroplasmaviruses 453
Addendum 456
References 456

Chapter 30 Viruses of *Drosophila* Sex-Ratio Spiroplasma

DAVID L. WILLIAMSON, KUGAO OISHI, AND DONALD F. POULSON

I. Introduction 465
II. The Sex-Ratio Spiroplasmaviruses 466
III. Structure and Morphogenesis 467
References 468

Index 473

List of Contributors

Numbers in parentheses indicate the pages on which the authors' contributions begin.

L. BAILEY (141), Rothamsted Experimental Station, Harpenden, Herts, England

J. B. BANCROFT (287), Department of Plant Sciences, The University of Western Ontario, London, Ontario, Canada

ROGER M. COLE (451), National Institute of Allergy and Infectious Diseases, National Institutes of Health, Laboratory of Streptococcal Diseases, Bethesda, Maryland

JYOTIRMOY DAS (439), Departments of Microbiology and of Radiation Biology and Biophysics, School of Medicine and Dentistry, University of Rochester, Rochester, New York

T. O. DIENER (431), ARS, U.S. Department of Agriculture, Plant Virology Laboratory, Agricultural Research Center, Beltsville, Maryland

R. I. B. FRANCKI (221), Department of Plant Pathology, Waite Agricultural Research Institute, The University of Adelaide, Glen Osmond, South Australia

R. W. FULTON (371), Department of Horticulture, University of Wisconsin, Madison, Wisconsin

SIMON GARZON (29, 67), Comparative Virology Research Group, Department of Microbiology and Immunology, Faculty of Medicine, University of Montreal, Montreal, Canada

E. L. HALK (371), Department of Horticulture, University of Wisconsin, Madison, Wisconsin

K. A. HARRAP (105), Natural Environment Research Council, Unit of Invertebrate Virology, Oxford, England

T. HATTA (221), Department of Plant Pathology, Waite Agricultural Research Institute, The University of Adelaide, Glen Osmond, South Australia

R. W. HORNE (287), Department of Ultrastructural Studies, John Innes Institute, Norwich, Norfolk, England

H. W. ISRAEL (363, 405), Department of Plant Pathology, Cornell University, Ithaca, New York

E. M. J. JASPARS (215), Department of Biochemistry, State University of Leiden, Leiden, The Netherlands

B. KASSANIS (281), Plant Pathology Department, Rothamsted Experimental Station, Harpenden, Herts, England

RENATE KOENIG (331), Biologische Bundesanstalt für Land- und Forstwirtschaft, Messeweg, Braunschweig, West Germany

EDOUARD KURSTAK (29, 67), Comparative Virology Research Group, Department of Microbiology and Immunology, Faculty of Medicine, University of Montreal, Montreal, Canada

PETER E. LEE (93), Department of Biology, Carleton University, Ottawa, Canada

D. E. LESEMANN (331), Biologische Bundesanstalt für Land- und Forstwirtschaft, Messeweg, Braunschweig, West Germany

L. S. LOESCH-FRIES (371), Department of Horticulture, University of Wisconsin, Madison, Wisconsin

JACK MANILOFF (439), Department of Microbiology, School of Medicine and Dentistry, University of Rochester, Rochester, New York

G. P. MARTELLI (181, 257), Universita degli Studi di Bari, Istituto di Patologia Vegetale, Bari, Italy

R. E. F. MATTHEWS (347), Department of Cell Biology, University of Auckland, Auckland, New Zealand

J. E. MELLEMA (167), Department of Biochemistry, State University of Leiden, Leiden, The Netherlands

KUGAO OISHI (465), Department of Biology, Kobe University, Kobe, Japan

R. D. PARES (311), Department of Agriculture, Rydalmere, New South Wales, Australia

C. C. PAYNE* (105), Natural Environment Research Council, Unit of Invertebrate Virology, Oxford, England

DONALD F. POULSON (465), Department of Biology, Yale University, New Haven, Connecticut

RESHA M. PUTZRATH (439), Department of Radiation Biology and Biophysics, University of Rochester, School of Medicine and Dentistry, Rochester, New York

A. QUACQUARELLI (257), Universita degli Studi di Bari, Istituto di Patologia Vegetale, Bari, Italy

W. F. ROCHOW (363), Department of Plant Pathology, Cornell University, Ithaca, New York

M. RUSSO (181, 257), Universita degli Studi di Bari, Istituto di Patologia Vegetale, Bari, Italy

K. N. SAKSENA (421), Mellon Institute, Carnegie–Mellon University, Pittsburgh, Pennsylvania

ROBERT J. SHEPHERD (159), Department of Plant Pathology, University of California, Davis, California

EISHIRO SHIKATA (237, 377), Department of Botany, Faculty of Agriculture, Hokkaido University, Sapporo, Japan

MAX D. SUMMERS† (3), The Cell Research Institute, The University of Texas at Austin, Austin, Texas

EDWARD S. SYLVESTER (131), Department of Entomological Sciences, University of California, Berkeley, California

D. S. TEAKLE (311), Department of Microbiology, University of Queensland, St. Lucia, Brisbane, Queensland, Australia

P. TIJSSEN (67), Comparative Virology Research Group, Department of Microbiology and Immunology, Faculty of Medicine, University of Montreal, Montreal, Canada

SUE A. TOLIN (303), Department of Plant Pathology and Physiology, Virginia Polytechnic Institute and State University, Blacksburg, Virginia

* Present address: Department of Entomology, Glasshouse Crops Research Institute, Rustington, Littlehampton, Sussex, England.

† Present address: Department of Entomology, College of Agriculture, Texas A & M University, College Station, Texas.

A. VAN KAMMEN (167), Department of Molecular Biology, Agricultural University, Wageningen, The Netherlands

M. H. V. VAN REGENMORTEL (323), Department of Microbiology, University of Cape Town, Rondebosch, South Africa

DAVID L. WILLIAMSON (465), Department of Anatomical Sciences, Health Science Center, SUNY at Stony Brook, Stony Brook, New York

H. J. WILSON (405), Department of Biology, University of Alabama in Huntsville, Huntsville, Alabama

R. D. WOODS (141), Rothamsted Experimental Station, Harpenden, Herts, England

Foreword

Comments on the earlier atlas on animal viruses and bacteriophages ("Ultrastructure of Animal Viruses and Bacteriophages: An Atlas." A. J. Dalton and F. Haguenau, editors. Academic Press, New York and London, 1973) were generally favorable, emphasizing that it filled a definite need in the particular field. However, several comments, both in reviews and in personal contacts, included the wish that we had gone further and had added viruses of insects and plants. Actually, we felt at the time that we had already gone a bit further than our knowledge and expertise had really allowed.

Fortunately, one of those who expressed the view that we should develop a second part to the atlas which would include insect and plant viruses was Dr. Karl Maramorosch. Our immediate reaction was "fine, if you will agree to be the primary editor." And so it came to pass. In our view, our immediate and only reaction has proved to be unquestionably the correct one.

In organizing the contents of this monograph the recommendations of the International Committee on the Nomenclature of Viruses in regard to classification have been followed. However, in the insect viruses, placing the viruses of bees in one group, regardless of classification, was an editorial decision in favor of the reader. A similar decision was applied to viruses of mycoplasma and spiroplasma.

Most major groups of plant viruses have been included, and only one group of "other viruses" has been omitted. This group, the viruses of protozoa, has been more than adquately dealt with elsewhere [Diamond, L. S., and Mattern, C. F. T. *Adv. Virus Res.* **20,** 87–112 (1976)].

As editors of the series on "Ultrastructure in Biological Systems," of which this atlas is number eight, we wish to thank personally all of the contributors and particularly Dr. Maramorosch for the time and effort they have given to produce what we truly believe will be a timely, useful, and successful addition to the literature of this rapidly advancing field.

ALBERT J. DALTON
Scientist Emeritus
National Cancer Institute
National Institutes of Health
Bethesda, Maryland

FRANÇOISE HAGUENAU
Sous-Directeur
Laboratoire de Médecine Expérimentale
Collège de France
Paris, France

Preface

"The Atlas of Insect and Plant Viruses" is a companion volume to "Ultrastructure of Animal Viruses and Bacteriophages: An Atlas," edited by Dr. Albert J. Dalton and Dr. Françoise Haguenau, published by Academic Press in 1973. It includes a description of viruses infecting insects, higher plants, fungi, mycoplasmas, and spiroplasmas. The smallest recognized disease agents, viroids, have also been included, although they are not classified as viruses. The contributors have presented the information available on the ultrastructural aspects of viral replication and the organization of the virion in the diverse groups.

Viruses of microorganisms in the families Mycoplasmataceae and Spiroplasmataceae, discovered recently, have been described in detail. Viruses of protozoa and algae have been omitted because information on them has been inadequate or unavailable. Among the taxonomically recognized groups of plant viruses, all but the following have been included: alfalfa mosaic virus (monotypic); pea enation mosaic virus (monotypic); Carlavirus (carnation latent virus) group; tomato spotted wilt virus (monotypic); and Hordeivirus (barley stripe mosaic virus) group.

Although the taxonomy of viruses has progressed since the publication of "Insect and Plant Viruses," the chapter in that volume on the classification and nomenclature of viruses by Melnick is recommended as a guideline to this book as well. Specific references to various groups of viruses give the sources of published information.

Those interested in the current status of virus taxonomy are referred to the description of viruses in the following publications: Peter Wildy (1971). *In* "Monographs in Virology Series" (J. L. Melnick, ed.), Volume V: "Classification and Nomenclature of Viruses." Karger, Basel; Frank Fenner (1976). *ASM News* **42,** 170; Frank Fenner (1976). *Virology* **71,** 371. The officially accepted families and genera of viruses are given in these publications.

It is hoped that this volume will not only serve those who work with viruses of invertebrates and plants, but also virologists interested in the comparative aspects of all viruses.

KARL MARAMOROSCH

PART I

Insect Viruses

Chapter 1

Baculoviruses (Baculoviridae)

MAX D. SUMMERS

I. Introduction 3
II. Baculovirus Structure 4
A. The Occlusion 4
B. Nucleocapsids and Enveloped Nucleocapsids 6
C. Virus Genomes 6
III. Baculovirus–Host Interactions 7
Virus Envelope–Membrane Associations 8
IV. Serology 9
V. Conclusion 9
References 9

I. INTRODUCTION

Baculoviruses are DNA, rod-shaped, enveloped nucleocapsids which are identified on the basis of enveloped nucleocapsid morphology and occlusion in a proteinic crystal (Summers, 1975; Wildy, 1971). Baculoviruses in the occluded form exhibit three morphological types. The granulosis viruses (GV's) contain one enveloped nucleocapsid per occlusion (Figs. 1, 5, and 6). The nuclear-polyhedrosis viruses (NPV's) usually exhibit a polyhedral shape and two distinct structural relationships with enveloped nucleocapsids: (a) NPV's in which enveloped single nucleocapsids are observed with many *singles* occluded (SNPV) (Figs. 3 and 4); and (b) *more than one* nucleocapsid common to a virus envelope with many bundles occluded (MNPV) (Figs. 2 and 9). The occluded NPV's range in size to 15 μm with the numbers of nucleocapsids, per virus bundle, up to 39 (Kawamoto and Asayama, 1975).

Based upon virus morphology and host susceptibility there have been approximately 284 NPV's and 65 GV's reported (David, 1975). Most NPV's are found in the order Lepidoptera (243 viruses) with reports of NPV's in Orthoptera, Neuroptera, Trichoptera, Coleoptera, Hymenoptera, and Diptera. All GV's have been from species of Lepidoptera.

Several books and reviews are available for a more comprehensive survey of the nature of baculovirus structure, invasion, and replication (Bulla, 1973; David, 1975; Gibbs, 1973; Maramorosch, 1968; Paschke and Summers, 1975;

Smith, 1967, 1971; Steinhaus, 1963; Summers *et al.*, 1975; Tinsley and Harrap, 1972; Vago and Bergoin, 1968). The nature of virus specificity, host immunity, and defenses to infection have also been given recent attention (Bulla and Cheng, 1975; Gibbs, 1973; Maramorosch and Shope, 1975; Summers *et al.*, 1975). Early reviews of excellent quality by Bergold (1953, 1958, 1959, 1963a, 1964) and Aizawa (1963), provide detailed accounts of chemical and biochemical information usually not cited in such detail by more recent reviews.

The development of invertebrate cell cultures has been followed by studies on the infection and replication of baculoviruses (Brown and Faulkner, 1975; Hink and Vail, 1973; Knudson and Tinsley, 1974; Knudson and Harrap, 1976; Faulkner and Henderson, 1972; MacKinnon *et al.*, 1974; Raghow and Grace, 1974; Volkman and Summers, 1975; McIntosh, 1975; McIntosh and Shamy, 1975).

The occlusion of the rod-shaped enveloped nucleocapsid into a crystalline protein matrix has been considered a unique feature of insect baculoviruses until recently. Couch (1974) described an occluded virus of the pink shrimp, *Panaeus duorarum* (Figs. 4 and 8). Also, enveloped nucleocapsids with structural similarities to NPV and GV nucleocapsids but which are not occluded are being considered as baculoviruses. Examples of these have been reported in *Gyrinis* (Gouranton, 1972), and *Oryctes* (Huger, 1966). The latter was considered nonoccluded, but it is possible that occlusion may occur at some stage in the life cycle of the virus–host relationship. In parasitoid Hymenoptera a fluid or secretion derived from the parasitoid ovary, which is involved in the ability of the host to mount a defense against the egg of the parasite, contains nonoccluded particles with structural properties similar to those of baculovirus enveloped nucleocapsids (Stoltz *et al.*, 1976) (Fig. 10).

II. BACULOVIRUS STRUCTURE

A. The Occlusion

The structure and function of the occlusion (proteinic crystal) is only partially characterized (Summers, 1975). It provides protection for the occluded virus in the natural environment thereby allowing certain of the baculoviruses to be utilized as alternatives for chemical pesticides. In the case of certain lepidopterous NPV's and GV's, there appears to be a tissue dependency related to the occlusion process. This has been described by following the invasion and infection sequence in susceptible larvae (Harrap, 1970; Summers, 1971). In midgut cells, the primary site of infection, large numbers of GV or NPV progeny are not occluded in the nucleus of infected cells but are apparently released to bud through the plasma membrane of the midgut cell into the hemolymph. Similar observations have been made in epidermal tissues (Cunningham, 1971). However, in fat body cells and other tissues during secondary infection, large numbers of enveloped particles are occluded, thus signaling the termination of the maturation process. Some nonoccluded viruses can be observed in cells of those tissues as well (Nappi and Hammill, 1975).

There is now adequate evidence documenting the presence of a distinct structure on the surface of the occlusion. This has been described as a polyhedral or capsule "membrane" (Harrap, 1972a). Summers and Arnott (1969) demonstrated the unique morphology of the "membrane" and its structural similarity with regards to the "membrane profiles" (Figs. 3 and 13). The "membrane profiles" were observed in intimate association with fibrillar elements (Figs. 13, 14, and 15). Similar observations and structures have been observed *in vitro*

(MacKinnon *et al.*, 1974; Knudson and Harrap, 1976). Harrap (1972a), using negative staining techniques, observed that the "membrane" exhibited holes 6 nm in diameter. His structural model suggested protein subunits arranged in a rhombic or as a hexagonal tessellation around the central hole.

Although the "membrane" has not been characterized in terms of chemical properties, the solubility of the structure is distinctly different as compared to granulin or polyhedrin, the crystalline form of that protein (other than the enveloped nucleocapsid), which constitutes the major component of the crystal for both NPV's and GV's. Although the structural composition and function are not known, the origin of the "membrane" as possibly related to cellular membranes has been noted (Knudson and Harrap, 1976; Summers and Arnott, 1969).

The major structural component of the protein occlusion (Figs. 1–9), occasionally described as matrix protein, is now referred to as granulin for GV's and polyhedrin for NPV's (Summers and Egawa, 1973). Early attempts to characterize granulin and polyhedrin were conducted by Bergold (1947, 1948). Recent studies have supported some of the generalities of Bergold's observations (Bergold, 1963b; Harrap, 1972a). Using a combination of negative staining and thin section electron microscopy, the structural organization of the crystal exhibits lattice arrangements varying from 4 to 7 nm with lattice angles of approximately 65°–85°. Occasionally 90° angles have been observed (Arnott and Smith, 1968a; Summers, 1975).

The physical and chemical properties of a few polyhedrins and granulins have been studied. The usual procedure for dissociation of crystal structure requires dilute alkaline saline (Bergold, 1947). However, several solvents and reagents have been utilized to solubilize the protein crystal at neutral conditions (Egawa and Summers, 1972; Kawanishi *et al.*, 1972a). It has been documented that the presence of a protease in the occlusion is capable of degrading granulin or polyhedrin into lower molecular weight components (Eppstein *et al.*, 1975; Eppstein and Thoma, 1975; Kozlov *et al.*, 1975a,b; Summers, 1975; Summers and Smith, 1975). For accurate comparisons of the structural properties of granulins and polyhedrins, protease activity must be inhibited.

Granulins and polyhedrins are similar in size and composition exhibiting a molecular weight of approximately 28,000–30,000. This suggests that the major structural component (Fig. 5) as visualized by electron microscopy must be an aggregate assembly of polypeptides of lower molecular weight (Arnott and Smith, 1968a; Harrap, 1972a). SDS-polyacrylamide gel electrophoresis (PAGE), peptide mapping, and amino acid analyses (Summers and Smith, 1975, 1976) have shown that polyhedrins and granulins apparently have related primary structure; however, each protein was different and apparently specific in association with each virus species studied. Granulins and polyhedrins are also phenol soluble and apparently phosphorylated.

There have been reports of RNA in the occlusion (Faulkner, 1963; Himeno and Onodera, 1969). However, Summers and Egawa (1973) demonstrated that a significant portion of the orcinol positive component of *Trichoplusia ni* GV could be removed by extensive washing. Also identified as a component of the occlusion of a GV is a synergistic factor responsible for the enhancement of a NPV infection (Hara *et al.*, 1976; Tanada *et al.*, 1975). This factor is believed to be a protein, perhaps an enzyme, with an estimated molecular weight of 152,000–163,000.

Unusual formations of polyhedra and capsules (Figs. 14–16) exhibit strikingly aberrant structures (Arnott and Smith, 1968b; Summers and Arnott, 1969).

Occlusion of enveloped nucleocapsids is known to vary according to the virus and host system studied. The crystalline form of the protein is deposited on the surface of a GV enveloped nucleocapsid, then, by an orderly process, the virus particle is occluded (Arnott and Smith, 1968a). Nuclear-polyhedrosis virus enveloped nucleocapsids are occluded in crystallizing polyhedra by a nonrandom process (Figs. 2, 3, 8, and 9) (Knudson and Harrap, 1976). In the NPV infection of *Rhynchosciara*, inclusion formation initiates on the inner nuclear membrane (Fig. 7; Stoltz *et al.*, 1973), then enveloped nucleocapsids are occluded at random. This mechanism of occlusion is not similar to that which occurs in lepidopterous NPV infections.

B. Nucleocapsids and Enveloped Nucleocapsids

The rod-shaped enveloped nucleocapsids are released from the crystalline occlusion as infectious particles, using the dilute alkaline saline procedure (Bergold, 1947). Extracted from the crystal and prepared in highly purified form (Arif and Brown, 1975; Harrap, 1972b; Kawanishi and Paschke, 1970a,b; Summers and Paschke, 1970; Summers and Volkman, 1976), the virus particles are observed as enveloped single nucleocapsids or bundles of nucleocapsids (Fig. 18). The envelope of particles extracted from occlusions, with the exception of Harrap's observations (1972b), is routinely observed without any distinct surface or internal structure. However, nonoccluded virus (Fig. 19), which buds from the plasma membranes of cells in culture or from infectious insect hemolymph, does exhibit modified surface structure (Summers and Volkman, 1976).

Enveloped nucleocapsids are apparently complex as analyzed by SDS-PAGE (Padhi *et al.*, 1974; Young and Lovell, 1973) consisting of 10–12 major proteins. Padhi *et al.* (1974) report the structure of the nucleocapsid to be complex, exhibiting approximately 10 major SDS-PAGE bands.

Negatively stained preparations of capsids (Fig. 17) show a regular structural arrangement (Harrap, 1972b; Khosaka *et al.*, 1971; Kozlov and Alexeenko, 1967; Summers and Paschke, 1970). Observed are a series of bands approximately 2.5 to 3.0 nm in width almost at 90° to the longitudinal axis of the capsid.

Purified nucleocapsids show a tendency to aggregate by the end structure (Kawanishi and Paschke, 1970b). The nucleocapsid shows a distinct structural polarity at opposite ends of the particle (Fig. 12) (Kawanishi and Paschke, 1970b; Teakle, 1969).

C. Virus Genomes

Estimates of the size of baculovirus DNA molecules as reported in the literature range from 2.0 to 100 $\times 10^6$ with the enveloped virus containing 7–15% DNA (Bergold, 1959; Summers, 1975). Sedimentation relative to bacteriophage T7 and T4 DNA standards demonstrated that, as isolated from highly purified virus preparations, baculovirus DNA existed in three forms: linear (dlDNA), relaxed circular (rcDNA), and covalently closed double-stranded molecules (ccDNA) (Fig. 29).

The nature of sedimentation of circular, relative to linear, DNA has been shown to be dependent on the speed of centrifugation and the ionic strength of the gradient. As estimated by relative sedimentation to bacteriophage DNA standards, NPV and GV DNA's have molecular weights of 90 to 100 $\times 10^6$. Estimates of size for *Trichoplusia ni* GV DNA measured by sedimentation have been supported by direct measurements on molecules observed by electron microscopy (Summers, 1975). Sedimentation did not adequately resolve differences between NPV and GV DNA's. However, the baculovirus DNA's have characteristic T_m values.

Under normal conditions of rate-zonal centrifugation in sucrose, it has been shown that ccDNA and rcDNA cosediment. However, the two different forms can be resolved in ethidium bromide–CsCl gradients, sucrose gradients with $I < 10^{-3}$, or by sedimentation in alkaline sucrose.

There have been reports of subviral infectivity (Zherebtsova *et al.*, 1972). The nature of the subviral components relative to the large DNA's described above is yet to be resolved. Regardless, the molecular biology of baculovirus DNA's shows considerable promise even though there is little factual information on several species of viruses reported.

III. BACULOVIRUS–HOST INTERACTIONS

The sequence of virus invasion and infection and a detailed consideration of virus replication has been adequately reviewed elsewhere (Dalgarno and Davey, 1973; Harrap, 1973; Hirumi *et al.*, 1975; Paschke and Summers, 1975). For both NPV's and GV's of lepidopterous species the infection cycle in larvae is apparently complex.

During infection, whether *in vivo* or *in vitro*, enveloped nucleocapsids can be observed to exhibit two significantly different structural and functional relationships: (a) the occluded virus (OV) (Figs. 1–9); and, (b) the nonoccluded virus (NOV) (Fig. 19). The OV is the most routinely observed form of the virus. The NOV of cell culture and infectious hemolymph of insects are known to have distinctly different physical, chemical, and infectious properties as compared to enveloped nucleocapsids purified from the occlusion (Dougherty *et al.*, 1975; Henderson *et al.*, 1974; Knudson and Tinsley, 1974; Summers and Volkman, 1976).

In general, the enveloped nucleocapsid is released from the crystal in the gut lumen of susceptible larvae to penetrate the midgut cells. Enveloped nucleocapsids as singles (Harrap, 1970; Summers, 1971), or as bundles (Fig. 24) (Kawanishi *et al.*, 1972b; Tanada *et al.*, 1975) apparently fuse with the microvillar membrane thus allowing entry of the nucleocapsid(s) to the gut cell (Figs. 24, 27, and 28). The nature of uncoating of the virus genome has not been clearly documented. Knudson and Harrap (1976) suggest that uncoating may occur in the cytoplasm, although a possible nuclear pore interaction may be involved thereby allowing release of the virus DNA directly through the pore into the nucleus (Figs. 20 and 22). Recent studies suggest by the presence of nucleocapsids in the nucleus within 3 to 4 hours postinfection, although rarely observed, that the intact nucleocapsid may be able to directly penetrate the nucleus (Hirumi *et al.*, 1975; Knudson and Harrap, 1976).

For comparison, virus entry into susceptible cells in culture (Fig. 25) appears to occur by viropexis of enveloped nucleocapsids (Hirumi *et al.*, 1975; Knudson and Harrap, 1976; Raghow and Grace, 1974). However, nucleocapsids have also been observed to enter culture cells by viropexis (Vaughn *et al.*, 1972). *In vivo*, viropexis has been observed by insect hemocytes (Kislev *et al.*, 1969).

Perhaps the mechanisms associated with baculovirus entry into susceptible cells, as in the area of vertebrate virology, cannot be generalized as of yet because of possible and potential differences between *in vivo* and *in vitro* mechanisms. In particular, when studying virus entry *in vivo* it is important to consider the tissue involved during invasion and subsequent interactions with host cells.

The cytopathology resulting from virus replication is similar regardless of whether studied *in vivo* or *in vitro*. A detailed account of those events has been reviewed (Hirumi *et al.*, 1975; Harrap, 1972b,c; Knudson and Harrap, 1976; MacKinnon *et al.*, 1974; Steinhaus, 1963). Quantitative studies using a plaque assay (Hink and Vail, 1973; Hink and Strauss, 1976; Volkman and Summers, 1975) and cells in culture have given a more accurate picture of the sequence of events. Dependent upon the virus or cell line employed, a few hours postinfection the nucleus of the infected cell undergoes progressive and distinct changes during eclipse phase. Characteristic of baculovirus infections is the presence of a virogenic stroma (Fig. 26), an electron dense structure or chromatin-like network which appears intimately involved in nucleocapsid assembly. With the exception of nonoccluded forms, the virus appears to be confined to the nucleus of the cell until cellular disruption occurs. Although GV infection in midgut cells appears confined to the nucleus, replication and occlusion occurs in the cytoplasm of fat body cells (Summers, 1971).

Virus Envelope–Membrane Associations

NPV's and GV's exhibit complex associations and behavior with membranes during the virus–host cycle (Robertson *et al.*, 1974; Stoltz *et al.*, 1973). Stoltz catagorized two types of virus membranes or envelopes based upon the fate of the virus as observed in the cell: (a) the virus envelope; and (b) "transport" membranes. The virus envelope represents that form or association which is found with, or derived from, the occluded virus (Figs. 3, 9, and 18). "Transport" membranes are associated with the nucleocapsid in a vesicular nature; the enveloped form is not occluded and can be observed to escape the cell by synhymenosis (Fig. 23). Both associations of nucleocapsids with membranes or envelopes can be observed *in vivo* and *in vitro*. If the unit membrane structure in association with the nucleocapsid is found to be virus specific, then it may be assumed that the membrane is an envelope for which specific factors, giving rise to the observed functional and structural differences, cannot be described at this time.

In NPV infections of lepidopterous larvae it is possible that the virus envelope is derived from the inner nuclear membrane (Summers and Arnott, 1969). In the NPV infection of *Rhynchosciara angelae* (Stoltz *et al.*, 1973), which shows a distinctly different process of occlusion as compared to NPV's of Lepidoptera, viral envelopes may be derived by *de novo* synthesis in the nucleus (Fig. 11). Granulosis viruses, aside from midgut cell infection, appear to obtain an envelope from cytoplasmic membranes by an apparently specific process (Arnott and Smith, 1968a). The end of the nucleocapsid first associates with a smooth membrane then acquires an envelope by budding. In fat body cells GV enveloped nucleocapsids are occluded in the cytoplasm. Granulosis virus progeny in infected midgut cells apparently acquire an envelope from other cellular sources and are not occluded.

The origin of the virus envelope or membrane associated with the NPV is not clear but may be derived from several possible sources (Summers, 1971). As studied in cell culture, the events involving the release of nucleocapsids from the nucleus into the cytoplasm to bud through the plasma membrane are a series of complex membrane-associated phenomena not yet clearly understood (Hirumi *et al.*, 1975; Knudson and Harrap, 1976). Regardless, the two different modes of envelopment likely result in infectious viruses with different fate(s) and/or purposes (Knudson and Harrap, 1976). Both mechanisms are required for persistence of the virus in nature. The source and function of the baculovirus envelope provides a very interesting series of problems related to the study of virus specificity associated with tissue tropisms.

IV. SEROLOGY

Early work confirmed by more recent studies have shown that the granulin, polyhedrin, and enveloped nucleocapsid antigenic systems are complex (Mazzone, 1975; Scott and Young, 1973). Several serological techniques have been used (Cunningham, 1968; Norton and Dicapua, 1975). The cross-reactivity of antisera to heterologous antigens for granulins, polyhedrins, and enveloped virus, correlated with contradictory reports from different sources, makes it difficult to compare results in a way that one can decipher the serological results for the identification of baculoviruses. Although some studies using more sophisticated serological techniques show good promise (Benton *et al.*, 1973; Kurstak and Kurstak, 1974), a more definitive evaluation of baculovirus serology must be momentarily delayed for the advent of more detailed chemical and biochemical characterization studies. These studies must be conducted relative to the search for virus group and specific antigens. This must be coupled with serological techniques which are quantitative in their use.

V. CONCLUSION

Baculoviruses are apparently complex structures about which little is known in terms of molecular biology, identification of virus species, and the molecular specificity of virus–host interactions. As presently characterized, the virus–host cycle is complex both *in vivo* and *in vitro*, an indication that a considerable amount of study is needed to adequately characterize these interactions in such a way as to follow the fate and function of any baculovirus in susceptible and apparently nonsusceptible systems.

REFERENCES

Aizawa, K. (1963). *In* "Insect Pathology: An Advanced Treatise" (E. A. Steinhaus, ed.), Vol. 1, pp. 381–412. Academic Press, New York.

Arif, B. M., and Brown, K. W. (1975). *Can. J. Microbiol.* **21,** 1224–1231.

Arnott, H. J., and Smith, K. M. (1968a). *J. Ultrastruct. Res.* **21,** 251–268.

Arnott, H. J., and Smith, K. M. (1968b). *J. Ultrastruct. Res.* **22,** 136–158.

Benton, C. V., Reichelderfer, C. F., and Hetrick, F. M. (1973). *J. Invertebr. Pathol.* **22,** 42–49.

Bergold, G. H. (1947). *Z. Naturforsch., Teil B* **2,** 122–143.

Bergold, G. H. (1948). *Z. Naturforsch., Teil B* **3,** 338–342.

Bergold, G. H. (1953). *Adv. Virus Res.* **1,** 91–139.

Bergold, G. H. (1958). *In* "Handbuch der Virusforschung" (C. Hallauer and K. F. Meyer, eds.), 2nd ed. Vol. 4, pp. 60–142. Springer-Verlag, Berlin and New York.

Bergold, G. H. (1959). *In* "Viruses" (F. M. Burnet and W. M. Stanley, eds.), Vol. 1, pp. 505–523. Academic Press, New York.

Bergold, G. H. (1963a). *In* "Insect Pathology: An Advanced Treatise" (E. A. Steinhaus, ed.), Vol. 1, pp. 413–456. Academic Press, New York.

Bergold, G. H. (1963b). *J. Ultrastruct. Res.* **8,** 360–378.

Bergold, G. H. (1964). *In* "Techniques in Experimental Virology" (R. J. C. Harris, ed.), pp. 111–144. Academic Press, New York.

Brown, M., and Faulkner, P. (1975). *J. Invertebr. Pathol.* **26,** 251–257.

Bulla, L. A., and Cheng, T. C. (1975). *Ann. N. Y. Acad. Sci.* **266,** 1–540.

Bulla, L. A., Jr., (1973). *Ann. N. Y. Acad. Sci.* **217,** 1–243.

Couch, J. A. (1974). *J. Invertebr. Pathol.* **24,** 311–331.

Cunningham, J. C. (1968). *J. Invertebr. Pathol.* **11,** 132–141.

Cunningham, J. C. (1971). *Can. J. Microbiol.* **17,** 69–72.

Dalgarno, L., and Davey, M. W. (1973). *In* "Viruses and Invertebrates" (A. J. Gibbs, ed.), Vol. 31, pp. 246–270. North-Holland Publ., Amsterdam.

David, W. A. L. (1975). *Annu. Rev. Entomol.* **20,** 97–117.

Dougherty, E. M., Vaughn, J. L., and Reichelderfer, C. F. (1975). *Intervirology* **5,** 109–121.

Egawa, K., and Summers, M. D. (1972). *J. Invertebr. Pathol.* **19,** 395–404.
Eppstein, D., and Thoma, J. A. (1975). *Biochem. Biophys. Res. Commun.* **62,** 478–484.
Eppstein, D., Thoma, J. A., Scott, H. A., and Young, S. Y. (1975). *Virology* **67,** 591–594.
Faulkner, P. (1963). *Virology* **16,** 477–484.
Faulkner, P., and Henderson, J. F. (1972). *Virology* **50,** 920–924.
Gibbs, A. J., ed. (1973). "Viruses and Invertebrates," Vol. 31. North-Holland Publ., Amsterdam.
Gouranton, J. (1972). *J. Ultrastruct. Res.* **39,** 281–294.
Hara, S., Tanada, Y., and Omi, E. M. (1976). *J. Invertebr. Pathol.* **27,** 115–124.
Harrap, K. A. (1970). *Virology* **42,** 311–318.
Harrap, K. A. (1972a). *Virology* **50,** 114–123.
Harrap, K. A. (1972b). *Virology* **50,** 124–132.
Harrap, K. A. (1972c). *Virology* **50,** 133–139.
Harrap, K. A. (1973). *In* "Viruses and Invertebrates" (A. J. Gibbs, ed.), Vol. 31, pp. 272–299. North-Holland Publ., Amsterdam.
Henderson, J. F., Faulkner, P., and MacKinnon, E. A. (1974). *J. Gen. Virol.* **22,** 143–146.
Himeno, M., and Onodera, K. (1969). *J. Invertebr. Pathol.* **13,** 89–90.
Hink, W. F., and Vail, P. V. (1973). *J. Invertebr. Pathol.* **22,** 168–174.
Hink, W. F., and Strauss, E. (1976). *J. Invertebr. Pathol.* **27,** 49–55.
Hirumi, H., Hirumi, K., and McIntosh, A. H. (1975). *Ann. N. Y. Acad. Sci.* **266,** 302–326.
Huger, A. M. (1966). *J. Invertebr. Pathol.* **8,** 38–51.
Kawamoto, F., and Asayama, T. (1975). *J. Invertebr. Pathol.* **26,** 47–55.
Kawanishi, C. Y., and Paschke, J. D. (1970a). *J. Invertebr. Pathol.* **16,** 89–92.
Kawanishi, C. Y., and Paschke, J. D. (1970b). *Proc. Colloq. Insect Pathol., 4th, 1970* pp. 127–146.
Kawanishi, C. Y., Egawa, K., and Summers, M. D. (1972a). *J. Invertebr. Pathol.* **20,** 95–100.
Kawanishi, C. Y., Summers, M. D., Stoltz, D. B., and Arnott, H. J. (1972b). *J. Invertebr. Pathol.* **20,** 104–108.
Khosaka, T., Himeno, M., and Onodera, K. (1971). *J. Virol.* **7,** 267–273.
Kislev, N., Harpaz, I., and Zelcer, A. (1969). *J. Invertebr. Pathol.* **14,** 245–257.
Knudson, D. L., and Harrap, K. A. (1976). *J. Virol.* **17,** 254–268.
Knudson, D. L., and Tinsley, T. W. (1974). *J. Virol.* **14,** 934–944.
Kozlov, E. A., and Alexeenko, I. P. (1967). *J. Invertebr. Pathol.* **9,** 413–419.
Kozlov, E. A., Sidorova, N. M., and Serebryani, S. B. (1975a). *J. Invertebr. Pathol.* **25,** 97–101.
Kozlov, E. A., Levitina, T. L., Sidorova, N. M., Radauski, Yu. L., and Serebryani, S. B. (1975b). *J. Invertebr. Pathol.* **25,** 103–107.
Kurstak, E., and Kurstak, C. (1974). *In* "Viral Immunodiagnosis" (E. Kurstak and R. Morisset, eds.), pp. 3–30. Academic Press, New York.
McIntosh, A. H. (1975). *In* "Baculoviruses for Insect Pest Control: Safety Considerations" (M. D. Summers *et al.*, eds.), pp. 63–69. Am. Soc. Microbiol., Washington, D. C.
McIntosh, A. H., and Shamy, R. (1975). *Ann. N. Y. Acad. Sci.* **266,** 327–331.
MacKinnon, E. A., Henderson, J. F., Stoltz, D. B., and Faulkner, P. (1974). *J. Ultrastruct. Res.* **49,** 419–435.
Maramorosch, K. (1968). *Curr. Top. Microbiol. Immunol.* **42,** 1–192.
Maramorosch, K., and Shope, R. E., eds. (1975). "Invertebrate Immunity." Academic Press, New York.
Mazzone, H. M. (1975). *In* "Baculoviruses for Insect Pest Control: Safety Considerations" (M. D. Summers *et al.*, eds.), pp. 33–37. Am. Soc. Microbiol., Washington, D. C.
Nappi, A. J., and Hammill, T. M. (1975). *J. Invertebr. Pathol.* **26,** 387–392.
Norton, P. W., and Dicapua, R. A. (1975). *J. Invertebr. Pathol.* **25,** 185–188.
Padhi, S. B., Eikenberry, E. F., and Chase, T., Jr. (1974). *Intervirology* **4,** 333–345.
Paschke, J. D., and Summers, M. D. (1975). *In* "Invertebrate Immunity" (K. Maramorosch and R. E. Shope, eds.), pp. 75–112. Academic Press, New York.
Raghow, R., and Grace, T. D. C. (1974). *J. Ultrastruct. Res.* **47,** 384–399.
Robertson, J. S., Harrap, K. A., and Longworth, J. F. (1974). *J. Invertebr. Pathol.* **23,** 248–251.
Scott, H. A., and Young, S. Y. (1973). *J. Invertebr. Pathol.* **21,** 315–317.
Smith, K. M. (1967). "Insect Virology." Academic Press, New York.
Smith, K. M. (1971). *In* "Comparative Virology" (K. Maramorosch and E. Kurstak, eds.), pp. 479–507. Academic Press, New York.
Steinhaus, E. A., ed. (1963). "Insect Pathology: An Advanced Treatise," Vol. 1. Academic Press, New York.
Stoltz, D. B., Pavan, C., and da Cunha, A. B. (1973). *J. Gen. Virol.* **19,** 145–150.
Stoltz, D. B., Vinson, S. B., and MacKinnon, E. A. (1976). *Can. J. Microbiol.* **22,** 1013–1023.
Summers, M. D. (1971). *J. Ultrastruct. Res.* **35,** 606–625.
Summers, M. D. (1975). *In* "Baculoviruses for Insect Pest Control: Safety Considerations" (M. D. Summers *et al.*, eds.), pp. 17–29. Am. Soc. Microbiol., Washington, D. C.

Summers, M. D., and Arnott, H. J. (1969). *J. Ultrastruct. Res.* **28,** 462–480.
Summers, M. D., and Egawa, K. (1973). *J. Virol.* **12,** 1092–1103.
Summers, M. D., and Paschke, J. D. (1970). *J. Invertebr. Pathol.* **16,** 227–240.
Summers, M. D., and Smith, G. E. (1975). *J. Virol.* **16,** 1108–1116.
Summers, M. D., and Smith, G. E. (1976). *Intervirology* **6,** 168–180.
Summers, M. D., and Volkman, L. E. (1976). *J. Virol.* **17,** 962–972.
Summers, M. D., Engler, R., Falcon, L. A., and Vail, P. V., eds. (1975). "Baculoviruses for Insect Pest Control: Safety Considerations." Am. Soc. Microbiol., Washington, D. C.
Tanada, Y., Hess, R. T., and Omi, E. M. (1975). *J. Invertebr. Pathol.* **26,** 99–104.
Teakle, R. E. (1969). *J. Invertebr. Pathol.* **14,** 18–27.
Tinsley, T. W., and Harrap, K. A., eds. (1972). "Moving Frontiers in Invertebrate Virology," Vol. 6. Karger, Basel.
Vago, C., and Bergoin, M. (1968). *Adv. Virus Res.* **13,** 247–303.
Vaughn, J. L., Adams, J. R., and Wilcox, T. (1972). *In* "Moving Frontiers in Invertebrate Virology" (T. W. Tinsley and K. A. Harrap, eds.), Vol. 6, pp. 27–35. Karger, Basel.
Volkman, L. E., and Summers, M. D. (1975). *J. Virol.* **16,** 1630–1637.
Wildy, P. (1971). "Monographs in Virology," Vol. 5. Karger, Basel.
Young, S. Y., and Lovell, J. S. (1973). *J. Invertebr. Pathol.* **22,** 471–472.
Zherebtsova, E. N., Strokvskaya, L. I., and Gudz-Gorban, A. P. (1972). *Acta Virol. (Engl. Ed.)* **16,** 427–431.

Fig. 1. GV of *Plodia interpunctella* (Arnott and Smith, 1968a). Bar = 0.5 μm.

Fig. 2. MNPV of *Heliothis armigera* (courtesy of C. Y. Kawanishi). Bar = 1.0 μm.

Fig. 3. SNPV of the cabbage looper, *Trichoplusia ni* (M. D. Summers). Bar = 0.5 μm.

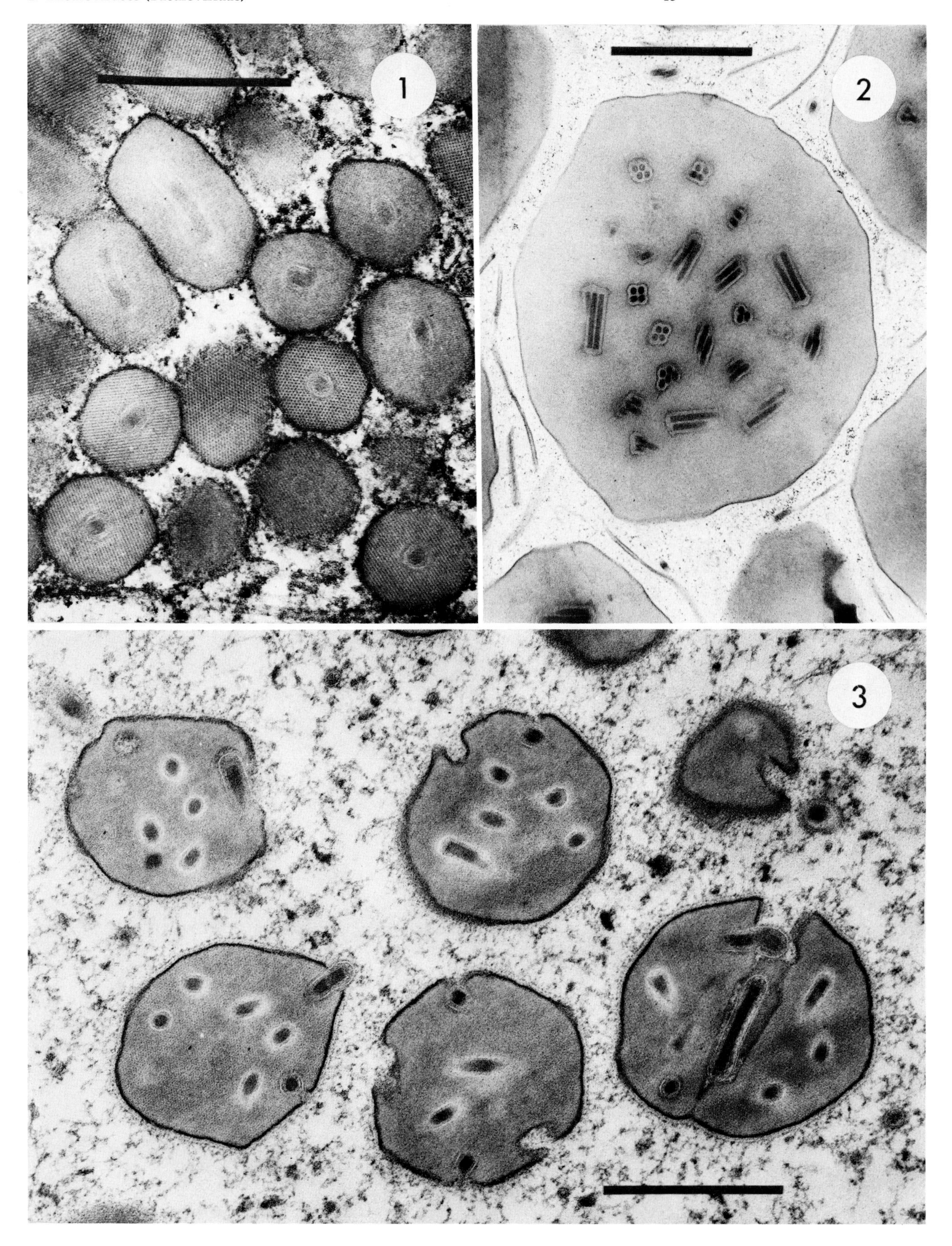
1
2
3

Fig. 4. SNPV of the pink shrimp, *Penaeus duorarum* (Couch, 1974). Bar = 1.0 μm.

Fig. 5. GV of *Plodia interpunctella* (Arnott and Smith, 1968a). Bar = 0.1 μm.

Fig. 6. GV of *Plodia interpunctella* (Arnott and Smith, 1968b). Bar = 0.1 μm.

Fig. 7. SNPV of *Rhynchosciara angelae* (courtesy of D. B. Stoltz). Bar = 1.0 μm.

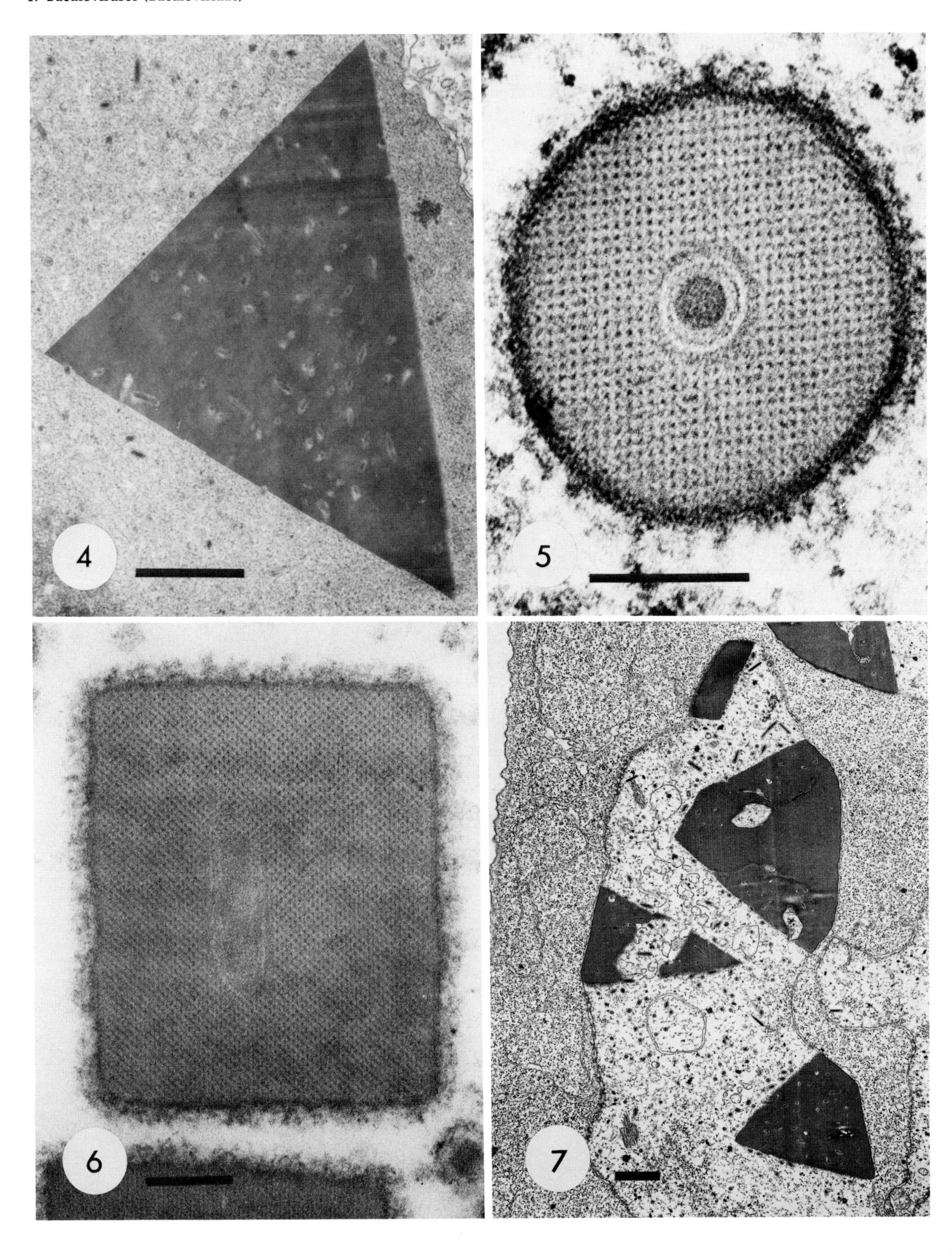
4
5
6
7

Fig. 8. SNPV of *Penaeus duorarum*. Note the subunit structure of the crystalline occlusion in the vicinity of enveloped nucleocapsid incorporation (Couch, 1974). Bar = 0.5 μm.

Fig. 9. Bundles of occluded enveloped nucleocapsids, *Heliothis armigera* (courtesy of C. Y. Kawanishi). Bar = 0.1 μm.

Fig. 10. Particles with baculovirus-like structure as observed in the secretion derived from a braconid parasitoid ovary, *Chelonus texanus* (Stoltz *et al.*, 1976). Bar = 0.1 μm.

Fig. 11. NPV nucleocapsids in process of envelopment by intranuclear membranes of possible *de novo* origin (Stoltz *et al.*, 1973). Bar = 0.5 μm.

Fig. 12. Nucleocapsid of the GV of *Trichoplusia ni* showing structural differentiation at opposite ends of the particle (courtesy of D. B. Stoltz). Bar = 0.1 μm.

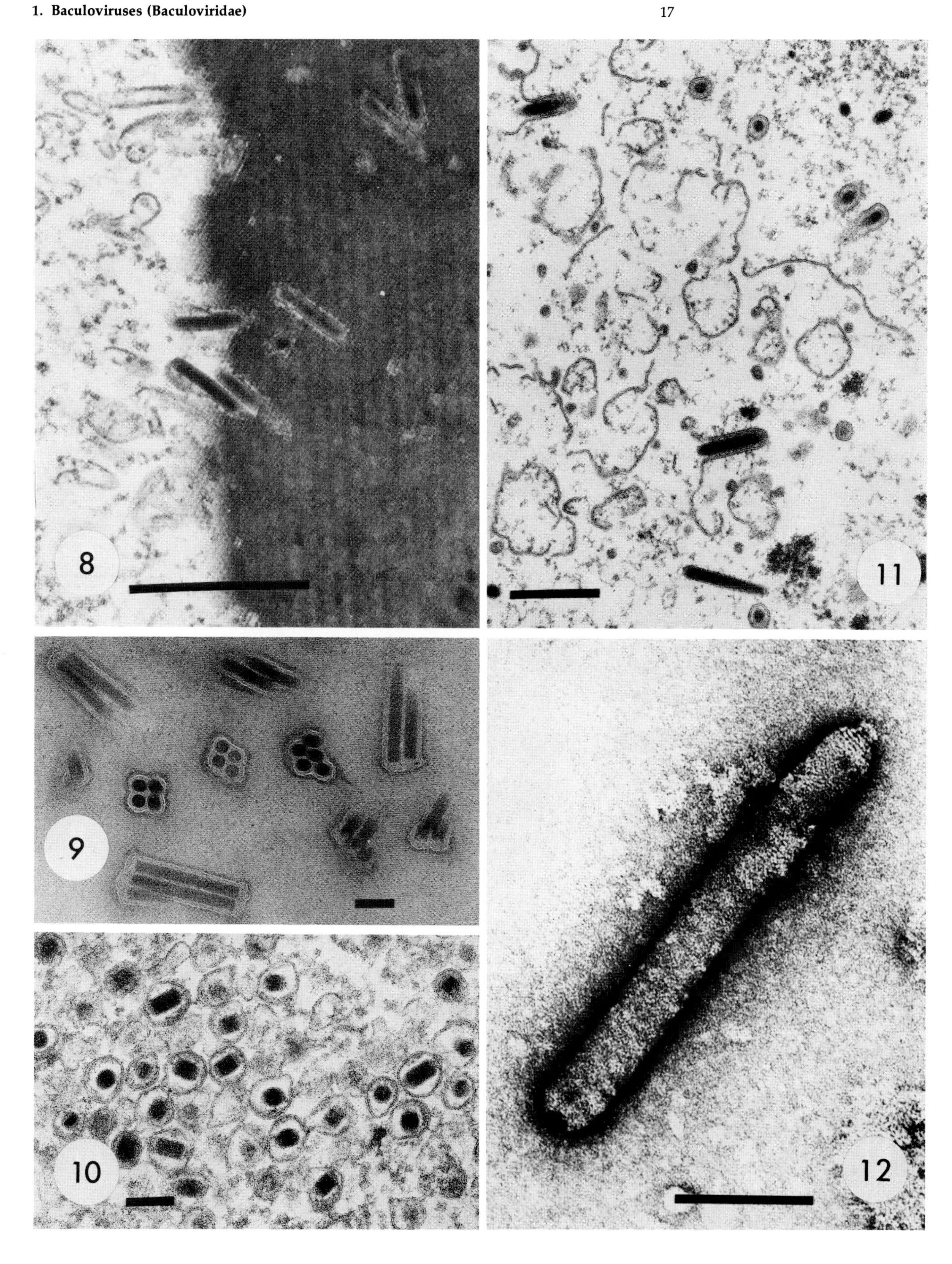
8
9
10
11
12

Fig. 13. "Membrane" profile and fibrillar elements observed in a MNPV infection. Note the structural similarity of one-half of the profile with the polyhedral "membrane" (Summers and Arnott, 1969). Bar = 0.2 μm.

Fig. 14. Fibrillar elements in association with MNPV polyhedra of *Rachiplusia ou* (Summers and Arnott, 1969). Bar = 1.0 μm.

Fig. 15. Fibrillar elements and "profiles" in a GV infection of fat body cells of *Trichoplusia ni*. Bar = 0.5 μm.

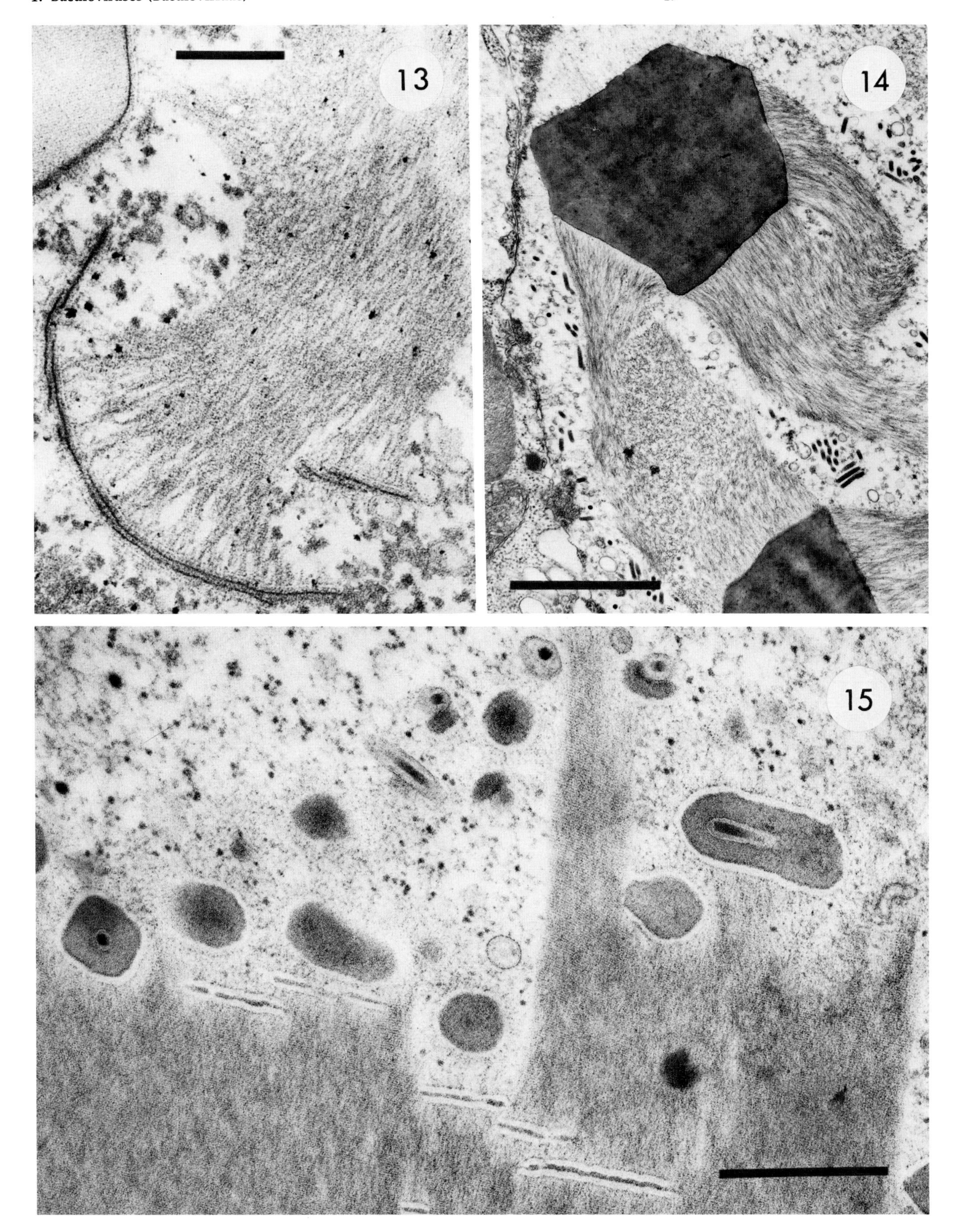
13
14
15

Fig. 16. Aberrant occlusion formation in the GV infection of *Plodia interpunctella* (Arnott and Smith, 1968b). Bar = 0.5 μm.

Fig. 17. Capsids of the MNPV of *Autographa californica*. Bar = 0.1 μm.

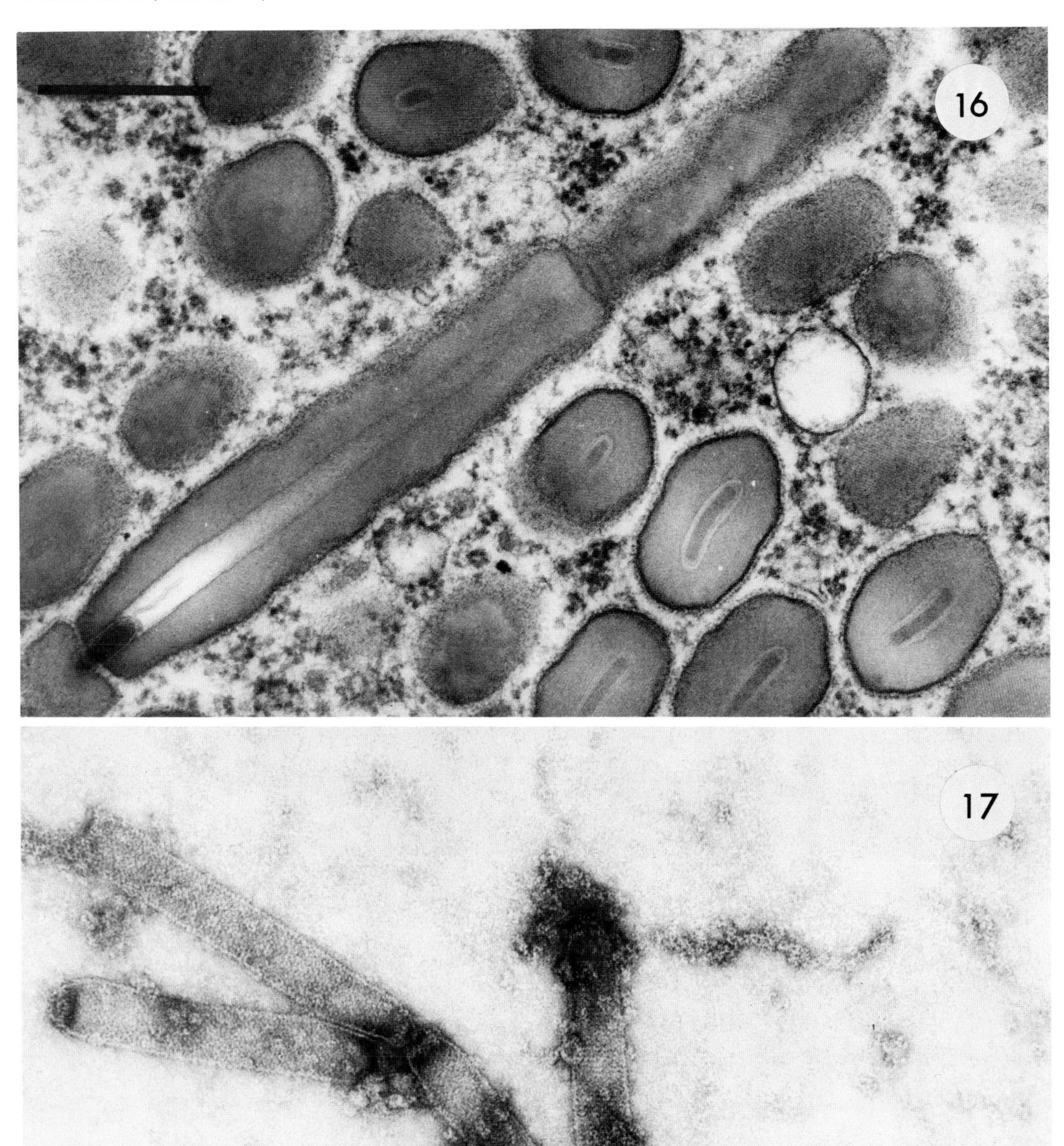
16
17

Fig. 18. Enveloped nucleocapsids of the GV of *Trichoplusia ni* after alkali solubilization of the occlusion and purification of the virus. Note the difference of observed structure where the envelope remains intact. Bar = 0.1 μm.

Fig. 19. Nonoccluded enveloped nucleocapsid of *Autographa californica* purified from cell culture medium (Summers and Volkman, 1976). Bar = 0.1 μm.

Figs. 20 and 22. Nuclear pore interaction by the GV of *Trichoplusia ni* (Summers, 1971). Bar = 0.5 μm.

Fig. 21. Disruption of the occlusion protein by exposure to 0.1 *M* Na_2CO_3, pH 10.9, to release the enveloped nucleocapsid of the GV of *Trichoplusia ni* (courtesy of C. Y. Kawanishi and D. B. Stoltz). Bar = 0.1 μm.

Fig. 23. Nonoccluded GV of *Trichoplusia ni* in infected midgut cell (Summers, 1971). Bar = 0.3 μm.

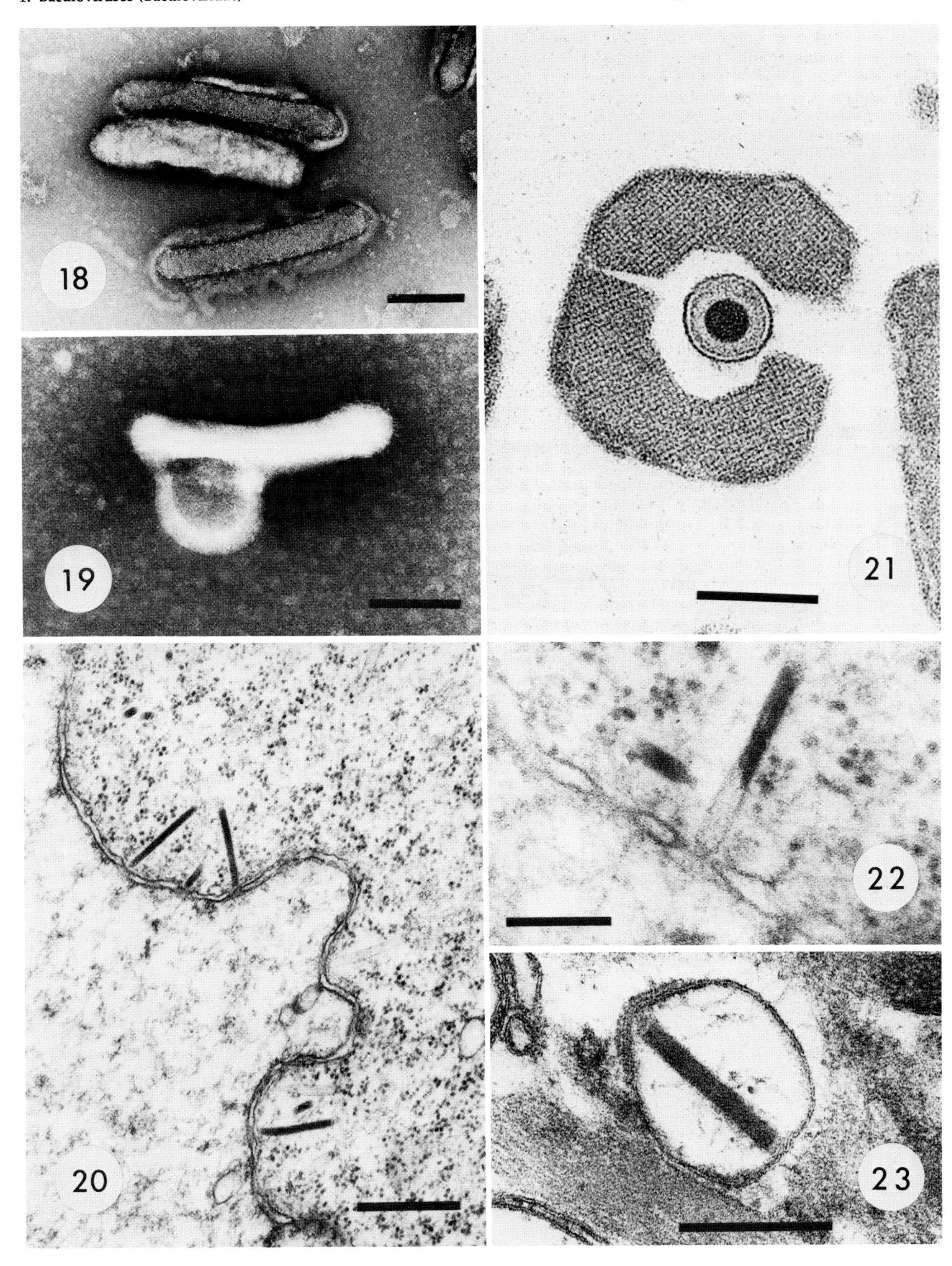
18
19
20
21
22
23

Fig. 24. Fusion of *Rachiplusia ou* MNPV with the microvillar membrane of *Trichoplusia ni* midgut cell (Kawanishi *et al.*, 1972b). Bar = 0.1 μm.

Fig. 25. Penetration of a bundle of nucleocapsids of nonoccluded *Spodoptera frugiperda* MNPV into *Spodoptera frugiperda* cells in culture (Knudson and Harrap, 1976). Bar = 0.3 μm.

Fig. 26. Cytopathology of *Rachiplusia ou* MNPV infection in tracheol cells of *Trichoplusia ni*. Note the virogenic stroma in the cell with occlusions. Bar = 0.2 μm.

Figs. 27 and 28. GV nucleocapsids in midgut cell microvilli (Summers, 1971). Bar = 0.2 μm.

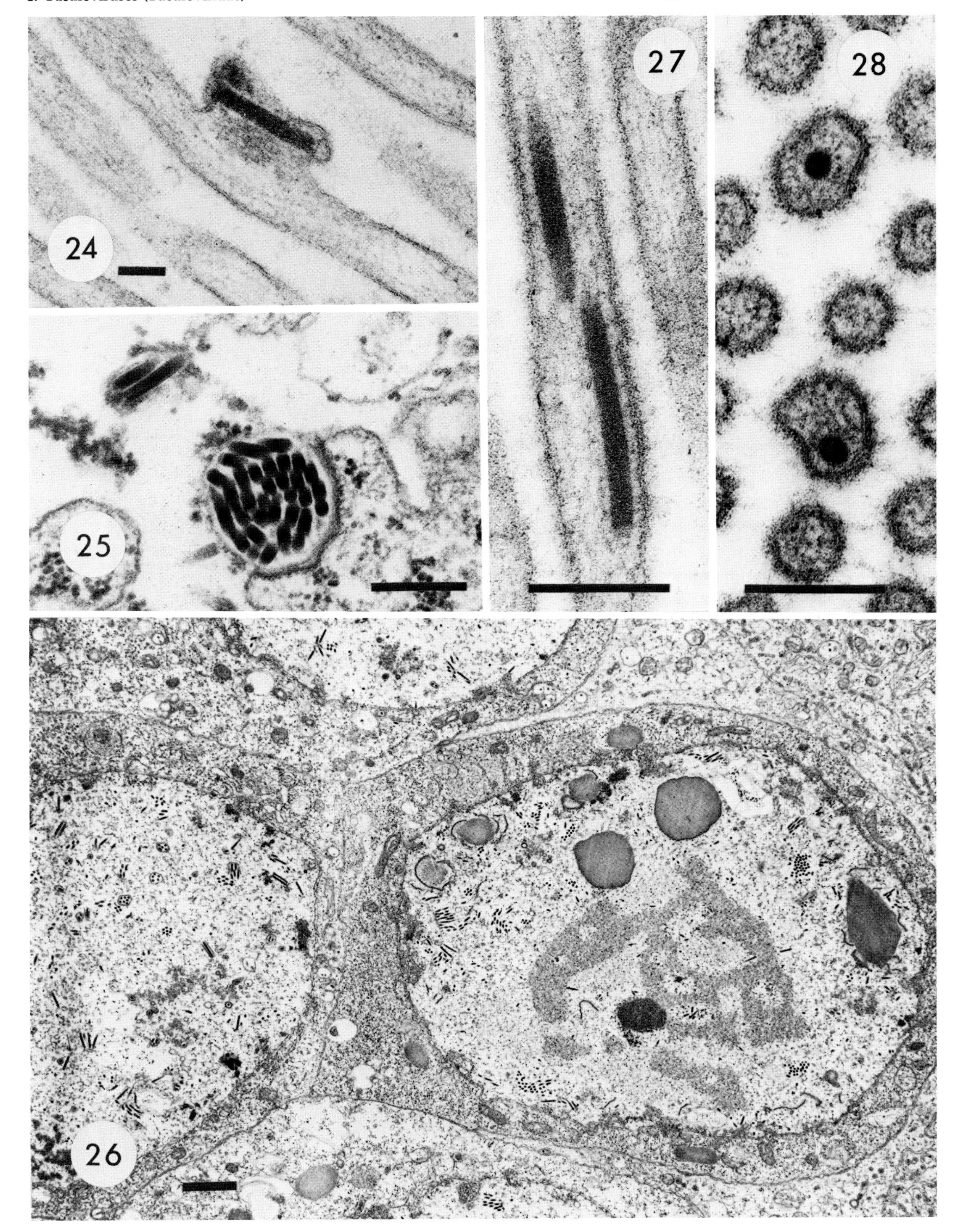
24
25
27
28
26

Fig. 29. Circular DNA molecule of the GV of *Trichoplusia ni*. Note the highly coiled form with an apparent break in the molecule. Bar = 1.0 μm.

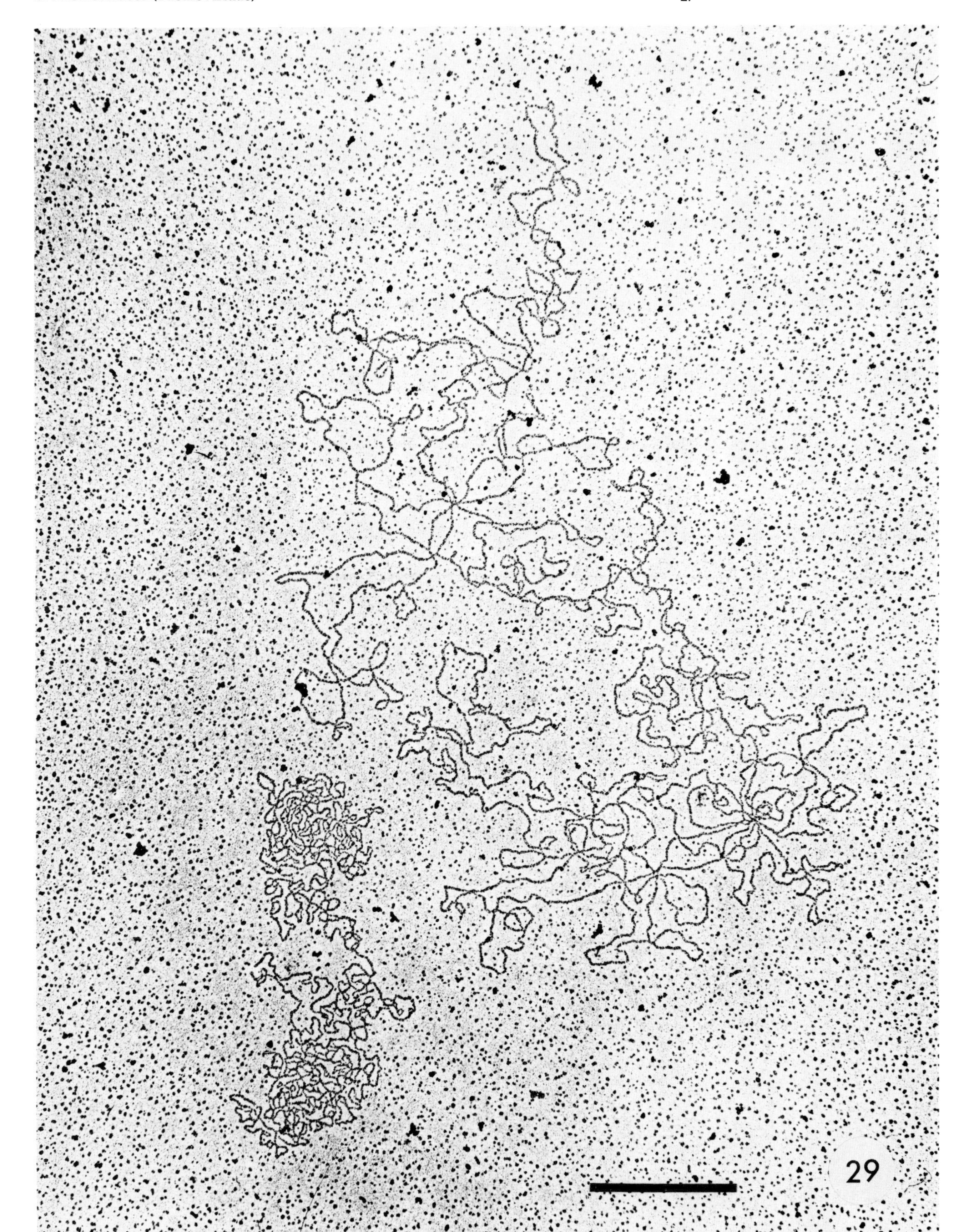
29

Chapter 2

Entomopoxviruses (Poxviruses of Invertebrates)

EDOUARD KURSTAK AND SIMON GARZON

I. Introduction 29
II. Taxonomy 30
III. The Virion 30
IV. Inclusion Bodies 31
A. Spheroids 31
B. Spindles 34
V. Pathology 34
VI. Morphogenesis 35
A. Penetration 35
B. Maturation 35
VII. Conclusion 37
References 37

I. INTRODUCTION

The poxviruses of invertebrates have been described only in insects and are known as entomopoxviruses. These viruses present morphological and physicochemical characteristics very similar to those of the poxviruses of vertebrates. However, all known entomopoxviruses (EPV) occur predominantly as occluded particles within proteinaceaous inclusion bodies, named spheroids (Goodwin and Filshie, 1969). Fusiform paracrystalline bodies (spindles) devoid of virus particles are usually also produced during the replicative cycle of most of these viruses.

Some entomopoxviruses were known before their isolation as the agents of the spindle diseases or viroses (Vago and Bergoin, 1968). A decade after the isolation of the first entomopoxvirus still very little work has been done on the characterization of this new genus of viruses. More is known on the pathology of these diseases, which are produced in several species of insects.

Since the first EPV disease was reported in the coleopteran larvae *Melolontha melolontha* (Hurpin and Vago, 1963), several new members of the entomopoxvirus genus have been discovered. Entomopoxviruses were found to infect several species of insects, belonging to Coleoptera, Lepidoptera, Diptera, and Orthoptera, with a widespread geographical distribution.

Three of these EPV, those infecting larvae of the Coleoptera *Melolontha melolontha,* the Lepidoptera *Estigmene acrea,* and more recently the Lepidoptera *Choristoneura fumiferana* have been studied extensively (Arif, 1976; Bergoin and Dales, 1971; Bergoin *et al.*, 1968a,b, 1969a,b, 1971; Bird, 1974; Devauchelle *et al.*, 1970, 1971; Granados, 1970, 1973a,b; Granados and Roberts, 1970; Granados and Naughton, 1975; McCarthy *et al.*, 1974, 1975a,b; Roberts and Granados, 1968). Complementary studies on the structure and morphogenesis of the *Choristoneura fumiferana* poxvirus were recently initiated in this laboratory.

II. TAXONOMY

The new genus *Entomopoxvirus* was included by the International Committee on Taxonomy of Viruses (ICTV) in the poxvirus group where the type species is the vaccinia virus (Wildy, 1971).

The genus *Entomopoxvirus* has the following cryptogram: D/2: 140-240/5-6:X/*:I/0 and includes three probable subgenera based on the morphology of the virions (Fenner, 1977). The best known species are *Entomopoxvirus melolontha* (subgenus A), *Entomopoxvirus amsacta* and *Entomopoxvirus choristoneura* (subgenus B) and *Entomopoxvirus chironomi* (subgenus C). Approximately 24 species or variants of entomopoxviruses are presently under study (Table I).

III. THE VIRION

The virions of entomopoxviruses are either oval or brick-shaped and vary from 150 to 470 nm in length and from 165 to 300 nm in width. They are largest in the Coleoptera and smallest in the Diptera (Table I).

Negatively stained virions show on their surface the presence of spherical units which appear to be formed by folds in the outer virus envelope (Fig. 1a). Depending on the virus species, these spherical units measure 22 nm for the Coleoptera *Melolontha* (Bergoin *et al.*, 1971) and 40 nm for the Lepidoptera *Amsacta* (Granados and Roberts, 1970) and give the virion a beaded surface analogous to the M-(mulberry) forms of the vertebrate poxviruses (Westwood *et al.*, 1964). The C-(capsule) forms where the beaded surface is not clearly visible are also present in negatively stained preparations (Fig. 1b) and may represent a stage of degradation of the virus particle (Westwood *et al.*, 1964; Easterbrook, 1966).

In fine sections, the mature virion consists of a beaded lipoprotein envelope, a plate-like core containing the double-stranded DNA viral genome, and one or two lateral bodies (Figs. 1c and d). The core contains a substance of high electron opacity delimited by a three-layered coat. Inside the core, a coiled rodlike, electron lucent structure, folded into 4 to 6 segments, is usually present (Figs. 2a and d). No rodlike structure is observed in the occluded dipteran poxviruses (Fig. 3). Between the core coat and the virus envelope, a substance of intermediate density forms one or two lateral bodies. Most of the occluded EPV seem to be surrounded by a diffuse proteinous layer of low density producing a "halo" (Bergoin *et al.*, 1968b). Removal of the "halo" by trypsin results in virions with a RNA-polymerase activity and a relatively high infectivity (McCarthy *et al.*, 1975a). Such layers do not appear to be associated with the *Choristoneura* poxvirus (Arif, 1976). On the basis of the size and morphology of the virion, 3 subgroups can be distinguished: (i) the EPV from Lepidoptera (Figs. 2a and b) and Orthoptera, generally smaller (350 × 250 nm) than the Coleoptera EPV, possessing a symmetrical cylindrical core and two discrete lateral bodies; (ii) the EPV infecting the Coleoptera (Figs. 2c and d), relatively large (450 × 250 nm) and having a unilaterally concave core with a single lateral body located in the concavity of the core; (iii) the Diptera EPV (Fig. 3), smallest (320 × 230 nm) and cuboidal in outline. They present a biconcave core and two well-developed lateral bodies. These viruses resemble the vertebrate poxviruses more closely than any other entomopoxvirus.

The virions are reported to have a buoyant density of 1.26 gm/cm for the *Amsacta* poxvirus and 1.31 gm/cm for the *Choristoneura* poxvirus as compared to 1.28 determined for some vertebrate poxviruses (Yau and Rouhandeh, 1972, 1973). The genetic material of this virus is a double-stranded DNA which constitutes approximately 5–6% of the particle. Four enzyme activities have been found in the *Amsacta* virions as part of their structural proteins: an RNA polymerase, a nucleotide phosphohydrolase, and neutral and acidic DNases (Pogo *et al.*, 1971; McCarthy *et al.*, 1974, 1975a). The viral DNA genome was estimated to have a mean length of 65 nm, a molecular weight of 132–160 × 10^6 daltons and a sedimentation coefficient of 59 S. The DNA melted at 61.6° C and had a density of 1.6849 gm/ml. The G + C content of the *Choristoneura* poxvirus DNA investigated by thermal denaturation and buoyant density analysis in CsCl gave values of 26.6% and 25.4%, respectively, which are lower than those reported for the DNA of vertebrate poxviruses (G + C = 32.5–39%). So far, no antigenic relationship has been demonstrated between entomopoxviruses and vertebrate poxviruses (Arif, 1976; Bergoin *et al.*, 1969a; Langridge and Roberts, 1975; McCarthy *et al.*, 1974, 1975a,b; Roberts *et al.*, 1975).

IV. INCLUSION BODIES

The inclusion bodies observed in EPV can be either spheroid or spindle-shaped.

A. Spheroids

These inclusion bodies appear as large (up to 20–24 μm in diameter) paracrystalline, proteinaceaous structures. The number of virions occluded per spheroid is highly variable in different hosts and even in the different cell types of the same host. The oval-shaped virions are occluded in the inclusion bodies, usually without any fissure in the crystalline lattice (Figs. 2b and d; Figs. 3c and d). The distribution of the virions within the crystalline lattice appears to be random or oriented radially in relation to the center of the inclusion. Spheroids that have completed their development usually have a marginal zone or cortex of varying thickness which contains no virus particles. These structures are enveloped in a distinctive epispheroid layer with associated fibrils (Fig. 17b). The size, form, and distribution of the spheroids depend on the EPV species. They range in shape from nearly spherical (coleopterous host) to nearly regular ellipsoidal (lepidopterous host). Some unusual shapes have also been noted such as in the case of *Demodena boranensis* virus and of *Melanoplus sanguinipes* virus which tend to be either rounded subcubical or ellipsoidal in sagittal section but almost square in cross section (Henry *et al.*, 1969). Generally, in the infected tissues of the Coleoptera (Fig. 4b) one or more large spheroids develop per cell but in the Lepidoptera (Fig. 4d) numerous smaller spheroids develop per cell.

TABLE I[a]

Morphological and Pathological Characteristics of Some Entomopoxviruses

Original host	Virus size (nm)	Inclusion body size (μm)	Virus shape	Virus core shape (vertical sections)	Virus-free spindles[b]	Tissue target	References
Coleoptera							
Melolontha melolontha (Scarabaeidae)	450 × 250	10–24	Oval	Unilaterally concave	Present, macrospindle, NO	Adipose tissue; hemocyte	Hurpin and Vago (1963); Bergoin *et al.* (1968a,b); Hurpin and Robert (1967)
Othnonius batesi (Scarabaeidae)	470 × 265	5–10	Oval	Unilaterally concave	Present, large, NO	Adipose tissue	Goodwin and Filshie (1969) Goodwin and Roberts (1975)
Demodena boranensis (Scarabaeidae)	420 × 230	7.8–11.0	Oval	Unilaterally concave	Present, large, NO	Adipose tissue	Vago *et al.* (1968a)
Geotrupes silvaticus (Scarabaeidae)	366–416 × 255–286	3.5–11	Oval	Unilaterally concave	Present, large, NO	Adipose tissue	Lipa and Bartkowski (1972)
Dermolepida alborhirtum (Scarabaeidae)	420–450 × 220–240	3–5	Oval	Unilaterally concave	Absent	Adipose tissue	Goodwin and Filshie (1975) Goodwin and Roberts (1975)
Aphodius tasmaniae (Scarabaeidae)	380–430 × 250–300	5–12	Oval	Unilaterally concave	Present, NO	Adipose tissue	Goodwin and Filshie (1975)
Anomala cuprea (Scarabaeidae)	440 × 250	5 × 8	Oval	Unilaterally concave	Present, large, NO	Adipose tissue; hemocyte	Katagiri *et al.* (1975)
Phyllopertha horticola (Rutelidae)	400 × 240	6–25	Oval	Unilaterally concave	Present, large, NO	Adipose tissue	Vago *et al.* (1969)
Figulus sublaevis (Lucanidae)	330 × 290	1–5	Oval	Unilaterally concave	Present, large, NO	Adipose tissue; hypoderm; and gut (rare)	Vago *et al.* (1968b)
Lepidoptera							
Amsacta moorei (Arctiidae)	350 × 250	1–4	Oval	Rectangular	Rare	Polytropic nature	Roberts and Granados (1968) Granados and Roberts (1970)
Oreopsyche angustella (Psychidae)	360 × 260	3–10 × 2–7	Oval	Rectangular	Present	Adipose tissue	Meynadier *et al.* (1968)
Operophtera brumata (Geometridae)	400 × 350	3–15	Oval	Rectangular	Present, large	Adipose tissue	Weiser and Vago (1966); Weiser *et al.* (1970)
Euxoa auxiliaris (Noctuidae)	260 × 165	4–5	Oval	Rectangular	Absent	Adipose tissue	Sutter (1972)
Choristoneura biennis (Tortricidae)	400 × 300	2.2–3.2	Oval	Rectangular	Present, O	Adipose tissue; midgut	Bird *et al.* (1971)
Choristoneura conflictana (Tortricidae)	273 × 235	7 × 4.3	Oval	Rectangular	Present, O	Adipose tissue	Cunningham *et al.* (1973)
Choristoneura diversana (Tortricidae)	280 × 220	7 × 5	Oval	Rectangular	Present, O		Katagiri (1973)

Orthoptera							
Melanoplus sanguinipes (Acrididae)	320–250	2–11	Oval	Rectangular or dumbbell shaped		Adipose tissue	Henry *et al.* (1969)
Diptera							
Chironomus luridus (Chironomidae)	320 × 230 × 110	4–7	Cuboidal	Dumbbell shaped	Absent	Polytropic nature	Götz *et al.* (1969); Huger *et al.* (1970)
Chironomus attenuatus (Chironomidae)	330 × 250 × 130	up to 6	Cuboidal	Dumbbell shaped	Absent	Hemocyte	Stoltz and Summers (1972)
Camptochironomus tentans (Chironomidae)	200–250 × 270–300 × 130–150	2.16 × 8.10	Cushion shaped	Dumbbell shaped	Absent	Adipose tissue	Weiser (1969)
Goeldichironomus holo-prasinus (Chironomidae)	346 × 300 × 160	3 × 5	Cushion shaped	Dumbbell shaped	Absent	Hemocyte	Federici *et al.* (1974)
Aedes aegypti (Culicidae)	320 × 230			Dumbbell shaped			Buchatsky (1974)

[a] Modified from Granados (1973).
[b] O, occluded; NO, nonoccluded.

B. Spindles

These paracrystalline inclusion bodies are devoid of virus particles and are produced in large numbers in the course of an infection with a variety of EPV. The molecular lattice spacing of the spindle is about 58–60 Å. It is to be noted that no spindles are associated with the infection with Diptera or Orthoptera EPV(Table I). Macrospindles (up to 25 μm) have been described in all EPV infections of Coleoptera (Fig. 12) with the exception of the *Dermolepida albohirtum*. However, most EPV of Lepidoptera are associated with the formation of microspindles and in this species they may be occluded randomly, together with the virions within the spheroids (Fig. 15a). Roberts (1970) reported that while spindles are not produced in *Estigmene acrea* infected by the *Amsacta* EPV they are produced in *Galleria mellonella* infected by the same virus. This observation suggests that the production of spindles could be a host-dependent phenomenon.

The spindle proteins seems to be antigenically and chemically different from the spheroid proteins (Bergoin and Veyrunes, 1970; Croizier and Veyrunes, 1971; Bergoin and Dales, 1971; Bergoin *et al.*, 1976) and react as a deoxyribonucleoprotein with acridine orange or by the EDTA staining method (Garzon and Kurstak, in preparation). On the other hand, spheroid proteins are not bleached by this latter method (Fig. 11c; Fig. 12f). The origin, function, and significance of the spindle proteins are unkown. They have no significance in the development and propagation of the virus and are even not present in some of the EPV infections.

The variations in the size and shape of the virions and spheroids as well as the presence or absence of spindles among the known EPV may indicate either different virus strains or species or may result from host-induced variations in virus morphology (Roberts, 1970; Goodwin and Roberts, 1975). Further work is needed, particularly in cross-infection experiments, to elucidate this question. In the case of the replication of the *Amsacta* EPV in *G. mellonella,* the replicative cycle is slowed down and is often incomplete, particularly at the level of the incorporation of the virions into the spheroids (Fig. 17c; Fig. 18). Fewer virions appeared also to be occluded in spheroids in the course of infection by *Amsacta moorei* EPV of primary ovarian and hemocytes tissue cultures prepared from *Estigmene acrea* larvae (Roberts, 1970; Granados and Naughton, 1975; Quiot *et al.*, 1975).

V. PATHOLOGY

Infection with EPV shows a range of cell tissue and host specificity (Table I). The host range seems to be relatively restricted. Even if some cross infection could be obtained in closely related species, no cross-infections were ever produced between the different insect families (Table II). The majority of the EPV replicates in the cytoplasm of the fat body cell and this replication is characterized by the presence of large spheroid inclusions accompanied in some cases by macrospindles. Microspindles are not visible under the light microscope (Fig. 4).

Some of the EPV are restricted to the fat body, others infect also the hemocytes or the epidermis, or produce systemic infection in their host. Hemocytes are the only apparent site of infection in larvae infected with the *Chironomus attenuatus* or *Goeldichironomus holoprasinus* EPV. Staining with acridine orange of EPV-infected hemocytes reveals an intense RNA synthesis in the infected cells which occurs before the appearance of spheroids and spindles (Fig. 4a). The

TABLE II

Cross-Transmission of Some Entomopoxviruses

EPV original host	EPV adapted hosts	References
Coleoptera	No success in cross-transmission	Hurpin and Robert (1967); Granados (1973b); Goodwin and Roberts (1975)
Lepidoptera		
Amsacta moorei	*Estigmene acrae; Galleria mellonella*	Granados and Roberts (1970)
Choristoneura biennis	*Choristoneura fumiferana*	Bird *et al.* (1971)
Choristoneura conflictana	*Choristoneura fumiferana*	Cunningham *et al.* (1973)
Choristoneura diversana	*Archippus isshikii*	Katagiri (1973)
Orthoptera		
Melanoplus sanguinipes	*Melanoplus bivatus; Melanoplus differentialis; Schistocerca americana*	Henry *et al.* (1969)

development of the disease in the tissues causes a hypertrophy of the cytoplasm and of the nucleus and a regression of the fat granules (Figs. 4c and d). In an advanced stage of the disease the tissue disintegrates (Fig. 4f). Weiser (1969) had also observed an increased division of fat body cells infected by the *Camptochironomus* EPV.

VI. MORPHOGENESIS

As the replicative cycle of EPV inoculated *in vivo* is asynchronous, its morphogenesis has to be deduced from vertebrate models and from a series of events already well studied with various EPV (Bergoin and Dales, 1971; Dales, 1973; Dales and Mosbach, 1968; Dales and Siminovitch, 1961; De Harven and Yohn, 1966; Ichihashi and Dales, 1973; Ichihashi *et al.*, 1971; Joklik, 1966).

A. Penetration

Penetration of EPV into a susceptible cell may depend on the route of infection. In an infection through the digestive system, which is the natural way of propagation of the virus, dissolution of the ingested spheroid and liberation of the occluded virus take place in the midgut of the infected larvae. The entry of the virion involves fusion of the viral envelope with the microvillus membrane of the gut epithelial cells, followed by the passage and migration of the core to the cell cytoplasm (Granados, 1973a) (Figs. 5a and b). In intrahemocoelic infection, however, penetration of the free virus is assured by a mechanism of viropexis (Devauchelle *et al.*, 1971). In this case, as for the vertebrate poxviruses, "denudation" of the virion takes place in the phagosome derived from the viropexis process (Figs. 5c and d; Fig. 6), followed by the release of the nucleoid in the cytoplasm where the "uncoating" and liberation of the DNA viral genome is completed (Fig. 6b).

B. Maturation

After the period of latency, the virus progeny is assembled in the cell cytoplasm, either in association with loci of electron dense amorphous viroplasm (viroplasm I) (Fig. 7) or with loosely aggregated matrices of granular material interspersed with numerous small spicules-coated spherical vesicles (viroplasm II) (Fig. 8) (Granados, 1973b). Immature viral particles are assembled at the periphery of one or the other viroplasm where membranes develop. Incomplete spherical envelopes encapsulate portions of the viroplasm before membrane closure (Fig. 7b; Figs. 8a and b). Envelopes of immature particles appear to consist of an inner unit membrane and an outer layer of spicules separated by a less dense layer (Fig. 8c; Fig. 9). The immature particles then undergo differentiation within the envelope leading to (i) the condensation of the inner material, (ii) the progressive differentiation of nucleoid and the lateral bodies, (iii) the development of the three-layered membrane of the nucleoid consisting of a unit membrane and an outer layer of spicules or subunits, (iv) the lateral compression of the particle, and (v) the modification of the outer envelope into the beaded mulberry outline (Fig. 1c; Fig. 2; Fig. 9; Fig. 17d).

Mature particles are usually dispersed away from the viroplasmic site. They migrate toward the cytoplasmic membrane for exocytosis (Fig. 10) or toward a cytoplasmic site where occlusion of viral particles into the paracrystalline spheroid occurs. Nonoccluded viral particles can acquire a modified cell membrane by budding on the ergastoplasm or the cytoplasmic membrane. The viruses liberated in such a manner may contribute to the cell-to-cell transmission of the infection (Figs. 10c and d). The process of viral occlusion must be very specific as only viral particles or microspindles (in the lepidopteran EPV) are occluded. Little is known about the process of nucleation and growth of spheroids. The spheroids differ in their nature and structure from the A-type inclusion bodies of vertebrate poxviruses. They are comparable, however, to the inclusion bodies of the polyhedrosis or granulosis viruses of insects, although the nature of their virions is completely different (Summers and Arnott, 1969). EPV spheroids first begin to appear only after the formation of some mature virions and microspindles (when present). A prerequisite of occlusion seems to be the presence of an outer membrane. The polyhedra of nuclear polyhedrosis viruses (NPV) also appear to nucleate on definite structures which may be originated from unit membrane derivatives (Summers and Arnott, 1969). A characteristic peripheral structure, probably not involved in the limitation of spheroid size, was observed on spheroids of *C. attenuatus* poxvirus (Stoltz and Summers, 1972) (Fig. 3d). The authors suggested that the deposition of spheroid protein may be initiated on the inner surface of this structure. In *C. fumiferana* we observed that nucleation takes place in different cytoplasmic sites and that a virion could be the initial site of nucleation (Fig. 9b). The spheroid growth occurs through continuous protein accumulation and crystallization and the random occlusion of virions and microspindles in the vicinity of the initial nucleation and/or through fusion of several small spheroids in the course of formation (Fig. 14b; Fig. 15b). It has been reported that immature viral particles may also be occluded, the maturation of which being completed in the course of occlusion. However, this postoccluded maturation process was not observed during the development of *Amsacta* EPV in *E. acrea* larvae (Granados and Roberts, 1970). Structural changes which occur within the spheroid for most EPV are (i) a reduction in size, (ii) the acquisition of the wavy appearance of the outer viral envelope, (iii) the development of a more electron-dense core with the appearance of a coiled rodlike structure, and (iv) the formation in some case of a "halo" of variable size (Figs. 11, 14, 15, 16, and 17). These changes were interpreted by some authors as representing artifacts due to fixation or resulting from the pressure, exerted by the molecular arrangement of spheroid proteins around the viral particles (Bergoin *et al.*, 1971; Granados, 1973b).

Parallel to the production of virions and to their occlusion into the spheroids, paracrystalline spindles are often produced in the cytoplasm. These spindles are formed by the accumulation or progressive synthesis of an electron dense material which is then followed by its crystallization inside the vesicles surrounded by a unit membrane (Fig. 12a). Membrane fragments are often observed inside these structures (Fig. 12b). The spindles induced in *M. melolontha* by infection with EPV are very large. First produced inside the vesicles, derived from degenerative mitochondria or from the endoplasmic reticulum, the constant growth of these spindles leads to the rupture of the membrane of the vesicle (Fig. 12c). A sponge-like, less electron dense substance envelopes the spindles (Fig. 12d). The chemical nature of this material seems to be different from that of the spindles (Figs. 12e and f). The spindles produced in *C. fumiferana* by infection with EPV have a smaller size and grow within the limits of an ergastoplasmic membrane (Fig. 15a). The spindles seem to be occluded together with their vesicular membrane (Fig. 15b). However, it is difficult to state whether or not the presence of this membrane is a prerequisite for the occlusion.

Hypertrophied mitochondria and fascicles of fibrils (Fig. 16b; Figs. 17a, b, and c) are frequently observed in the cytoplasm of infected Lepidoptera cells, often in relation with the spheroids in formation (Fig. 17a). Tubules with a helicoidal structure, similar to the one described in the *in vitro* infection with cowpox (Ichihashi *et al.*, 1971) have been observed in the EPV infected larvae of *M. melolontha* (Fig. 13). Accumulation of concentric lamellae are often seen in the cytoplasm of EPV infected Diptera cells (Huger *et al.*, 1970; Stoltz and Summers, 1972). The nature and the role of these structures remain to be determined.

Simultaneously with the formation of spheroids in the cytoplasm, the nucleus also shows some changes: margination and fragmentation of the chromatin, formation of dense, amorphous masses, bundles of fibrils, and/or paracrystalline inclusions (Fig. 14a). Some observations suggest that the nucleus is involved in the synthesis of the spindles (Bergoin *et al.*, 1976).

VII. CONCLUSION

In recent years the entomopoxviruses have been extensively studied by several workers as a new genus of poxviruses. The *in vitro* inoculation of continuous cell lines of insects was of considerable importance in advancing our knowledge of Entomopoxviruses. It *permitted* the comparative study of ultrastructural and biochemical changes of the multiplication cycle. The results obtained so far show that poxviruses of invertebrates represent a relatively homogeneous group. The possible cross-infections of different species of insects by entomopoxviruses brings further stimulus for additional research on their specificity for invertebrates and possibly for vertebrates, including man.

ACKNOWLEDGMENTS

This work was supported by grants A-3746 from the National Research Council of Canada and MA-2385 from the Medical Research Council of Canada.

We wish to thank Drs. F. T. Bird and S. Sohi (Insect Pathology Research Institute, Sault Ste-Marie, Ontario) and Dr. D. W. Roberts (Boyce Thompson Institute, Yonkers, New York) for providing *C. fumiferana* and *Amsacta* poxviruses, respectively.

REFERENCES

Arif, B. M. (1976). *Virology* **69,** 626.

Bergoin, M., and Dales, S. (1971). *In* "Comparative Virology" (K. Maramorosch and E. Kurstak, eds.), pp. 169–205. Academic Press, New York.

Bergoin, M., and Veyrunes, J. C. (1970). *Virology* **40,** 760.
Bergoin, M., Devauchelle, G., Duthoit, J. L., and Vago, C. (1968a). *C. R. Hebd. Seances Acad. Sci.* **266,** 2126.
Bergoin, M., Devauchelle, G., and Vago, C. (1968b). *C. R. Hebd. Seances Acad. Sci.* **267,** 382.
Bergoin, M., Devauchelle, G., and Vago, C. (1969a). *Arch. Gesamte Virusforsch.* **28,** 285.
Bergoin, M., Veyrunes, J. C., and Vago, C. (1969b). *C. R. Hebd. Seances Acad. Sci.* **269,** 1464.
Bergoin, M., Devauchelle, G., and Vago, C. (1971). *Virology* **43,** 453.
Bergoin, M., Devauchelle, G., and Vago, C. (1976). *J. Ultrastruct. Res.* **55,** 17.
Bernhard, W. (1968). *C. R. Hebd. Seances Acad. Sci.* **267,** 2170.
Bird, F. T. (1974). *J. Invertebr. Pathol.* **23,** 325.
Bird, F. T., Sanders, C. J., and Burke, J. M. (1971). *J. Invertebr. Pathol.* **18,** 159.
Buchatsky, L. P. (1974). *Mikrobiol. Zh. (Kiev)* **36,** 797.
Croizier, G., and Veyrunes, J. C. (1971). *Ann. Inst. Pasteur (Paris)* **120,** 709.
Cunningham, J. C., Burke, J. M., and Arif, B. M. (1973). *Can. Entomol.* **105,** 767.
Dales, S. (1973). In "Ultrastructure of Animal Viruses and Bacteriophages: An Atlas" (A. J. Dalton and F. Haguenau, eds.), pp. 109–129. Academic Press, New York.
Dales, S., and Mosbach, E. H. (1968). *Virology* **35,** 564.
Dales, S., and Siminovitch, L. (1961). *J. Biophys. Biochem. Cytol.* **10,** 475.
De Harven, E., and Yohn, D. (1966). *Cancer Res.* **26,** 995.
Devauchelle, G., Bergoin, M., and Vago, C. (1970). *C. R. Hebd. Seances Acad. Sci.* **271,** 1138.
Devauchelle, G., Bergoin, M., and Vago, C. (1971). *J. Ultrastruct. Res.* **37,** 301.
Easterbrook, K. B. (1966). *J. Ultrastruct. Res.* **14,** 484.
Federici, B. A., Granados, R. R., Anthony, D. W., and Hazard, E. I. (1974). *J. Invertebr. Pathol.* **23,** 117.
Fenner, F. (1977). *Intervirology* **7,** 4–105.
Goodwin, R. H., and Filshie, B. K. (1969). *J. Invertebr. Pathol.* **13,** 317.
Goodwin, R. H., and Filshie, B. K. (1975). *J. Invertebr. Pathol.* **25,** 35.
Goodwin, R. H., and Roberts, R. J. (1975). *J. Invertebr. Pathol.* **25,** 47.
Götz, P., Huger, A. M., and Krieg, A. (1969). *Naturwissenschaften* **56,** 145.
Granados, R. R. (1970). *Proc. Int. Colloq. Insect Pathol., 4th, 1970* pp. 110–114.
Granados, R. R. (1973a). *Virology* **52,** 305.
Granados, R. R. (1973b). *Misc. Publicat. Entomol. Soc. Am.* **9,** 73–94.
Granados, R. R., and Naughton, M. (1975). *Intervirology* **5,** 62.
Granados, R. R., and Roberts, D. W. (1970). *Virology* **40,** 230.
Henry, J. E., Nelson, B. P., and Jutila, J. W. (1969). *J. Virol.* **3,** 605.
Huger, A. M., Krieg, A., Einschermann, P., and Götz, P. (1970). *J. Invertebr. Pathol.* **15,** 253.
Hurpin, B., and Robert, P. (1967). *Entomophaga* **12,** 175.
Hurpin, B., and Vago, C. (1963). *Rev. Pathol. Veg. Entomol. Agric. Fr.* **42,** 115.
Ichihashi, Y., and Dales, S. (1973). *Virology* **51,** 297.
Ichihashi, Y., Matsumoto, S., and Dales, S. (1971). *Virology* **46,** 507.
Joklik, W. K. (1966). *Bacteriol. Rev.* **30,** 33–66.
Katagiri, K. (1973). *J. Invertebr. Pathol.* **22,** 300.
Katagiri, K., Kushida, T., Kasuga, S., and Ohba, M. (1975). *Jpn. J. Appl. Zool.* **19,** 243 (Engl. Summ.).
Langridge, W. H. D., and Roberts, D. W. (1975). *Proc. Int. Congr. Virol., 3rd, 1975* Abstract, p. 96.
Lipa, J. J., and Bartkowski, J. (1972). *J. Invertebr. Pathol.* **20,** 218.
McCarthy, W. J., Granados, R. R., and Roberts, D. W. (1974). *Virology* **59,** 59.
McCarthy, W. J., Granados, R. R., Sutter, G. R., and Roberts, D. W. (1975a). *J. Invertebr. Pathol.* **25,** 215.
McCarthy, W. J., Neser, C. F., and Roberts, D. W. (1975b). *Intervirology* **5,** 169.
Meynadier, G., Fosset, J., Vago, C., Duthoit, J. L., and Bres, N. (1968). *Ann. Epiphyt.* **19,** 703.
Pogo, B. G. T., Dales, S., Bergoin, M., and Roberts, D. W. (1971). *Virology* **43,** 306.
Quiot, J. M., Bergoin, M., and Vago, C. (1975). *C. R. Hebd. Seances Acad. Sci.* **280,** 2273.
Roberts, D. W. (1970). *Bacteriol Proc.* p. 177 (abstr.).
Roberts, D. W., and Granados, R. R. (1968). *J. Invertebr. Pathol.* **12,** 141.
Roberts, D. W., Langridge, W. H. R., McCarthy, W. J., and Granados, R. R. (1975). *Proc. Int. Congr. Virol., 3rd, 1975* Abstract, p. 173.
Stoltz, D. B., and Summers, M. D. (1972). *J. Ultrastruct. Res.* **40,** 581.
Summers, M. D., and Arnott, H. J. (1969). *J. Ultrastruct. Res.* **28,** 462.
Sutter, G. R. (1972). *J. Invertebr. Pathol.* **19,** 375.
Vago, C., and Bergoin, M. (1968). *Adv. Virus Res.* **13,** 247–303.
Vago, C., Amargier, A., Hurpin, B., Meynadier, G., and Duthoit, J. L. (1968a). *Entomophaga* **13,** 373.
Vago, C., Monsarrat, P., Duthoit, J. L., Amargier, A., Meynadier, G., and van Waerebeke, D. (1968b). *C. R. Hebd. Seances Acad. Sci.* **266,** 1621.

Vago, C., Robert, P., Amargier, A., and Duthoit, J. L. (1969). *Mikroskopie* **25,** 378.
Weiser, J. (1969). *Acta Virol. (Engl. Ed.)* **13,** 549.
Weiser, J., and Vago, C. (1966). *J. Invertebr. Pathol.* **8,** 314.
Weiser, J., Tchubianishvili, C., and Zizka, Z. (1970). *Acta Virol. (Engl. Ed.)* **14,** 314.
Westwood, J. C. N., Harris, W. J., Zwartouw, H. T., Titmus, D. H. J., and Appleyard, G. (1964). *J. Gen. Microbiol.* **34,** 67.
Wildy, P. (1971). *Monogr. Virol.* **5,** 1.
Yau, T. M., and Rouhandeh, H. (1972). *Arch. Gesamte Virusforsch.* **39,** 140.
Yau, T. M., and Rouhandeh, H. (1973). *Biochim. Biophys. Acta* **299,** 210.

Fig. 1. Ultrastructure of the Lepidoptera poxviruses. (a) Whole mount of negatively stained *Amsacta* virion released from the spheroid by reducing alkaline solution. The envelope of the oval-shaped particles shows prominent globular protrusions of 40 nm mean diameter that give the virion a mulberrylike appearance. Potassium phosphotungstate staining. (×115,000.) (b) Whole mount of degraded *Amsacta* virions after prolonged contact with the reducing alkaline solution used to dissolve the spheroid protein. These structurally altered virions (C-forms) are penetrated by the potassium phosphotungstate and show enlargement of the envelope accompanied by the attenuation of the globular protrusions. The lateral body and the core (►) are separated from the envelope. Tubular or finger-like structures (→) are evident on the surface of a viral envelope ghost. (×52,000.) (c) Thin section of cytoplasmic nonoccluded *C. fumiferana* virions. The mulberry-like surface is evident in the transversal sections through the globular surface units (→). These globular surface units in longitudinal section appear to be protrusions of the outer layer of the virion envelope. Between the core (vc) and the outer beaded envelope, the lateral body (lb) of high electron opacity is located. The core, rectangular in outline, contains filamentous material surrounded by a three-layer envelope. (×83,000.) (d) Occlusion of a mature *C. fumiferana* poxvirus in a spheroid. The protrusions of the envelope in the part already occluded are no longer visible. vc, Virus core; lb, lateral body. (×92,000.)

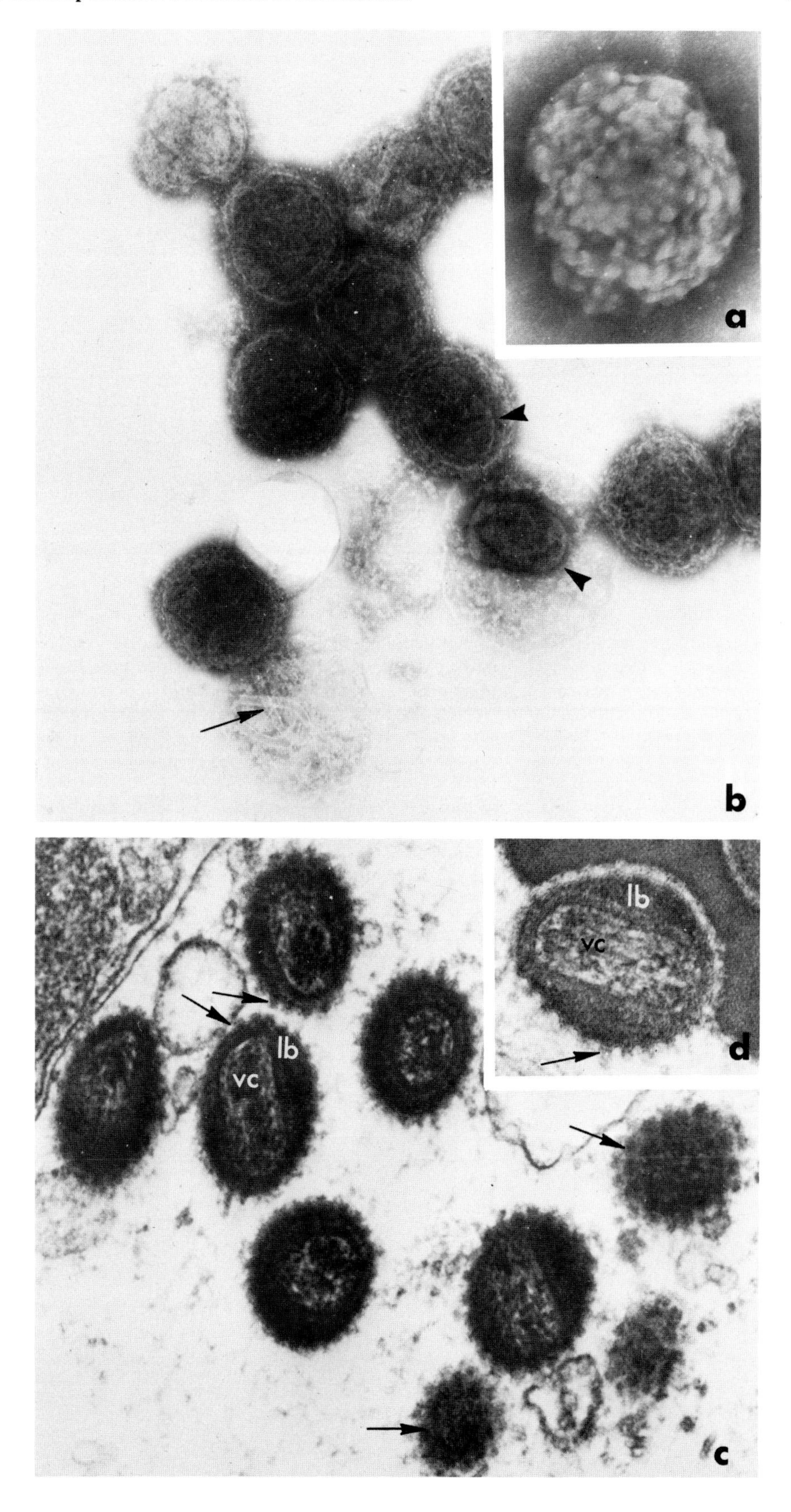
a
b
lb
vc
d
lb
vc
c

Fig. 2. Comparative ultrastructure of Lepidoptera and Coleoptera poxviruses. (a) Thin section of nonoccluded Lepidoptera *Amsacta* poxvirus in infected *G. melonella* adipose tissue. The lateral body (lb) surrounds a cylindrical virus core (vc) which appears rectangular in sagittal or horizontal sections and circular in transverse sections. Inside the core, a dense and coiled ropelike structure is present (➤). Six folds can be seen in a cross section of the core (→). (× 45,000.) (b) Section through occluded Lepidoptera *C. fumiferana* virions showing the beaded appearance of the envelope surrounded by an electron-lucent "halo" (►). The crystalline lattice is interrupted at the "halo" interface. (× 82,000.) (c and d) Ultrastructure of the Coleoptera *Melolontha melolontha* poxvirus. This virus has a unilaterally concave core (vc), and a lateral body (lb) is located at the hilus of the concavity. Whereas in nonoccluded virions the content of the core appears composed of filamentous substance, the occluded virions present a coiled ropelike structure (➤) within a dense core. (c, × 112,000; d, × 97,000.)

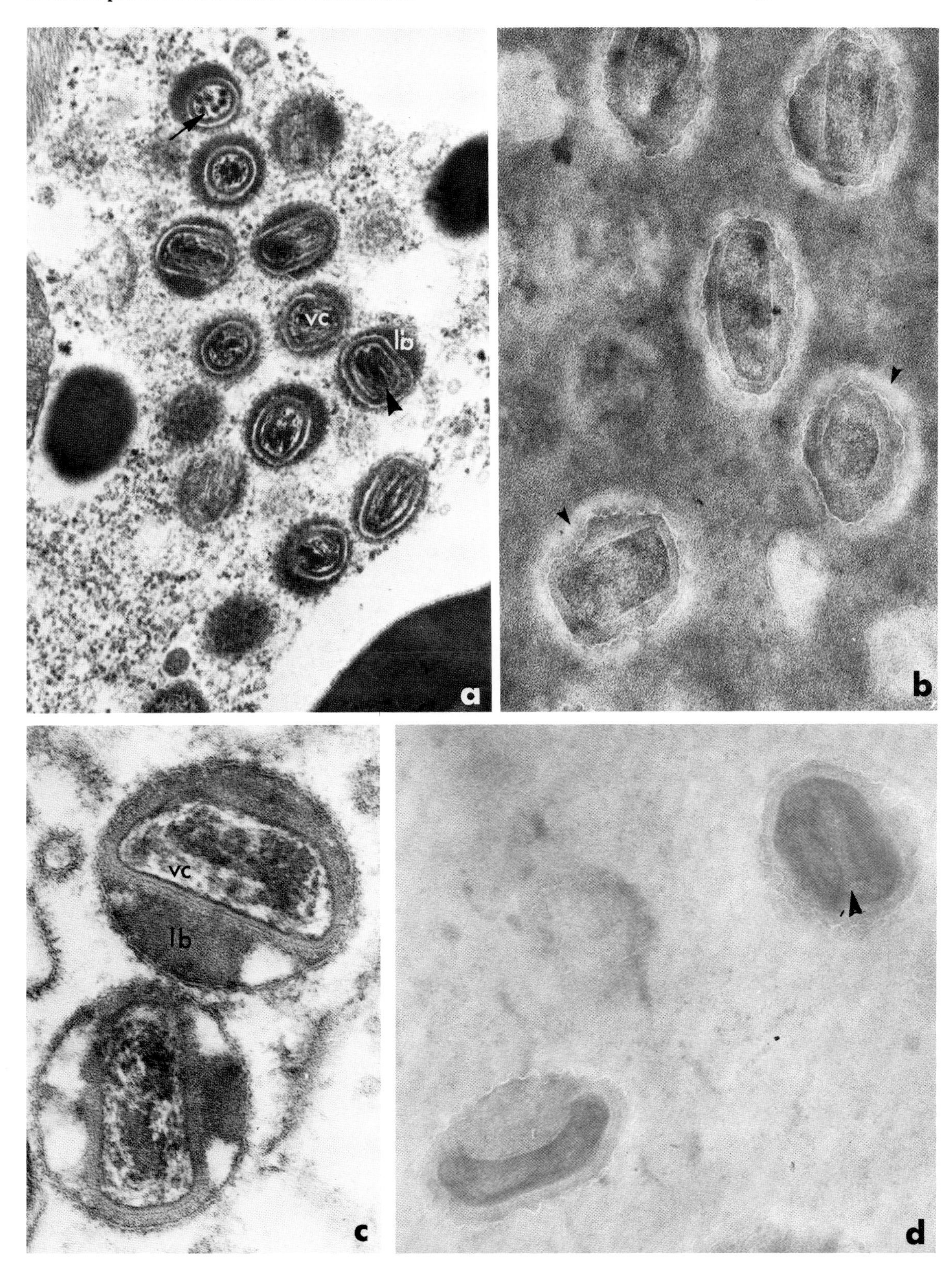
vc
lb
a
b
vc
lb
c
d

Fig. 3. Ultrastructure of the Diptera *Chironomus attenuatus* poxvirus (courtesy of Dr. D. B. Stoltz and Dr. M. D. Summers). (a and b) The mature virion, cuboidal in horizontal section, is dumbbell-shaped in transverse sections. Two lateral bodies are located on either side of the biconcave core. The core envelope consists of both a trilaminar "unit" membrane and a layer of subunits (►). (× 98,000.) (c) Section of a spheroid. The virions are randomly oriented within the crystalline lattice. Note that the core of the deeply occluded particles is more markedly biconcave than that of the more peripheral newly occluded particles. (× 70,000.) (d) Higher magnification of an occluded mature virion. The dumbbell-shaped core and the lateral body are well defined. Note the distinct lattice lines and a distinctive peripheral structure (⤚) of the spheroid. (× 98,000.)

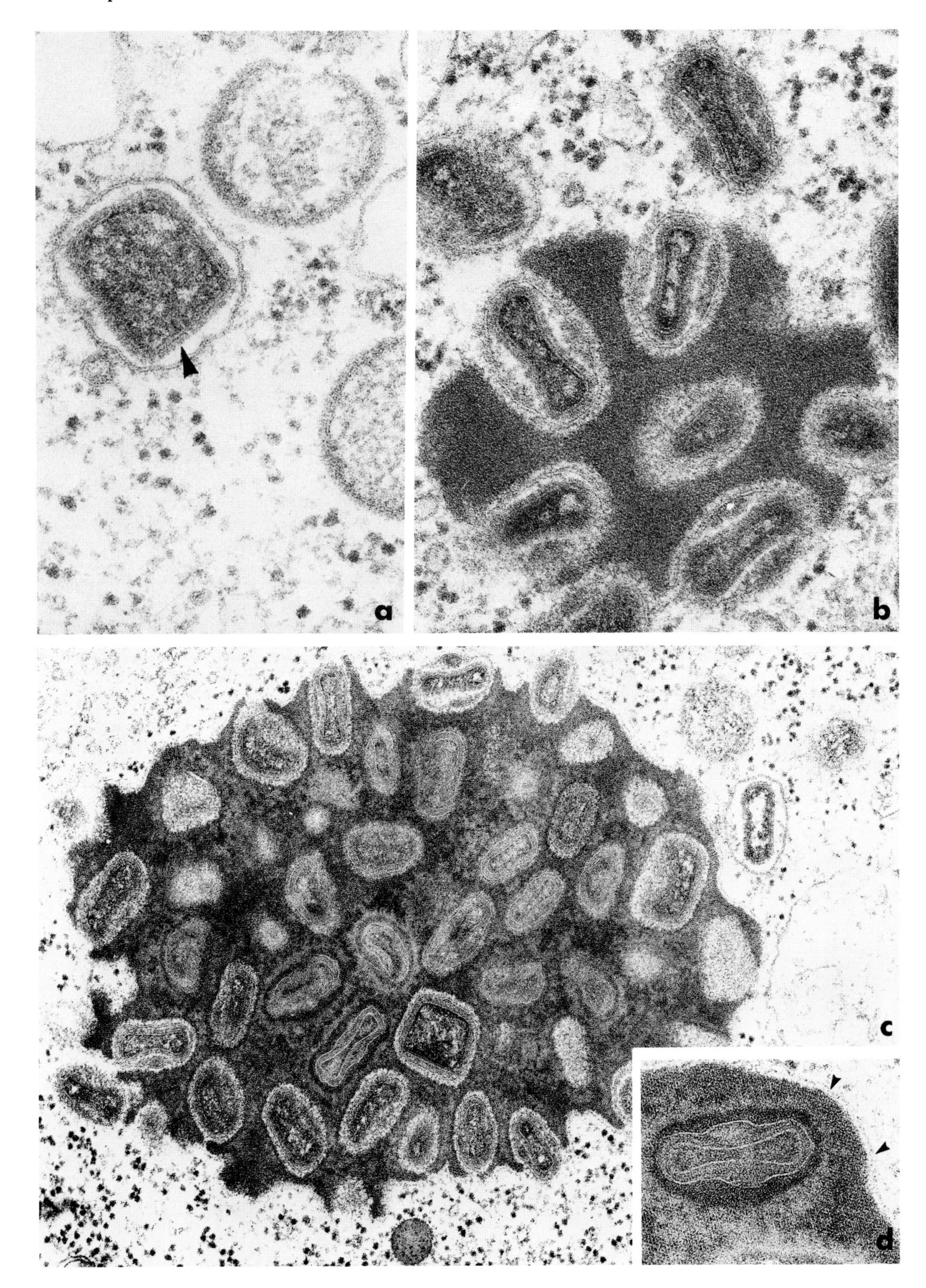
a
b
c
d

Fig. 4. Histopathology of EPV infections. (a and b) EPV-infected cells of the grub *M. melolontha* (Coleoptera). (a). Acridine orange coloration of infected hemocytes presenting a high content of RNA (bright red fluorescence, →) and the presence of large spindles (green fluorescents, ►). (×2,000.) (b) Hamm coloration of section of infected adipose tissue characterized by numerous macrospindles (►) of variable size and usually one large ovoid inclusion body (spheroid) per cell (►). (×4,000.) (c–e) Different stages in the pathology of the 25-day, EPV-infected cells of the Lepidoptera *C. fumiferana* lobe of adipose tissue indicating the asynchronous replication of the virus in the tissue. The infected cells lose their fat granules, which are rapidly replaced by foci of virogenic stroma (B-type inclusions of vertebrate poxviruses) (⇉). In a second step, numerous spheroids are formed in the infected cells (►). No spindles are apparent. Long cytoplasmic fascicles of fibrils are present in close relationship with the spheroids in formation (→). These structures were originally described by Weiser and Vago (1966) and are apparently characteristic for infected Lepidoptera species. Nuclear hypertrophy and some dense nuclear inclusions are also observed (►). (f) In a more advanced stage of viral replication the cells undergo lysis and liberate a great number of spheroids into the general cavity. (×2000.)

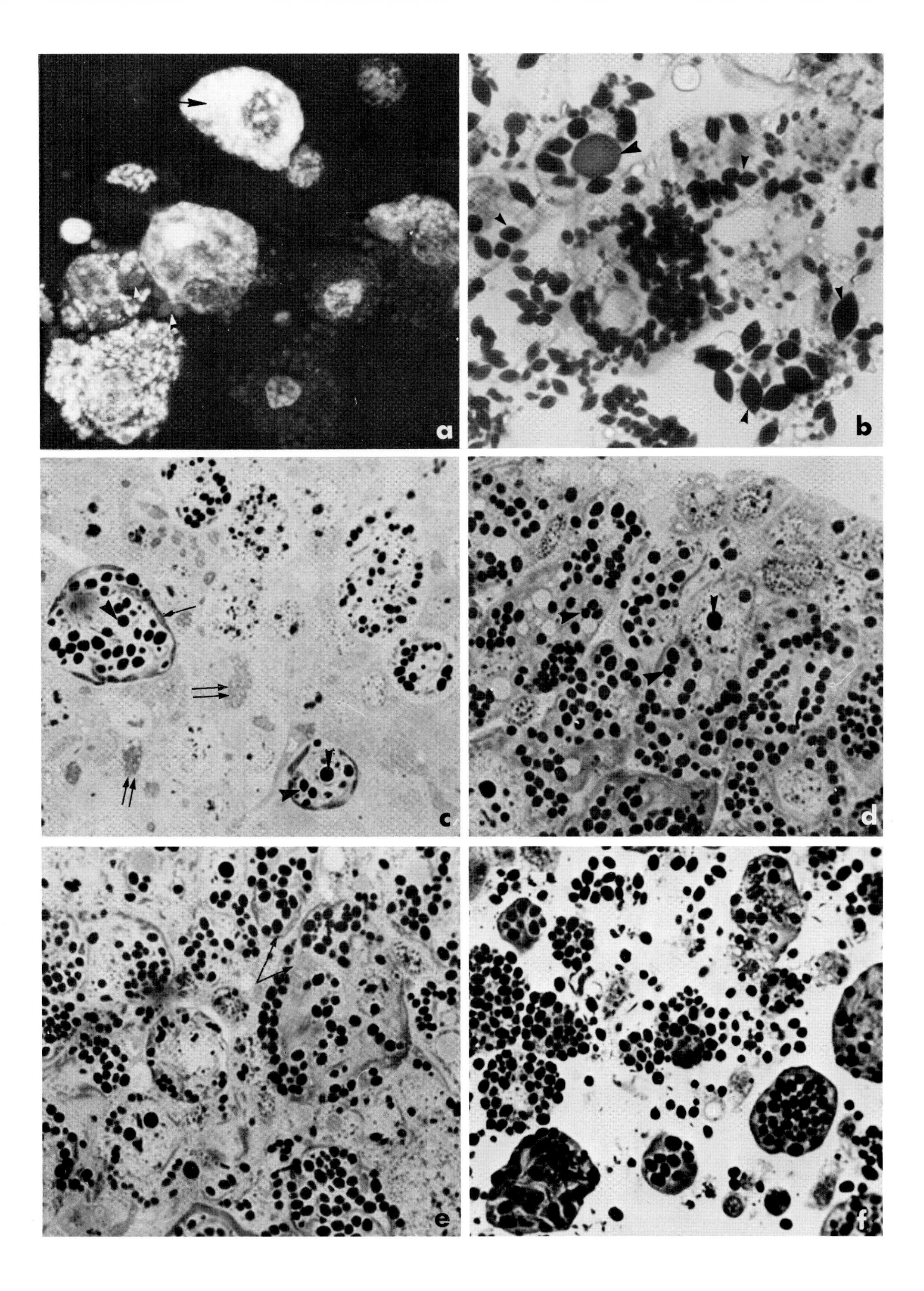

a
b
c
d
e
f

Fig. 5. Illustration of the two entry processes of *Amsacta* EPV in infected *Estigmene acrea* larvae (courtesy of Dr. R. R. Granados). (a and b) Entry of the viral core (vc) and the lateral body (lb) into the cytoplasm by fusion of the virus envelope and a microvillus (M) membrane of the midgut epithelium. The core structure released from the invading particle migrates down the microvillus into the cell cytoplasm. (×81,000.) (c and d) Purified virions infecting the hemocoel enter the host hemocytes by viropexis of the intact virion. (c) The virus attachment to the plasma membrane is followed by the invagination of the plasma membrane and the engulfment of the viral particle. (d) Viral particles contained within a phagocytic vacuole in the cell cytoplasm. (×81,000.)

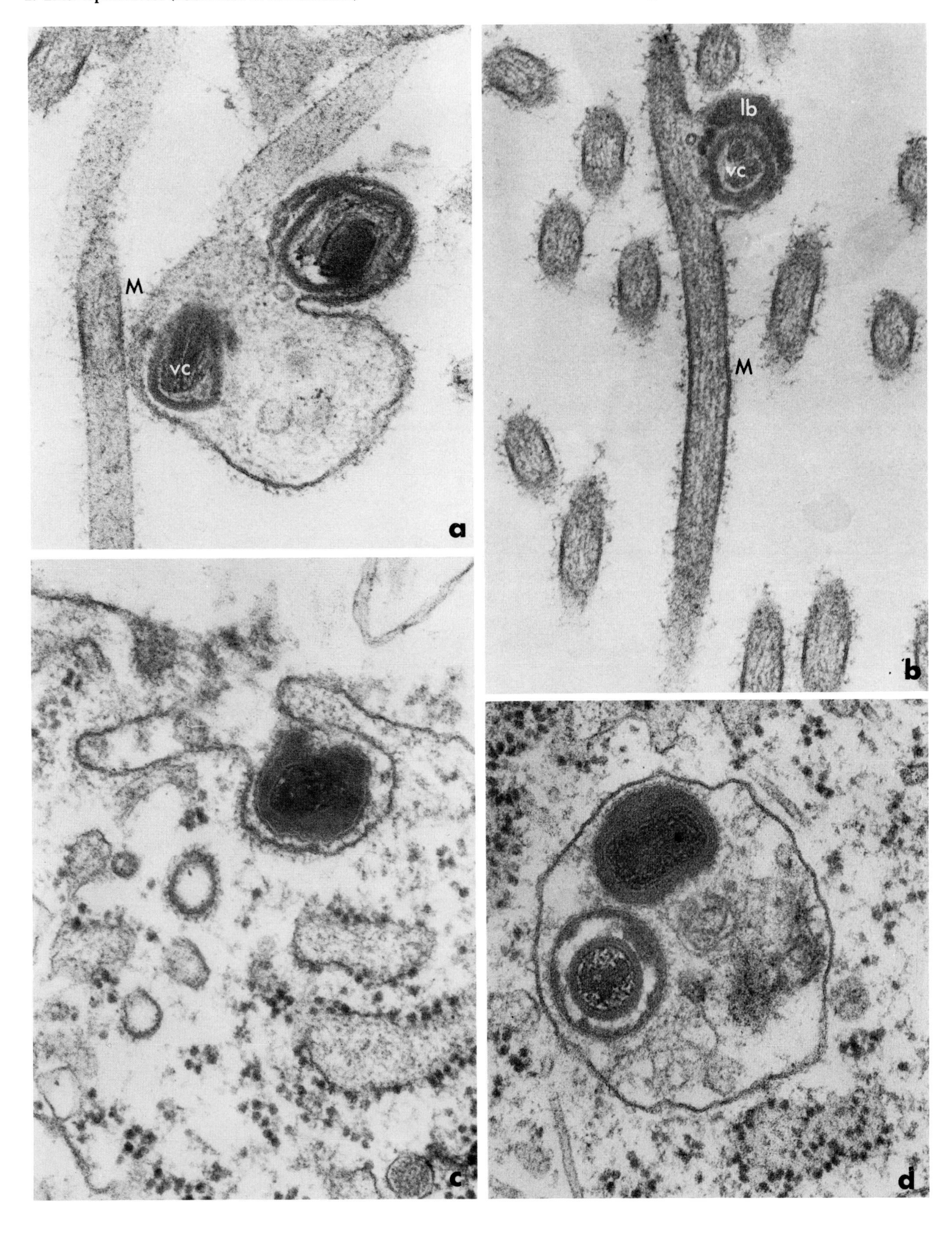
M
vc
a
lb
vc
M
b
c
d

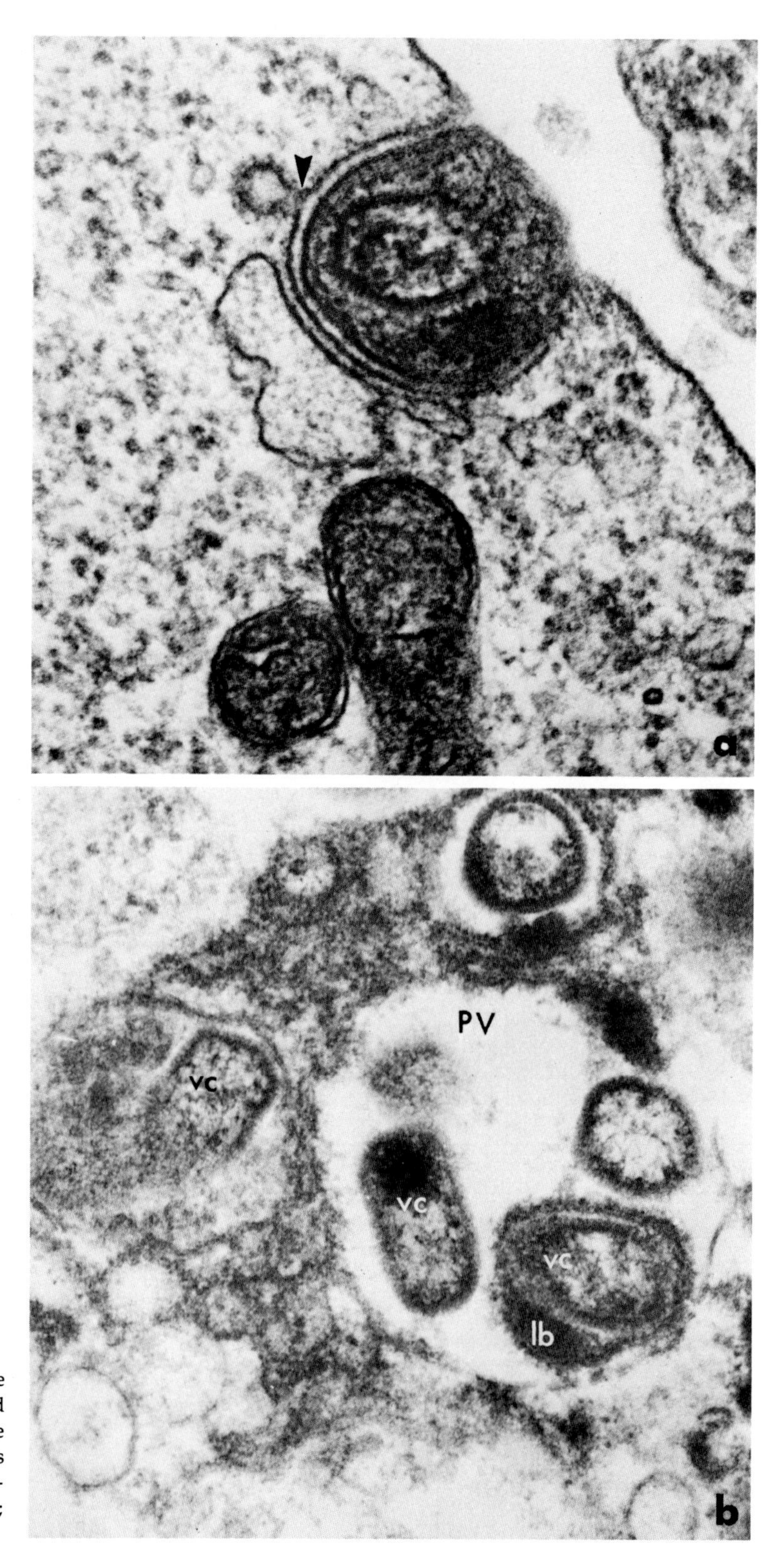

Fig. 6. Viropexis of *C. fumiferana* EPV by cell of adipose tissue. (a) Engulfment of the attached virion in the invaginated plasma membrane. The formation of the phagocytic vacuole begins before closure of the vacuole by fusion with vesicles (➤). (×140,000.) (b) Virus particles at different stages of "denudation" within a phagocytic vacuole (PV). vc, Virus core; lb, lateral body. (×90,000.)

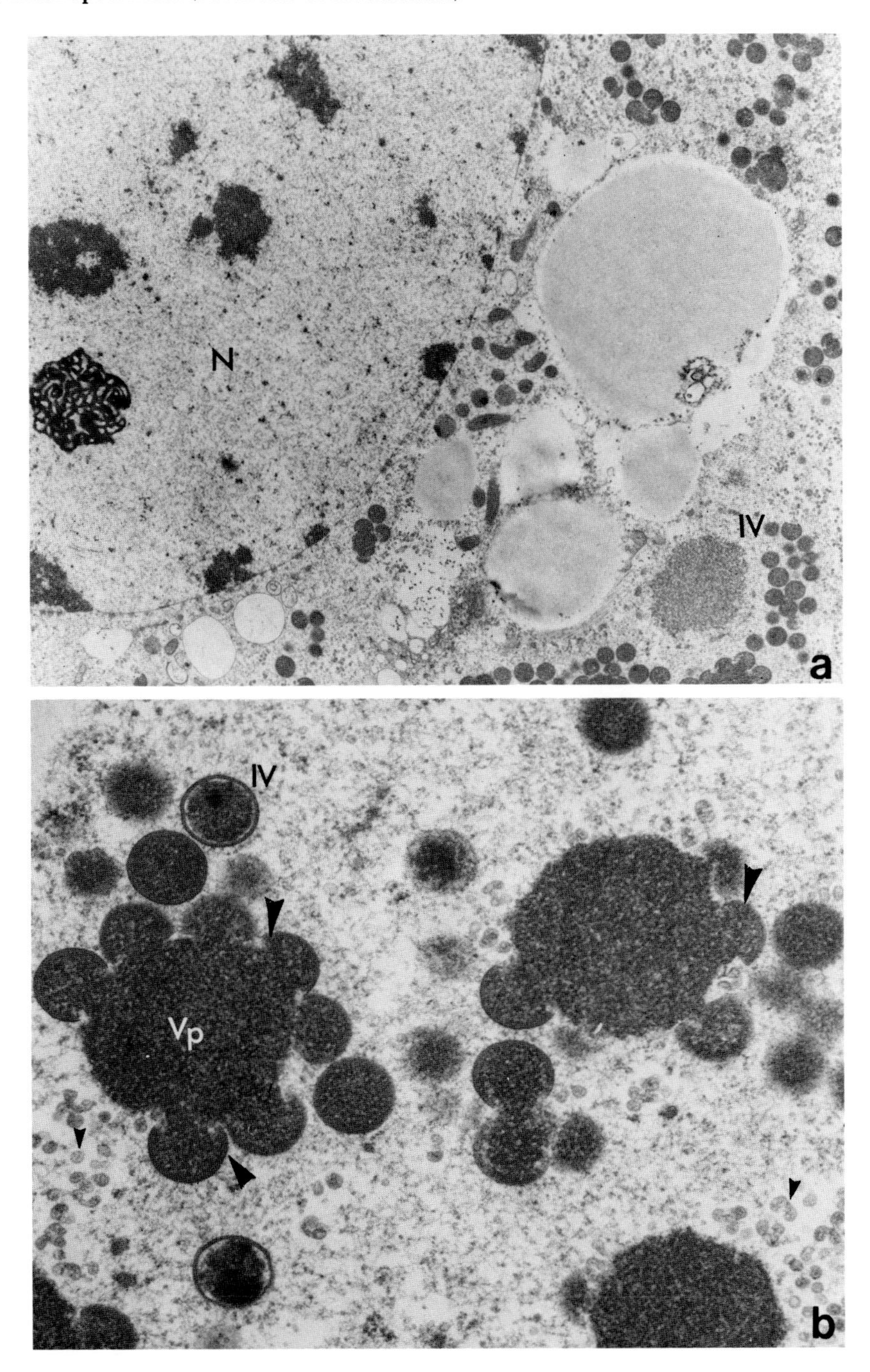

Fig. 7. Development of *M. melolontha* EPV in the cytoplasm of fat cell in association with loci of electron dense viroplasm (Vp) in the vicinity of the nucleus (N). Immature viral particles (IV) are assembled at the periphery of this viroplasm where the incomplete spherical envelope (➤) encapsulates portions of the viroplasm before membrane closure. Numerous small vesicles (➤) are usually found near the viroplasmic areas. (a, ×9000; b, ×29,000.)

Fig. 8. Development of *C. fumiferana* EPV in association with two distinct, sometimes adjacent, viroplasm foci. (a) The viroplasm I (Vpl) consists of electron dense amorphous material whereas viroplasm II (Vp2 seen in Fig. b) consists of a diffused, slightly electron lucent area containing numerous spherical vesicles. The vesicles appear to have a unit membrane structure with some spicules at their surface (►). Incomplete virus envelopes (➤) apparently increase their surface area while enclosing material from one or the other viroplasm, giving rise to spherical immature virus particles (IV) of different densities. (×58,000.) (b and c) Some immature viral particles (IV) present a condensation of the inner dense material at one pole and the formation of a surrounding envelope (→) structurally similar to the outer viral envelope (➤). The viral envelope (➤) of the immature virion consists of an inner unit membrane and outer layer spicules separated by a less dense layer. (b, ×34,000; c, ×78,000.)

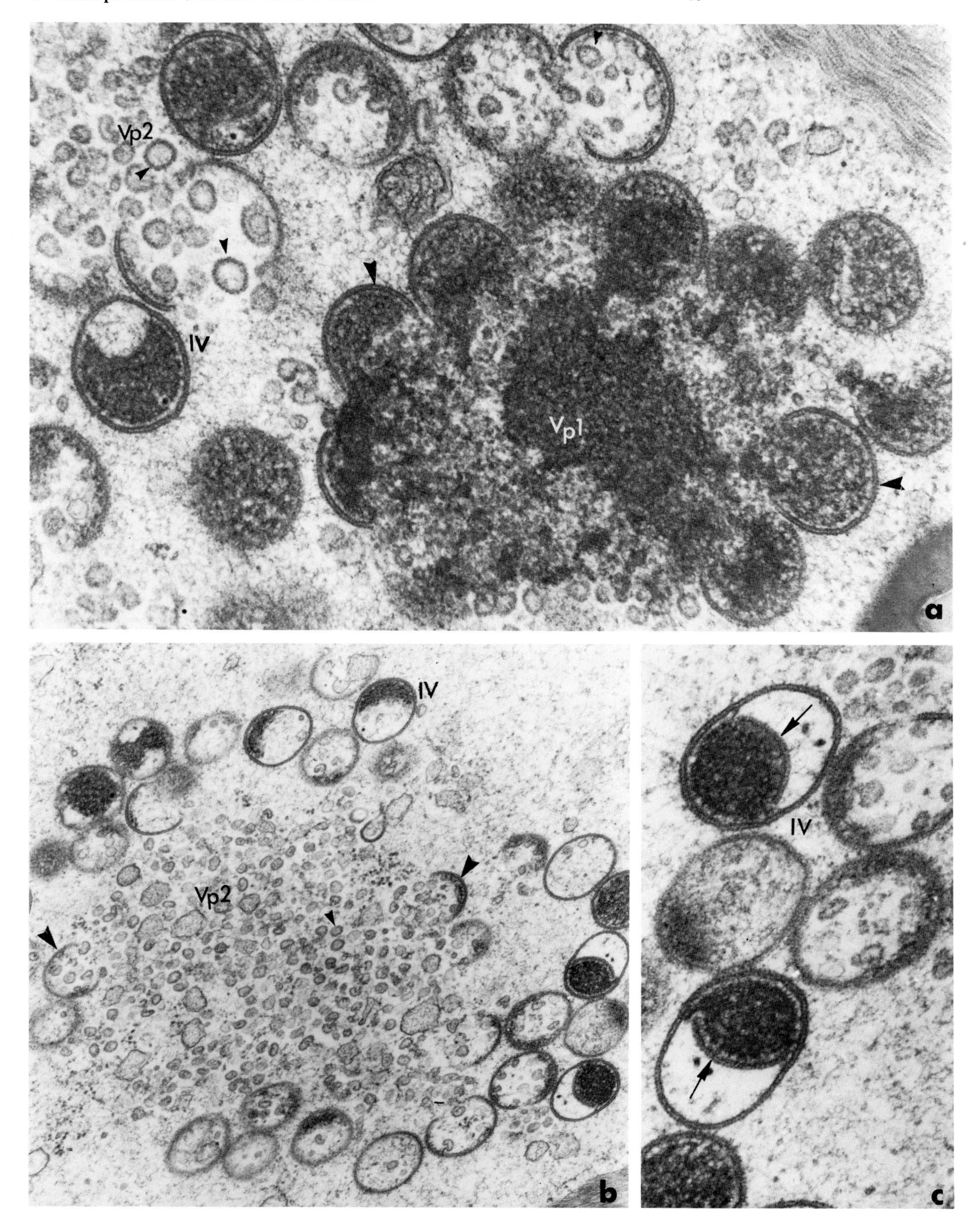
Vp2
IV
Vp1
a
IV
Vp2
b
IV
c

Fig. 9. Higher magnification of immature particles (IV) of *C. fumiferana* EPV. (a and b) Two distinct membranes, each of about 75 Å, separated by a less dense layer of about 100 Å are involved in the structure of the viral envelope. The trilaminar structure of the inner membrane (→) is typical of a "unit" membrane. The outer membrane shows an arrangement of radially disposed spicules (⇉). Note in (b) the closure of virus envelope by the addition of structural subunits (➤) or by adjustment of the envelope length (➤➤) and also the initiation of the spheroid formation by deposition and crystallization of protein (⇾) around an immature virion (IV). (a, ×84,000; b, ×110,000.)

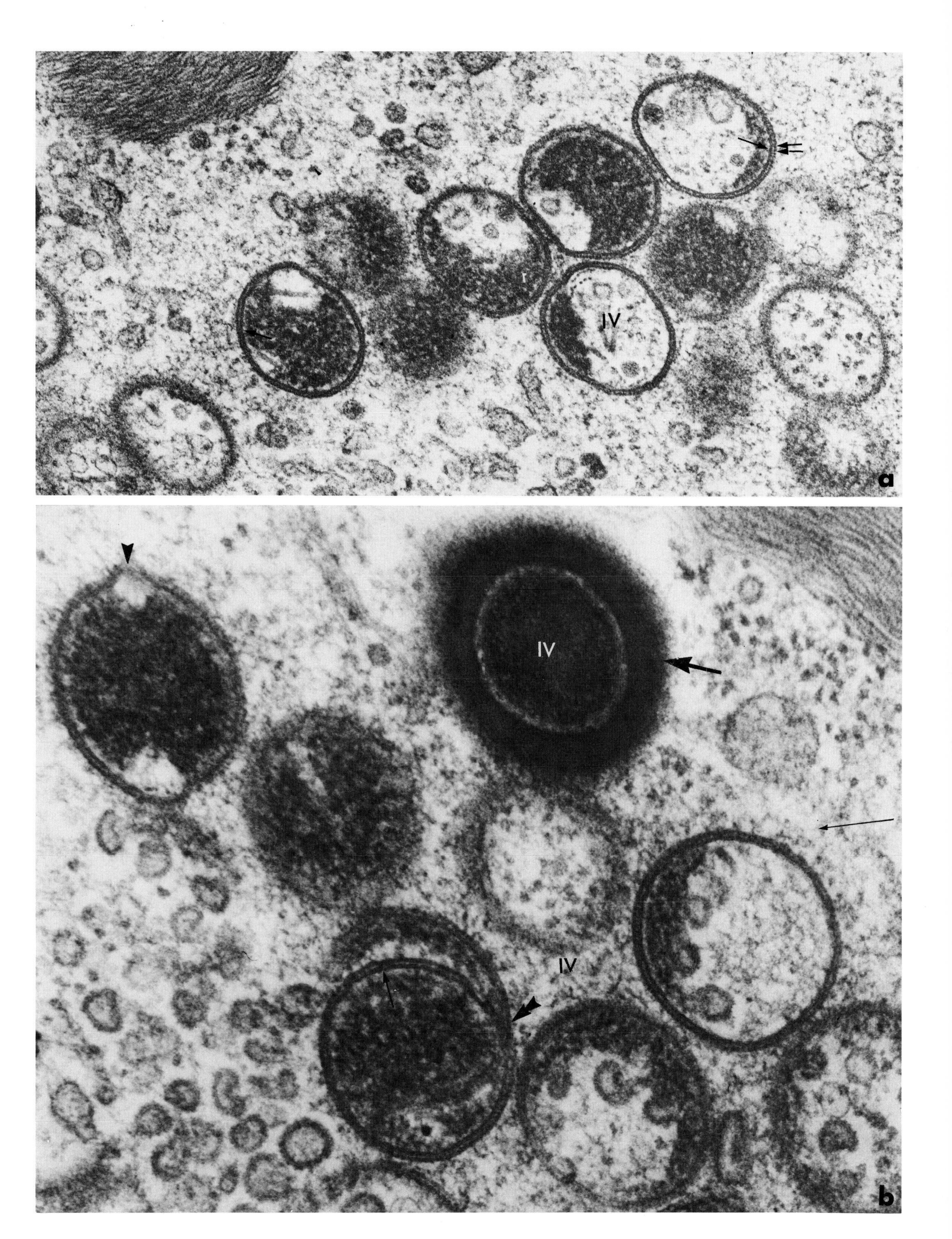
IV
r
a
IV
IV
b

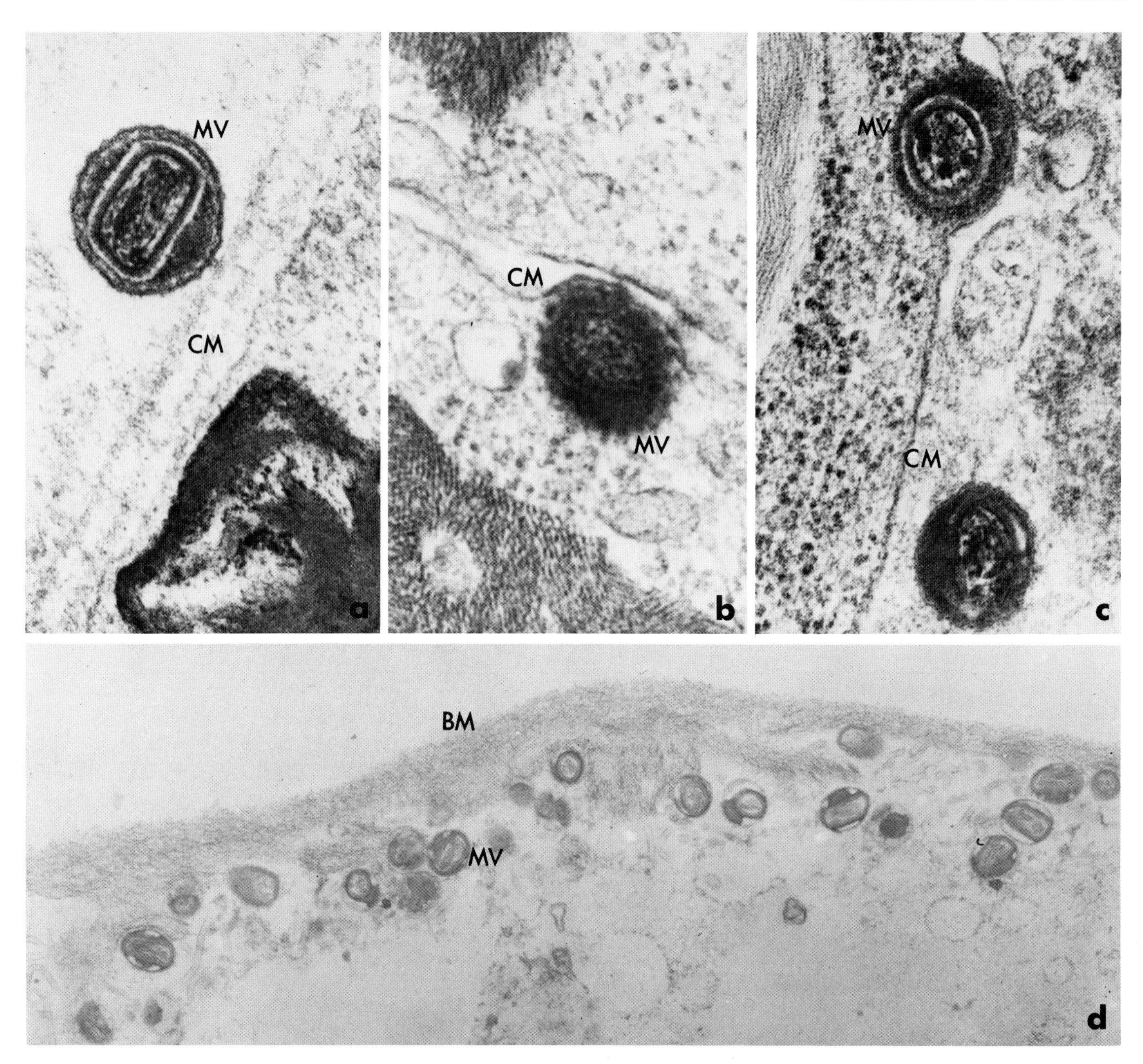

Fig. 10. Exocytosis and possible cell-to-cell transmission in EPV-infected adipose tissue of *C. fumiferana* (a–c) and *M. melolontha* (d). MV, mature virion; CM, cell membrane, BM, basal membrane. (a) Attachment of the mature virion to the cell membrane. (×81,000.) (b) Budding of the virion by evagination of the cell membrane. (×81,000.) (c and d) Liberation of the virion in the intercellular space or along the basal membrane. (c, ×81,000; d, ×24,000.)

Fig. 11. Stages in the development of *M. melolontha* spheroids. (a) Small immature spheroid occluding immature virions. (×47,000.) (b and insert) Mature spheroid showing the occlusion of virions without fissure in the crystalline lattice. The virions present some structural changes, such as an electron lucent "halo" surrounding the virus particle (►), a wavy appearance of the outer virus envelope (→) and a more electron dense unilaterally concave core with the appearance of a coiled ropelike structure within the core (⟶). The occluded virions present 5 to 6 profiles of the ropelike structure. lb, lateral body. (×120,000.) (c) Regressive staining with EDTA (Bernhard, 1958). The viral core (vc) of the nonoccluded virions (MV), the occluded virions (ov), and the spindles (S) are "bleached" by this method demonstrating the deoxyribonucleoprotein nature of these structures. The early immature viral particles iv, the lateral bodies (lb), the viral envelopes, the spheroid proteins (Sp), and the ribosomes keep their usual contrast (Epon-EDTA, 30 minutes). (×43,000.)

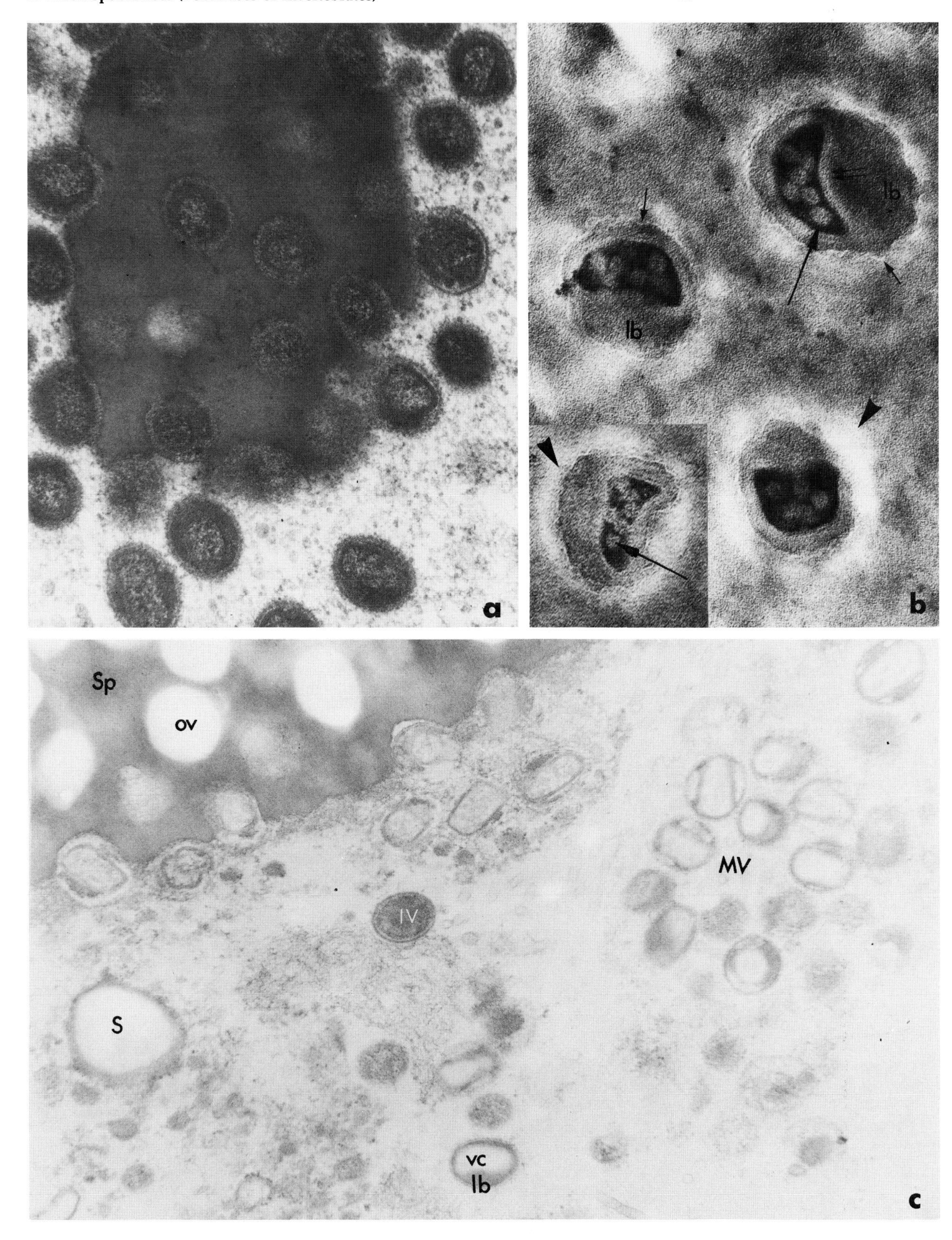

a
lb
lb
b
Sp
ov
MV
IV
S
vc
lb
c

Fig. 12. Stages in the development of *M. melolontha* spindles. (a) Note the condensation and accumulation (→) of a dense material inside the mitochondria. The continued accumulation and crystallization of this material gives rise to small paracrystalline inclusions within the membrane (⇉). (×30,000.) (b, c, and d) Continuous growing of these inclusions out of the limit of the cellular membrane produces large paracrystalline spindles (S). A spongelike material (➤) surrounds the spindles in formation. (b and c, × 52,000; d, × 17,000.) (e) Acridine orange coloration, (f) and regressive staining with EDTA of the spindles (S). The spindle fluoresces green by acridine orange coloration and is bleached by EDTA, demonstrating the deoxyribonucleoprotein nature of the crystallized proteins. The spongelike material (➤) surrounding the spindles reacts as does a ribonucleoprotein to the two methods. (e, × 6000; f, × 42,000.)

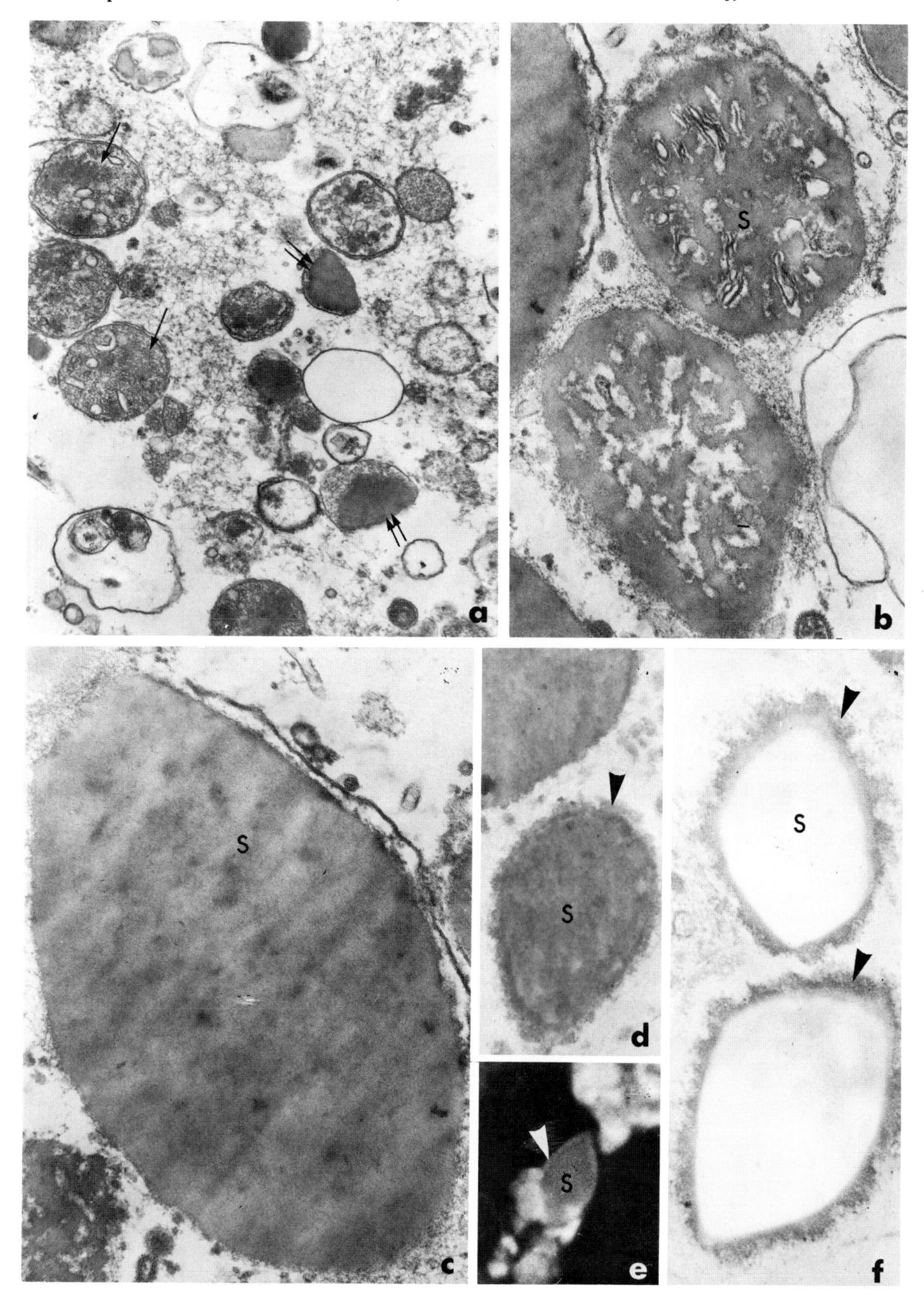
S
S
S
S
S
a
b
c
d
e
f

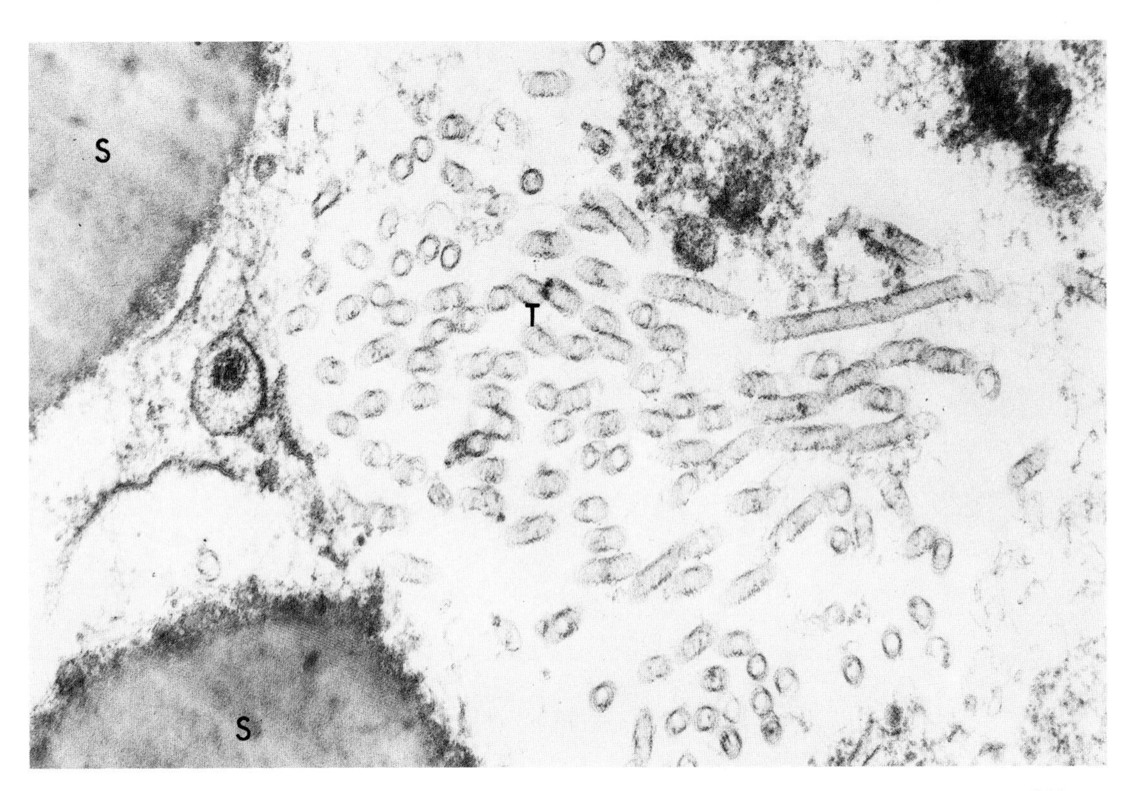

Fig. 13. Fascicle of helicoidal tubular structures (T) associated with the replication of *M. melolontha* EPV. Note the crystalline lattice of the spindles (S). (× 60,000.)

Figs. 14. Stages in the morphogenesis of *C. fumiferana* EPV. (a) In an early stage of viral replication, immature particles (IV) develop around the virogenic stroma (Vp) in the vicinity of the nucleus (N). Near this virogenic stroma, few spheroids (Sp) are under formation, occluding mature virions (MV) and enveloped spindles (S). Some fascicles of fibrils (ff) are present. The nucleus shows some alterations: condensation of chromatin; fascicle of fibrils; and granular (⇉) or dense (→) inclusions. (× 6400.) (b) The spheroid proteins deposit around the mature virion (MV) or enveloped spindles (S) as initial sites of nucleation (►). By subsequent deposition of spheroid proteins, they occlude randomly other virions and spindles present in the area of occlusion. Bigger spheroids (Sp) are produced by fusion of many growing small paracrystalline inclusions. ff, fascicles of fibrils; Vp, virogenic stroma. (× 13,000.)

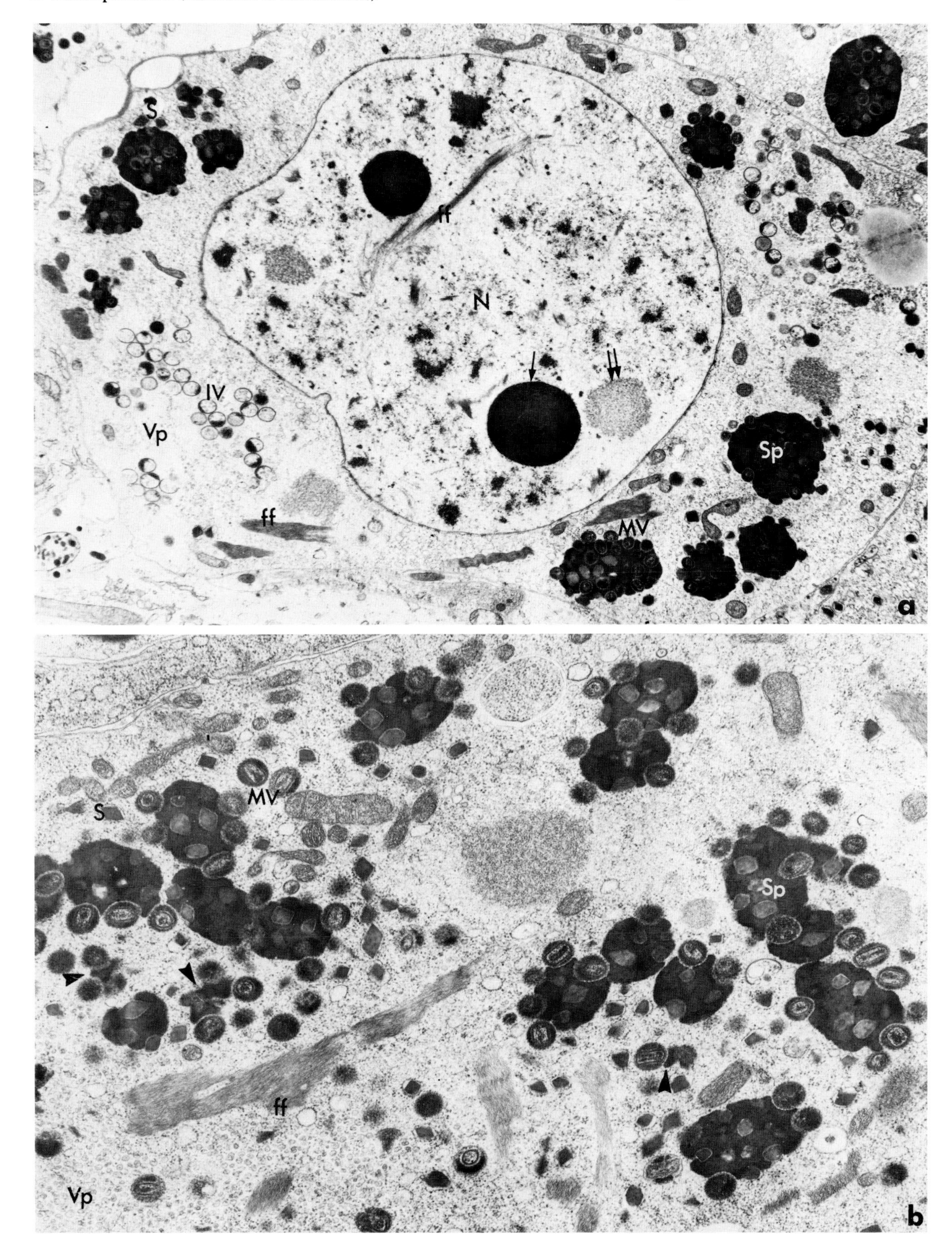
S
ff
N
IV
Vp
Sp
MV
ff
a
MV
S
Sp
ff
Vp
b

Fig. 15. Stages in the morphogenesis of *C. fumiferana* EPV. (a) Note the ergastoplasmic membrane (→) enveloping the spindle (S) and the occlusion of mature virions (MV) in the crystalline lattice of the spheroid (Sp). (×78,000.) (b) Growing spheroid by fusion of two small inclusions. The spindles (S) are occluded with their membrane (→) which lose their electron density (⇉) after complete occlusion. Note the random distribution of virions and spindles inside the spheroid. (×47,000.)

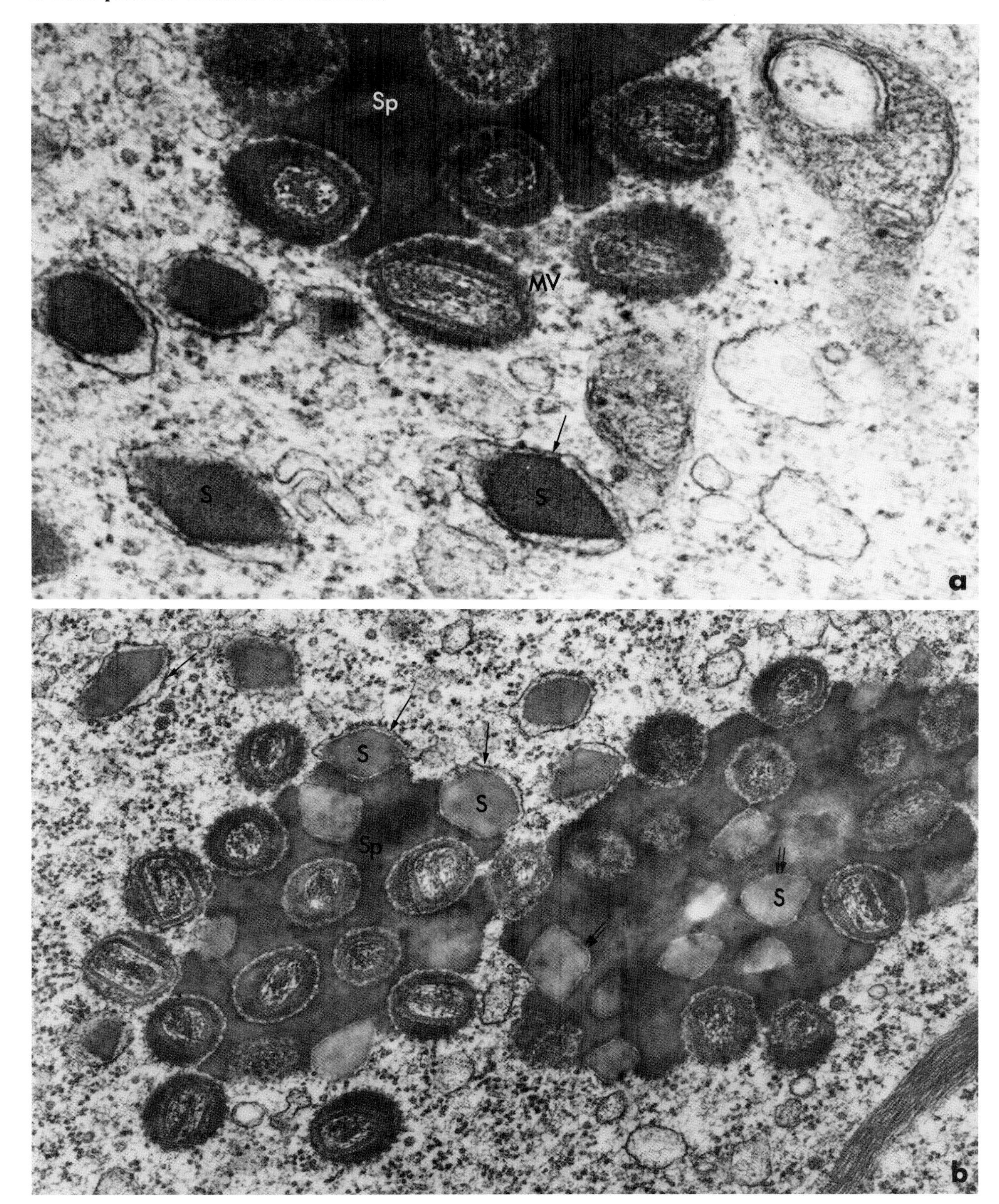
Sp
MV
S
S
a
S
S
Sp
S
b

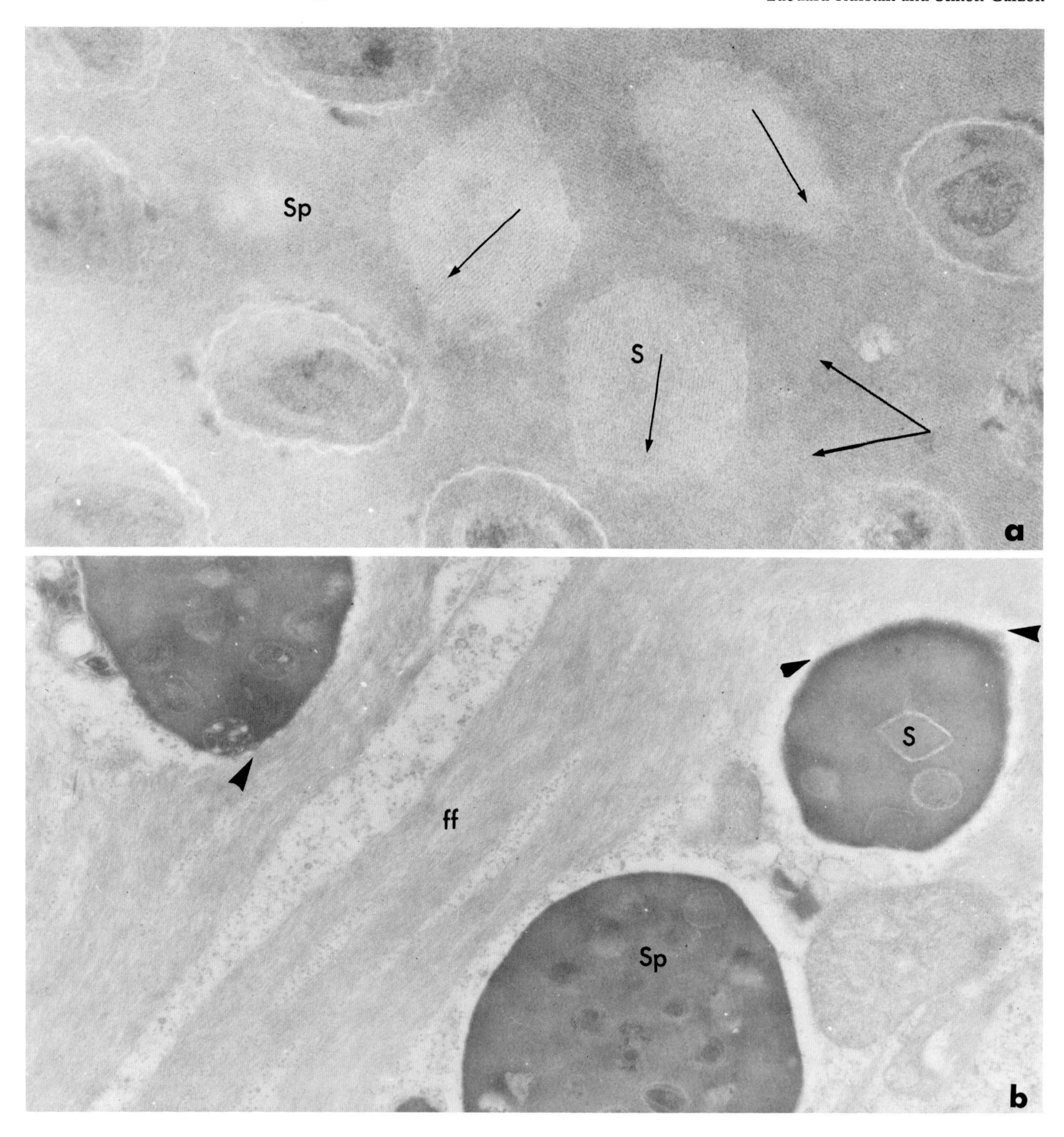

Fig. 16. Stages in the morphogenesis of *C. fumiferana* EPV. (a) Mature spheroid. Patterns of the lattice of the two proteins, spheroid (Sp) and occluded spindle (S), are oriented (⟶) in different directions as expected from a random positioning of the spindles. Neither the spindle nor the virion interferes with the protein lattice of the spheroid. (× 110,000.) (b) Many long fascicles of fibrils (ff) are associated with the multiplication of *C. fumiferana* EPV in adipose cells. They sometimes show a close relationship with the spheroids in formation (►). (× 27,000.)

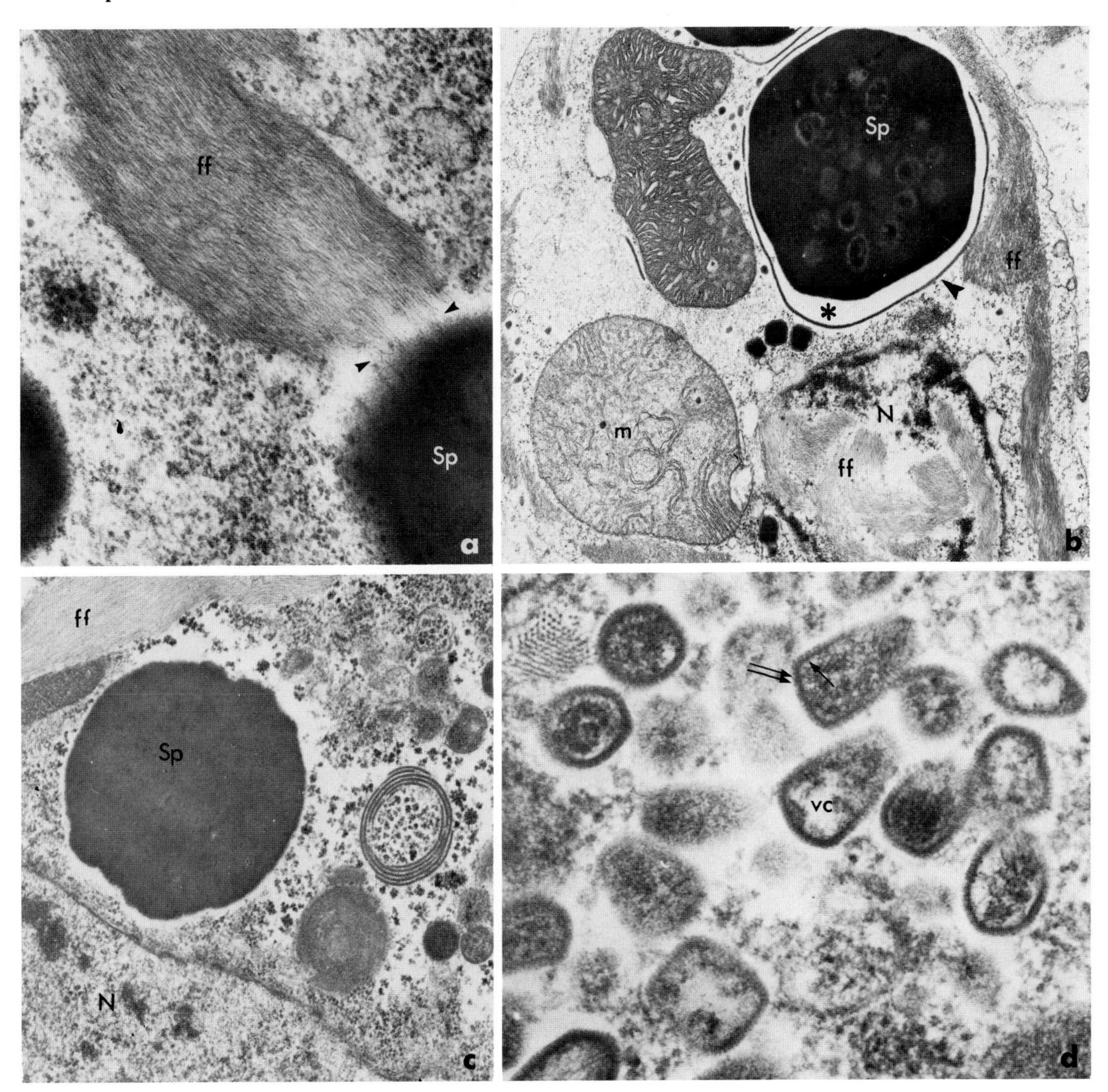

Fig. 17. Some aspects of the morphogenesis of Lepidoptera EPV. (a and b) *C. fumiferana* EPV morphogenesis. (a) The continuity of the fascicles of fibrils (ff) and the spheroid (Sp) lattice proteins is clearly visible (►). (×55,000.) (b) Hypertrophied mitochondria (m) near a spheroid which has completed its development. The spheroid (Sp) has enclosed numerous virions and spindles in its center, however, the cortical region of the spheroid contains no occluded structures. The spheroid is enveloped in a distinctive epispheroid dense layer (►). This layer is separated from the cortical zone by the procedure of fixation, embedding, and sectioning (*). ff, fascicle of fibrils; N, nucleus. (×20,000.) (c and d) Stages in the morphogenesis of *Amsacta* EPV replicating in *G. mellonella* larvae. In this replicative cycle few or no virions are occluded in the spheroids (Sp). Many viral cores (VC), apparently "mature," are produced in the cytoplasm, the envelope of which is composed of an inner "unit" membrane (→) and outer layer spicules (⇉). ff, fascicles of fibrils; N, nucleus. (c, ×30,000; d, ×90,000.)

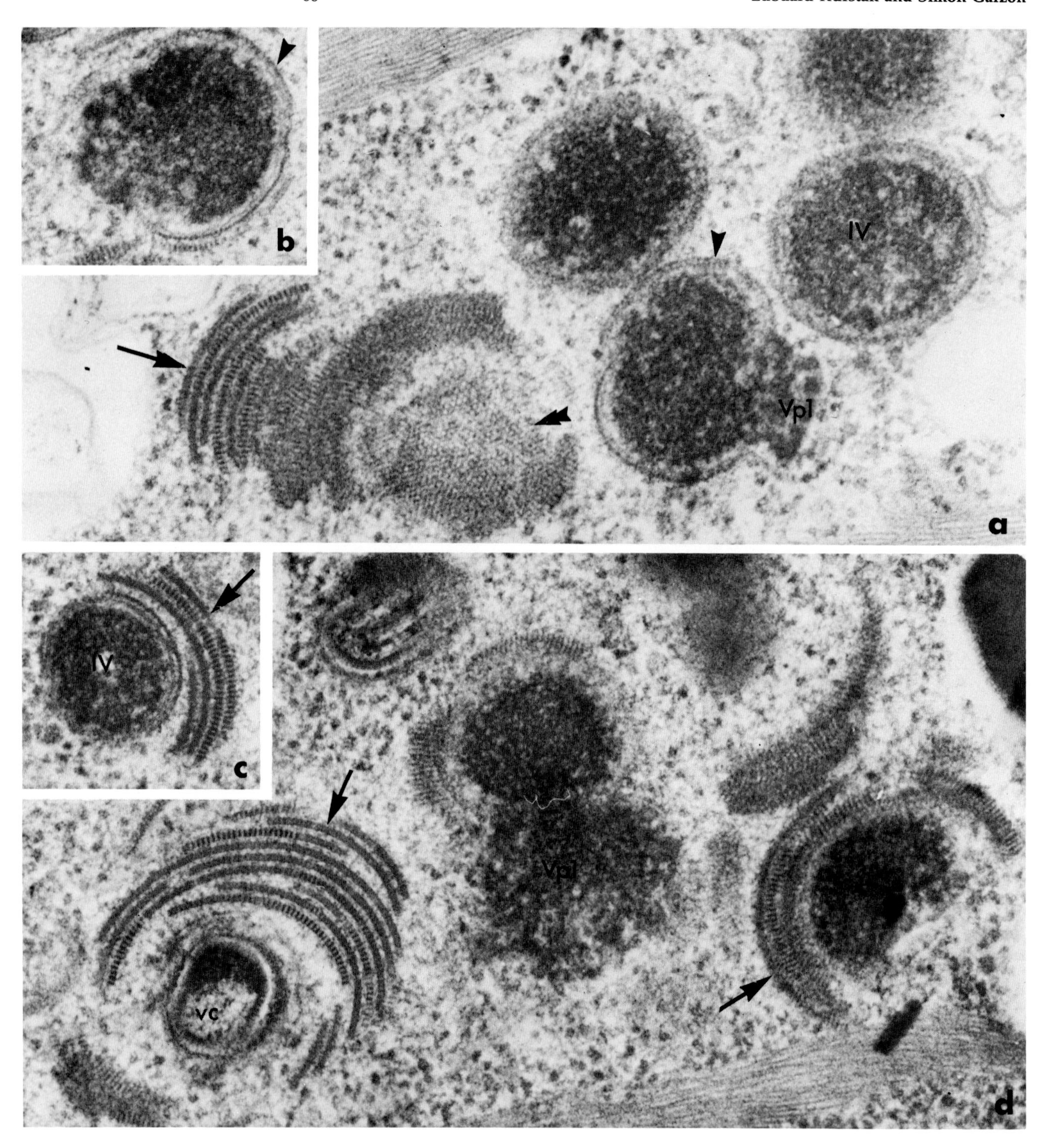

Fig. 18. Stage in the development of the immature virions (IV) of *Amsacta* EPV in *G. mellonella* larvae. The incomplete viral envelope (➤) encloses material from a viroplasmic focus type I (Vpl), before closure. (a and b) In the vicinity of the viroplasm, envelopes or tubules with periodic lateral structures are assembled (→). These structures are found around portions of the viroplasmic stroma or mature nucleoid (c and d). Tangential sections of these structures show a honeycomb-like arrangement (➤➤) of subunits (a). It is not known if these structures correspond to the outer layer spicules of the viral or core envelopes. vc, Viral core. (×90,000.)

Chapter 3

Densonucleosis Viruses (Parvoviridae)

E. KURSTAK, P. TIJSSEN, AND S. GARZON

I. Introduction 67
II. Classification and Nomenclature 68
III. Physicochemical Properties of Virions and Subviral Components 68
A. Nucleic Acid 68
B. Proteins 69
C. Virions 71
IV. Morphology 72
V. Biological Properties 73
A. Host Range and Pathogenicity 73
B. Multiple Infections and Interference 73
C. Carrier States 74
VI. Replication and Morphogenesis 74
VII. Conclusion 75
References 75

I. INTRODUCTION

All viruses containing linear single-stranded DNA belong to the parvoviruses. These viruses have very similar physicochemical properties. Although the vertebrate parvoviruses are not within the scope of this book, parvoviruses of invertebrates have many similarities with the former group.

Among the various invertebrate or insect parvoviruses known, such as *Aedes* densonucleosis virus (Lebedeva *et al.*, 1973), *Junonia* densonucleosis virus (Tinsley and Longworth, 1973), and other insect viruses probably belonging to the parvoviruses (Hirumi *et al.*, 1976), only the densonucleosis virus (DNV) (Fig. 1) responsible for the fatal disease of the larvae of *Galleria mellonella* (Lepidoptera) has been studied in some detail (Kurstak, 1972). Recently, a DNV-like virus containing single-stranded DNA was found in the silkworm (Dr. S. Kawase, Nagoya University, personal communication).

In general, parvoviruses are ubiquitous in nature. In spite of their wide distribution, they have only been discovered during the last 15 years, probably due to their extremely small size (22–25 nm; the smallest DNA viruses known). For recent reviews the reader is referred to Hoggan (1971); Kurstak (1972), Mayor, (1973), Mayor and Kurstak (1974), Rose (1974), Berns (1974), and Bachmann *et al.* (1975).

II. CLASSIFICATION AND NOMENCLATURE

All viruses with the cryptogram D/1: 1.4–2/20–40:s/s:I,V/0 are parvoviruses. The name parvovirus (latin, parvus = small) was first proposed by Lwoff and Tournier (1966), and the generic name was accepted by the International Committee on the Nomenclature of Viruses in 1970 (Andrewes, 1970). Earlier, the name "picodnavirus" (Mayor and Melnick, 1966) was sometimes used. In a report by the study group on Parvoviridae, from the Coordinating Subcommittee of the International Committee on Taxonomy of Viruses (ICTV), three genera were established within the family of Parvoviridae (Bachmann *et al.*, 1975), namely, (1) *Parvovirus,* (2) Adeno-associated virus (AAV) and (3) *Densovirus.* The differentiation into these 3 genera is not very well defined, since it is based mainly on differences in the mode of replication which is still poorly understood.

The dependence of some viruses (AAV serotypes) on helper viruses has been emphasized as a reason for classifying them into a separate group. However, the replication rate of several other so-called autonomous parvoviruses is strongly enhanced by the presence of helper viruses. On the other hand, it can be argued that an undiscovered permissive host or host cell might exist for the defective viruses.

It seems most likely that parvoviruses can be classified into two genera, based on the nature and physiochemical properties of the nucleic acid in the virion. The first genus would be characterized by the presence of only the minus strand of DNA in the virion; the viruses of the second genus have the complementary (plus and minus) DNA strands separately encapsidated. This striking feature could be responsible for a different mode of replication of these two genera. Together with AAV, densonucleosis virus belongs to the latter genus. Bachmann *et al.* (1975) reported that other arthropod paroviruses may also contain the complementary strands separately encapsidated, but this statement has not yet been substantiated with experimental facts.

III. PHYSICOCHEMICAL PROPERTIES OF VIRIONS AND SUBVIRAL COMPONENTS

A. Nucleic Acid

Early studies on parvoviruses suggested that these viruses contained DNA because they all produced intranuclear inclusions in susceptible cells (Rabson *et al.*, 1961). Crawford (1966) demonstrated that the DNA of the parvovirus MVM had a single-stranded structure. This feature distinguishes the parvoviruses from the papovaviruses (the latter have about the same genome content but their DNA is double-stranded).

Although the single-stranded nature of the DNA of parvoviruses is well established today, initial studies on several parvoviruses (AAV-serotypes, DNV) indicated double-stranded structures (Atchison *et al.*, 1965; Truffaut *et al.*, 1967; Rose *et al.*, 1968). In a study of Crawford *et al.* (1969) the physicochemical prop-

TABLE I

Physicochemical Properties of DNA Molecules of Parvoviruses

Genus	Virus	CsCl (gm/cm³)	T_m (°C)	Sedimentation rate (s)	Length (μm)	MW (× 10^{-6})	Reference[b]
Only "−" strands in virions	H-1	1.720		27.8	—	1.7	(1)
	KBSH	1.724		24	—	1.4	(2)
	KRV	1.726		27	1.5 ± 0.2	1.6	(3)
	MVM	1.722		—	1.2	1.5	(4)
"+" and "−" strands in virions	AAV-1	1.717[a]	91.5	15.5	1.38 + 0.05		(5)–(9)
	AAV-2	1.729		—		1.35	(5)–(9)
		1.714[a]	90.4	15	1.38 ± 0.06		(5)–(9)
		1.726		24		1.35	(5)–(9)
	AAV-3	1.715[a]	90.8	15	1.39 ± 0.06		(5)–(9)
		1.727		—		1.35	(5)–(9)
	AAV-4	1.720[a]	93.0	15.7	1.5 ± 0.21		(5)–(9)
		1.728		—		1.50	(5)–(9)
	DNV	1.701[a]	85.0	17	1.69		(10–(13)
		1.711		16		1.60	(10)–(13)

[a] Double-stranded molecules.

[b] Key to references: (1) McGeoch *et al.* (1970); (2) Siegl (1972); (3) Salzman *et al.* (1971); (4) Crawford *et al.* (1969); (5) Koczot *et al.* (1973); (6) Parks *et al.* (1967); (7) Rose (1974); (8) Rose *et al.* (1966, 1968, 1969); (9) Rose and Koczot (1971); (10) Barwise and Walker (1970); (11) Kurstak *et al.* (1971); (12) Truffaut *et al.* (1967); (13) Kurstak (1972).

erties of the virion and DNA of AAV and MVM were compared with those of ϕX174 (this bacteriophage is not a parvovirus; it has circular DNA and different protein composition) and the DNA in the AAV virion was revealed to be single-stranded. To explain this paradox, it was suggested that there were two types of AAV virions which contained either the plus or the minus strand and that these complementary strands would base pair upon extraction to form double-stranded DNA. This hypothesis was supported by the experimental findings of Rose *et al.* (1969) for AAV and Barwise and Walker (1970) and Kurstak *et al.* (1971, 1973) for DNV (Fig. 2). The physicochemical properties of the DNA molecules of different parvoviruses are given in Table I. The low buoyant density of the DNV–DNA duplex molecule suggests a high adenine content. The base composition of some of the parvoviruses have been reported recently (Rose, 1974). However, with the exception of MVM, these data do not agree with the values of their buoyant densities when verified with the theoretical formula for the mathematical relationship between these two parameters (Riva *et al.*, 1969).

Recent studies on AAV DNA revealed the presence of a limited number of nucleotide sequence permutations and also of terminal nucleotide sequence repetition (either inverted, natural, or both; Berns, 1974).

B. Proteins

Proteins make up about 60–80% of the weight of the virion. Dissociation with sodium dodecyl sulfate (SDS) or urea and subsequent electrophoresis on polyacrylamide gel showed the presence of 3 or 4 polypeptides. Among all arthropod parvoviruses, only the proteins of DNV have been studied in some detail.

DNV proteins behave differently in SDS-polyacrylamide gels than do the marker proteins. The widely employed technique of Weber and Osborn (1969) for the estimation of molecular weights of proteins proved to be unreliable in this system (Tijssen *et al.*, 1976). On the other hand, the molecular weights of these proteins could be estimated from the retardation of their electrophoretic mobility in gels with increasing polyacrylamide concentrations (Tijssen *et al.*, 1976). With this method, four structural proteins were found with molecular weights of 49,000, 58,500, 69,000, and 98,000 and were designated as p49, p59, p69, and p98, respectively. There are some indications that p98 might be a dimer of p49.

Tinsley and Longworth (1973) cited in their review on parvoviruses a conference presentation on the structural proteins of DNV, indicating that three different proteins are detectable in 5% polyacrylamide gels, whereas four are found in 10% gels. This phenomenon was never observed in this laboratory. An explanation for this observation would be that DNV structural proteins do not have interprotein but intraprotein disulfide bridges. Omission of β-mercaptoethanol used for the dissociation of the virions would lead to the formation of a double band for p49, i.e., p49 and p49′, the latter corresponding to the protein with the intact intraprotein disulfide bridge. This component behaves as a protein with a lower molecular weight, its dissociation after reduction lowers the R_f so that the mobility of p49′ becomes the same as that of p49.

The range of the molecular weights of the structural proteins (50,000–100,000) is the same for all the parvoviruses thus far analyzed (Table II).

The number of protein molecules per virion can be estimated by densitometry or by the use of radioactive precursors. The mass of the main protein in DNV (p49) and in other parvoviruses is equivalent to about 60 protein molecules per virion, the other proteins being present in significantly lower numbers (<10) (Tijssen *et al.*, 1976).

It has been reported for almost all the parvoviruses that the aggregate molecular weights of their proteins exceed the coding capacity of the viral genome. Several possible alternatives could account for this paradox: (1) some of the capsid proteins could be cleaved during virion maturation or purification; (2) one of the structural proteins could be coded for by the host; and (3) larger precursor proteins could be synthesized which could vary either at the point of initiation or of cleavage in the subsequent processing. The first alternative seems to hold for MVM (Clinton and Hayashi, 1975), where an intermediate-sized protein is cleaved to lower molecular weight proteins. This conversion has not been observed for other parvoviruses, although Johnson *et*

TABLE II

Comparison of the Molecular Weights of the Structural Proteins of the Parvoviruses and of DNV[a,b]

Protein	DNV	Kilham rat virus	H-1	AAV	AAV-3	Haden	MVM
I	*49,000*	52,000	56,000	*62,000*	*65,900*	*66,800*	69,000[c]
II	58,500	*62,000*	*72,000*	73,000	79,300	76,750	72,000[c]
III	69,000	72,000	92,000	87,000	91,000	85,500	92,000
IV	98,000	–	–	–	±[d]	–	–

[a] The italic figures indicate the main protein for each virus.

[b] References: DNV: Tijssen *et al.* (1976); KRV: Salzman and White (1970); H-1: Kongsvik and Toolan (1972); AAV: Rose *et al.* (1971); AAV-3: Johnson *et al.* (1971); Haden: Johnson and Hoggan (1973); MVM: Clinton and Hayashi (1975).

[c] II (main protein) of MVM is converted to I by proteolytic action.

[d] ±, sometimes a fourth band was obtained (117,000).

al. (1972) reported that antisera prepared against the SDS-dissociated II protein of AAV cross-reacted with the I protein of AAV.

Unpublished data from this laboratory of the amino acid compositions of the DNV proteins showed a resemblance between p49 and p59 but not p69. Amino acid compositions of the proteins of other parvoviruses are not yet available, although the total amino acid compositions for AAV-2 and H-1 have been reported (Rose *et al.*, 1971; Kongsvik and Toolan, 1972).

So far only one messenger RNA (MW: 900,000) appears to be synthesized for AAV *in vivo* (Carter and Rose, 1974). This mRNA would be just sufficient to code for the largest polypeptide. If no other mRNA's are found, then the precursor protein must be cleaved to give rise to the smaller proteins. Posttranslational formative cleavage is sometimes necessary since the protein synthesizing machinery of animal cells, in contrast to that of bacteria, is incapable of carrying out internal initiation in the translation of polycistronic viral messages. Much more widespread are the morphogenetic proteolytic scissions which are characterized to be specific at certain intermediary stages in viral assembly, such as in the packaging of DNA.

Enzymatic activity of one of the structural proteins has been detected by Salzman (1971) for rat virus. This DNA polymerase activity is associated with the high molecular weight protein (Salzman and McKerlie, 1975). An attempt to confirm this finding with H-1 virus failed (Rhode, 1973).

C. Virions

Parvoviruses are among the viruses with the highest buoyant densities in CsCl (about 1.40 gm/cm^3; Hoggan, 1971), owing to the high nucleic acid content of the particles (20–40%). Other viruses, e.g., rhinoviruses, with this density are unstable in acidic solutions, whereas parvoviruses are stable within the pH range of 3–9.

The resistance of parvoviruses to inactivation with ether or chloroform is consistent with an absence of essential lipids in the virion. Moreover, there is very little or no carbohydrate in the DNA virion.

Parvoviruses are among the most stable animal viruses. Depending on the salt conditions, they withstand heating at 56° for at least 1 hour and remain viable for years at room temperature (Rose, 1974). They also resist inactivation by DNase (Vasquez and Brailovsky, 1965; Siegl *et al.*, 1971).

TABLE III

Physical Properties of Some Parvovirus Particles

Virus	CsCl gm/cm^3	Sedimentation rate(s)	Particle weight ($\times 10^{-6}$)	References[a]
KRV	1.400	110, 122	6.6	2, 4, 6
KBSH	1.395	105	5.3	7, 10
H-1	1.422	110	—	2, 4
AAV-1	1.395	104, 125	5.4	1, 2, 5
AAV-4	1.445	137	5.4	2, 3
DNV-I	1.400	111	5.6	8, 9
DNV-II	1.440	89	5.0	2, 8

[a] Key to references: (1) Crawford *et al.* (1969); (2) Hoggan (1971); (3) Mayor *et al.* (1969); (4) McGeoch *et al.* (1970); (5) Rose *et al.* (1971); (6) Salzman and Jori (1970); (7) Siegl *et al.* (1971); (8) Tijssen *et al.* (1977); (9) Truffaut *et al.* (1967); (10) Siegl (1972).

The density values and sedimentation rates of different parvoviruses are given in Table III. For DNV, two different buoyant densities have been reported, 1.40 gm/cm^3 (Truffaut *et al.*, 1967) and 1.44 gm/cm^3 (Hoggan, 1971). It was demonstrated recently that virions of both densities existed and that their physicochemical properties were slightly different (Tijssen *et al.*, 1977) (Fig. 3). The difference between these two types is possibly due to two reasons: (1) in type II (with buoyant density of 1.44 gm/ml) about one-half of the minor proteins are missing and (2) type II has a different quaternary structure than type I. The specific absorption coefficients were $A^{0.1}_{260\text{ nm}, 1\text{ cm}} = 6.5$ for DNV-I and 7.1 for DNV-II. The absorption spectra, corrected for light scattering, showed the following characteristic ratios: $A_{260\text{ nm}}/A_{280\text{ nm}} = 1.59$ for DNV-I and 1.64 for DNA-II, whereas the $A_{260\text{ nm}}/A_{240\text{ nm}}$ ratio was 1.64 and 1.74 for DNV-I and DNV-II, respectively.

The balance of forces in the quaternary structure is maintained both for the DNA and for the protein components, as well as for the interprotein and protein-DNA interactions. These values reflect the minimum free energy conformation of their constituent components as they occur in their integrated state. It is suggested that protein-DNA interactions are more important in DNV-II than in DNV-I (Tijssen *et al.*, 1977).

IV. MORPHOLOGY

Some variation exists in the size of parvoviruses, although they all seem to have a diameter of the order of 20–25 nm (Bachmann *et al.*, 1975). These variations may be due either to technical errors, to buffer and staining effects (staining with uranyl acetate seems to yield larger virions than with PTA; Vernon and Rubin, 1973–1974), or they may reflect real differences in the size of parvoviruses. The latter alternative seems to be supported by several studies. Payne *et al.* (1963) found considerable variation in size (18–24 nm) of the X14 virus by negative staining with PTA, whereas Breese *et al.* (1964) found RV to vary in size between 15 to 25 nm. DNV was found to have a diameter of 21–23 nm (Amargier *et al.*, 1965), whereas Kurstak and Côté (1969) reported an average diameter of 20 ± 1.5 nm. These discrepancies can, at least partially, be explained by the recent finding that densonucleosis virus is heterogeneous and consists of two types with mean diameters of 24 nm and 21 nm for type I and II, respectively (Tijssen *et al.*, 1977). Mixing of these types on the same grid and measuring the diameters of these particles gave a two peak histogram. Similar observations were also made for AAV and Haden virus (Hoggan, 1971).

Most studies on the fine structure of parvoviruses have been hindered by their small size and their tight close-packed capsids. This close-packing of the capsid is probably due to the fact that the interprotein interactions are of a secondary character, acting through oriented complementary surfaces on the protein subunits, similar to antigen–antibody interactions. Such a capsid subunit would have 4 complementary surfaces, since there are 60 subunits per virion (triangulation number of 1, suggesting a dodecahedron; Caspar and Klug, 1962). The molecular weight of such a subunit suggests a Stokes radius of 30–35 Å (Andrews, 1970) if its protein is globular. The report of the study group on Parvoviridae (Coordinating Subcommittee, ICTV; Bachmann *et al.*, 1975; see also reviews given in Section I) that capsomers have a diameter of only 3–4 nm seems interpretable with an elongated structure of such capsomers. Calculations reveal that the protein of the capsomer would have a length of 14 nm and would not cover the virions' surface. Vernon and Rubin (1973–1974) studied the virus subunits after disruption by lyophilization followed by rehydration in the presence of SDS or treatment with lithium thiocyanate. A mean diameter of 4.1–4.6 nm was found for the virus depending on the conditions of the negative staining. In order to cover the surface of a virion of a diameter of 24 nm by means of a 12-pentamer structure (dodecahedron), the theoretical diameter of a capsomer would be ± 10 nm (Fig. 4). The dodecahedron structure seems to be supported by the dissociation of DNV into pentamers. This finding would also suggest that two types of secondary interactions exist, an intrapentamer and interpentamer, and that selective breakage of these interactions produces clusters of 3 and 5 subunits, respectively. The capsomers of DNV have the same appearance as the pentamers of ϕX174 (also a dodecahedron of the same size; Horne, 1974). Earlier studies in this laboratory suggested a 42-capsomer model (Kurstak and Côté, 1969).

V. BIOLOGICAL PROPERTIES

A. Host Range and Pathogenicity

Only a few parvoviruses are known to be the cause of natural diseases, although they are often found as tenacious contaminants of cell lines, virus stocks, and certain vaccines. The insect parvoviruses, however, are responsible for fatal diseases of their hosts. DNV kills the larvae of *Galleria mellonella* in 2–14 days, depending on the concentration of the virus in the inoculum. No other host is known for DNV. DNV is polytropic; all tissues, with the exception of the midgut, are infected (Amargier *et al.*, 1965; Kurstak and Vago, 1967; Garzon and Kurstak, 1968; Kurstak, 1972) (Figs. 5 and 6). The small genome of the parvoviruses might explain their considerable host-specificity, since they will depend for many of their functions on the host cells. *In vitro*, this host range may not be as restricted. It was shown that DNV was capable of replicating *in vitro* in the ovarian cells of *G. mellonella* and *Bombyx mori* (Vago *et al.*, 1966) and even in mammalian cells, such as L cells (Kurstak *et al.*, 1969). However, extensive experiments attempting to infect mice *in vivo* did not meet with success (Charpentier and Kurstak, in preparation).

B. Multiple Infections and Interference

Numerous examples of both homologous and heterologous interference are well known for the parvoviruses (Rose, 1974). For example, the yield of AAV was found to decrease drastically if more than the optimal multiplicity is used for inoculation (Rose and Koczot, 1972). Kurstak and Garzon (1975) and Odier (1975) showed a selective inhibition of the production of DNV in adipose cells if the DNV inoculation was preceeded by the inoculation of nuclear polyhedrosis virus (NPV) by at least 16 hours. The silk glands, however, which are not susceptible to infection by NPV show a typical DNV cytopathogenicity. Simultaneous inoculation of DNV with NPV gave a characteristic DNV infection in all cells. Odier (1975) observed the two viruses in the same nucleus: virogenic stroma of DNV and bacilloformic NPV virions in different stages of development. This phenomenon has not been observed in our laboratory. The primary infection of DNV suppresses mostly the nuclear development of NPV in a later stage. In another study (Kurstak and Garzon, 1975), single cells were dually infected with tipula iridescent virus (TIV; a cytoplasmic DNA virus) and DNV. Under these conditions a simultaneous production of both viruses was observed, without any apparent mutual effect other than "exhausting" the cell by metabolic competition.

C. Carrier States

Possible activations of latent parvoviruses may occur by certain manipulations, such as shipping, freeze-storage cells, or X irradiation (Hallauer *et al.*, 1971; Payne *et al.*, 1964). Hoggan (1970) showed that kidney cells, immunologically negative for AAV, released infectious AAV when exposed to purified helper virus. Detroit 6 cells exposed to AAV without helper virus remained positive for the ability to release AAV when exposed to helper virus for at least 100 passages (Hoggan *et al.*, 1973). Subsequent studies revealed that these AAV-carrier clones contained about 3–5 AAV genome equivalents per cell and that 10^4–10^5 AAV particles per cell were produced upon exposure to adenovirus (Hoggan, 1975). It is questionable, however, whether or not this important biological feature plays a role in nondefective viruses. The ubiquitous latency of parvoviruses is not well understood, in spite of the fact that parvoviruses are very potent contaminants of biological systems and as such can interfere easily with experiments in the laboratory.

VI. REPLICATION AND MORPHOGENESIS

The degree of the dependence of a virus on the host's replicative apparatus is related to its size. The single-stranded DNA bacteriophages have 8 to 10 genes of which several are implicated in replication (at least genes, 2, 3, or 5 for the filamentous phage M13 and genes *A*, *C*, or *D* for ϕX174; both phages code for at least 5 nonstructural proteins; (Kornberg, 1974).

The picture for parvoviruses is far less clear. As discussed earlier, the aggregate molecular weights of the structural proteins already exceed the coding capacity of the viral genome. The finding that only 50–60% of the viral genome of AAV is transcribed *in vivo* (Carter and Rose, 1972) makes the understanding of the mechanism of virus replication even more complex.

The mode of replication of parvovirus DNA is virtually unknown. From a theoretical point of view, the replication of single-stranded viral DNA should include 3 stages: (1) conversion of viral DNA to a replicative form (RF); (2) multiplication of RE; and (3) synthesis of progeny single strands. The first stage probably relies on the host's replicative system and, thus, would not require virus-coded proteins. There is an increasing body of evidence that parvoviruses appropriate the replicative assembly used by the cell for the duplication of its chromosome, since their replication is associated with late S phase gene functions (Rhode, 1973). The fact that parvoviruses replicate shortly after cellular DNA replication would support this view. It is not yet understood, however, by what mechanism(s) this small viral DNA is recognized inside the cell and can compete with the excess of cellular DNA. Similarly, the mechanism which prevents the mitosis of the infected cells is also unknown. Available evidence suggests a rolling circle mechanism for the second stage in the replication for the single-stranded DNA bacteriophages (Kornberg, 1974), but no data support such a mechanism for parvoviruses. Parvoviral DNA is linear and AAV DNA shows terminal repetition (Koczot *et al.*, 1973; Gerry *et al.*, 1973), suggesting that multiplication of the RF would be possible by concatemeric intermediates as described for linear bacteriophage DNA's. In a morphogenetic study (Garzon and Kurstak, 1976), it was demonstrated that the nucleolus of the infected cells undergoes a hypertrophy, which is accompanied by the segregation of its fibrillar and granular components. The development of the granular portion coincided with the synthesis of the double-stranded DNA of the RF in the virogenic stroma. As the infection progresses, the granular portion of the nucleolus regresses in favor of the fibrillar portion and DNV seems to stimulate the formation of intranuclear bodies associated with the virogenic stroma (Figs. 7–11). The third stage, the synthesis of progeny single strands from the replicative intermediate is a particular stage since the viral strands are separated at a temperature which is about 50° below the melting point of the duplex. This step might involve unwinding proteins (Alberts *et al.*, 1972) or the binding of structural and other proteins (Fidanian and Ray, 1974). The appearance of empty DNV capsids in dually infected cells (Kurstak and Garzon, 1975) seems to exclude the monomeric structural proteins in strand separation. On the other hand, DNA does not seem necessary for the assembly of the viral capsid. The virions are assembled inside the virogenic stroma and may be in relation with nucleolus and the intranuclear bodies (Garzon and Kurstak, 1976) (Fig. 9).

VII. CONCLUSION

All parvoviruses have been discovered during the last 1–2 decades, their small size rendering their discovery difficult. Vertebrate parvoviruses are ubiquitous in nature and it is expected that this is the case also for invertebrate parvoviruses. Each parvovirus species seems to be restricted to its specific host. The infection is characterized by the formation of dense, DNA-positive inclusions in the hypertrophied nuclei of the infected cells.

The resemblance of the physiochemical properties of all parvoviruses is striking and leaves no doubt in the classification of a virus as a parvovirus. The classification of the parvoviruses into three genera by the ICTV study group on parvoviridae seems premature, since this classification is mainly based on the not yet understood mode of replication.

Our knowledge on the morphology of parvoviruses is restricted, due to the lack of structure in their closely-packed capsids. The way of clustering of the 60 subunits in the virion seems only interesting for studies on the stabilizing forces in the virion and not for the morphological identification of a parvovirus, since there are indications that viral proteins obscure these capsomers.

ACKNOWLEDGMENTS

This work was supported by grants MA-2385 from the Medical Research Council of Canada and A-3746 from the National Research Council of Canada.

REFERENCES

Alberts, B., Frey, L., and Delius, H. (1972). *J. Mol. Biol.* **68,** 139.
Amargier, A., Vago, C., and Meynardier, G. (1965). *Arch. Gesamte Virusforsch.* **15,** 659.
Andrewes, C. (1970). *Virology* **40,** 1070.
Andrews, P., (1970). *Methods Biochem. Anal.* **18,** 1.
Atchison, R. W., Casto, B. C., and Hammon, W. McD. (1965). *Science* **194,** 754.
Bachmann, P. A., Hoggan, M. D., Melnick, J. L., Pereira, H. G., and Vago, C. (1975). *Intervirology* **5,** 83.
Barwise, A. H., and Walker, I. O. (1970). *FEBS Lett.* **6,** 13.
Bernhard, W. (1968). *C. R. Hebd. Seances Acad. Sci.* **267,** 2170.
Berns, K. I. (1974). *Curr. Top. Microbiol. Immunol.* **65,** 1.
Breese, S. S., Howatson, A. F., and Chany, C. (1964). *Virology* **24,** 598.

Carter, B. J., and Rose, J. A. (1972). *J. Virol.* **10,** 9.
Carger, B. J., and Rose, J. A. (1974). *Virology* **61,** 182.
Caspar, D. L. D., and Klug, A. (1962). *Cold Spring Harbor Symp. Quant. Biol.* **27,** 1.
Clinton, G. M., and Hayashi, M. (1975). *Virology* **66,** 261.
Crawford, L. V. (1966). *Virology* **29,** 605.
Crawford, L. V., Follett, E. A. C., Burden, M. G., and McGeoch, D. J. (1969). *J. Gen. Virol.* **4,** 37.
Fidanian, H. M., and Ray, D. S. (1974). *J. Mol. Biol.* **83,** 63.
Garzon, S., and Kurstak, E. (1968). *Nat. Can.* **95,** 1125.
Garzon, S., and Kurstak, E. (1976). *Virology* **70,** 517.
Gerry, H. W., Kelly, T. J., Jr., and Berns, K. I. (1973). *J. Mol. Biol.* **79,** 207.
Hallauer, C., Kronauer, G., and Siegl, G. (1971). *Arch. Gesamte Virusforsch.* **35,** 80.
Hirumi, H., Hirumi, K., Speyer, G., Yunker, C. E., Thomas, L. A., Cory, J., and Sweet, B. H. (1976). *In Vitro* **12,** 83.
Hoggan, M. D. (1970). *Prog. Med. Virol.* **12,** 211.
Hoggan, M. D. (1971). *In* "Comparative Virology" (K. Maramorosch and E. Kurstak, eds.), pp. 43–79. Academic Press, New York.
Hoggan, M. D. (1975). *Abstr. Int. Cong. Virol., 3rd, 1975* p. 181.
Hoggan, M. D., Thomas, G. F., and Johnson, F. B. (1973). *Proc. Lepetit Int. Colloq. Biol. Med., 4th, 1972:* Possible Episomes in Eukaryotes. p. 243.
Horne, R. W. (1974). "Virus Structure." Academic Press, New York.
Johnson, F. B., and Hoggan, M. D. (1973). *Virology* **51,** 129.
Johnson, F. B., Ozer, H. L., and Hoggan, M. D. (1971). *J. Virol.* **8,** 860.
Johnson, F. B., Blacklow, N. R., and Hoggan, M. D. (1972). *J. Virol.* **9,** 1017.
Koczot, F. J., Carter, B. J., Caron, C. F., and Rose, F. A. (1973). *Proc. Natl. Acad. Sci. U.S.A.* **70,** 215.
Kongsvik, J. R., and Toolan, H. W. (1972). *Proc. Soc. Exp. Biol. Med.* **139,** 1202.
Kornberg, A. (1974). "DNA Synthesis." Freeman, San Francisco, California.
Kurstak, E. (1972). *Adv. Virus Res.* **17,** 207.
Kurstak, E., and Côté, J. R. (1969). *C. R. Hebd. Seances Acad. Sci.* **268,** 616.
Kurstak, E., and Garzon, S. (1975). *Ann. N. Y. Acad. Sci.* **266,** 232.
Kurstak, E., and Vago, C. (1967). *Rev. Can. Biol.* **26,** 311.
Kurstak, E., Belloncik, S., and Brailovsky, C. (1969). *C. R. Hebd. Seances Acad. Sci.* **269,** 1716.
Kurstak, E., Vernoux, J. P., Niveleau, A., and Onji, P. A. (1971). *C. R. Hebd. Seances Acad. Sci.* **272,** 762.
Kurstak, E., Vernoux, J. P., and Brakier-Gingras, L. (1973). *Arch. Gesamte Virusforsch.* **40,** 274.
Lebedeva, O. P., Kusnetsova, M. A., Zelenko, A. P., and Gudz-Gorban, A. P. (1973). *Acta Virol. (Engl. Ed.)* **17,** 253.
Lwoff, A., and Tournier, P. (1966). *Annu. Rev. Microbiol.* **20,** 45.
McGeoch, O. J., Crawford, L. V., and Follett, E. A. C. (1970). *J. Gen. Virol.* **6,** 33.
Mayor, H. D. (1973). *Methods Cancer Res.* **8,** 203.
Mayor, H. D., and Kurstak, E. (1974). *In* "Viruses, Evolution and Cancer" (E. Kurstak and K. Maramorosch, eds.), pp. 55–78. Academic Press, New York.
Mayor, H. D., and Melnick, J. L. (1966). *Nature (London)* **210,** 331.
Mayor, H. D., Torikai, K., Melnick, J. L., and Mandel, M. (1969). *Science* **166,** 1280.
Odier, F. (1975). *C. R. Hebd. Seances Acad. Sci.* **280,** 2277.
Parks, W. P., Green, M., Piña, M., and Melnick, J. L. (1967). *J. Virol.* **1,** 980.
Payne, F. E., Shellabarger, G. F., and Schmidt, R. W. (1963). *Proc. Am. Assoc. Cancer Res.* **4,** 51.
Payne, F. E., Beals, T. F., and Preston, R. E. (1964). *Virology* **23,** 109.
Rabson, A. S., Kilham, L., and Kirchstein, R. L. (1961). *J. Natl. Cancer Inst.* **27,** 1217.
Rhode, S. L. (1973). *J. Virol.* **11,** 856.
Riva, S., Barrai, I., Cavalli-Sforza, L., and Falaschi, A. (1969). *J. Mol. Biol.* **45,** 367.
Rose, J. A. (1974). *Compr. Virol.* **3,** 1–61.
Rose, J. A., and Koczot, F. J. (1971). *J. Virol.* **8,** 771.
Rose, J. A., and Koczot, F. J. (1972). *J. Virol.* **10,** 1.
Rose, J. A., Hoggan, M. D., and Shatkin, A. J. (1966). *Proc. Natl. Acad. Sci. U.S.A.* **56,** 86.
Rose, J. A., Hoggan, M. D., Koczot, F. M., and Shatkin, A. J. (1968). *J. Virol.* **2,** 999.
Rose, J. A., Berns, K. I., Hoggan, M. D., and Koczot, F. M. (1969). *Proc. Natl. Acad. Sci. U.S.A.* **64,** 863.
Rose, J. A., Maizel, J. V., Jr., Inman, J. K., and Shatkin, A. J. (1971). *J. Virol.* **8,** 766.
Salzman, L. A. (1971). *Nature (London), New Biol.* **231,** 174.
Salzman, L. A., and Jori, L. A. (1970). *J. Virol.* **5,** 114.
Salzman, L. A., and McKerlie, L. (1975). *J. Biol. Chem.* **250,** 5583.
Salzman, L. A., and White, W. L. (1970). *Biochem. Biophys. Res. Commun.* **41,** 1551.
Salzman, L. A., White, W. L., and Kakefuda, T. (1971). *J. Virol.* **7,** 830.
Siegl, G. (1972). *Arch. Gesamte Virusforsch.* **37,** 267.

Siegl, G., Hallauer, C., Novak, A., and Kronauer, G. (1971). *Arch. Gesamte Virusforsch.* **35,** 91.
Tijssen, P., van den Hurk, J., and Kurstak, E. (1976). *J. Virol.* **17,** 686.
Tijssen, P., Tijssen-van der Slikke, T., and Kurstak, E. (1977). *J. Virol.* **21,** 225.
Tinsley, T. W., and Longworth, J. F. (1973). *J. Gen. Virol.* **20,** suppl., 7.
Truffaut, N., Berger, G., Niveleau, A., MAay, P., Bergoin, M., and Vago, C. (1967). *Arch. Gesamte Virusforsch.* **21,** 469.
Vago, C., Quoit, J. M., and Luciani, L. (1966). *C. R. Hebd. Seances Acad. Sci.* **263,** 799.
Vasquez, C., and Brailovsky, C. (1965). *Exp. Mol. Pathol.* **4,** 130.
Vernon, S. K., and Rubin, B. A. (1973–1974). *Intervirology* **2,** 114.
Weber, K., and Osborn, M. (1969). *J. Biol. Chem.* **244,** 4406.

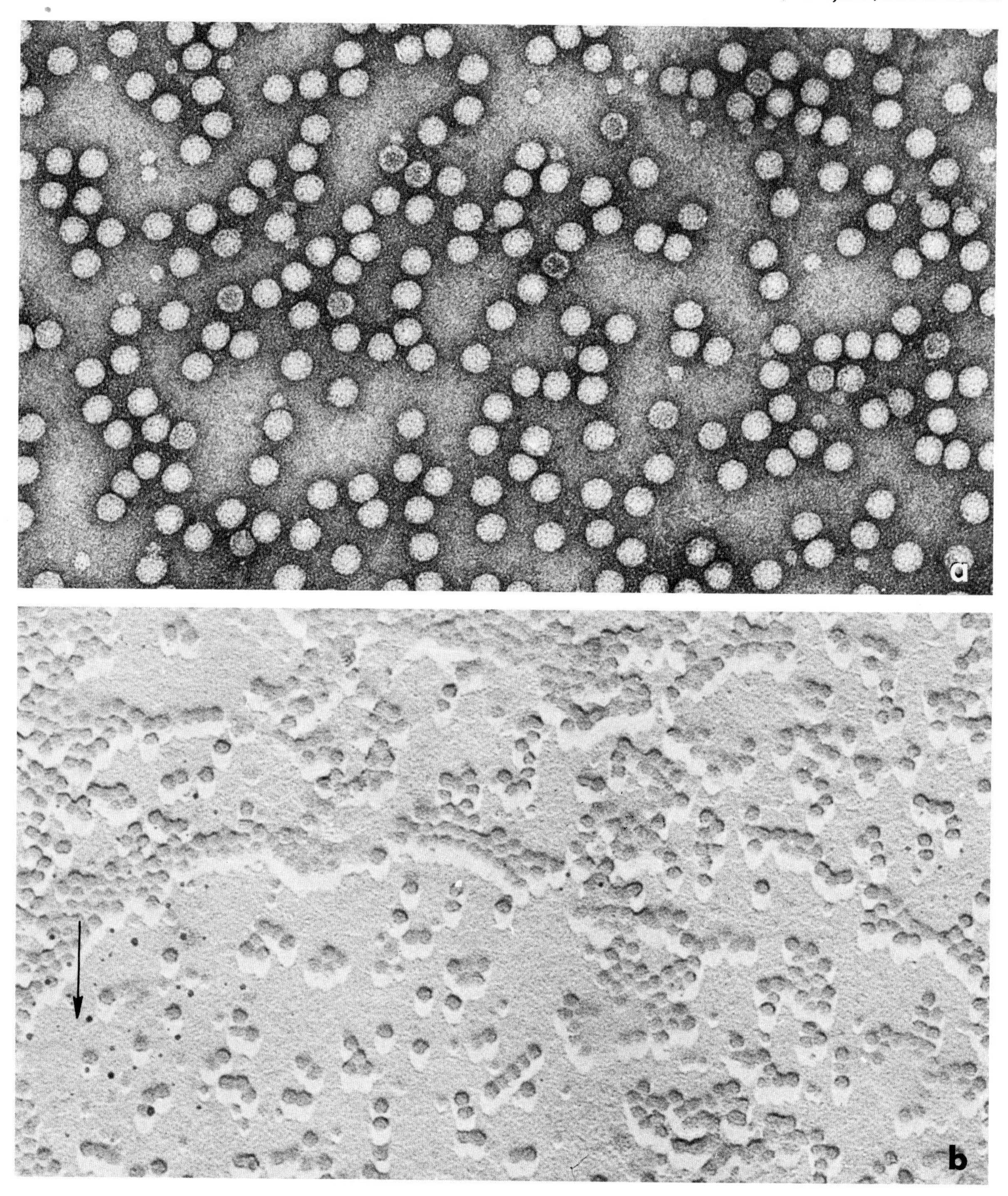

Fig. 1. DNV particles isolated from infected *G. mellonella* larvae. Purification by ultracentrifugation on sucrose gradient. (a) Negatively stained DNA particles, (2% PTA). Full and empty capsids are present. (×215,000.) (b) Platinum shadowing at 30° angle. The arrow (→) indicates the direction of shadowing. (×100,000.)

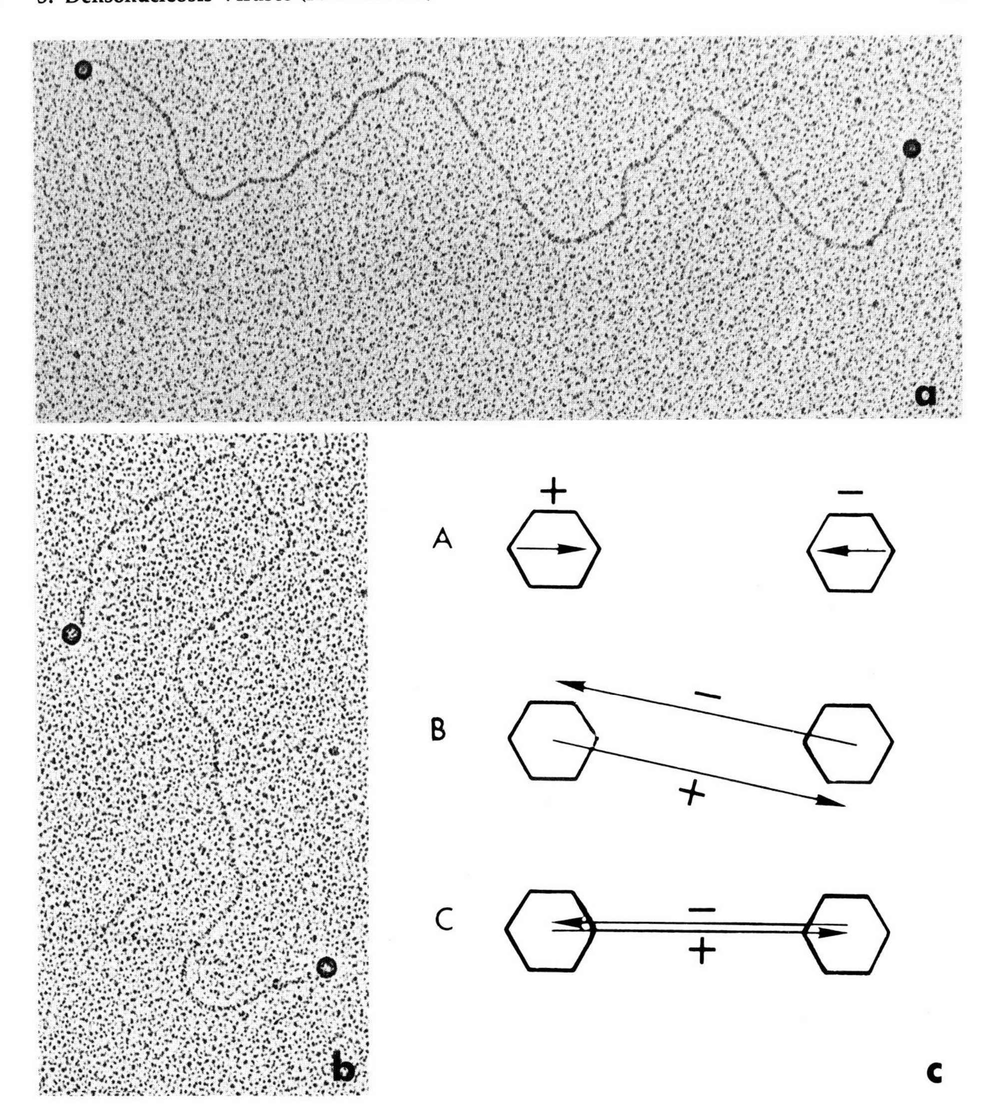

Fig. 2. (a and b) Viral DNA molecules from DNV. Kleinschmidt spreading method. The DNA molecules reassociated with empty capsids at the two extremities with a double-stranded appearance. Uranium oxide shadowing. (×76,000.) (c) Representation of the complementarity of single-stranded DNA chains with inverse polarities (+ and −) encapsidated separately (A); the absence of association in solution at low ionic strength (B); and association by complementarity in solution at high ionic strength (C). From Kurstak *et al.* (1971).

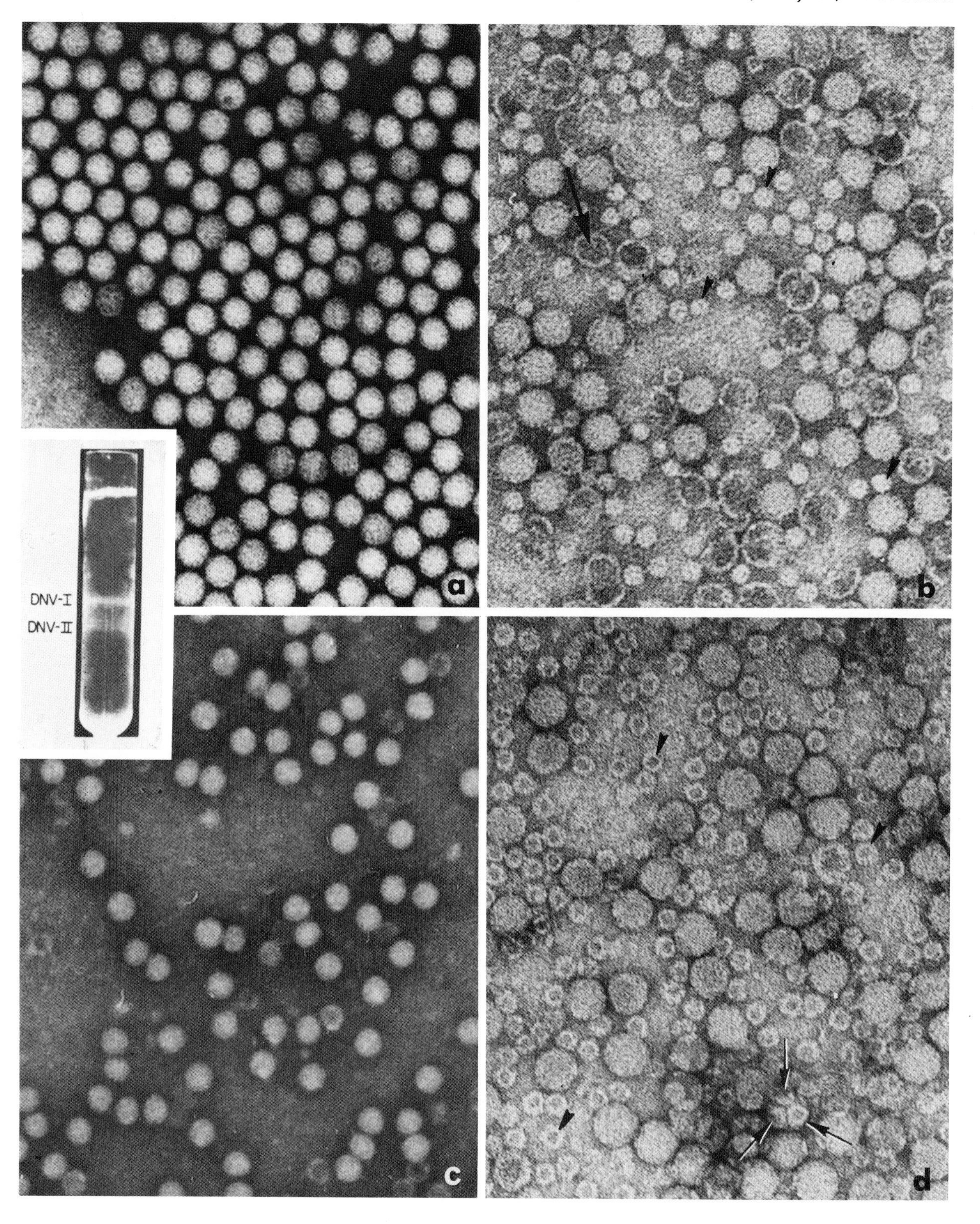
DNV-I
DNV-II
a
b
c
d

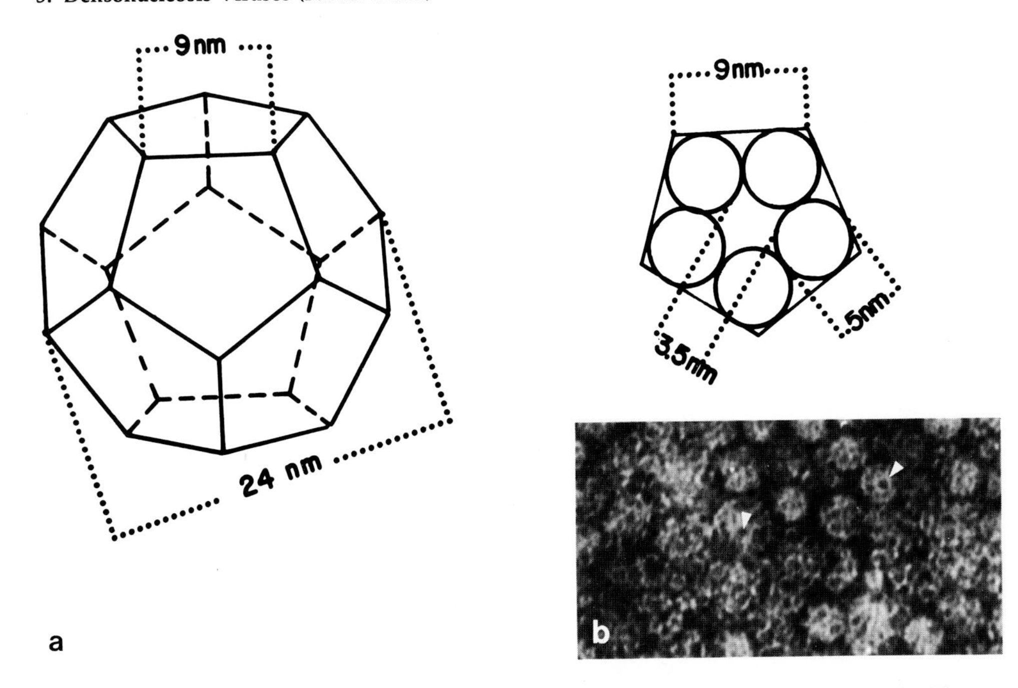

Fig. 4. (a) Model of DNV particle based on biochemical data and electron micrographs. The capsid, 21–24 nm in diameter, is icosahedral in outline, consisting of 12 capsomers with a 9 nm "edge." Each hollow-centered capsomer is a pentamer of p49. The dimensions, based on theoretical considerations, fit the electron microscopic observations. (b) Negatively stained (2% PTA) DNV particles showing different stages of disruption. Note some swollen capsids with clear hollow capsomers (➤). (×332,000.)

Fig. 3. Purification of DNV on CsCl gradient allows the separation of two populations of particles: DNV-1 and DNV-2 (Inset). (a) Negatively stained preparation of purified DNV-I in 0.05 *M* Tris-HCl, pH 8, and 0.15 *M* NaCl. The particles of 24 nm mean diameter are complete and icosahedral in outline, 2% PTA staining. (×332,000.) (b) Negatively stained preparation of purified DNV-I in 0.05 *M* Tris-HCl, pH 8, stained with 1% uranyl acetate. Under this condition many of the particles are disrupted on the collodion-carbon-coated grid into numerous full capsomers (➤) and empty shells (→). (×332,000.) (c) Negatively stained preparation of purified DNV-II in 0.075 *M* Tris-HCl and stained with 2% PTA. Most of the particles appear complete and of 21 nm mean diameter. (×300,000.) (d) Negatively stained preparation of purified DNV-II in 0.05 *M* Tris-HCl, pH 8, and stained with 1% uranyl acetate. Many particles are disrupted under these conditions. Free capsomers (➤), 8–10 nm in diameter, appear like rings in contrast with the capsomers from disrupted DNV-I. One particle (→) in disruption clearly shows an arrangement of three capsomers.

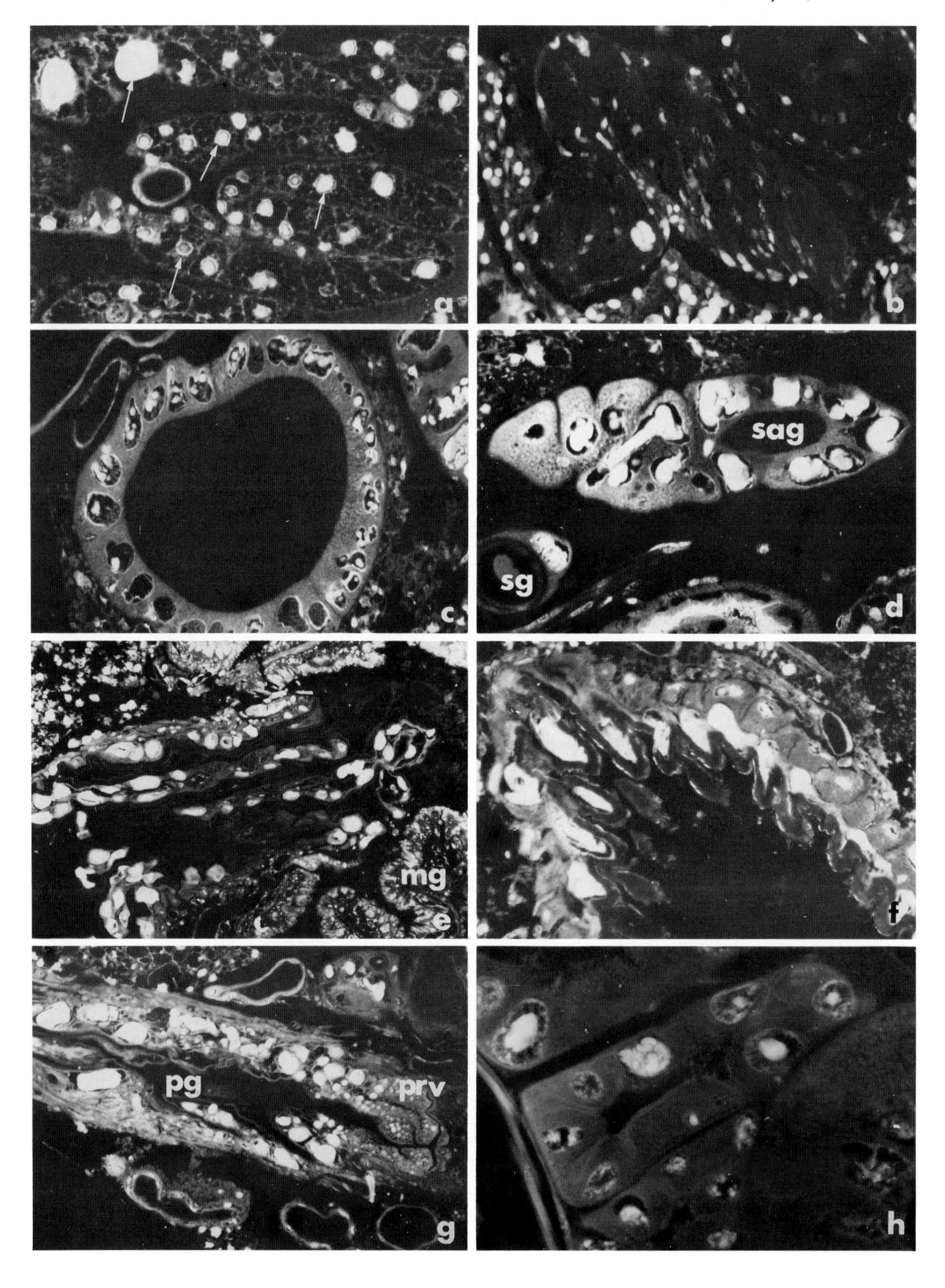
a
b
sag
sg
c
d
mg
e
f
pg
prv
g
h

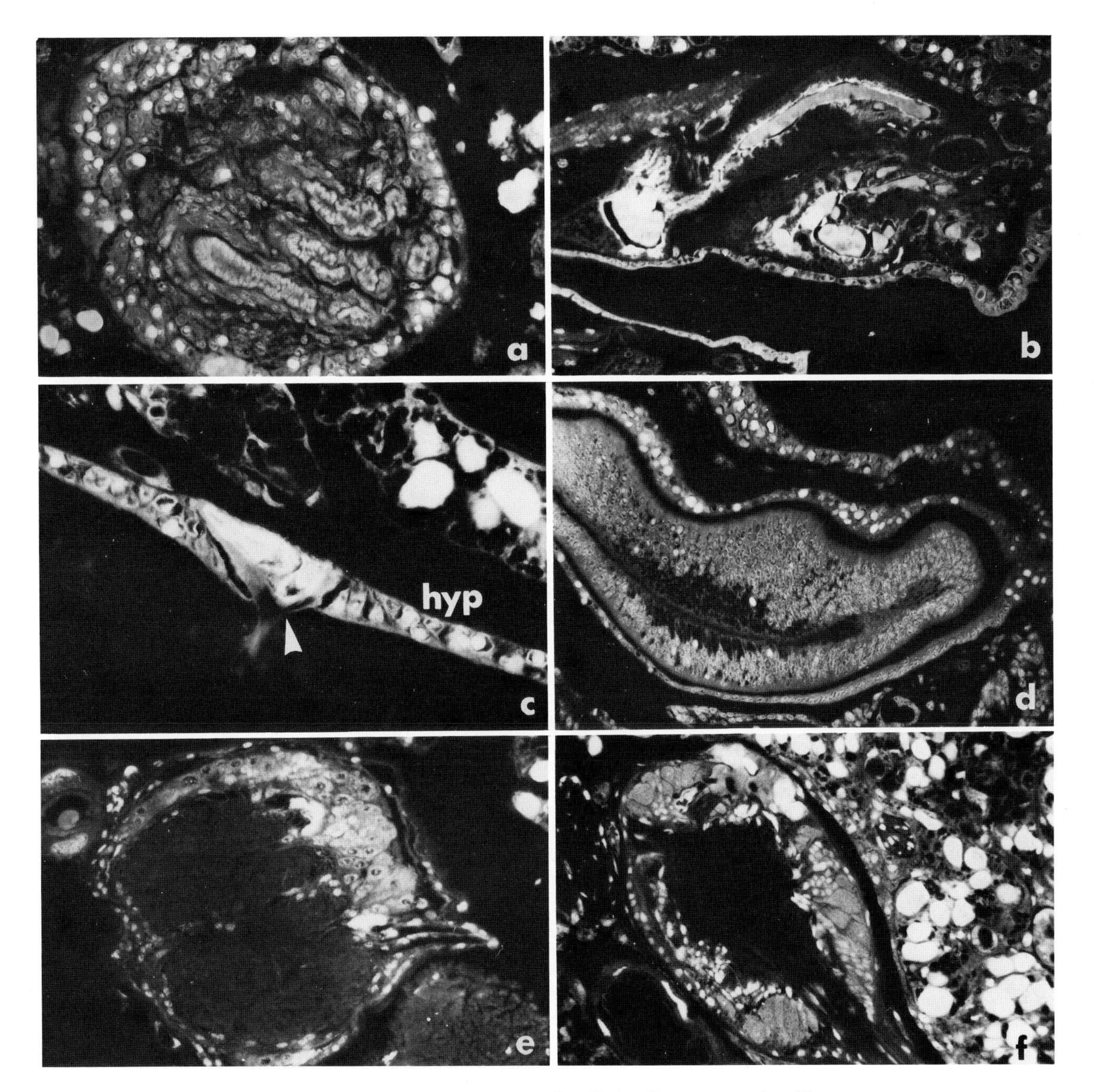

Figs. 5 and 6. Spectrum of virulence and polytropic nature of DNV. Acridine orange stain, pH 3.8. Nuclear lesions (nuclear hypertrophy and dense virogenic stroma) in the cells of various tissues of the infected larvae of *G. mellonella.* Note the asynchronous replication of DNV, reflecting the progression of the viral infection. At the beginning of the infection, the cells show a loss of the nuclear internal organization (green fluorescence) and the appearance of the perinuclear halo and of the yellow–green fluorescence of dense intranuclear masses. These dense masses, the virogenic stroma, develop progressively until they occupy the whole nuclear surface (→). The continued synthesis of viral material leads to the hypertrophy of the nucleus which can reach monstrous proportions, pushing the cytoplasm to the periphery. At the same time, the intensity of the fluorescence of the nucleus increases considerably, producing orange staining with occasional red islets.

Fig. 6. (a) Gonad. The epithelial sheath cells and the intersticial cells show marked nuclear alterations. The germinal cells are insensible to DNV (× 800); (b) molting gland in longitudinal section (× 600); (c) dermal gland (➤) and hypodermis (hyp) (× 1200); (d) wing bud (× 800); the apicale cells appear noninfected in contrast to the cells of the peripodium and of the median part of the bud; (e and f) nodules of the abdominal nerve chain showing infection of the gland cells and of some nerve cells (× 1200).

Fig. 5. (a) Adipose tissue (× 1200); (b) muscle (× 800); (c) silk gland (× 500); (d) salivary gland (sag) and silk gland canal (sg) (× 1600); (e) stomodeal valve (× 1000); the dense and hypertrophied nuclei of the foregut epithelium contrast with those of the midgut (mg); (f) foregut (× 800); (g) hindgut (pg) and proctodeal valve (prv) (× 800); (h) malpighian tubes (× 900).

Fig. 7. First stages of DNV replication (12 hours of infection). (a) Cell of adipose tissue showing the accumulation of small particles of 17–20 nm in diameter, probably ribosomes, condensing into a paracrystalline array inside a vesicle. Different plans of crystallization can be recognized (→). Many of these structures are present in this cell. (×35,000.) (b) Nucleus of a hypodermal cell. Note the hypertrophy and segregation of the nucleolus. f, Fibrillar component; g, granular component; nu, nucleolus. (×30,000.)

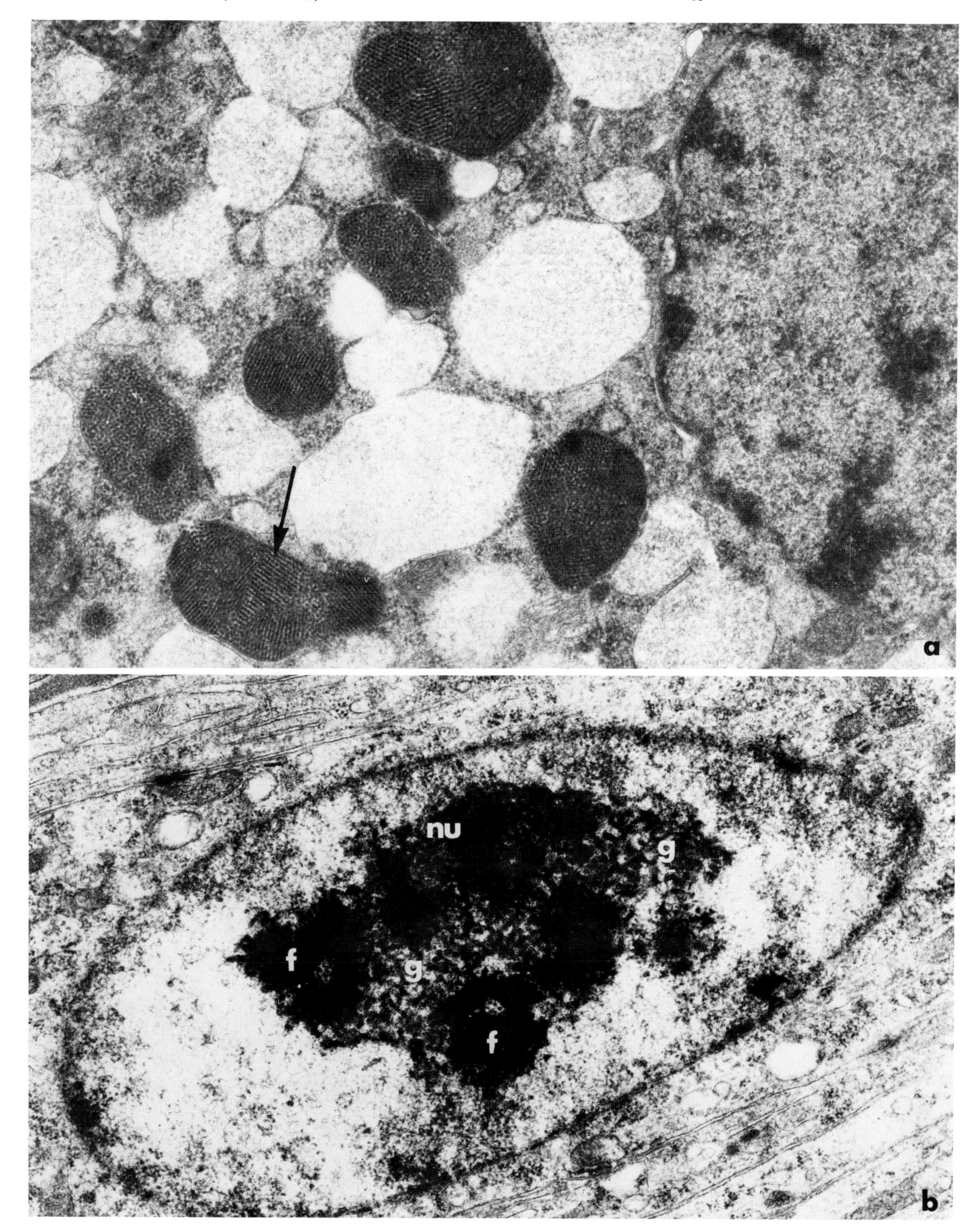
a
nu
g
f
g
f
b

Fig. 8. Early infected hypodermal cells. (a) Two of the hypodermal cells, 24 hours after infection, present a virogenic stroma (vs) which is in close relation with the nucleolus. The latter shows hypertrophy and segregation of its two components. The cytoplasm is rich in free ribosomes and shows the swelling of the endoplasmic reticulum (er) and mitochondria (m). The third cell which presents no signs of infection has a compact and uniformly dense nucleolus. ch, cellular chromatin; nu, nucleolus; f, fibrillar component; g, granular. (×16,000.) (b) Nuclei of hypodermal cells after 48 hours of infection: the virogenic stroma (vs) increases and pushes the chromatin and nucleolus to the periphery. In the center of the virogenic stroma, an intranuclear body is visible (➤). Inside and around the virogenic stroma mature virions (v) are released. (×18,800.) (c) Regressive staining with EDTA (Bernhard, 1968) which preferentially differentiate ribonucleoproteins from deoxyribonucleoproteins. The granular (g) and fibrillar (f) components of the nucleolus, the intranuclear body (➤) and the ribosomes (r) retain their contrast. The virogenic stroma which contains the viral replicative form (double-stranded DNA) as well as the mature virions (single-stranded DNA) are completely "bleached." (×18,800.)

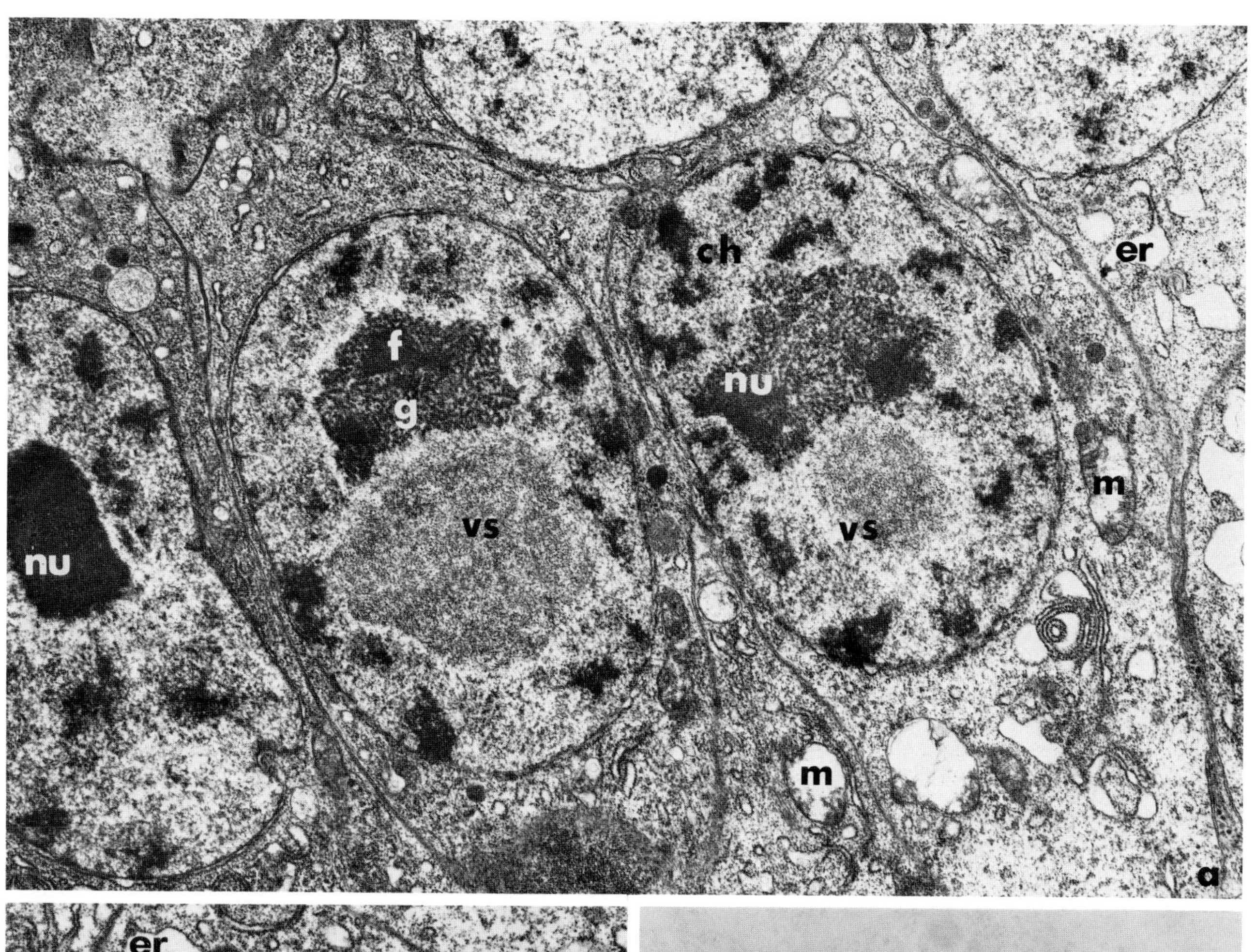

ch
er
f
nu
g
m
nu
vs
vs
m
a

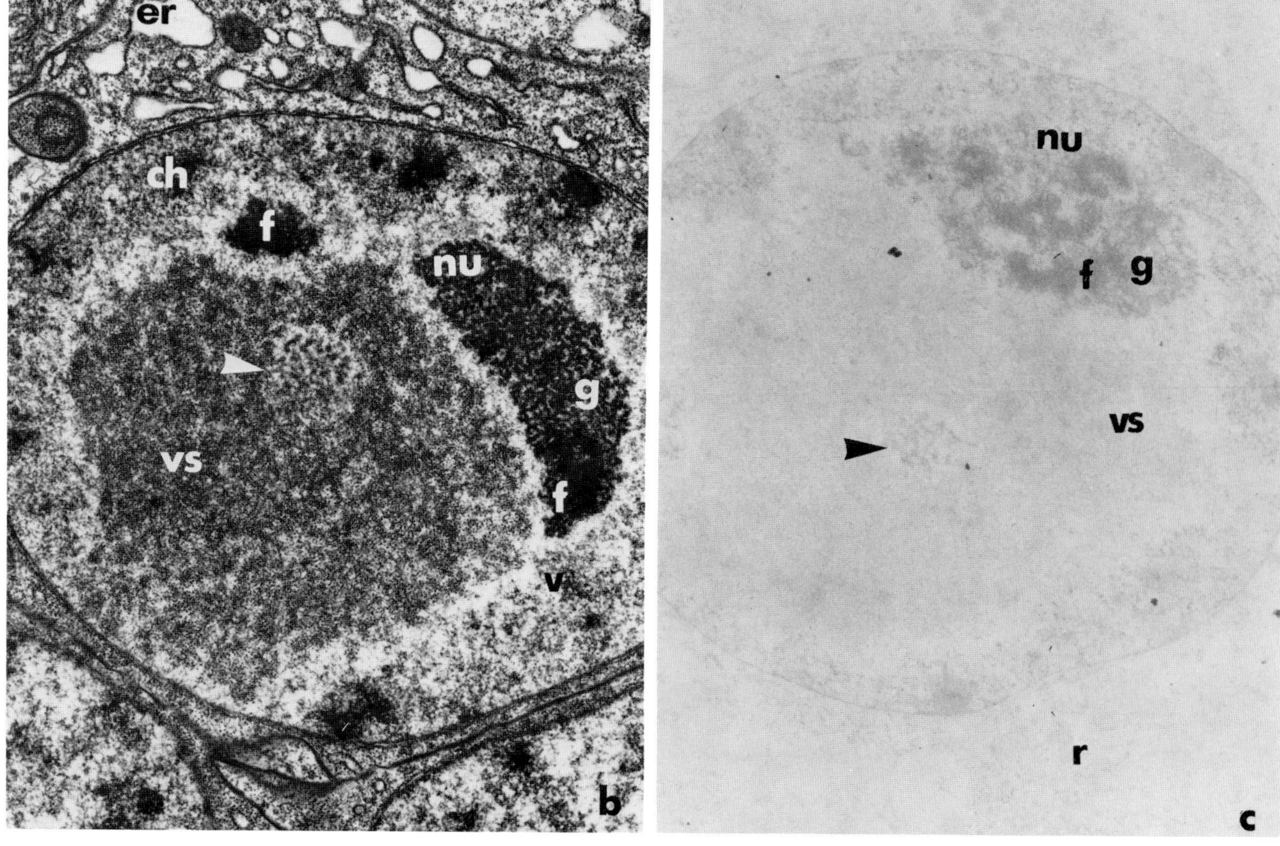

er
ch
f
nu
g
vs
f
v
b
nu
f
g
vs
r
c

Fig. 9. Late stages of infection. (a and b) Hypodermal cells after 4 days of infection; the viral inclusion occupies almost the entire nucleus. (a) Mature virions (v) are grouped together between the nucleolus (nu) and the virogenic stroma (vs). (× 14,500.) (b) The virions are assembled inside the virogenic stroma and grouped into islets. The nucleolus is pushed to the periphery of the nucleus and is reduced to the fibrillar component. (×10,500.) (c and d) Hypodermal cells after 6 days of infection. (c) The virogenic stroma (vs) is progressively replaced by plaques of virus particles (v). (× 19,500.) (d) The membrane of the nucleus which is hypertrophied by the production of virions is ruptured, allowing the passage of the virions to the cytoplasm (→). Note the paracrystalline formations of DNV in the nucleus (➤). (×26,000.)

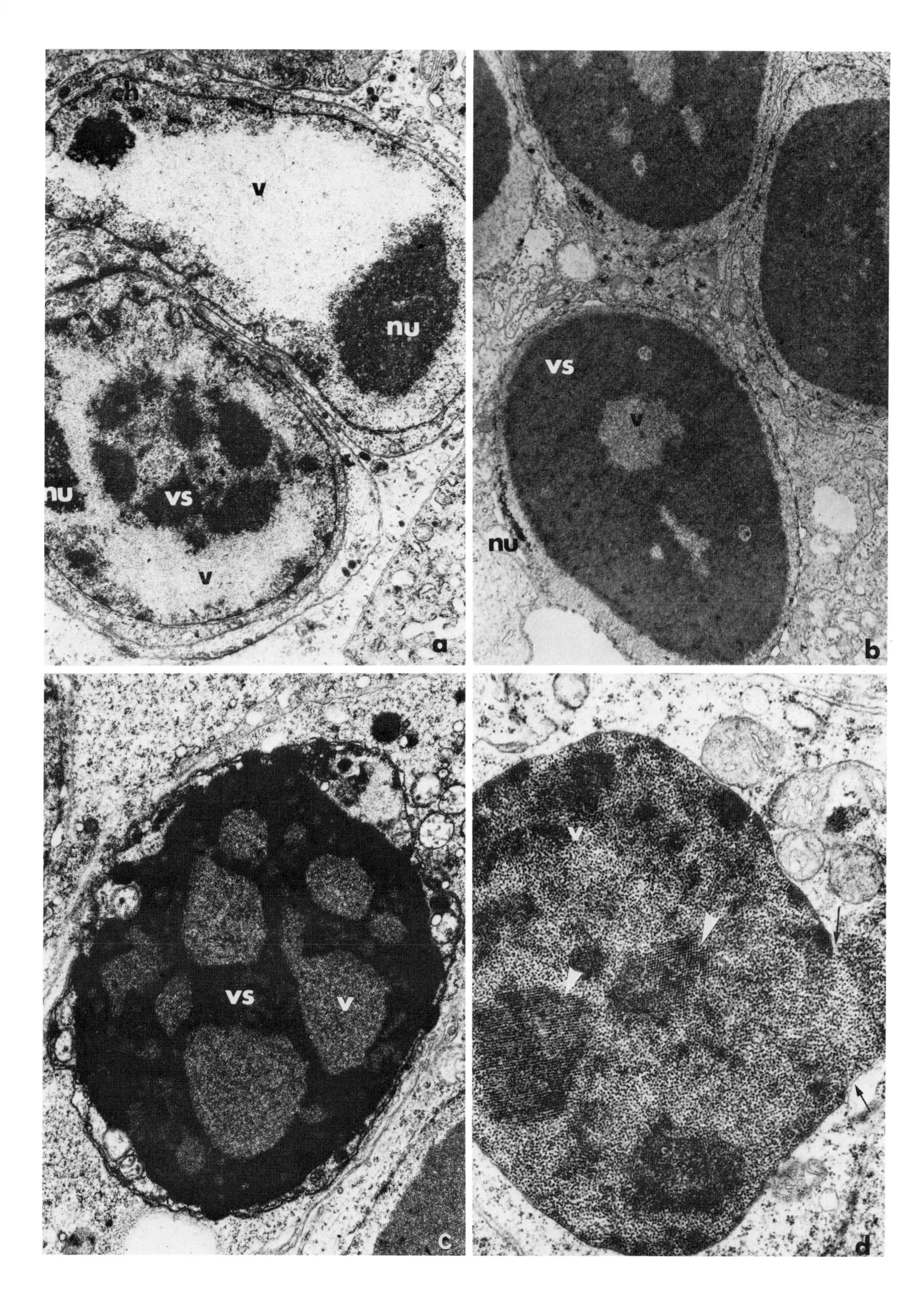

ch
v
nu
nu
vs
v
a
vs
v
nu
b
vs
v
c
v
d

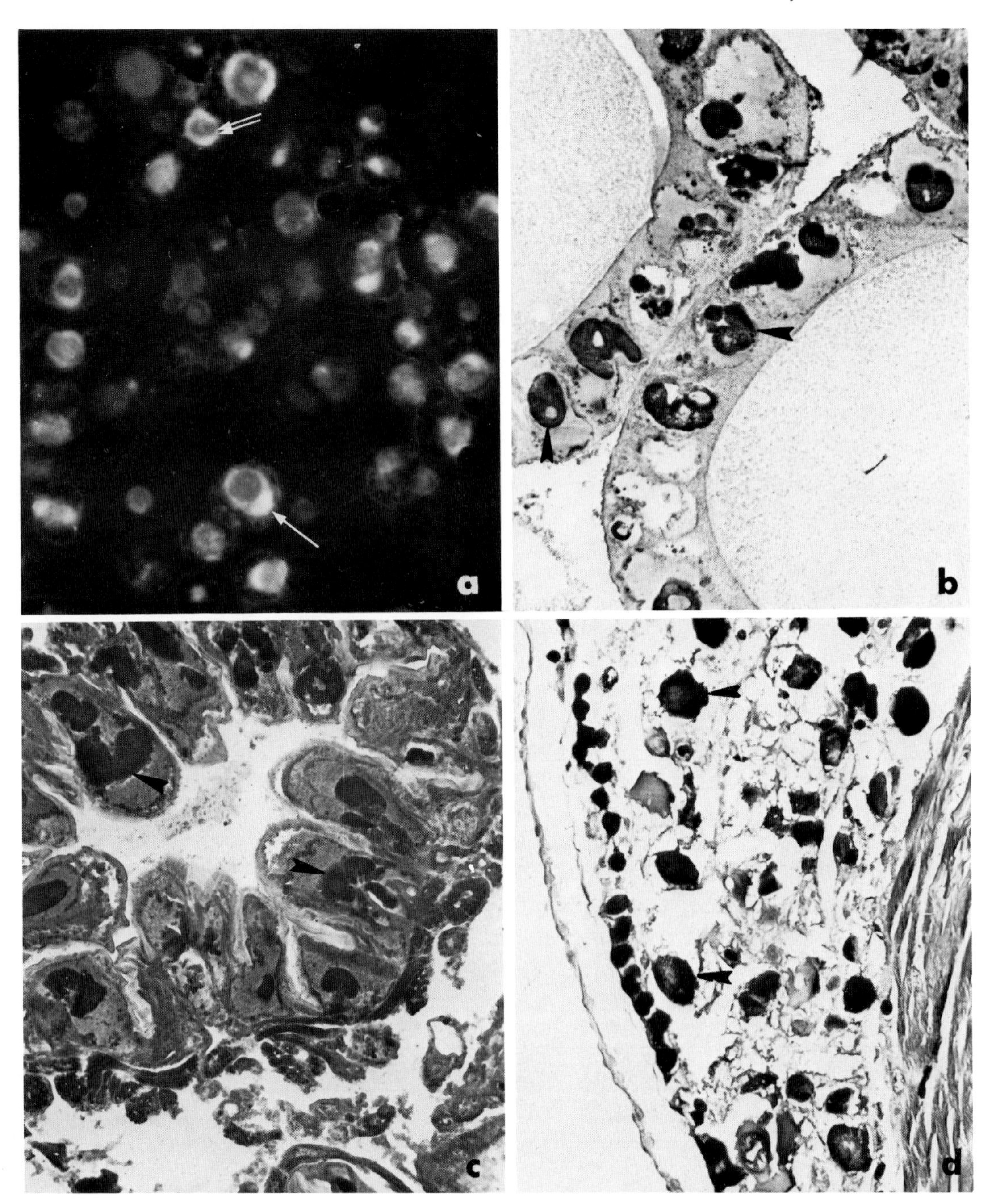

Fig. 10. Synthesis and localization of DNV viral antigens detected by immunohistochemical techniques. (a) Indirect immunofluorescence; 13-hour infected hemocytes of *G. mellonella* showing different stages of infection. The resulting fluorescence is mainly intracytoplasmic (→) and perinuclear (⇇). (× 660.) (b, c, and d) Direct immunoperoxidase. The anti-DNV antibodies conjugated with peroxidase and fixed to the viral antigens are revealed by staining with benzidine and H_2O_2; dark brown reaction (➤). Specific confirmation of the polytrophic nature of DNV. (b) silk gland; (c) posterior gut; (d) hypodermis and adipose tissue. (× 1200.)

Fig. 11. Localization of DNV antigens (⟶) and virions (v) in the nucleus of DNV infected *G. mellonella* cells (a and b) by the ultrastructural immunoperoxidase procedure. Fixation with paraformaldehyde–glutaraldehyde and osmium tetroxide, embedding with Epon 812. (× 26,500.)

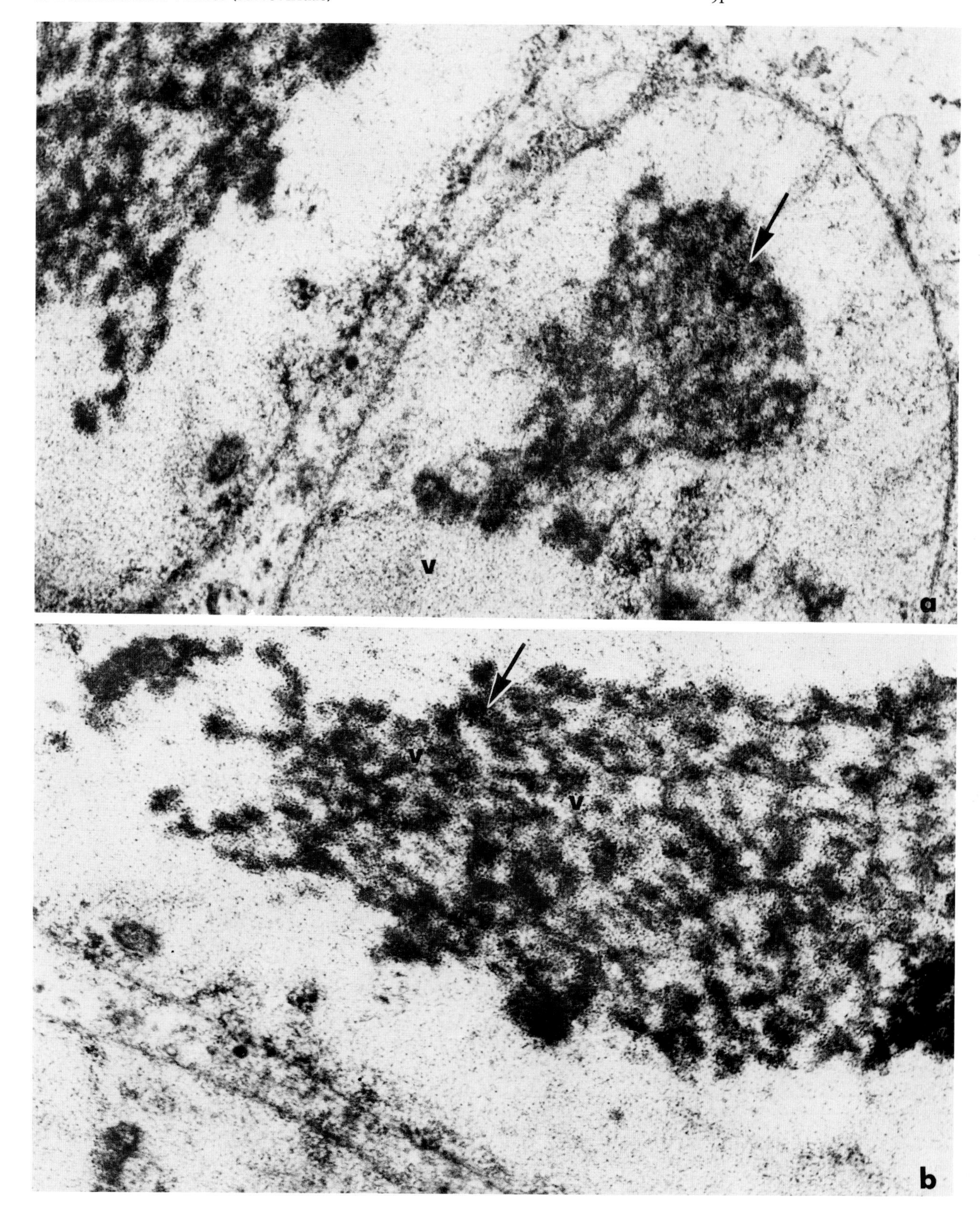
v
a
v
v
b

Chapter 4

Iridoviruses (Iridoviridae)

PETER E. LEE

I. Introduction 93
II. Fine Structure of Purified Virions 93
III. Virus–Cell Relationships 94
References 95

I. INTRODUCTION

The iridoviruses contain deoxyribonucleic acid (DNA), are icosahedral in shape, and purified virus suspensions or pellets iridesce in visible light due to the periodic spacings between virions. The iridoviruses mentioned here are confined to the class Insecta; they replicate in the cytoplasm of infected cells and infection does not appear to be limited to particular cell types or tissues. For a detailed account of these viruses, the reader is referred to Bellet (1968). The micrographs presented are from 3 members of the group: *Tipula* iridescent virus (TIV), *Sericesthis* iridescent virus (SIV), and *Chironomus* iridescent virus (*Chironomus* ICDV).

II. FINE STRUCTURE OF PURIFIED VIRIONS

The iridoviruses are large virions with a diameter ranging from 130 nm to 190 nm (Williams and Smith, 1957, 1958; Mercer and Day, 1965; Stoltz, 1971). They possess a dense irregular central core or nucleoid containing double stranded DNA. In negatively stained preparations, particles from the same sample have one or two surrounding membranes (Bellet, 1968; Wrigley, 1969; Stoltz, 1971). Wrigley (1969) examined purified suspensions of SIV by negative staining. He found that storage in distilled H_2O at 4° for a few weeks resulted in triangular arrays of subunits. Each triangle having 55 capsomeres (Fig. 1). If SIV was treated with Afrin, a nasal decongestant, prior to negative staining, capsomeres of the capsid were resolved, and with the use of the Goldberg diagram, it was suggested that the virion surface was composed of probably 1562 capsomeres. When TIV was similarly treated, the same results were obtained, as shown in Fig. 2 (Wrigley, 1970).

Using polyacrylamide gel electrophoresis (PAGE), SIV, *Chilo* iridescent virus (CIV), and TIV have 20, 19, and 28 polypeptides respectfully (Kelly and Tinsley, 1972; Krell and Lee, 1974). Stoltz (1971) found, using electron microscopy of negatively stained preparations of TIV and *Chironomus* ICDV, that virions of the former had a peripheral amorphous fuzz while those of the latter had a fringe of long fibrils. Pronase treatment of *Chironomus* ICDV disrupted the virion capsid into triangles composed of capsomeres but the long fibrils remained attached to the triangles and measured 150 ± 20 nm in length (Fig. 3).

III. VIRUS–CELL RELATIONSHIPS

TIV is the type virus of the iridescent group (Gibbs *et al.*, 1966) and extensive work has been conducted on its mode of infection and replication in susceptible cells and tissues (Xeros, 1954, 1964; Bird, 1961; Hukuhara and Hashimoto, 1966; Younghusband and Lee, 1969, 1970; Yule, 1971; Yule and Lee, 1973). Because of numerous similarities among the iridescent group it seems plausable to assume that the pattern of viral infection and synthesis in susceptible cells may be similar to that of TIV. TIV was initially found in larvae of the cranefly *Tipula paludosa* (M.), and replicates in several insect species (Xeros, 1954; Smith and Rivers, 1969; Smith *et al.*, 1961).

When primary cultures of hemocytes from *Galleria mellonella,* maintained in physiological saline, were inoculated with TIV, electron microscopic observation of the cells revealed several virions at the plasma membrane (Fig. 4), free in the cytoplasm, or in vesicles one hour postinoculation; but there was not a clear indication of viral replication at this or a later time (Younghusband, 1969). On the other hand, injection of larvae with TIV showed that hemocytes from such insects contained within their cytoplasm small virogenic stroma with a few particles suggesting that replication had commenced (P. E. Lee, unpublished). Using the Gomori assay for acid phosphatase activity in hemocytes from inoculated larvae, single-membraned structures with virions reacted positively; this was taken as an indication that the infecting particles are engulfed by lysosomes (Younghusband and Lee, 1970).

Thin sections of TIV in the cytoplasm of hemocytes also appear to have one or two surrounding membranes (Figs. 5 and 6) with apparent "budding," thus suggestive of the virion becoming membrane-bound: Figures 5 and 6 also show the end product of a virion completely enclosed by a membrane. The enclosure of particles by membranes was not limited to the plasma membrane region of infected cells. Figure 6 shows particles in close association to intracytoplasmic membranes.

The cytoplasmic virogenic centers (Fig. 7), in addition to containing new virions, are composed of a ground matrix unlike that of the cytoplasm and somewhat similar in appearance to the chromatin of the nucleus. These centers have been shown by autoradiography to incorporate ^{3}H-thymidine (Younghusband and Lee, 1970; Yule, 1971). Developmental viral forms were observed by Yule and Lee (1973) in virogenic stroma of infected hemocytes. Based on ferritin-tagged viral antibodies (Fig. 8), a plausable sequence of TIV assembly (Figs. 9 and 10) was made and is briefly presented: In Fig. 9 (a_2–f_2) it appears that the shell of the virion has just been formed, since 1-, 2-, 3-, 4-, 5-, and 6-sided faces of the icosahedron are seen specifically tagged with ferritin. These developing shells are in intimate contact with the viroplasm, and viroplasmic material can be seen with partial shells (Fig. 9c_2 and d_1). The final stage in shell assembly is the 5-sided or 6-sided pentagonal form. Figure 9e_1 and the corresponding ferritin-tagged form in Fig. 9e_2 (insert) supports the concept that the shell is formed first. Later stages of virion assembly, for example, the gradual buildup of core content are shown in Figs. 9f and 10g. The nucleoid material probably is introduced through a small opening left in the empty capsid; this appears feasible since some virions contained core material attached to one side of the capsid (Fig. 10g). The ferritin-tagged forms which resemble nucleoids may be glancing sections through the shell of the particle (Fig. 10j). The particle shell and nucleoprotein core do not appear to be assembled concurrently as proposed by Xeros (1964).

ACKNOWLEDGMENTS

The author wishes to express his thanks to Dr. G. B. Stoltz and Dr. N. G. Wrigley for permission to use their electron micrographs. Work of the author described here was supported by Grants A2911 and E2134 from the National Research Council of Canada.

REFERENCES

Bellet, A. J. D. (1968). *Adv. Virus Res.* **13,** 225.
Bird, F. T. (1961). *Can. J. Microbiol.* **7,** 827.
Gibbs, A. J., Harrison, B. D., Watson, D. H., and Wildy, P. (1966). *Nature (London)* **209,** 450.
Hukuhara, T., and Hashimoto, Y. (1966). *Appl. Entomol. Zool.* **1,** 166.
Kelly, D. C., and Tinsley, T. W. (1972). *J. Invertebr. Pathol.* **19,** 273.
Krell, P., and Lee, P. E. (1974). *Virology* **60,** 315.
Mercer, E. H., and Day, M. F. (1965). *Biochim. Biophys. Acta* **102,** 590.
Smith, K. M., and Rivers, C. F. (1969). *Virology* **9,** 140.
Smith, K. M., Mills, G. J., and Rivers, C. F. (1961). *Virology* **13,** 233.
Stoltz, D. B. (1971). *J. Ultrastruct. Res.* **37,** 219.
Williams, R. C., and Smith, K. W. (1957). *Nature (London)* **179,** 119.
Williams, R. C., and Smith, K. W. (1958). *Biochem. Biophys. Acta* **28,** 464.
Wrigley, N. G. (1969). *J. Gen. Virol.* **5,** 123.
Wrigley, N. G. (1970). *J. Gen. Virol.* **6,** 169.
Younghusband, H. B. (1969). M. Sc. Thesis, Carleton University Archives, Ottawa.
Younghusband, H. B., and Lee, P. E. (1969). *Virology* **38,** 247.
Younghusband, H. B., and Lee, P. E. (1970). *Virology* **40,** 757.
Yule, G. B. (1971). M.Sc. Thesis, Carleton University Archives, Ottawa.
Yule, G. B., and Lee, P. E. (1973). *Virology* **51,** 409.
Xeros, N. (1954). *Nature (London)* **174,** 562.
Xeros, N. (1964). *J. Insect Pathol.* **6,** 261.

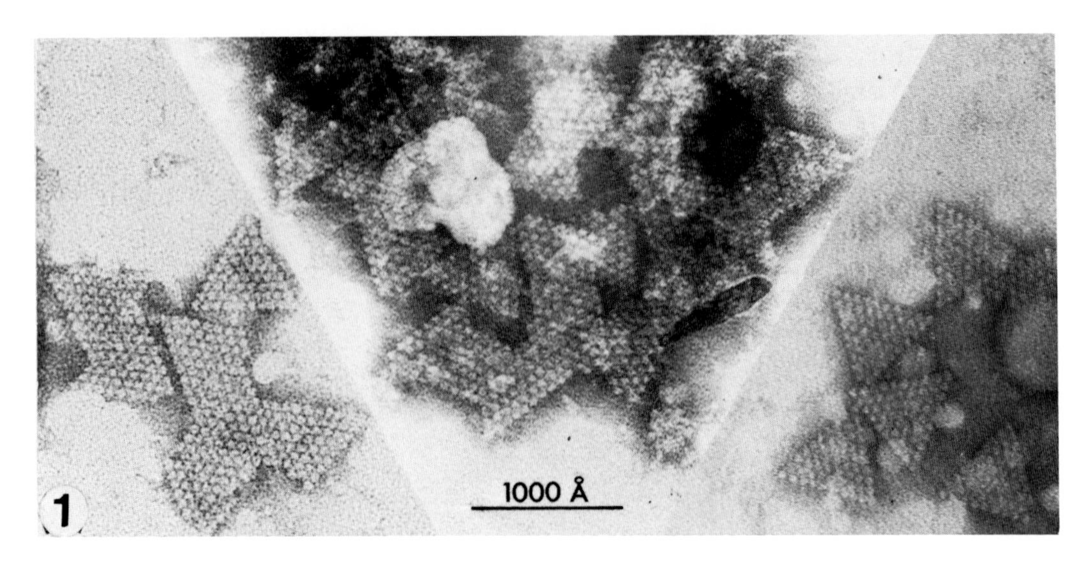

Fig. 1. Three fields showing "triangular" fragments of SIV, with their 55 subunits clearly resolved. Reproduced from Wrigley (1969), courtesy of Cambridge University Press, London and New York.

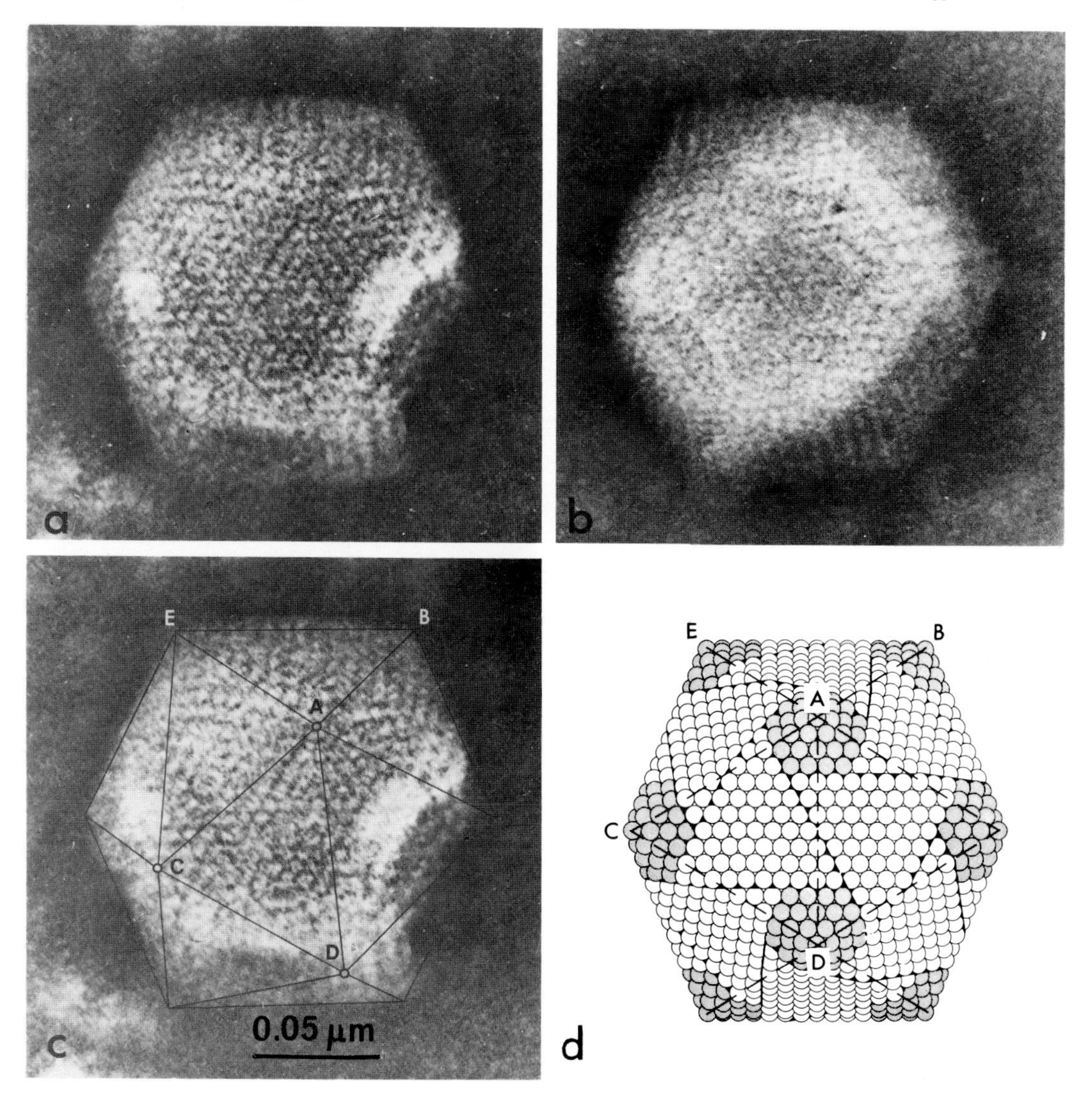

Fig. 2. (a) and (b) Two examples of TIV particles after 24-hour treatment with Afrin. Rows of subunits can be clearly seen in parts of the particles. (c) The same particle as in (a) but with a possible interpretation of the icosahedron edges superimposed. (d) 1562-subunit model proposed for SIV, for comparison with (c). The size scale for these pictures is given by the icosahedron edge-length 824 Å (lengths A to B, B to E, etc.) determined for TIV. Reproduced from Wrigley (1970), courtesy of Cambridge University Press, London and New York.

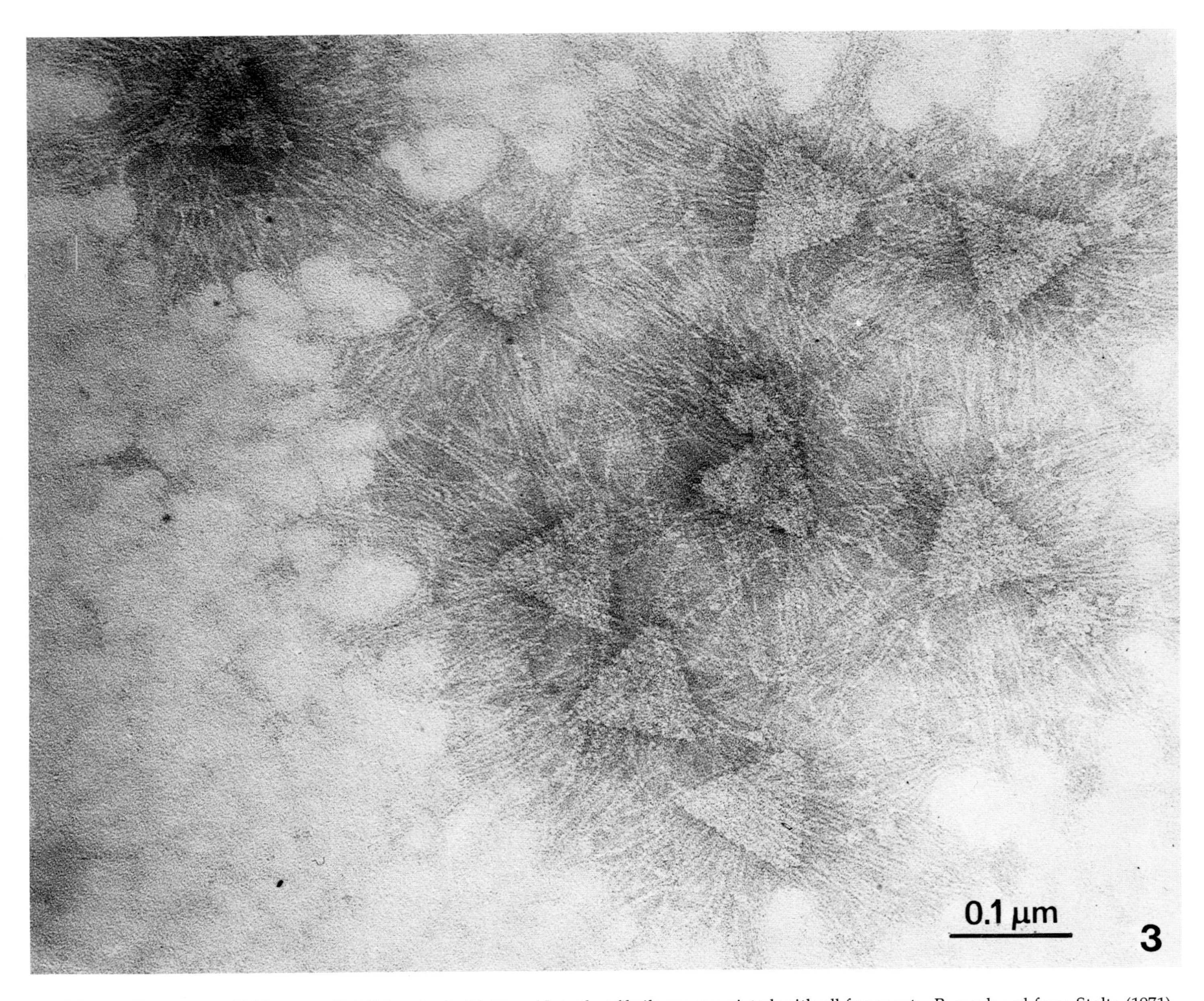

Fig. 3. Fragments of *Chironomus* ICDV icosahedral lattice. Note that fibrils are associated with all fragments. Reproduced from Stoltz (1971), courtesy of Academic Press, New York.

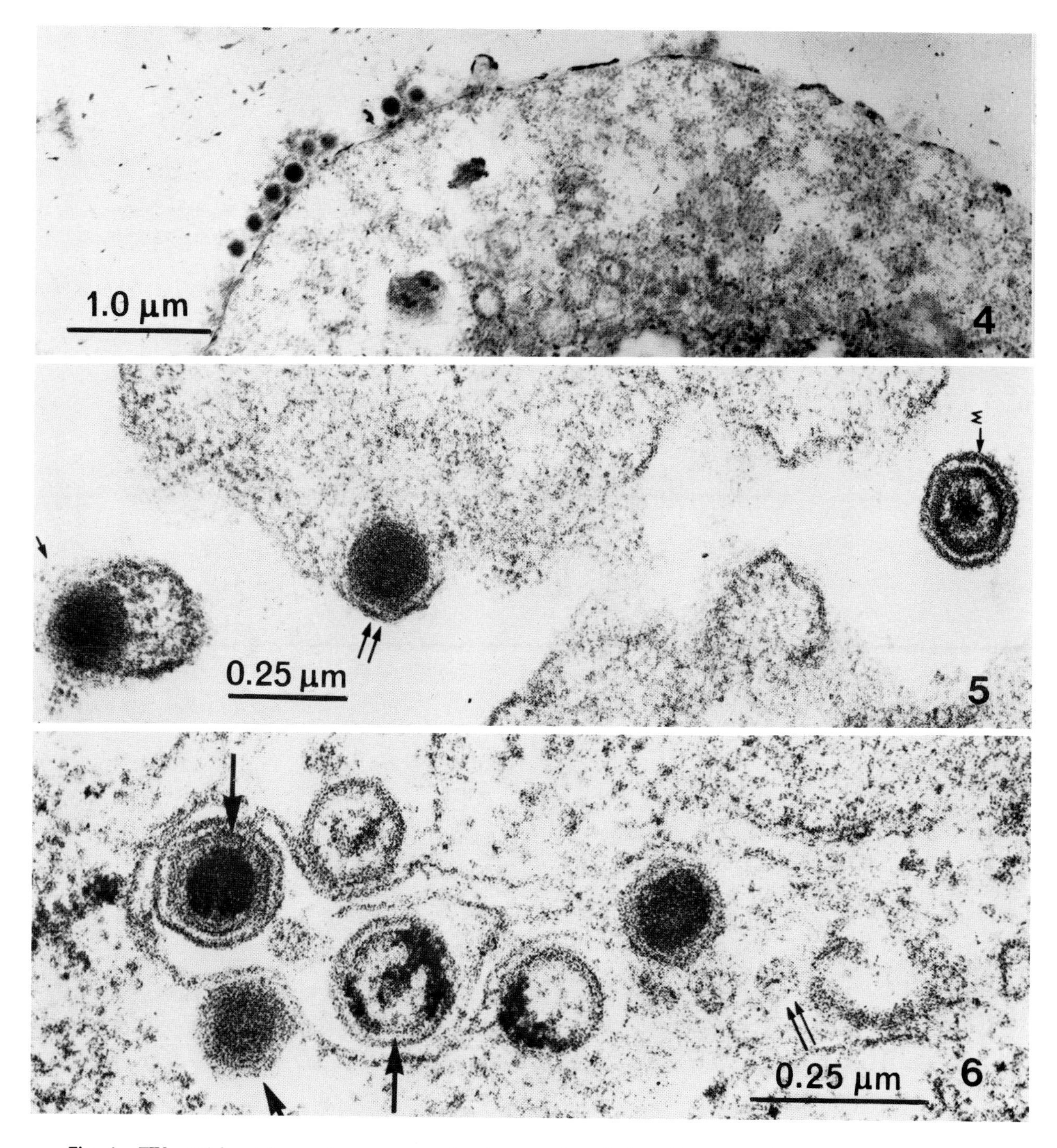

Fig. 4. TIV particles at hemocyte membrane. One-hour postinoculation of hemocytes in physiological saline.

Fig. 5. Virions of TIV in different stages of being "coated" with a membrane. The single arrow indicates a free virion with partial membrane. The double arrow shows a part of the host-cell membrane being utilized by a virion. There is also a virion surrounded by a fully formed membrane (M).

Fig. 6. Membrane bound group of virus particles, 24 hours after inoculation. The single arrows indicate outer edges of normal virus particles. The double arrow indicates a small particle of an undetermined nature. Fixed in aldehyde, treated with digitonin and postfixed in aldehyde and OsO_4; stained with uranyl acetate and lead citrate. ($\times$130,000.)

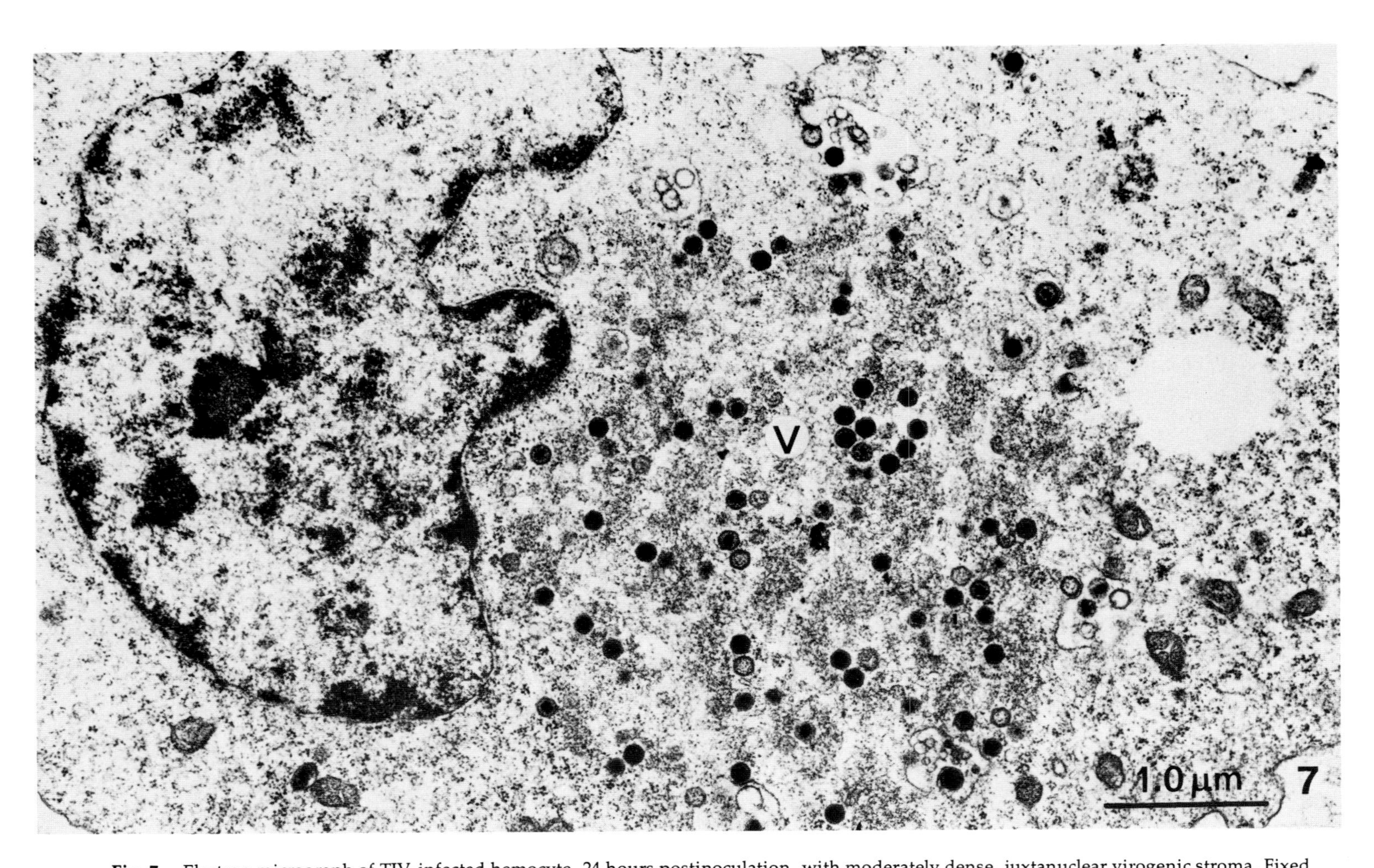

Fig. 7. Electron micrograph of TIV-infected hemocyte, 24 hours postinoculation, with moderately dense, juxtanuclear virogenic stroma. Fixed in aldehyde, permeabilized with digitonin and postfixed in aldehyde and OsO_4. Stained with uranyl acetate and lead citrate. V, virogenic center. (×26,900.)

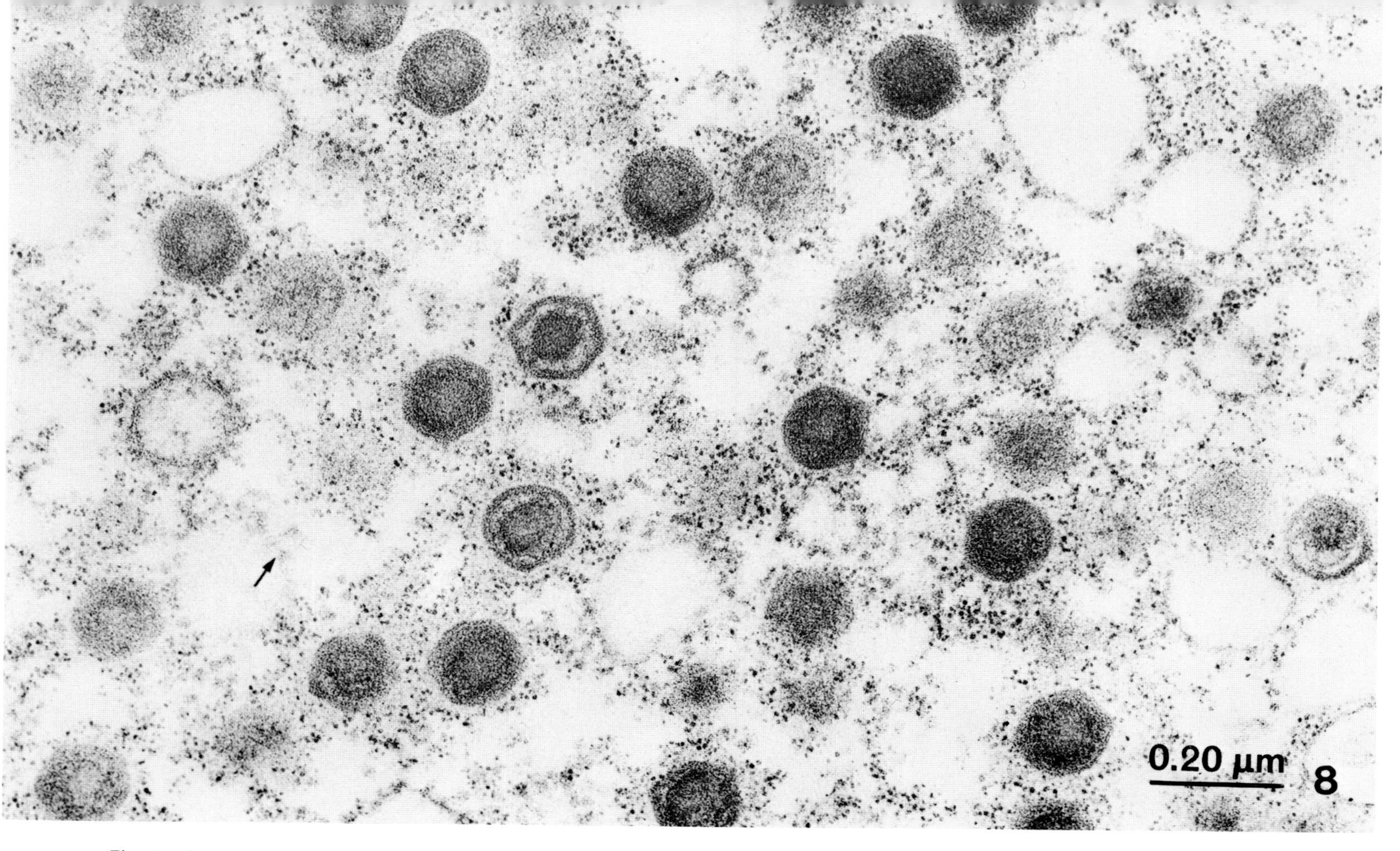

Fig. 8. Electron micrograph of viroplasmic center from TIV-infected hemocyte, 48 hours postinoculation, treated indirectly with immunoferritin. The arrow points to small spherical structures which are probably ribosomes. TIV particles and these structures in the viroplasmic center are ferritin-tagged. Fixed in aldehyde, permeabilized with digitonin, postfixed in aldehyde and OsO_4. Stained with uranyl acetate. (×108,000.)

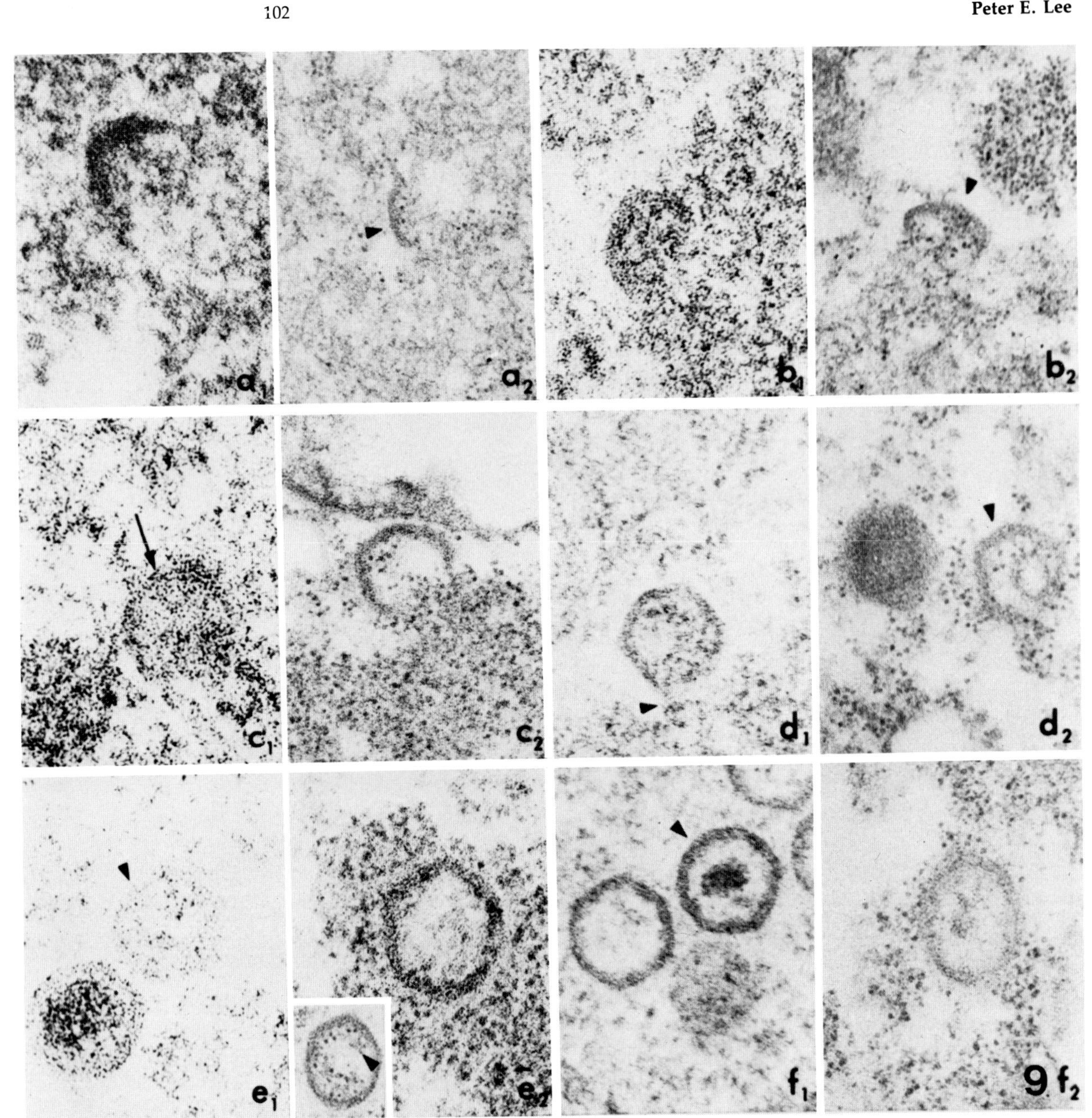

Fig. 9. Montage of *Tipula* iridescent virus (TIV) arranged in a possible sequence of assembly. a_1–f_1, untagged viral developmental stages; a_2–f_2, ferritin-tagged viral developmental stages. $a_{1,2}$, 2-sided particle; arrow (a_2) indicates tagging of inner edge of shell. (a_1, ×132,240; a_2, ×114,000.) $b_{1,2}$, 3-sided shell; arrow (b_2) indicates tagging of inner edge of shell. (b_1, ×132,240; b_2, ×114,000.) $c_{1,2}$, 4-sided shell (arrow, c_1). Note ferritin-tagged material filling shell in c_2. (c_1, ×132,240; c_2, ×114,000.) $d_{1,2}$, 5-sided shell. Arrow (d_1) indicates attachment of particle to stroma. Arrow (d_2) indicates ferritin tagging inside shell ($d_{1,2}$, ×114,000). $e_{1,2}$, 6-sided shell (arrow, e_1). (e_2 and insert) ferritin tagging inside completed shell (arrow). (e_1, ×131,000; e_2, ×148,200; insert e_2, ×82,800.) $f_{1,2}$, particles with beginning of core development (arrow). ($f_{1,2}$, ×114,000.) a_1–f_1 stained with uranyl acetate and lead citrate; a_2–f_2 stained with uranyl acetate.

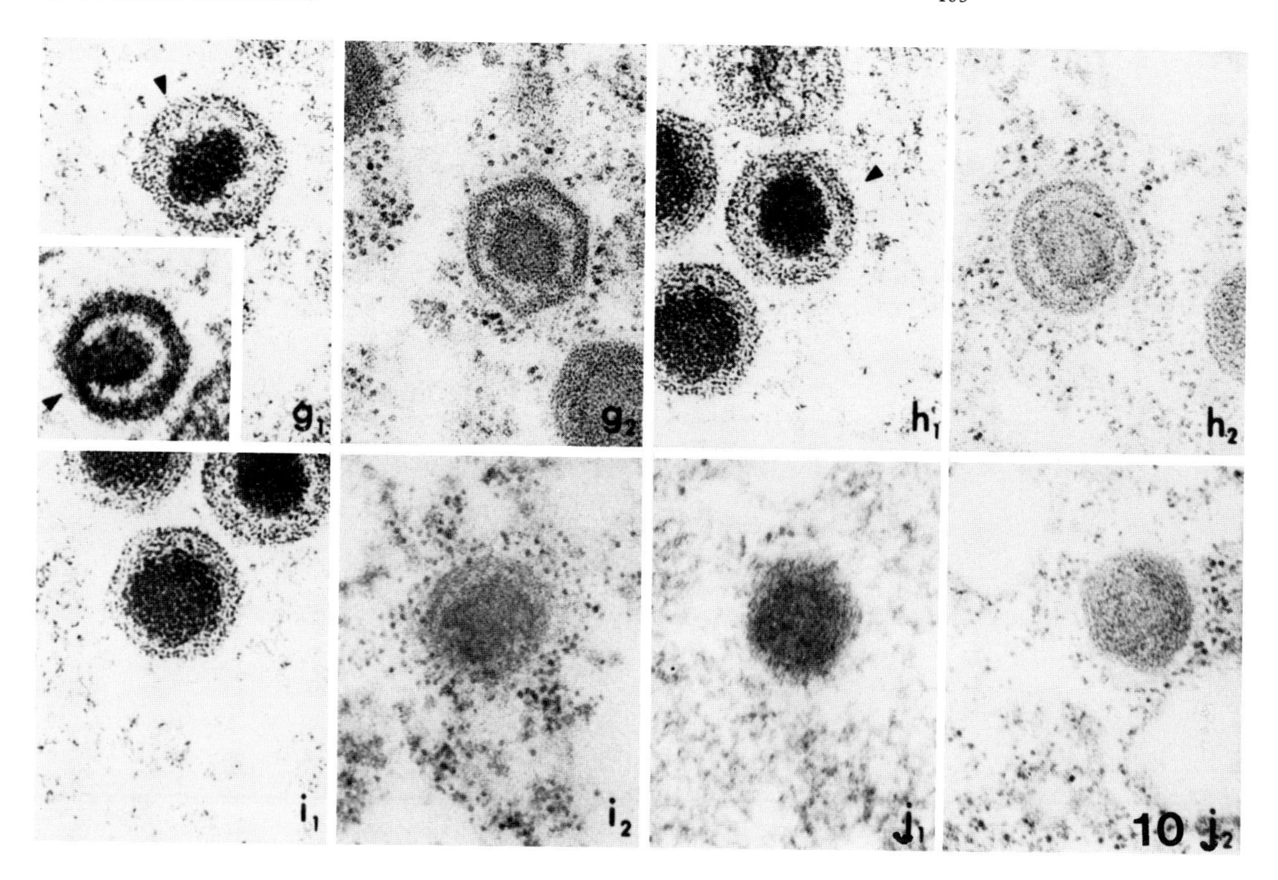

Fig. 10. Continuation of *Tipula* iridescent virus (TIV) possible sequence of assembly. g_1–j_1, Untagged viral developmental stages; g_2–j_2 ferritin-tagged developmental stages. $g_{1,2}$, Partially filled shells (arrow). Insert g_1 indicates attachment point of core to shell. (g_1, ×131,000; g_1 insert, ×114,000; g_2, ×124,260.) $h_{1,2}$, Almost filled viral shells. (h_1, ×131,000; h_2, ×120,840.) $i_{1,2}$, Mature virus. (i_1, ×120,840; i_2, ×114,000.) $j_{1,2}$, Particles believed to be glancing sections through shell, masking core content. Ferritin tagging of this form indicates that it is not a naked icosahedral core. ($j_{1,2}$, ×106,020.) Fixed in aldehyde, treated with digitonin and postfixed in aldehyde and OsO_4. g_1–j_1 stained with uranyl acetate and lead citrate; g_2–j_2 stained with uranyl acetate.

Chapter 5

Cytoplasmic Polyhedrosis Viruses

C. C. PAYNE AND K. A. HARRAP

I. Introduction 105
II. Members of the Group 106
III. The Structure of CPV's 106
A. Polyhedra 106
B. Virus Particles 107
IV. Virus Replication 109
References 111

I. INTRODUCTION

Cytoplasmic polyhedrosis viruses (CPV's) are characterized by the production of polyhedral inclusion bodies ("polyhedra") usually in the cytoplasm of virus-infected cells. These polyhedra contain isometric virus particles which can be released from the surrounding matrix protein ("polyhedron protein") by treatment with dilute alkali (Hills and Smith, 1959; Hosaka and Aizawa, 1964). The viral genome is a segmented double-stranded (ds) RNA (Miura *et al.*, 1968; Kalmakoff *et al.*, 1969; Payne and Tinsley, 1974).

Viruses with these properties have been isolated from the larvae of many species of Lepidoptera (Smith, 1963; Aruga and Tanada, 1971) in which their replication appears to be confined to the gut epithelial tissue. Viruses of this morphological type have also been observed in Diptera (Kellen *et al.*, 1966; Stoltz and Hilsenhoff, 1969; Bailey *et al.*, 1975), Hymenoptera (Longworth and Spilling, 1970), and, more recently, in the freshwater crustacean *Simocephalus expinosus* (Federici and Hazard, 1975). Until this last report, CPV's were generally considered to be restricted to the class Insecta.

The International Committee for the Taxonomy of Viruses is considering the inclusion of CPV's as a genus within the Reoviridae. They are distinguishable from reoviruses by host range, morphology of the virus particle, and the production of polyhedra in infected cells.

II. MEMBERS OF THE GROUP

Many CPV isolations have been reported from different insect species (Aruga and Tanada, 1971), but extensive studies have been restricted to a small number and, in particular, to a virus of the silkworm *Bombyx mori.* (Miura *et al.*, 1968; Kalmakoff *et al.*, 1969; Lewandowski and Traynor, 1972; Payne and Kalmakoff, 1975). As a result, it is not clear whether or not each isolate represents a distinct virus type. However, just as reovirus types 1, 2, and 3 can be distinguished by differences in the distribution of the RNA genome segments (Shatkin *et al.*, 1968) so it is possible to distinguish several CPV types by the same method (Fig. 1). Such gel profiles provide a means for the provisional classification of the large number of isolates of these viruses. Using this method, eleven distinct CPV "types" have been distinguished among 33 isolates (Payne and Rivers, 1976). Some viruses share the same RNA gel profile and may therefore be very similar or identical. Differences in RNA's are also reinforced by differences in the sizes and antigenicity of virus structural proteins (Payne, 1976; Payne and Rivers, 1976). Such results suggest that considerable variation exists between members of the CPV group, and it is likely that many more "types" remain to be identified.

III. THE STRUCTURE OF CPV'S

A. Polyhedra

Polyhedra of CPV's and other occluded insect viruses differ from the inclusion bodies observed in most virus infections. They contain apparently mature virus particles packed within a matrix protein and represent a terminal stage of virus synthesis (Fig. 2). Electron microscopic examination of polyhedra reveals a crystalline array of protein arranged in a cubic lattice (Figs. 2 and 22), the center-to-center spacing of the lattice varying between 41 Å (Arnott *et al.*, 1968), and 74 Å (Bergold and Suter, 1959). The virus particles comprise only a small amount (approximately 5%) by weight of the polyhedron (Hukuhara and Hashimoto, 1966), and the preponderance of protein is reflected in the relatively low buoyant density (1.279 gm/cm^3) of the inclusions (Martignoni, 1967). The lattice structure is only slightly, or not at all, disturbed by the presence of the virus particles (Bergold and Suter, 1959; Arnott *et al.*, 1968). Surface cavities have been observed in polyhedra, which probably arise from the loss of virus particles at the surface (Bergold and Suter, 1959). This suggests that there is no limiting layer or "membrane" at the periphery of the polyhedron.

The alkali-soluble polyhedron protein has a molecular weight of 25,000–30,000, as measured by electrophoresis on polyacrylamide gels (Payne and Rivers, 1976). In the CPV from *Bombyx mori,* this protein is glycosylated (Payne and Kalmakoff, 1975) and is antigenically unrelated to the structural proteins of the virus particles (Hukuhara and Hashimoto, 1966). This suggests that the matrix of the polyhedron is not produced by excessive synthesis of a virus structural protein but rather by *de novo* synthesis of a distinct host or viral protein. Although it is likely that the synthesis of polyhedron protein is virus-directed the evidence is, at best, circumstantial. For example, many CPV's contain an RNA segment which, from its size, can theoretically contain the genetic information for a protein of the size of polyhedron protein (Lewandowski and Traynor, 1972; Payne, 1976). In addition, Aruga *et al.* (1961) have obtained CPV strains from *Bombyx mori,* with different inclusion body shapes. The distinctive shape is retained after several passages, implying that it is controlled by the virus genome.

B. Virus Particles

1. *Morphology.* "Occluded" virus particles can be extracted from polyhedra by dissolution in alkaline solutions under controlled conditions (Hills and Smith, 1959; Hukuhara and Hashimoto, 1966; Payne, 1976). "Nonoccluded" particles have also been extracted from homogenates of larval midguts (Hayashi and Bird, 1968). It appears that both occluded and nonoccluded particles have similar morphological and biochemical features (Hayashi and Bird, 1968; Payne and Kalmakoff, 1975). The virus from *Bombyx mori* can be used as a model for the group.

Hosaka and Aizawa (1964) demonstrated by shadowing techniques that the virus particles were icosahedral and showed a hexagonal or circular outline. In thin sections of polyhedra (Fig. 2) the virus particles appear to consist of an electron dense core surrounded by an outer shell (Lewandowski and Traynor, 1972). In negatively stained preparations (Figs. 3–5) the particles have an average diameter of 60–70 nm and are characterized by the presence of projections or "spikes," probably twelve in number, at the vertices of the particles (Hosaka and Aizawa, 1964; Miura *et al.*, 1969; Lewandowski and Traynor, 1972). These distinct spikes, which are 20 nm in length (Miura *et al.*, 1969) appear to be hollow and to originate within the core of the particle. Asai *et al.* (1972) have described spherical structures, 12 nm in diameter, attached to the top of the spikes. As the virus is known to hemagglutinate, it has been suggested that these structures may serve as the site of adsorption on to erythrocytes or host cells (Miyajima and Kawase, 1969; Asai *et al.*, 1972). Surface projections have also been observed in CPV particles isolated from insects other than *Bombyx mori* (Figs. 6 and 7) and appear to be common features of the group.

Although an early model for the structure of CPV particles proposed that the virus particles contained two parallel protein coats (Fig. 8; Hosaka and Aizawa, 1964), the electron micrographs show no evidence for the presence of a distinct double capsid structure of the reovirus type. In fact, many of the properties of CPV particles are similar to those of subviral particles or "cores" of reovirus (Lewandowski and Traynor, 1972; Payne and Tinsley, 1974). The number of capsomeres on the surface of the particle has not been resolved. However, 20 peripheral subunits can be counted on certain particles, apparently on their fivefold axis of symmetry [i.e., particles not showing a clearly hexagonal outline (Fig. 3)]. If it is assumed that each peripheral subunit represents one side of a hollow roughly cylindrical morphological unit (Fig. 4), ten of these can be envisaged on the periphery. On theoretical grounds this would yield a particle with a triangulation number (T) of 4 and 42 such morphological units or capsomeres on its surface. However, it is difficult to relate the position of the spikes with such a model.

2. *Biochemical Properties.* The virus particles of *Bombyx mori* CPV have a sedimentation coefficient of 370–440 S (Hukuhara and Hashimoto, 1966; Miura *et al.*, 1968; Lewandowski *et al.*, 1969), a density of approximately 1.44 gm/cm^3 (Lewandowski and Millward, 1971; Payne and Kalmakoff, 1975), and contain RNA comprising approximately 30% of the particle weight (Nishimura and Hosaka, 1969). Base composition analysis suggested that the RNA exists in a base-paired structure (Hayashi and Kawase, 1964), and other studies have confirmed its double-stranded nature (Miura *et al.*, 1968) (Fig. 9). Like reovirus RNA, CPV RNA exists as ten discrete segments present in equimolar amounts comprising a genome with a total molecular weight of approximately 15×10^6 daltons (Fujii-Kawata *et al.*, 1970). Further evidence that this is the approximate size of the genome ($15–21 \times 10^6$ daltons) comes from chemical analysis and an examination of the length of the RNA in the electron microscope (Nishimura and Hosaka, 1969). When RNA was examined by the protein monolayer

method, the longest filaments observed were 6.8 μm (Fig. 10). This is equivalent to an RNA molecular weight of 14–18 $\times 10^6$ daltons. The detection of these long RNA molecules also suggests that the genome segments are attached to one another in the virion. However, the attachment cannot involve covalent phosphodiester bonds between RNA segments, as the RNA within the virus particle has the same number of 3′-termini as does the segmented RNA which has been extracted from the virion with phenol (Lewandowski and Millward, 1971).

In a more recent study, Kavenoff *et al.* (1973) describe CPV RNA molecules, many of which have a circular conformation. The contour length of these molecules (15 μm) corresponds to an RNA molecular weight of 35 $\times 10^6$ daltons. They suggest that each circular molecule contains all 10 segments of double-stranded RNA plus a considerable fraction which may be "spacer" RNA. Clearly these results are incompatible with data from several studies where values of 15–21 $\times 10^6$ daltons are considered to represent the total RNA content of the virus particle.

Five structural polypeptides have been observed in virus particles of *Bombyx mori* CPV (Lewandowski and Traynor, 1972; Payne and Kalmakoff, 1975). The molecular weights of these are in close agreement with the theoretical coding capacity of five of the genome RNA segments and may therefore represent primary gene products of these segments. Recent studies have shown that many CPV's from different insect species also contain 10 RNA segments (Payne, 1976; Payne and Rivers, 1976). Even though these have size distributions different from *Bombyx mori* CPV (Fig. 1), the total genome size is approximately 15 $\times 10^6$ daltons. However, as yet, there is no apparent consistency in the number of structural proteins present in particles of the different CPV types (Payne and Rivers, 1976).

Purified virus particles of CPV have been shown to possess several enzymatic activities which provide insight into the probable means of replication. The presence of an RNA transcriptase activity in intact virions provides a means of transcribing the dsRNA viral genome (Lewandowski *et al.*, 1969). In contrast to the viral transcriptase of many viruses, including reovirus, the enzyme activity is expressed by virus particles which have not been heat-treated or partially disrupted by proteolytic enzymes (Lewandowski *et al.*, 1969; Hayashi and Donaghue, 1971). Attempts to further purify the transcriptase activity by disruption of the virion have failed, and it may be that several structural proteins form a multicomponent enzyme similar to the prokaryotic and eukaryotic DNA–RNA polymerases (Lewandowski and Traynor, 1972).

The enzyme reaction is dependent on the presence of all four ribonucleoside triphosphates and a divalent cation (Mg^{2+} is optimal) (Lewandowski *et al.*, 1969; Furuichi, 1974). As is to be expected, actinomycin D has no effect on the incorporation of triphosphates. However, when a methyl group donor [e.g., *S*-adenosylmethionine (SAM)] is added to the assay mixture, there is a dramatic increase in the rate of the reaction (Furuichi, 1974). The *in vitro* products are 10 single-stranded RNA's which represent complete transcripts of one strand of each of the 10 viral genome segments (Furuichi, 1974). The terminal nucleotide structures of all 10 genome RNA segments are identical, and transcription appears to start from the side where the 5′-terminus of the viral RNA carries a methylated adenylic acid residue (A*-G), using the complementary 3′-terminal sequence (U-C) as a template (Shimotohno and Miura, 1974; Miura *et al.*, 1974). If the transcripts are complete copies of the genome RNA, the 5′-terminus should end with a phosphorylated adenosine residue. This is observed in assays where SAM is omitted from the reaction mixture (Shimotohno and Miura, 1974). However, when SAM is included, the 5′-termini of the *in vitro* products are blocked by a methylated guanylic acid residue linked by a pyrophosphate bond to methylated adenylic acid (Furuichi and Miura, 1975). Such terminal structures have since been observed in many messenger RNA's and are considered to be important for translation (Rottman *et al.*, 1974). The fact that methylation occurs *in vitro* in the presence of *S*-adenosylmethionine also implies that the CPV particle contains a methylase enzyme.

Two additional activities have also been reported in CPV particles. Storer *et al.* (1973) observed a nucleotide phosphohydrolase activity, which preferentially converts adenosine triphosphate (ATP) to the diphosphate. This result explains an earlier finding that ATP is required in higher concentration than other ribonucleoside triphosphates in the transcriptase assay (Lewandowski *et al.*, 1969) and that the 5′-terminal nucleotide in the mRNA molecules synthesized *in vitro* lack the γ-phosphate (Shimotohno and Miura, 1974). Similarly the presence of an exonucleolytic ribonuclease (Storer *et al.*, 1973) could account for the initial failure to detect intact messenger RNA molecules in the products of the *in vitro* transcriptase assay (Lewandowski *et al.*, 1969).

IV. VIRUS REPLICATION

Most studies of CPV replication have been restricted to an examination of tissues removed from insect larvae infected with uncharacterized virus isolates. As yet no detailed studies have been made with synchronously infected cells, although there are several reports of the growth of these viruses in cell culture. The replication of CPV in primary cultures has been observed in cells derived from both ovarian tissue (Vago and Bergoin, 1963) and midgut epithelium (Kobayashi, 1971). In the first study with a continuous insect cell line, Grace (1962) reported a fortuitous CPV infection in *Antherea eucalypti* cells. More recently, Granados *et al.* (1974) produced evidence in the form of light and electron micrographs (Figs. 11 and 12) that a CPV from *Trichoplusia ni* would replicate and produce polyhedra in a *T.ni* cell line. No lysis of infected cells was observed and the cell-free medium was not infectious. However, when the cells were mechanically disrupted, infectious virus was released and could be passaged. In the only other study with a continuous cell line, Kawarabata and Hayashi (1972) studied viral RNA synthesis in *Aedes aegypti* cells infected with *Malacosoma disstria* CPV. They showed that ^{3}H-uridine was incorporated into components of the same size as intact virus particles. However, there is now some doubt as to the identity of the cell line used by these workers (Krywienczyk and Sohi, 1973).

Whereas "nonoccluded" or "occluded" virus particles are used to initiate infection in cell culture, the normal means by which CPV infection is transmitted from one insect to another is the oral ingestion of polyhedra. Virus particles are apparently not infectious when injected per os, although they can cause infection when injected into the hemocoel (Faust and Cantwell, 1968). The pH of the gut fluids of lepidopterous larvae is alkaline (Dadd, 1970), and the polyhedra are dissolved in the gut in a way mimicked by the *in vitro* solubilization of polyhedra in alkaline solutions. It has been suggested that the liberated virus particles then adsorb to the cell surface by their projections or spikes and the nucleic acid or nucleoprotein core is injected into the cytoplasm of the cell (Kobayashi, 1971). However, two pieces of evidence suggest that this is unlikely: (a) isolated viral RNA is not infectious (Lewandowski *et al.*, 1969); and (b) the viral transcriptase (which is required, presumably, for the production of virus-specific messenger RNA in the infected cell) appears unable to function if the virus particle is degraded in any way (Lewandowski and Traynor, 1972).

In reovirus-infected cells intact virus particles are taken up into lysosomes at the onset of infection. Part of the outer capsid is degraded by proteolysis, and the viral transcriptase is activated (Silverstein and Dales, 1968). Unlike reovirus, the CPV transcriptase apparently requires no activation (Lewandowski *et al.*, 1969) although treatment with organic solvents may enhance its activity (Shimotohno and Miura, 1973). However, CPV virus particles have been observed associated with lysosomes, though whether this occurs at the beginning of infection is not clear (Bird, 1965; Kobayashi, 1971).

Studies of RNA synthesis in CPV infections are not as far advanced as the *in vitro* studies on the functioning of the transcriptase might suggest. However, like the *in vitro* synthesis of mRNA, viral RNA synthesis is not inhibited by concentrations of actinomycin D, which suppress host DNA-dependent RNA transcription (Hayashi and Kawarabata, 1970; Payne, 1972). Up to 75% of the single-stranded RNA synthesized in the presence of the inhibitor is virus-specific, does not self-anneal, and probably represents mRNA (Payne and Kalmakoff, 1973; Furusawa and Kawase, 1973). Several studies have suggested that most of the RNA synthesis in infected cells occurs in the nucleus although all other aspects of replication usually appear to be confined to the cytoplasm (Hayashi and Kawase, 1965; Watanabe, 1967; Hayashi and Retnakaran, 1970).

One of the first signs of infection in the gut epithelial cells of diseased insects is the development of micronet structures within the cytoplasm (Fig. 13) usually referred to as the "virogenic stroma" (Xeros, 1956; 1966). It appears that the virus capsids are synthesized within these stroma (Kobayashi, 1971). However, there is some difference in interpretation of the process of virus particle assembly. Bird (1965) proposed that "core" particles develop first and subsequently are encapsidated. In contrast, Arnott *et al.* (1968) observed numerous empty particles in infected cells and suggested that assembly occurred by the reverse process. Although both observations may be correct, it is possible that the presence of empty particles (Fig. 16) may represent aberrant assembly and the formation of defective virus.

Polyhedron protein is first synthesized, as early as 9 hours after per os infection of *Bombyx mori* larvae, at sites near the striated border of the epithelial cells (Kawase and Miyajima, 1969). Arnott *et al.* (1968) observed tightly-packed fibrils ("crystallogenic matrix") presumably composed of uncrystallized polyhedron protein forming a denser micronet structure than the stroma, adjacent to developing polyhedra (Fig. 14). The precise mechanism of polyhedron formation is unknown, but it starts with the crystallization of protein around groups of virus particles (Figs. 15 and 16). There may be some relatively specific recognition (such as exists between enzyme and substrate) between virus particles and polyhedron protein, which initiates the crystallization and excludes components in the cell other than virus particles. In a CPV infection of *Arctia caja*, two serologically distinct viruses (types 2 and 3; Fig. 1) can replicate, although polyhedral protein characteristic of only one of these viruses (type 3) is produced (Payne, 1976). In polyhedra derived from such infections, virus particles of both types are found, indicating that any interaction between polyhedron protein and virus protein is not completely exclusive. However, proportionally fewer virus particles of type 2 are observed within these polyhedra, implying that there is some specificity in the process of occlusion.

The virus particles are usually occluded in large numbers, apparently at random (Fig. 2). Arnott *et al.* (1968) estimated that as many as 10,000 particles, 50–100 nm apart, were occluded within a single polyhedron of a CPV from *Danaus plexippus*. However, in some CPV types, only one particle may be occluded (Fig. 17) although separate inclusions may later fuse (Fig. 18) (Stoltz and Hilsenhoff, 1969). As occlusion proceeds, the virogenic stroma gradually regresses and remains as small remnants within large masses of polyhedra (Figs. 19 and 20).

In some infections, a significant proportion of the virus particles may not be occluded within polyhedra and can be isolated from the infected tissue without exposure to alkali. These "nonoccluded" virus particles may represent more than 70% of the total number of particles in an infected insect at an advanced stage of infection (Hayashi, 1970). A comparison of some of the properties of occluded and nonoccluded virus particles has not revealed any fundamental differences between the two types (Payne and Kalmakoff, 1975), and the reason why so many particles are not occluded is therefore unlikely to be the result of major structural differences. Whereas polyhedra are the vehicle for transmission of infection between insects, the occurrence of large numbers of nonoccluded particles would ensure that virus multiplication can continue within infected individuals.

During infection, the most pronounced effect on the cell is the degradation of the rough endoplasmic reticulum (Xeros, 1956; Arnott *et al.*, 1968; Kobayashi, 1971). Ultimately cell lysis may occur as the plasma membrane disintegrates (Kobayashi, 1971), and then polyhedra are released into the lumen of the midgut and are excreted. However, more epithelial cells may develop from regenerative cells within the tissue, and this continual replacement of susceptible cells may account for the chronic, rather than lethal, nature of CPV infections.

Although the nucleus may be involved in viral RNA synthesis, it does not usually undergo great morphological change during infection (Kobayashi, 1971). However, Kawase *et al.*, (1973) have reported the development of a CPV infection in *Bombyx mori,* where polyhedra are formed in the nucleus. The virus strain producing these inclusions was indistinguishable by several criteria from the strain producing only cytoplasmic inclusions (Kawase and Yamaguchi, 1974). This evidence, combined with the studies of a CPV (Figs. 21 and 22) in a crustacean (Federici and Hazard, 1975), indicates that these viruses are not as restricted in histopathological effect or host susceptibility as was originally thought.

ACKNOWLEDGMENTS

We are particularly grateful to the following for supplying photographs: Dr. B. Federici, Dr. R. R. Granados, Miss R. Rubinstein, Mr. P. F. Entwistle, and Dr. Y. Hosaka for allowing us to reproduce Figs. 8a, 9, and 10. We also thank Miss M. K. Arnold, Mr. C. D. Hatton, and Dr. J. Kalmakoff for help in the preparation of this chapter.

REFERENCES

Arnott, H. J., Smith, K. M., and Fullilove, S. L. (1968). *J. Ultrastruct. Res.* **24,** 479.
Aruga, H., and Tanada Y., eds. (1971). "The Cytoplasmic Polyhedrosis Virus of the Silkworm." Univ. of Tokyo Press, Tokyo.
Aruga, H., Hukuhara, T., Yoshitake, N., and Ayudhya, I. (1961). *J. Insect Pathol.* **3,** 81.
Asai, J., Kawamoto, F., and Kawase, S. (1972). *J. Invertebr. Pathol.* **19,** 279.
Bailey, C. H., Shapiro, M., and Granados, R. R. (1975). *J. Invertebr. Pathol.* **25,** 273.
Bergold, G. H., and Suter, J. (1959). *J. Insect Pathol.* **1,** 1.
Bird, F. T. (1965). *Can. J. Microbiol.* **11,** 497.
Dadd, R. H. (1970). *Chem. Zool.* **5,** 117–145.
Faust, R. M., and Cantwell, G. E. (1968). *J. Invertebr. Pathol.* **11,** 119.
Federici, B. A., and Hazard, E. I. (1975). *Nature (London)* **254,** 327.
Fujii-Kawata, I., Miura, K., and Fuke, M. (1970). *J. Mol. Biol.* **51,** 247.
Furuichi, Y. (1974). *Nucleic Acids Res.* **1,** 809.
Furuichi, Y., and Miura, K. (1975). *Nature (London)* **253,** 374.
Furusawa, T., and Kawase, S. (1973). *J. Invertebr. Pathol.* **22,** 335.
Grace, T. D. C. (1962). *Virology* **18,** 33.

Granados, R. R., McCarthy, W. J., and Naughton, M. (1974). *Virology* **59,** 584.
Hayashi, Y. (1970). *J. Invertebr. Pathol.* **16,** 442.
Hayashi, Y., and Bird, F. T. (1968). *J. Invertebr. Pathol.* **11,** 40.
Hayashi, Y., and Donaghue, T. P. (1971). *Biochem. Biophys. Res. Commun.* **42,** 214.
Hayashi, Y., and Kawarabata, T. (1970). *J. Invertebr. Pathol.* **15,** 461.
Hayashi, Y., and Kawase, S. (1964). *Virology* **23,** 611.
Hayashi, Y., and Kawase, S. (1965). *J. Seric. Soc. Jpn.* **34,** 171.
Hayashi, Y., and Retnakaran, A. (1970). *J. Invertebr. Pathol.* **16,** 150.
Hills, G. J., and Smith, K. M. (1959). *J. Insect Pathol.* **1,** 121.
Hosaka, Y., and Aizawa, K. (1964). *J. Insect Pathol.* **6,** 53.
Hukuhara, T., and Hashimoto, Y. (1966). *J. Invertebr. Pathol.* **8,** 234.
Kalmakoff, J., Kewandowski, L. J., and Black, D. R. (1969). *J. Virol.* **4,** 851.
Kavenoff, R., Klotz, L. C., and Zimm, B. H. (1973). *Cold Spring Harbor Symp. Quant. Biol.* **38,** 1.
Kawarabata, T., and Hayashi, Y. (1972). *J. Invertebr. Pathol.* **19,** 414.
Kawase, S., and Miyajima, S. (1969). *J. Invertebr. Pathol.* **13,** 330.
Kawase, S., and Yamaguchi, K. (1974). *J. Invertebr. Pathol.* **24,** 106.
Kawase, S., Kawamoto, F., and Yamaguchi, K. (1973). *J. Invertebr. Pathol.* **22,** 266.
Kellen, W. R., Clark, T. B., Lindegren, J. E., and Sanders, R. D. (1966). *J. Invertebr. Pathol.* **8,** 390.
Kobayashi, M. (1971). *In* "The Cytoplasmic Polyhedrosis Virus of the Silkworm" (H. Aruga and Y. Tanada, eds.), pp. 103–128. Univ. of Tokyo Press, Tokyo.
Krywienczyk, J., and Sohi, S. S. (1973). *In Vitro* **8,** 495.
Lewandowski, L. J., and Millward, S. (1971). *J. Virol.* **7,** 434.
Lewandowski, L. J., and Traynor, B. L. (1972). *J. Virol.* **10,** 1053.
Lewandowski, L. J., Kalmakoff, J., and Tanada, Y. (1969). *J. Virol.* **4,** 857.
Longworth, J. F., and Spilling, C. R. (1970). *J. Invertebr. Pathol.* **15,** 276.
Luftig, R. B., Kilham, S. S., Hay, A. J., Zweerink, H. J., and Joklik, W. K. (1972). *Virology* **48,** 170.
Martignoni, M. E. (1967). *J. Virol.* **1,** 646.
Miura, K., Fujii, I., Sakaki, T., Fuke, M., and Kawase, S. (1968). *J. Virol.* **2,** 1211.
Miura, K., Fujii-Kawata, I., Iwata, H., and Kawase, S. (1969). *J. Invertebr. Pathol.* **14,** 262.
Miura, K., Watanabe, K., and Sugiura, M. (1974). *J. Mol. Biol.* **86,** 31.
Miyajima, S., and Kawase, S. (1969). *Virology* **39,** 347.
Nishimura, A., and Hosaka, Y. (1969). *Virology* **38,** 550.
Payne, C. C. (1972). *Monogr. Virol.* **6,** 11–15.
Payne, C. C. (1976). *J. Gen. Virol.* **30,** 357.
Payne, C. C., and Kalmakoff, J. (1973). *Intervirology* **1,** 34.
Payne, C. C., and Kalmakoff, J. (1975). *Intervirology* **4,** 354.
Payne, C. C., and Rivers, C. F. (1976). *J. Gen. Virol.* **33,** 71.
Payne, C. C., and Tinsley, T. W. (1974). *J. Gen. Virol.* **25,** 291.
Rottman, F., Shatkin, A. J., and Perry, R. P. (1974). *Cell* **3,** 197.
Shatkin, A. J., Sipe, J. D., and Loh, P. (1968). *J. Virol.* **2,** 986.
Shimotohno, K., and Miura, K. (1973). *Virology* **53,** 283.
Shimotohno, K., and Miura, K. (1974). *J. Mol. Biol.* **86,** 21.
Silverstein, S. C., and Dales, S. (1968). *J. Cell Biol.* **36,** 197.
Smith, K. M. (1963). *In* "Insect Pathology" (E. A. Steinhaus, ed.), Vol. 1, pp. 457–497. Academic Press. New York.
Stoltz, D. B., and Hilsenhoff, W. L. (1969). *J. Invertebr. Pathol.* **14,** 39.
Storer, G. B., Shepherd, M. G., and Kalmakoff, J. (1973). *Intervirology* **2,** 87.
Vago, C., and Bergoin, M. (1963). *Entomophaga* **8,** 253.
Watanabe, H. (1967). *J. Invertebr. Pathol.* **9,** 480.
Xeros, N. (1956). *Nature (London)* **178,** 412.
Xeros, N. (1966). *J. Invertebr. Pathol.* **8,** 79.

Fig. 1. Electrophoretic separation of the RNA genome segments of six isolates of cytoplasmic polyhedrosis viruses (CPV's) (Payne and Rivers, 1976). Direction of migration is from left to right. (a) *Bombyx mori* (type 1) CPV; (b) *Inachis io* (type 2) CPV; (c) *Spodoptera exempta* (type 3) CPV; (d) *Actias selene* (type 4) CPV; (e) *Trichoplusia ni* (type 5) CPV; (f) *Triphena pronuba* (type 7) CPV. In these examples it has not been possible to resolve all the genome segments, but molar proportion measurements have confirmed the existence of 10 segments in each virus type, with a total molecular weight of approximately 15×10^6 daltons.

Fig. 2. Thin section of polyhedron of *Bombyx mori* CPV showing the large number of isometric virus particles embedded, apparently at random, within the matrix protein. In this virus strain, the polyhedra are characteristically cubic in shape. (Bar = 200 nm.) Inset, polyhedron of *Anoplonyx destructor* CPV showing the lattice structure of the polyhedron protein. Bar = 200 nm.

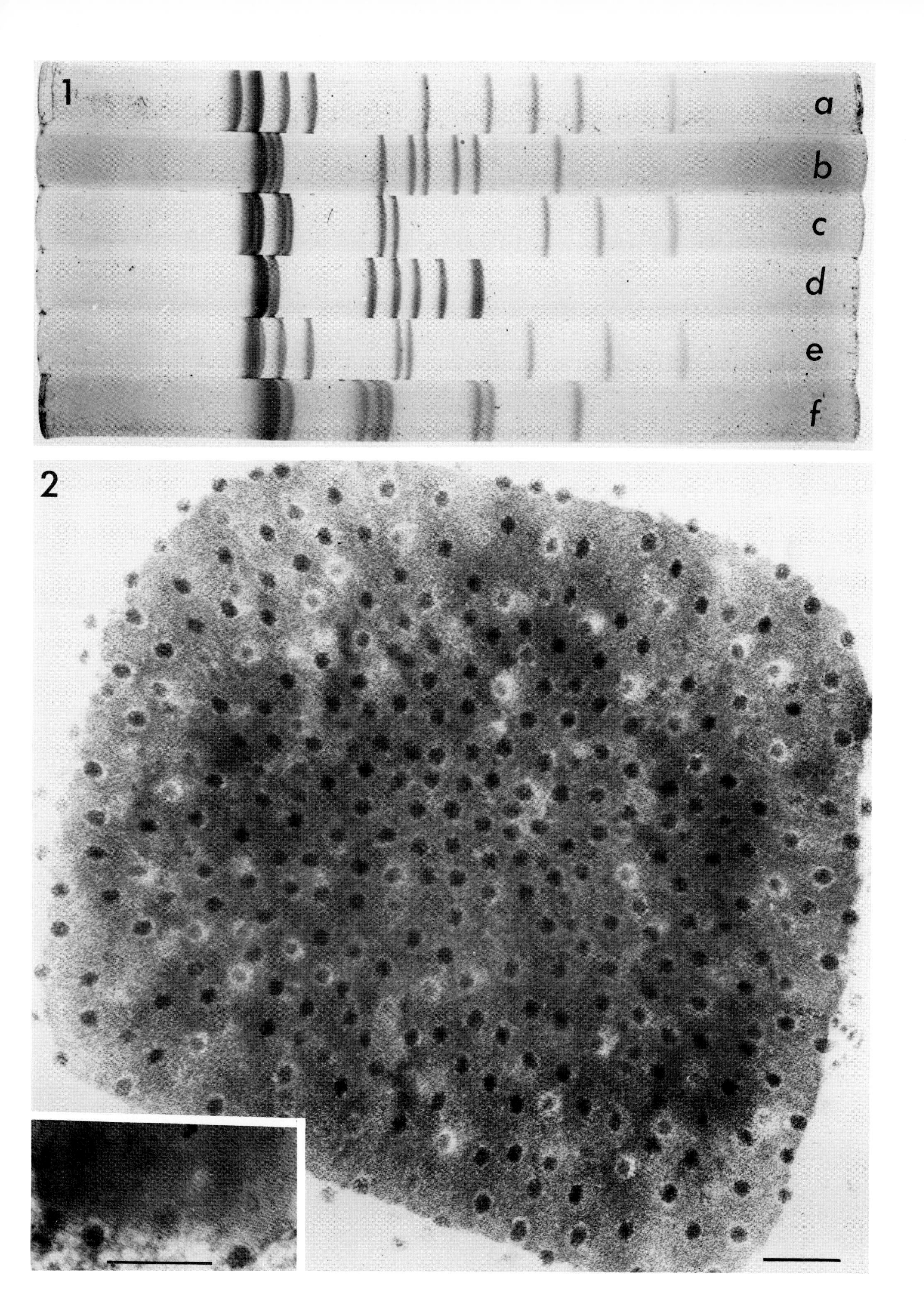
1
a
b
c
d
e
f
2

Figs. 3–5. Virus particles of *Bombyx mori* CPV released from polyhedra by treatment with dilute alkali and negatively stained with phosphotungstic acid.

Fig. 3. Preparation showing particles with "spikes," apparently empty particles, and (arrowed) a particle with a beaded arrangement of 20 surface subunits, apparently on its fivefold axis of symmetry. A similar arrangement of subunits has been observed in reovirus and reovirus cores (Luftig *et al.*, 1972). Bar = 100 nm.

Fig. 4. Particle showing apparently hollow morphological units (arrow) arranged probably on a fivefold axis of symmetry. Bar = 100 nm.

Fig. 5. Sixfold optical rotation (b) of a virus particle (a) demonstrating hexagonal symmetry and discontinuities in the capsid which probably represent the base of the hollow projections or spikes.

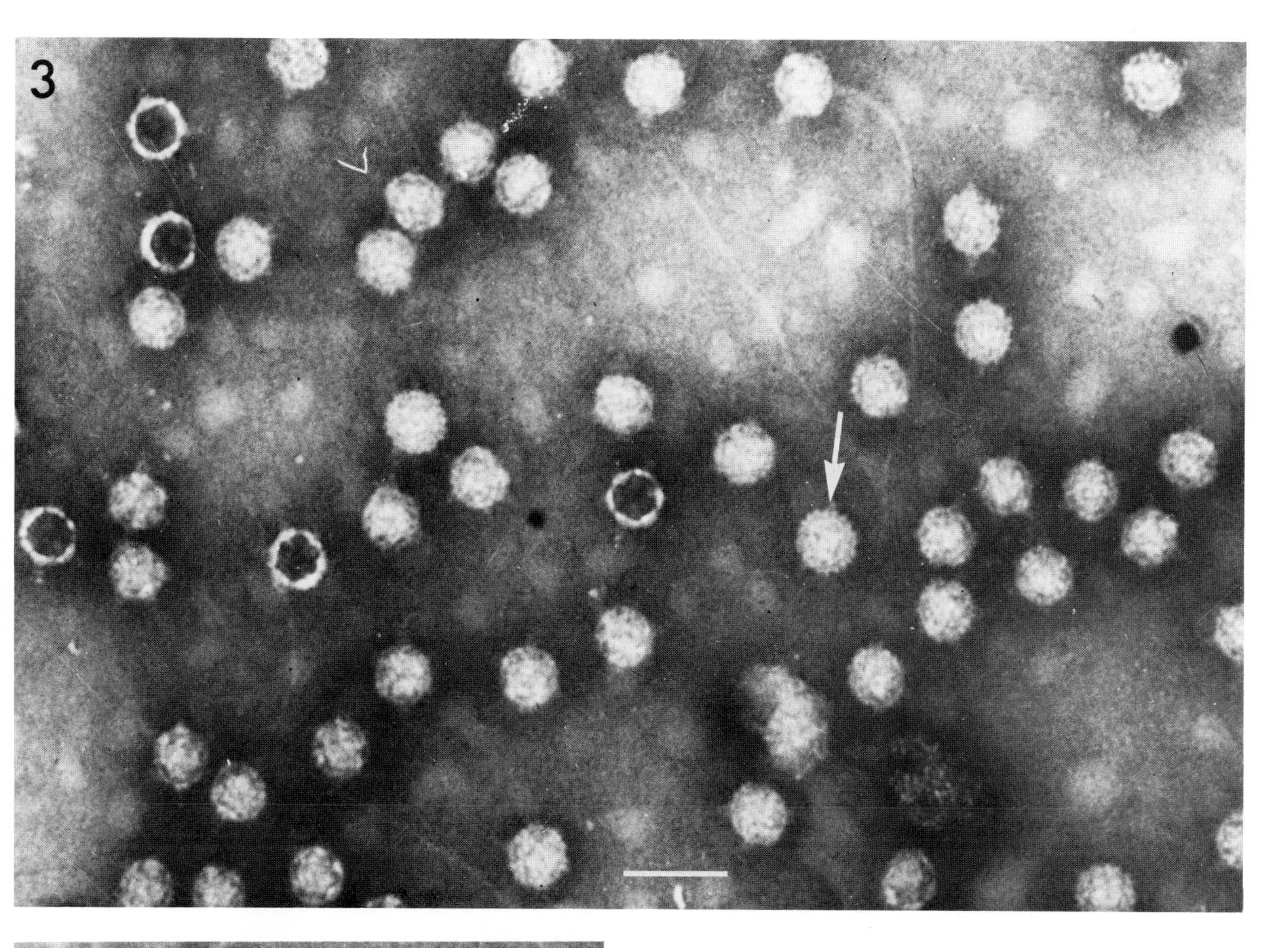
3

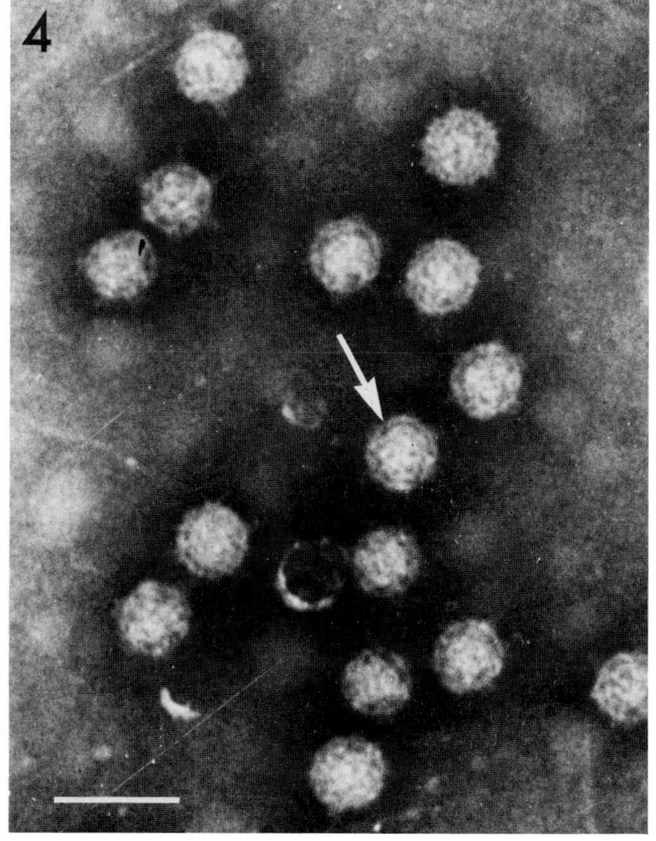
4

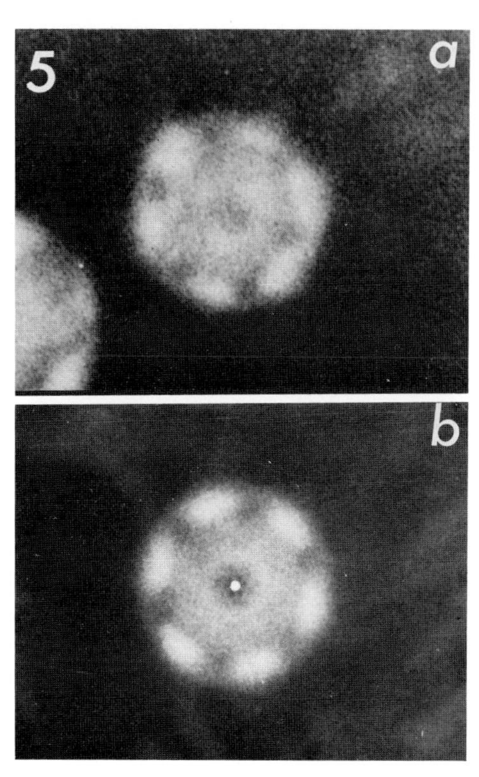
5
a
b

Figs. 6 and 7. Negatively stained particles of CPV's from *Phalera bucephala* (Fig. 6) and *Heliothis armigera* (Fig. 7). Despite differences in the sizes of the RNA genome segments of these viruses and *Bombyx mori* CPV (Payne and Rivers, 1976), the virus particles of the different types show considerable morphological similarities, such as size and the presence of spikes (particularly clear in Fig. 6. insert). Although no distinct double capsid structure is evident in the negatively stained preparations it appears that particles consist of a central core region which is surrounded by an outer shell or capsid. In "empty" particles, only this outer layer is visible.

Fig. 6 Bar = 100 nm; inset bar = 100 nm.

Fig. 7. Bar = 100 nm; Photograph courtesy of Miss R. Rubinstein.

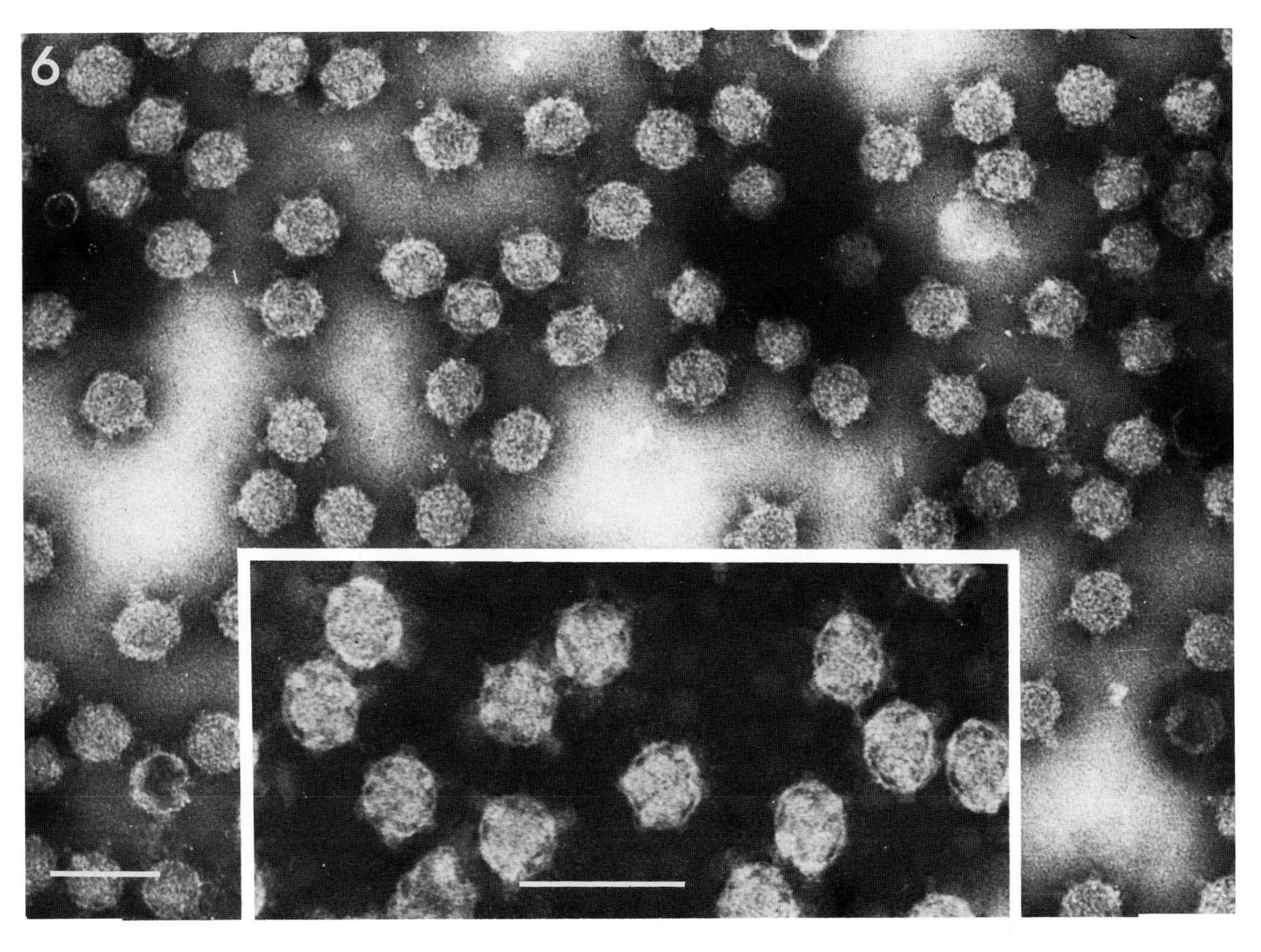
6

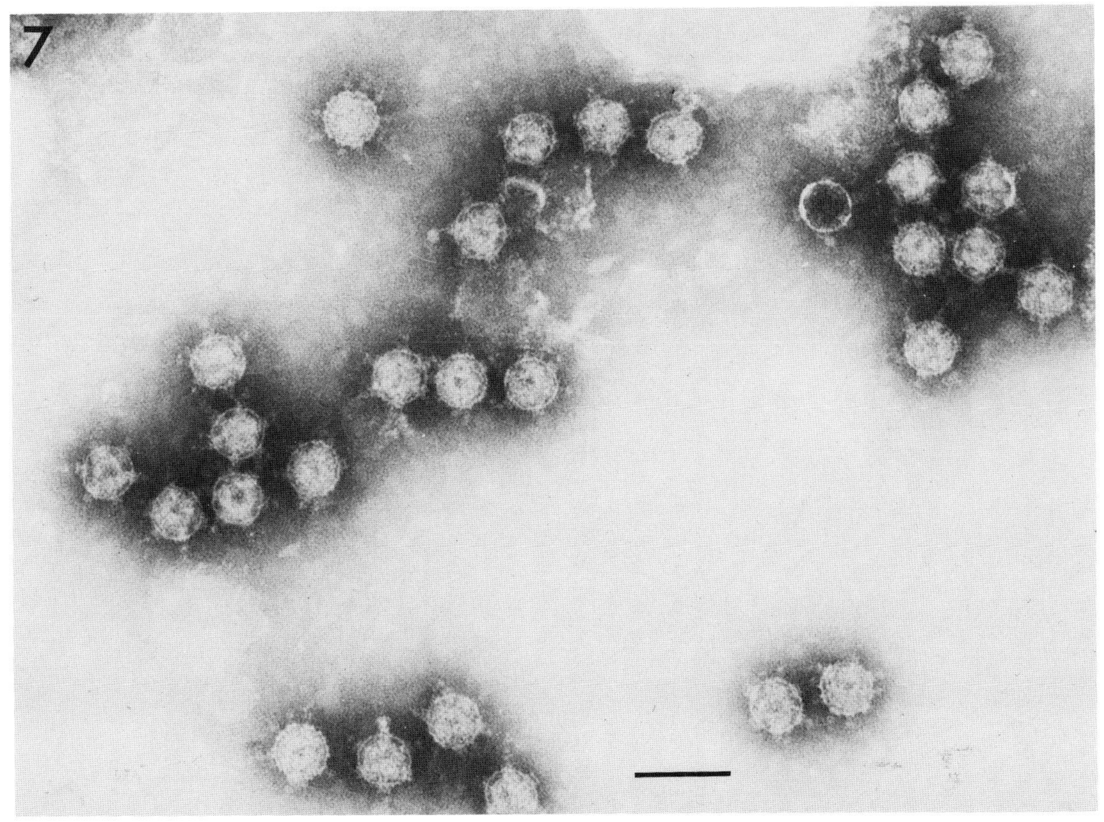
7

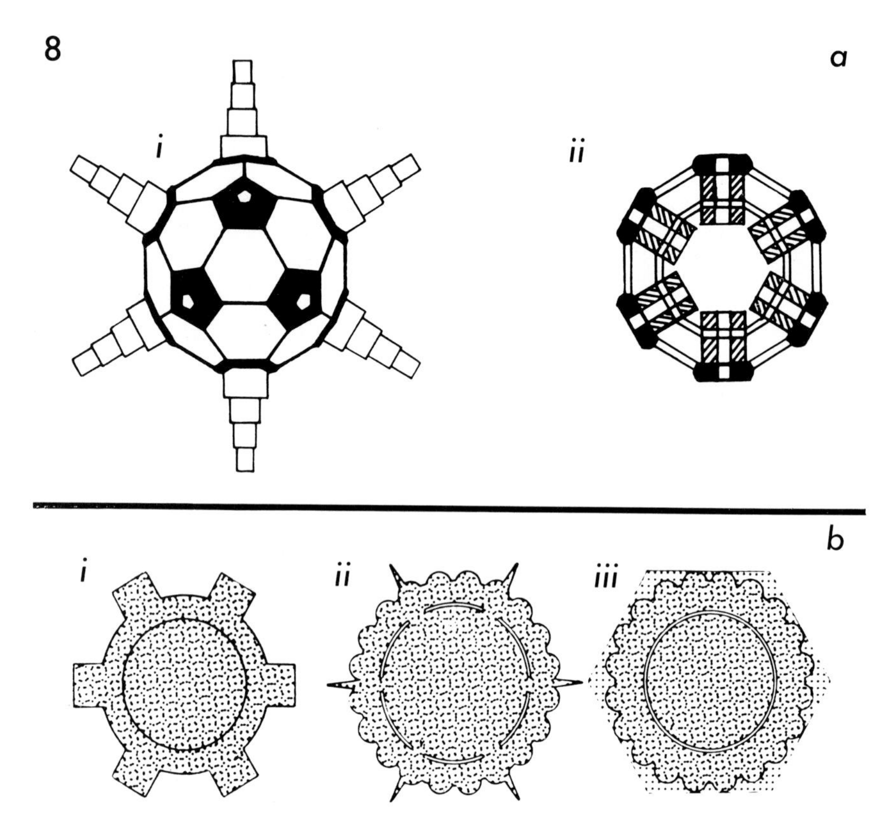

Fig. 8. Structural models for *Bombyx mori* CPV virus particles. (a) The model proposed by Hosaka and Aizawa (1964) consists of two concentric shells connected by tubular projections or spikes at the vertices of the particles (i) surface structure and (ii) internal structure. (Reproduced by kind permission of Dr. Y. Hosaka.) (b) Diagrammatic presentation of the structure of (i) reovirus core, (ii) CPV, and (iii) wound tumor virus, as proposed by Lewandowski and Traynor (1972). (Reproduced by kind permission of the American Society for Microbiology.)

Although both sets of workers agree that the virus particle appears to have a dense core surrounded by an outer shell, Lewandowski and Traynor (1972) suggest that the structure is not analogous to the double-capsid of reovirus. Both in structure and biological properties, CPV particles show considerable similarities with the subviral particles, or cores, of reovirus.

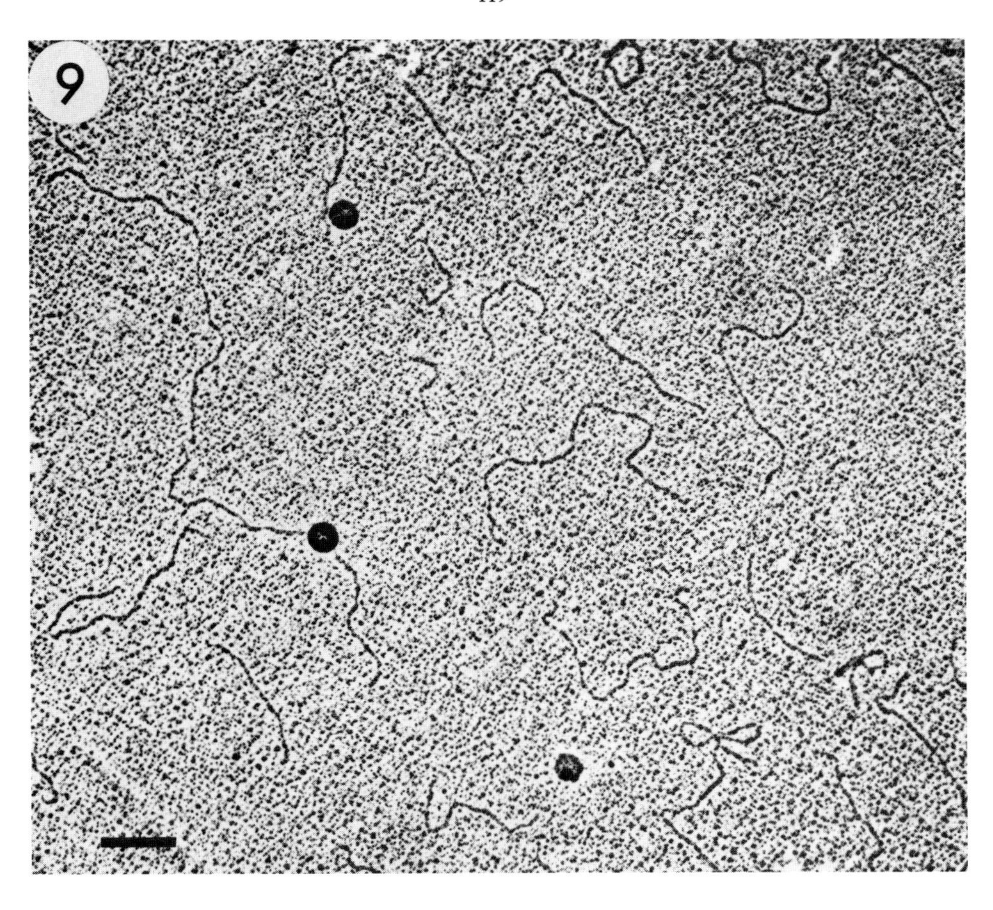

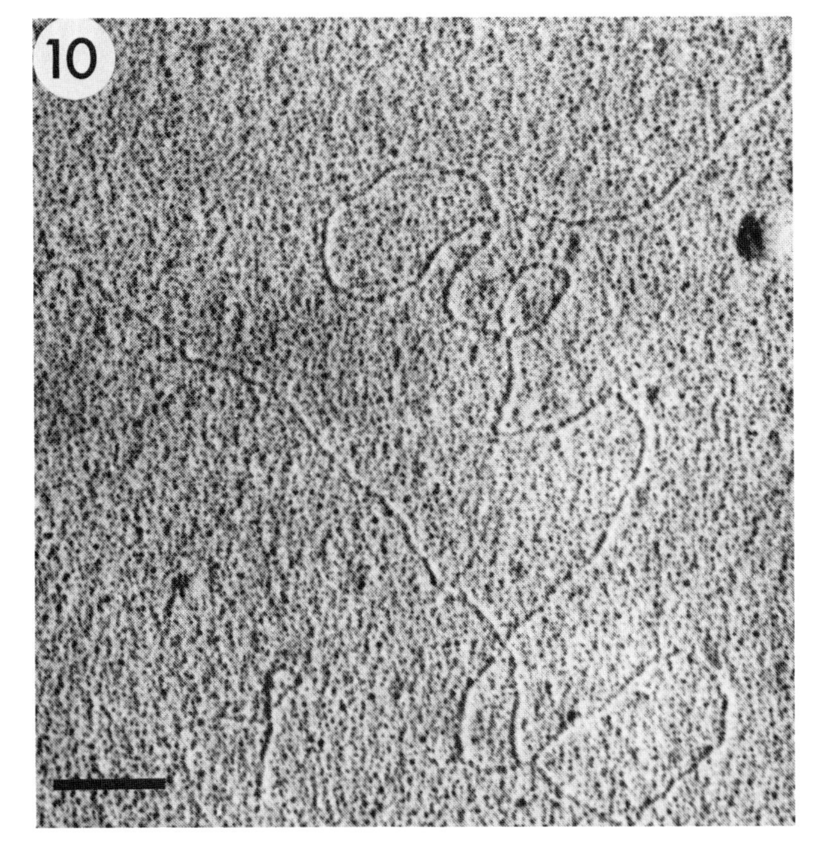

Figs. 9 and 10. Electron microscopy of the RNA of *Bombyx mori* CPV.

Fig. 9. RNA released from virus particles after treatment with sodium perchlorate. Linear molecules of nucleic acid, characteristic of a double-stranded structure, can be observed. These molecules are of different lengths and can be related to the sizes of the major classes of RNA segments which can be resolved by electrophoresis on polyacrylamide gels. Some RNA is still associated with the capsids. Bar = 200 nm.

Fig. 10. One molecule of RNA released by urea treatment of CPV virus particles. This molecule has a contour length of 6.8 μm and may represent the complete viral genome of 10 RNA segments linked end to end. Bar = 200 nm. [Photographs reproduced by kind permission of Dr. Y. Hosaka, from Nishimura and Hosaka (1969).]

Figs. 11 and 12. The replication of *Trichoplusia ni* CPV in a *T.ni* cell line.

Fig. 11. Phase contrast micrograph of cultured cells of *T.ni* 7 days after infection with *T.ni* CPV. Numerous polyhedra are visible in the cytoplasm of infected cells. Bar = 10 μm.

Fig. 12. An infected cell of *T.ni* showing polyhedra developing at the periphery of cytoplasmic micro-net structures ("virogenic stroma"). Numerous virus particles, consisting of a dense central core surrounded by a less dense shell (insert) can be observed in the stroma. Bar = 1 μm; inset bar = 200 nm. (Photographs courtesy of Dr. R. R. Granados.)

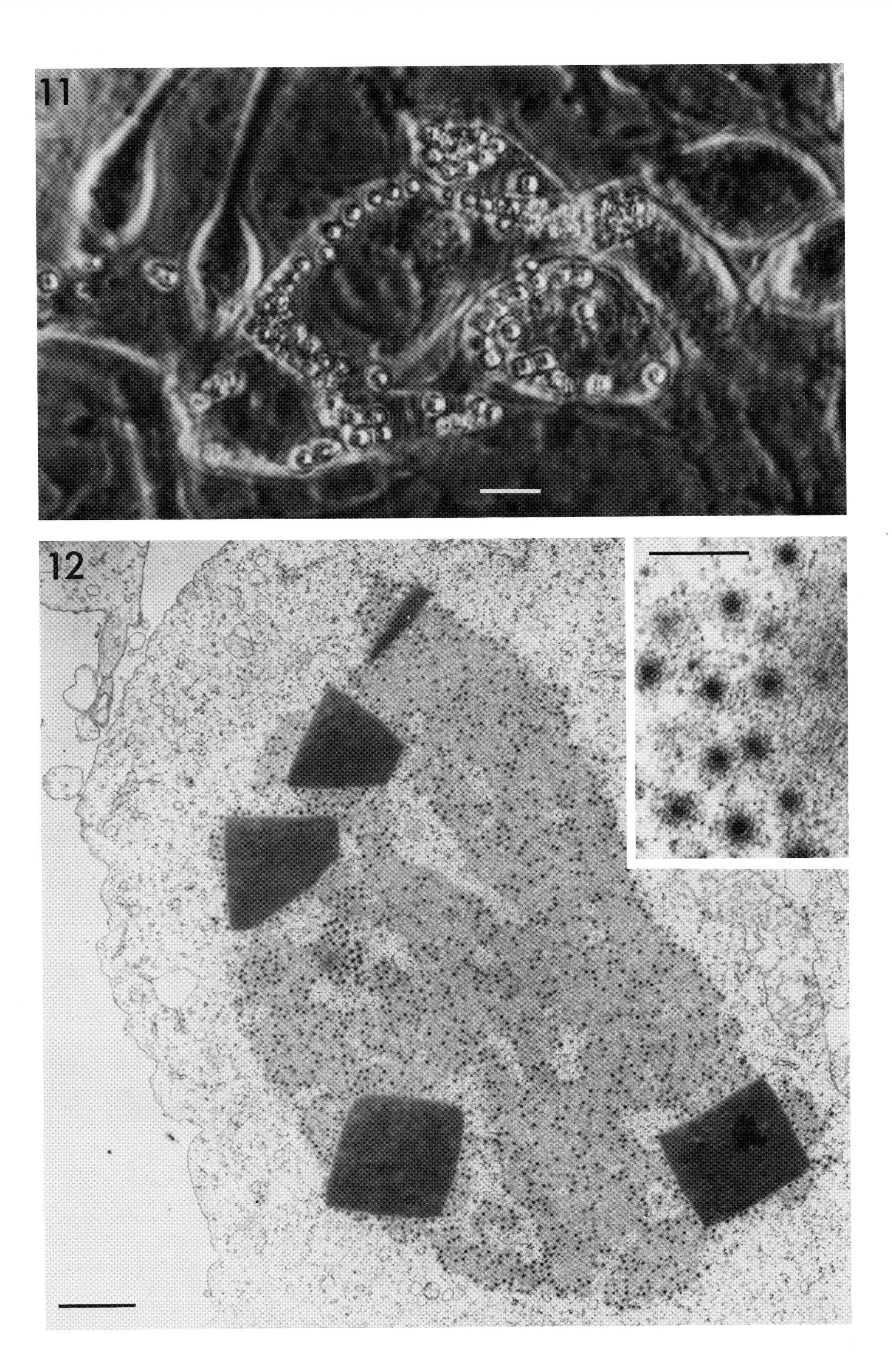
11
12

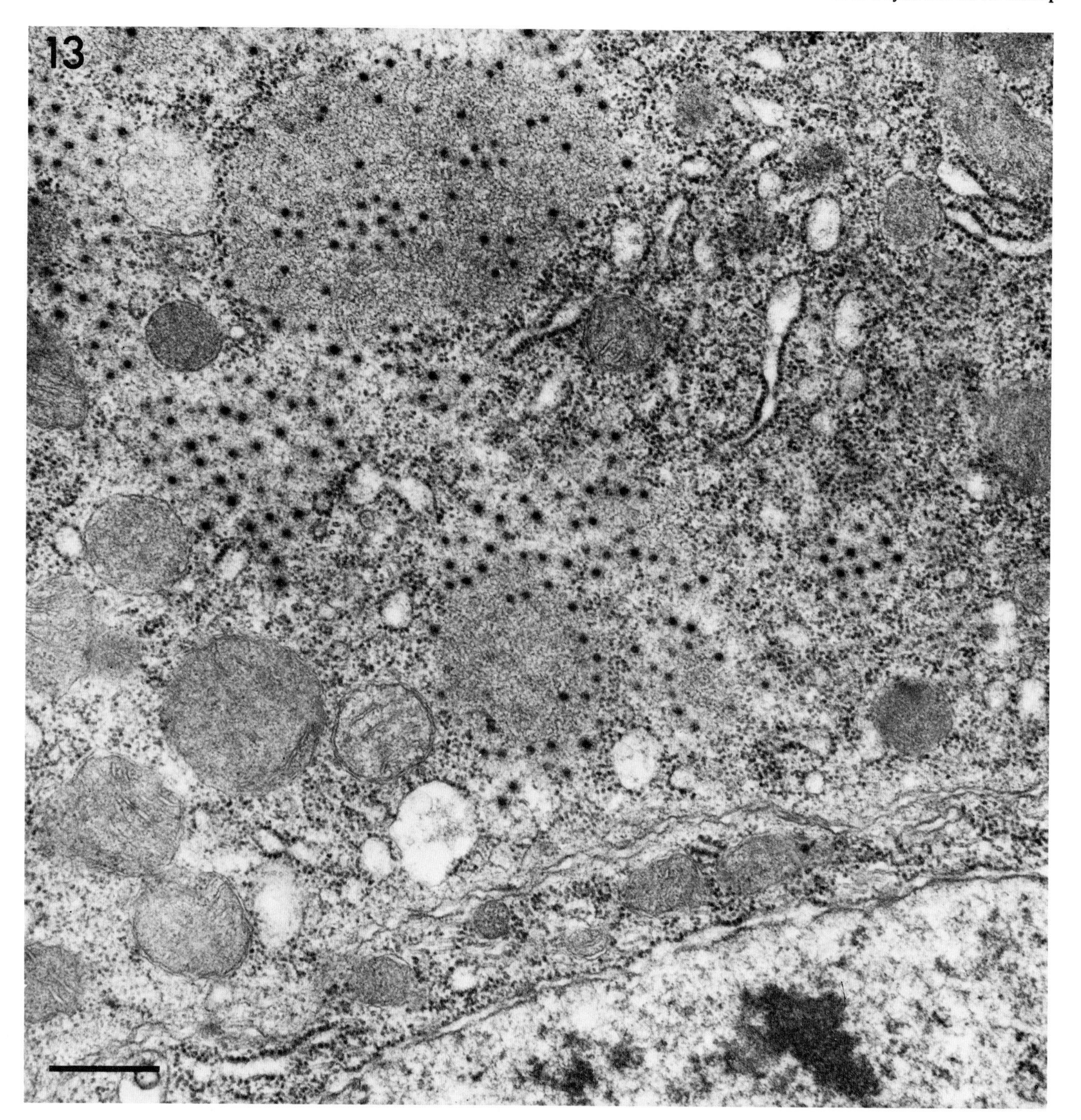

Fig. 13. CPV virus particles of *Aglais urticae* developing within small areas of virogenic stroma in a midgut epithelial cell. Later in infection individual stroma may coalesce. Bar = 500 nm.

Figs. 14 and 15. CPV infection in the midgut epithelium of *Bupalus piniarius.*

Fig. 14. The development of a polyhedron in association with the virogenic stroma (vs) and a more dense micro-net structure, possibly the "crystallogenic matrix" (cm) of Arnott *et al.* (1968). Bar = 200 nm.

Fig. 15. Immature polyhedra at a later stage of development showing the crystallization of polyhedron protein around groups of particles and the random occlusion of large numbers of virus particles. Bar = 200 nm. (Photographs courtesy of Mr. P. F. Entwistle.)

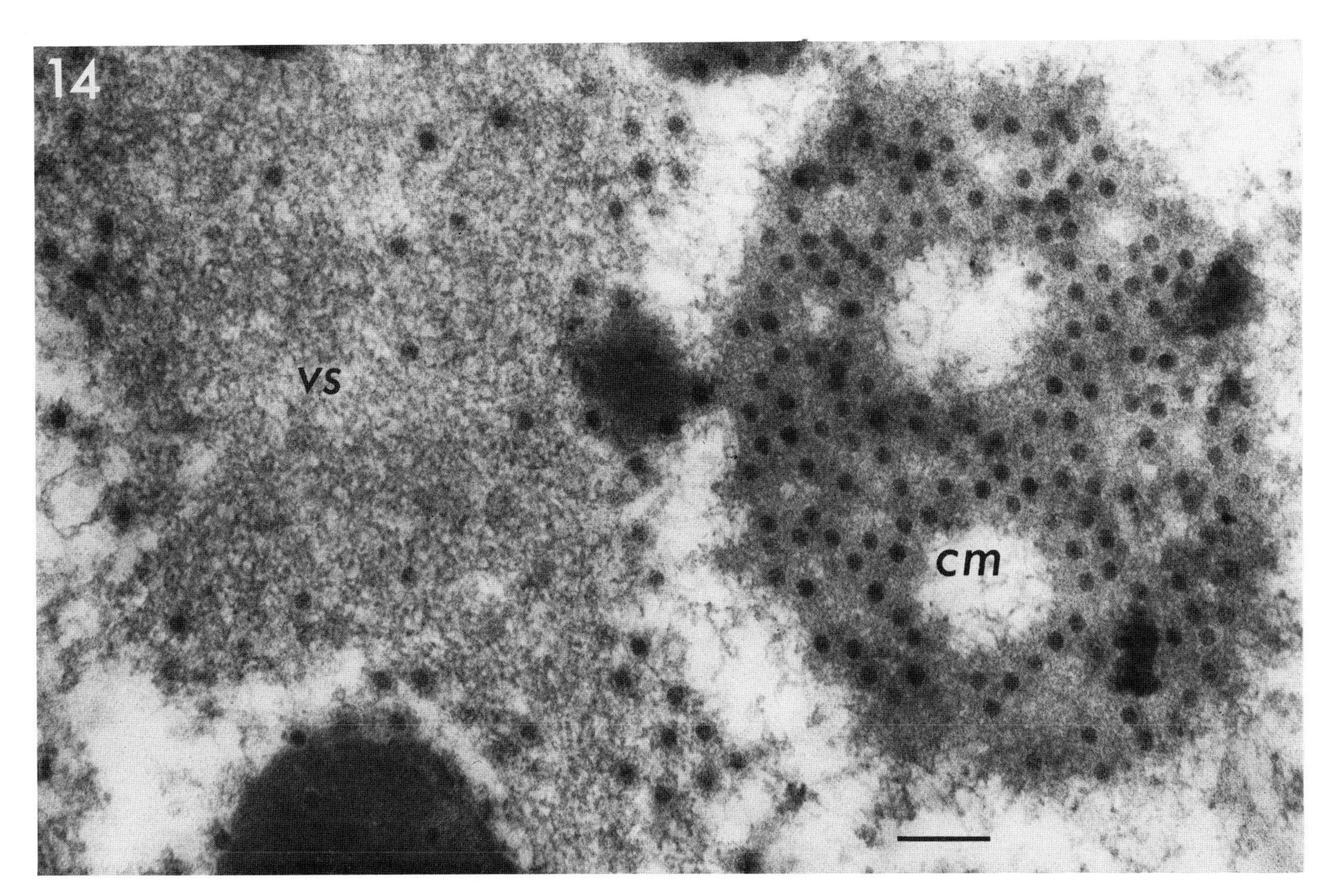
14
vs
cm

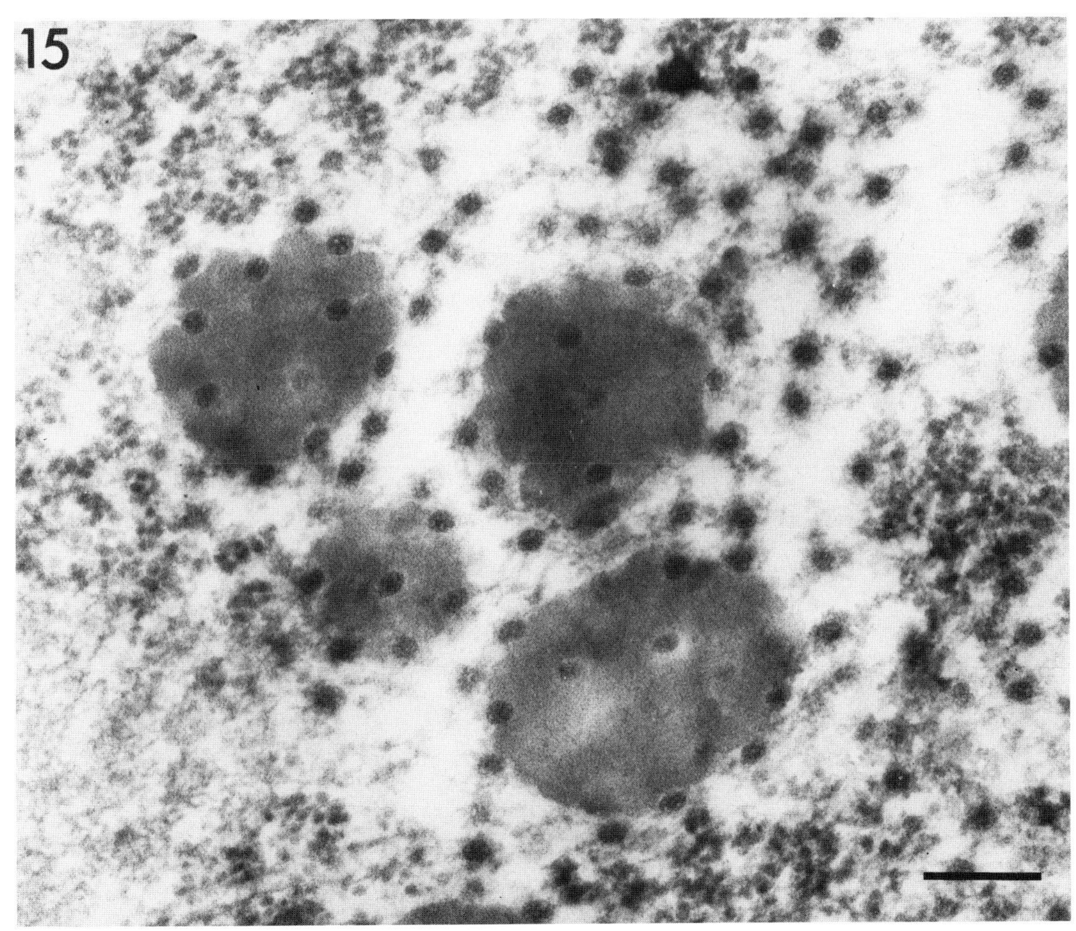
15

Fig. 16. Developing polyhedra in CPV-infected larvae of *Simulium vittatum* (Diptera) showing large numbers of apparently empty particles or virus capsids. Bar = 500 nm. (Photograph by kind permission of Dr. B. Federici.)

Fig. 17. Polyhedra in the midgut epithelium of larvae of *Aedes quadrimaculatus*. In these inclusions the virus particles are occluded singly. Bar = 500 nm. (Photograph by kind permission of Dr. B. Federici.)

Fig. 18. Polyhedra in the midgut epithelium of larvae of *Aedes taeniorhynchus* which may have resulted from the coalescence of singly embedded virus particles as described for *Chironomus plumosus* CPV by Stoltz and Hilsenhoff (1969). Bar = 500 nm. (Photograph by kind permission of Dr. B. Federici.)

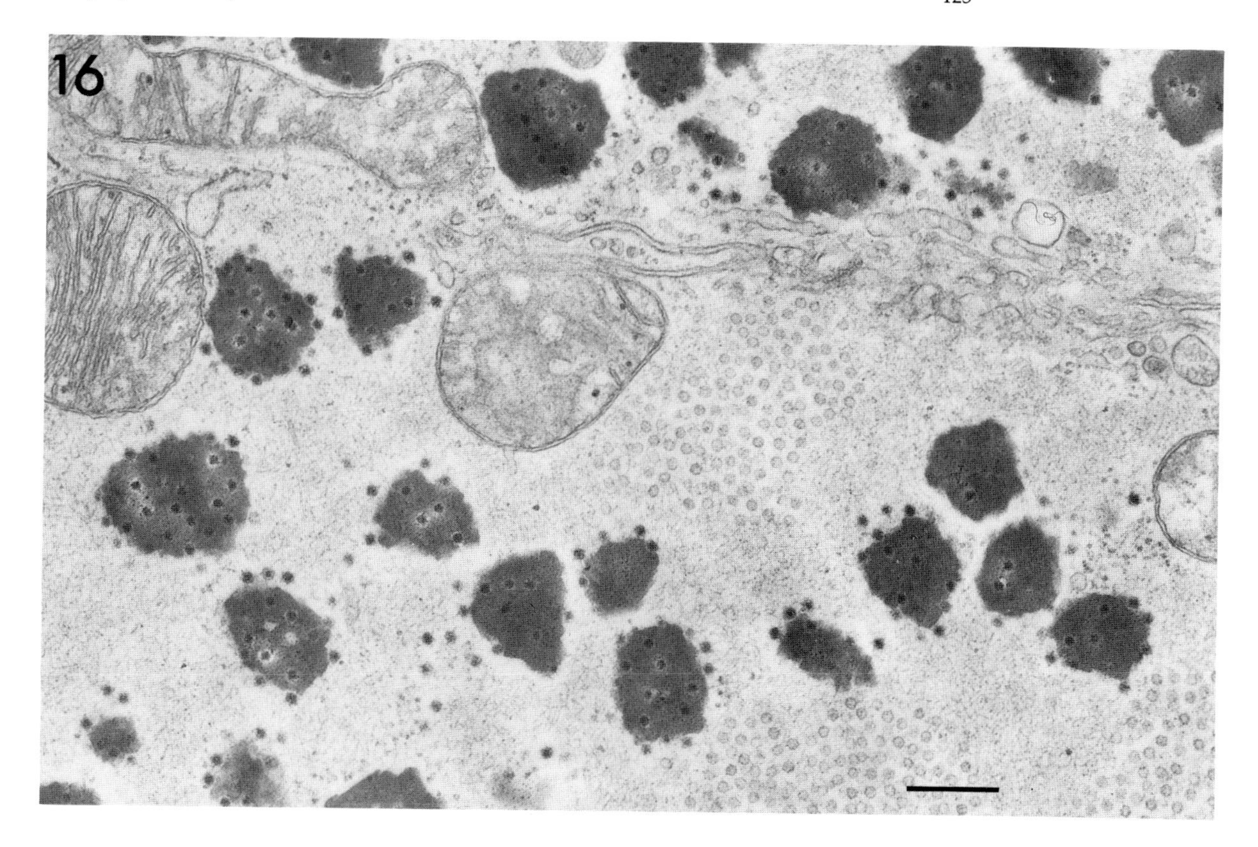
16

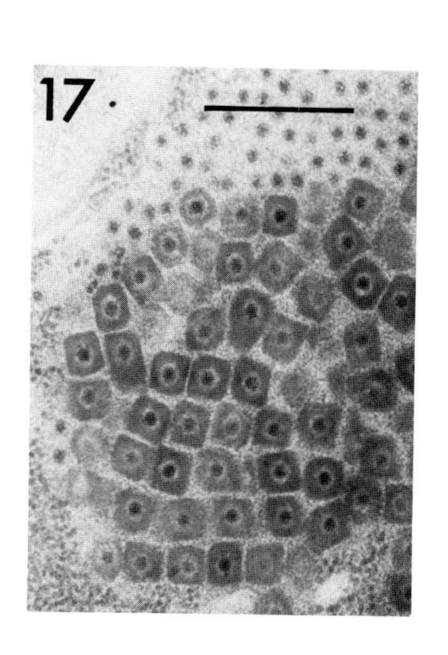
17

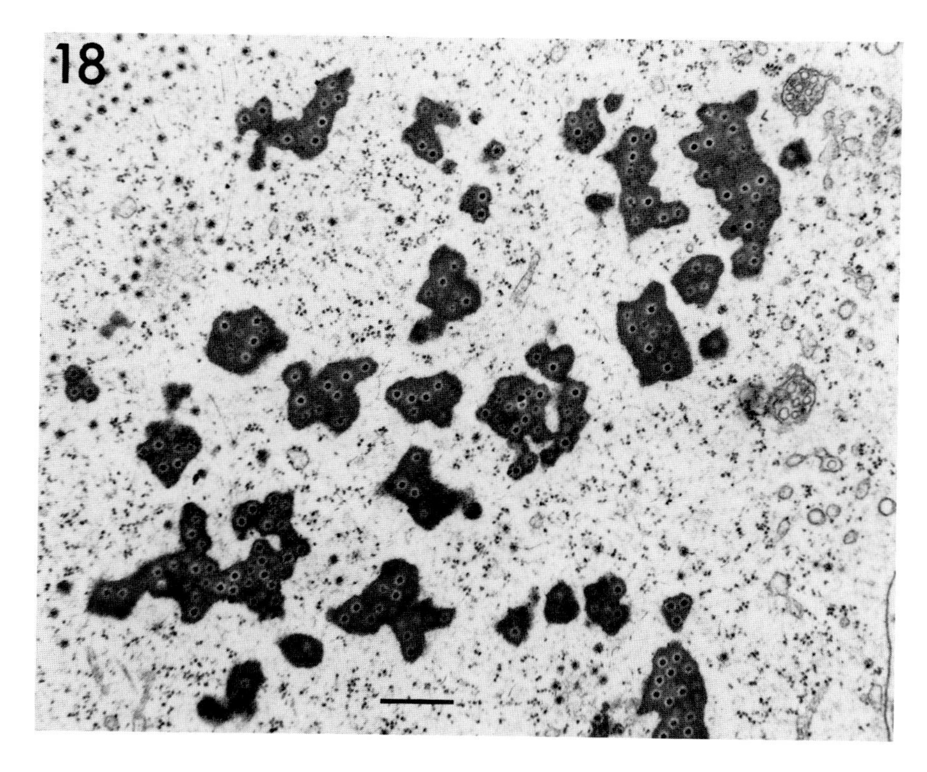
18

Figs. 19 and 20. Terminal stages of infection with CPV showing the production of large numbers of polyhedra within the cytoplasm of infected epithelial cells.

Fig. 19. *Anoplonyx destructor,* showing numerous polygonal polyhedra. Bar = 1 μm.

Fig. 20. *Bombyx mori,* showing apparently normal (cubic) and abnormal (multifaceted) polyhedra within the same cell. Bar = 1 μm.

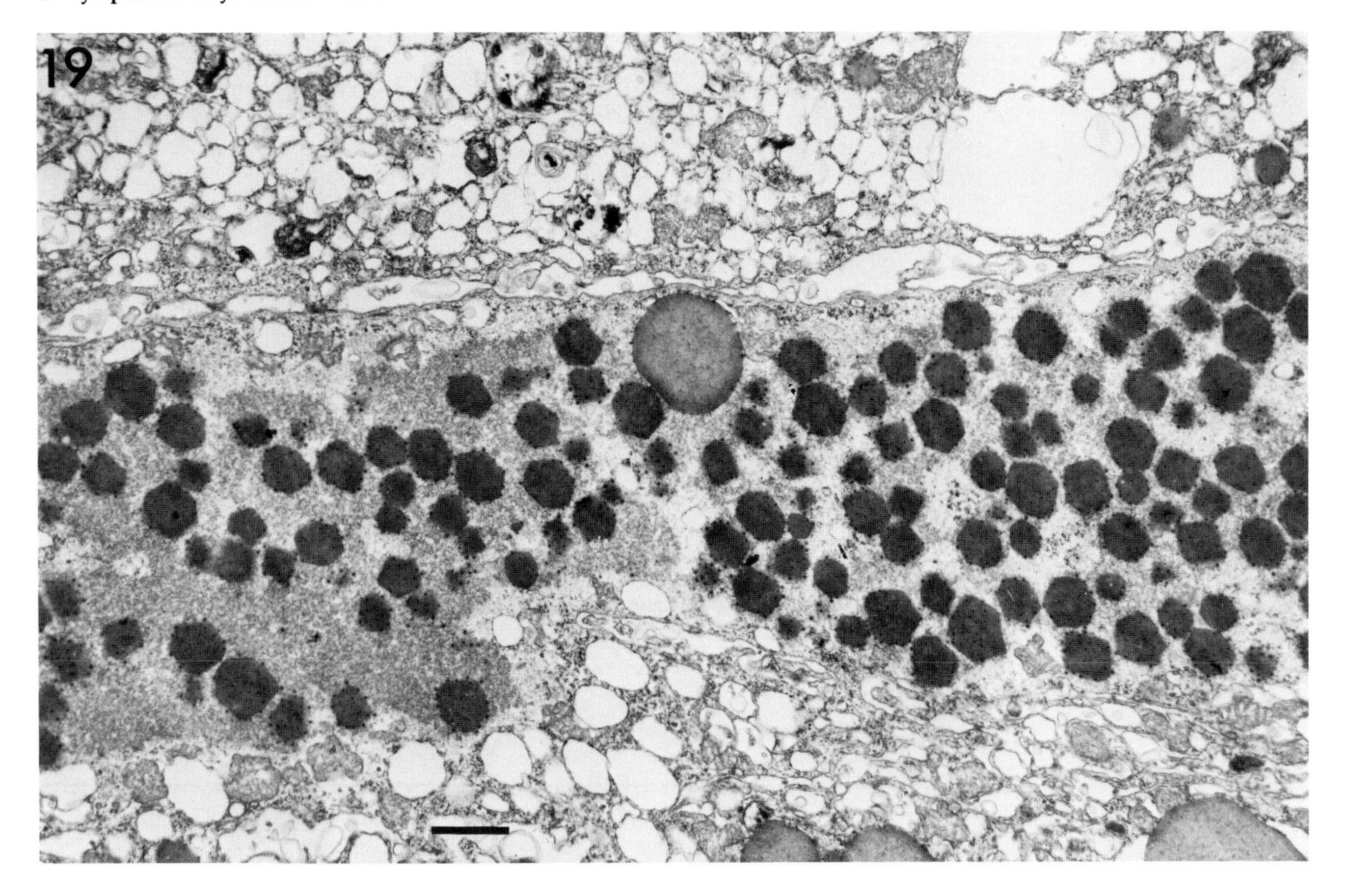
19

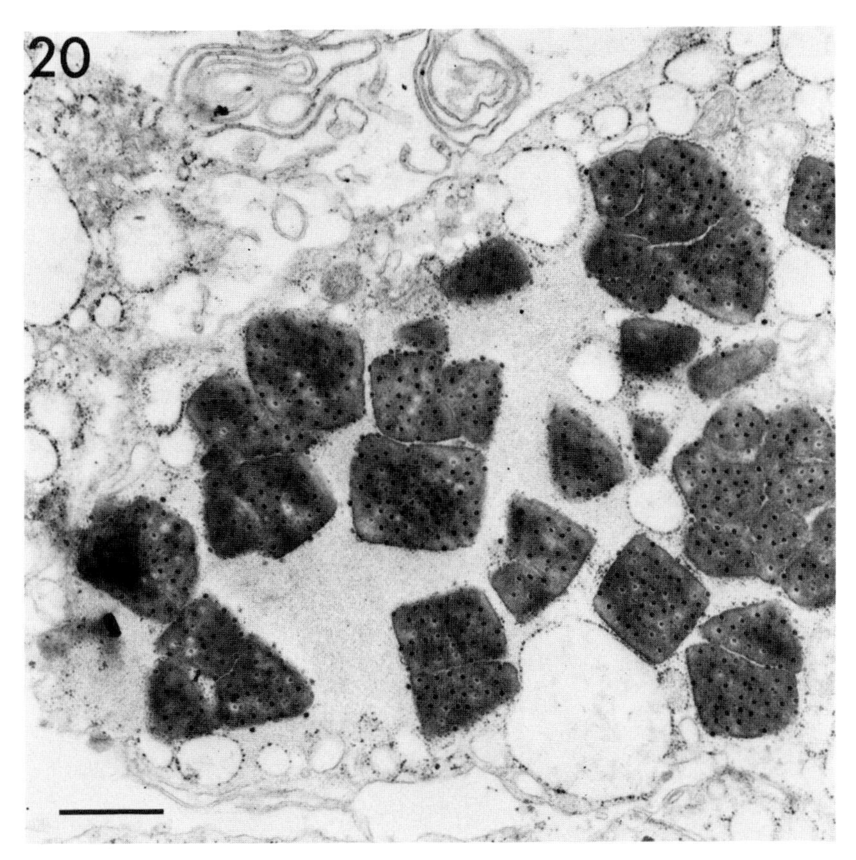
20

Figs. 21 and 22. Cytoplasmic polyhedrosis virus of the crustacean *Simocephalus expinosus*. The morphological features of polyhedra and virus particles of this crustacean virus are similar to those observed in virus isolations from the Insecta. (Photographs courtesy of Dr. B. Federici.)

Fig. 21. Oblong polyhedral inclusion bodies developing within the cytoplasm of gut epithelial tissue. Bar = 200 nm.

Fig. 22. Detail of polyhedron showing the lattice structure within the polyhedron protein, and virus particles composed of electron dense cores and a less dense outer shell. Bar = 200 nm.

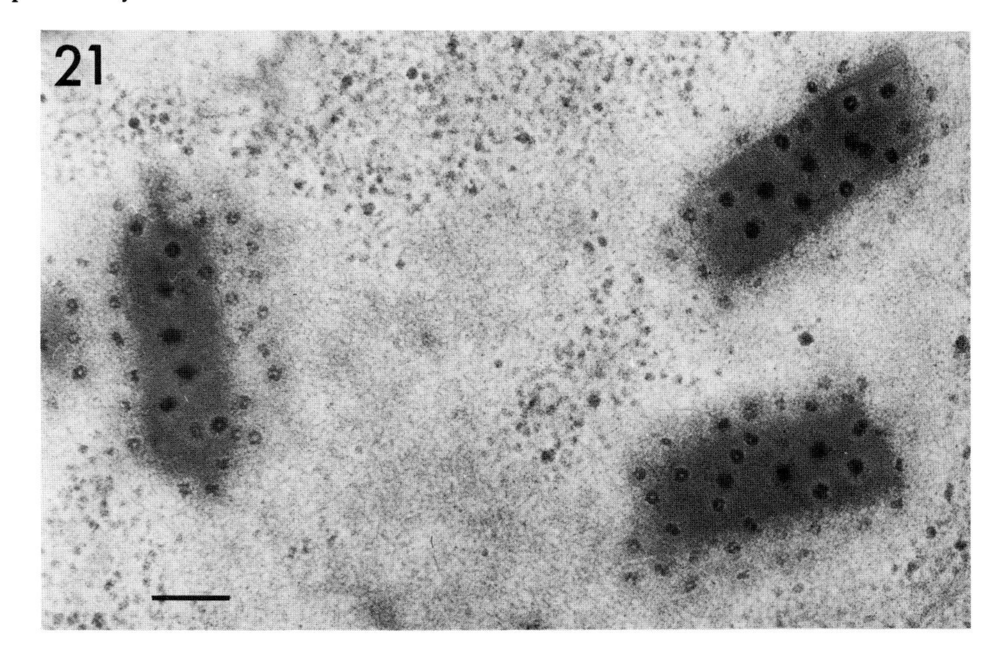
21

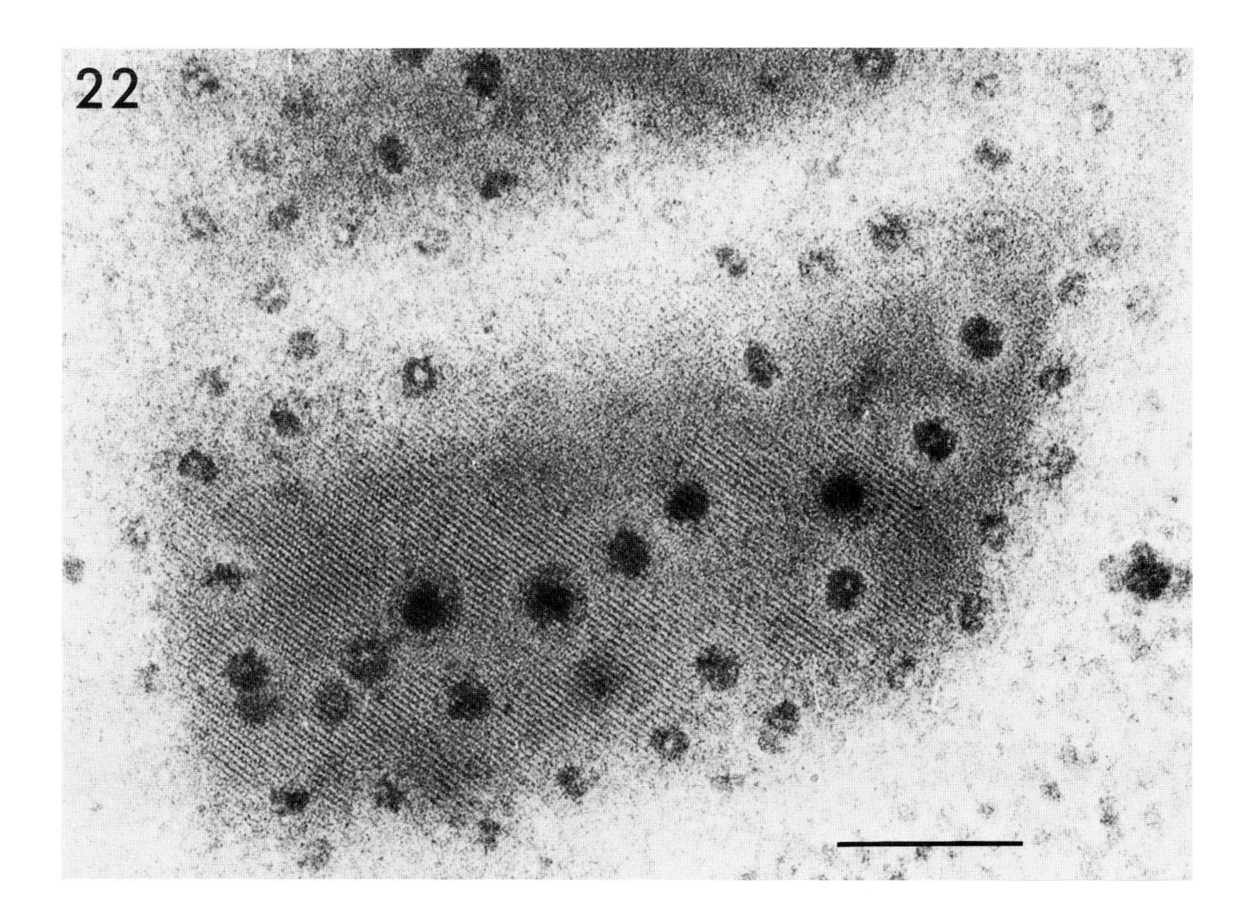
22

Chapter 6

Rhabdoviruses of Insects (Sigma Virus of *Drosophila*)

EDWARD S. SYLVESTER

I. Introduction 131
II. Sigma Virus of Drosophila 131
A. Morphology 132
B. Morphogenesis 133
References 134

I. INTRODUCTION

Sigma virus of *Drosophila* is included in the rhabdovirus group by the International Committee on Nomenclature of Viruses (Wildy, 1971). The prototype of the group is vesicular stomatitis virus (VSV). Characteristically the virions of the rhabdoviruses are large bacilliform or bullet-shaped particles, consisting of a low pitch internal helical structure (appearing as a series of transverse striations in some negatively stained preparations) enclosed by a membranous envelope whose outer surface is studded with numerous projections. The lipoprotein nature of the outer envelope accounts for the ether sensitivity of the rhabdoviruses, and most models interpret the internal component as one molecule of single-stranded RNA associated with protein. The nature of the nucleic acid of sigma virus is unknown, but recent evidence (Herforth, 1973) suggests it to be RNA. To date, among the rhabdoviruses, no cases of infectious RNA have been reported, presumably because unless it is accompanied by an RNA-dependent RNA polymerase (one of the proteins of the virion) the genetic information of the viral genome cannot be transcribed into messenger RNA.

II. SIGMA VIRUS OF DROSOPHILA

L'Héritier and Teissier (1937) first described the fatal paralysis of *Drosophila melanogaster* following exposure of the flies to CO_2. The factor responsible for this sensitivity to CO_2 later was found to be an inheritable infectious agent that could be transmitted by injection (L'Héritier and Hugon de Scoeux, 1947). A virus, designated by the Greek letter sigma (σ), was presumed to be the etiological agent, and in 1965, Berkaloff *et al.* published electron micrographs suggesting that sigma virus particles were bullet-shaped with a morphology strikingly similar to that of VSV.*

Sensitivity to CO_2 has been detected in wild populations of *D. melanogaster* from various regions of the world (Kalmus *et al.*, 1954). An infectious CO_2 sensitivity also has been found in two other species of *Drosophila* (Williamson, 1961).

Both male and female gametes can carry sigma virus, and artificially the virus can be moved by injection or by transplantation of infected organs. Two types of infected flies have been designated: *stabilized* and *nonstabilized*. Since the phenotypic expression of the infection in both situations is sensitivity to CO_2, progeny tests are needed to determine the status of any individual.

In the stabilized state, all offspring from infected females usually are both infected and stabilized. An exception is the occasional noninfected fly resulting from an accidental loss of virus from germ cells. Some spermatozoa from stabilized males may infect an egg at the time of fertilization, but genetic continuity of the stabilized condition is never maintained by paternal inheritance.

The nonstabilized state results if flies are infected by artificial means, i.e., injection or transplantation of infected tissue or if a virus-free female is mated with a stabilized male. Nonstabilized males will not transmit the virus, but nonstabilized females may transmit virus to some of their progeny. In such cases, infection occurs at the beginning of the formation of the ovarian cysts, thus oogonia usually are not infected and the progeny are nonstabilized. Occasionally, however, an infected ovarian cyst in a nonstabilized female produces a large quantity of virus and the oogonium is infected and a stabilized offspring is produced. Such progeny are called *neostabilized* and the event is termed "germinal passage" (Ohanessian-Guillemain, 1959). Only virus strains carrying the *g*+ gene can induce the neostabilized condition, and germinal passage can be prevented by heat treatment.

Strains of inheritable sigma virus (ultra rho) exist that yield no extractable infectious material, and infected flies are not CO_2 sensitive. However, flies infected with such latent virus are immune to superinfection by other strains of sigma virus. The most general hypothesis used to explain the unusual facts of sigma-virus infection is to assume that the female germ cells carry a noninfectious viral genome (nucleic acid or incomplete particle) while invaded somatic tissues produce infectious virions. Presumably, however, the male germ cells carry infectious virus.

A. Morphology

Berkaloff *et al.* (1965) and Teninges (1968) published electron micrographs of presumed sigma virions in ovarian cysts of stabilized female *D. melanogaster*, and in testicular tissue of stabilized males, respectively.

The particles were bullet-shaped (Fig. 3), approximately 70 nm in diameter, and with a usual length varying from 140 to 180 nm. An occasional particle with a length of 210 nm was found. Negatively stained preparations revealed the typical cross striations (internal helix) of a rhabdovirus surrounded by an outer envelope.

Particles in cross section consisted of two electron opaque rings surrounding an electron lucent core. The inner ring was the thicker, and it had a

* Strains of VSV have been found that, when injected, will multiply in *D. melanogaster* and will induce CO_2 sensitivity. Such virus, however, is not inherited and is serologically distinct from sigma virus (Printz, 1973).

greater electron opaqueness than did the outer ring. There was evidence of projections on the outer envelope.

Purification of sigma virus has not been accomplished. When this is done, a greater detail of the morphological characteristics will become available.

B. Morphogenesis

The assay of infectious material uses the dilution end point and an assumed Poisson distribution to define an infectious unit (IU) as that dose which produces sensitivity in approximately 63% of the flies inoculated (Plus, 1954). The incubation period (the time between inoculation and the development of CO_2 sensitivity) is proportional to the logarithm of the number of IU injected. Titration curves are developed for experimental work using flies of a standard genotype.

The growth curve of sigma virus, and the titer reached, (other factors such as temperature and the strain of the virus and of the fly used being equal) depends upon the mode of transmission. Virus inherited via infected oogonia increases exponentially from a base of less than 10^2 IU to a peak of about 10^4 IU in 9 days (third instar larvae). This is followed by a decrease to about 10^2 IU in early pupae and another increase in late pupae and early imagos to a final plateau of about 10^4 IU/fly. The plateau is thought to be due to a dynamic equilibrium between virus formation and destruction, and the titer of virus attained at the plateau tends to remain constant for the remainder of the adult's life.

If sigma virus is transferred via sperm, very few IU can be detected in embryos. However, in comparison to the oogonial passage, the exponential rise is more rapid with a higher plateau being reached in the early pupal stage of approximately 5×10^4 IU/fly.

When the virus is injected at 25° and with a multiplicity of infection rate of 10^3 IU/fly, a plateau of something in excess of 10^5 IU/fly can be reached in 6 to 7 days after inoculation (Printz, 1973). In all cases, sensitivity to CO_2 and maximum titer occur at about the same time (Seecof, 1968). The slower rate of multiplication and lower virus yield associated with oogonial infection as compared to the yield obtained by injection or by paternal transmission has not been explained. A common suggestion is that there is an inhibitory humoral factor involved in maternal germ line infection.

Knowledge of particle morphogenesis of sigma virus is limited. Study of the virus in cell culture is a relatively recent innovation (Ohanessian and Echalier, 1967; Richard-Moland, 1975), and only germinal tissues from stabilized flies have yielded sufficient particles for electron microscopy. The preference of the virus for reproductive tissue is illustrated by the fact that up to 50% of the infectious material in a female fly can reside in ovarian cysts (Bregliano, 1965; L'Héritier, 1970).

An electron micrograph of Berkaloff *et al.* (1965) showed a small collection of bullet-shaped particles in a space between follicular cells of an ovarian cyst. It was suggested that the particles matured, as do other rhabdoviruses, by budding from cellular membranes (Howatson and Whitmore, 1962). Additional details of the morphogenesis of sigma virions were provided by Teninges (1968, 1972a). She found virions in germinal cells of young spermatids (immature sperm cells) (Figs. 1 and 2) but not in spermatozoa. Dense, homogeneous areas, lacking the ribosomal content of surrounding cytoplasm, were found in some of the spermatids, and such areas were interpreted to be viroplasms. Micrographs (Fig. 3) showing the continuity between cellular membranes and virus particles provided strong evidence of the budding mode of assembly that is used for particle maturation and release. Virus particles were not found in the surrounding somatic tissues of the testicles. The number of IU in extracts of testes has been found to be independent of the presence of virus particles. This has led to the suggestion that the bullet-shaped virions found budding off plasmic membranes of young spermatids may have little, if anything, to do with the infectious form of sigma virus and indeed may be noninfectious (Teninges, 1972b).

REFERENCES

Berkaloff, A., Bregliano, J. C., and Ohanessian, A. (1965). *C. R. Hebd. Seances Acad. Sci.* **260,** 5956.
Bregliano, J. C. (1965). *Ann. Inst. Pasteur, Paris* **109,** 638.
Herforth, R. S. (1973). *Virology* **51,** 47.
Howatson, A. F., and Whitmore, G. F. (1962). *Virology* **16,** 466.
Kalmus, H., Kerridge, J., and Tattesfield, F. (1954). *Nature (London)* **173,** 1101
L'Héritier, P. (1970). *Evol. Biol.* **4,** 185.
L'Héritier, P., and Hugon du Scoeux, F. (1947). *Bull. Biol. Fr. Belg.* **81,** 70.
L'Héritier, P., and Teissier, G. (1937). *C. R. Hebd. Seances Acad. Sci.* **205,** 1099.
Ohanessian, A., and Echalier, G. (1967). *Nature (London)* **213,** 1049.
Ohanessian-Guilleman, A. (1959). *Ann. Genet.* **1,** 59.
Plus, N. (1954). *Bull. Biol. Fr. Belg.* **88,** 249.
Printz, D. (1973). *Adv. Virus Res.* **18,** 143.
Richard-Moland, C. (1975). *Arch. Virol.* **43,** 139.
Seecof, F. (1968). *Curr. Top. Microbiol. Immunol.* **42,** 59.
Teninges, D. (1968). *Arch. Gesamte Virusforsch.* **23,** 378.
Teninges, D. (1972a). *Ann. Inst. Pasteur, Paris* **122,** 541.
Teninges, D. (1972b). *Ann. Inst. Pasteur, Paris* **122,** 1183.
Wildy, P. (1971). *Monogr. Virol.* **5,** 51.
Williamson, D. (1961). *Genetics* **46,** 1053.

Fig. 1. Section through a testicular "cyst" of a 13-day-old *Drosophila melanogaster* pupa at the level of the zone of spermatid elongation. Throughout the section (marked by rectangles), sigma virions can be seen to be budding from plasma membranes of the spermatids into cisternae formed among the individual spermatic cells. A, Axonema; M, mitochondria. Micrograph and interpretation courtesy of D. Teninges (1972a), with permission of the Annales de L'Institut Pasteur, Paris. (×16,000.)

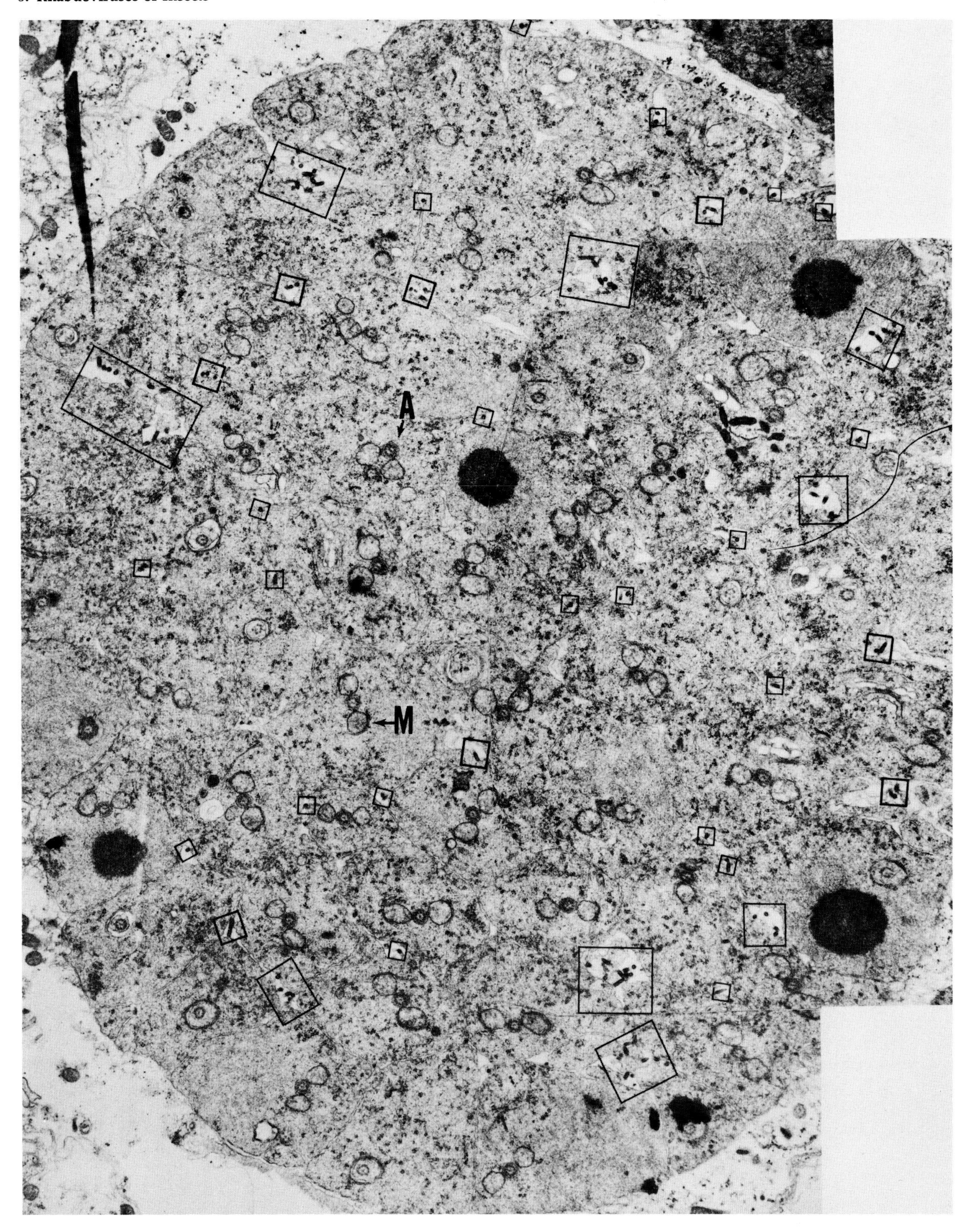
A
M

Fig. 2. Section through a testicular "cyst" of a 13-day-old *Drosophila melanogaster* pupa at the level of the "tail" part of the spermatids in a late stage of their development. Paracrystalline material has been deposited in the paired mitochondria (M), which have become unequal in size. Again, sigma virions (arrows) can be seen associated with, and budding from, plasma membranes of the spermatids. A, Axonema. Micrograph and interpretation courtesy of D. Teninges (1972a), with permission of the Annales de L'Institut Pasteur, Paris. (×30,000.)

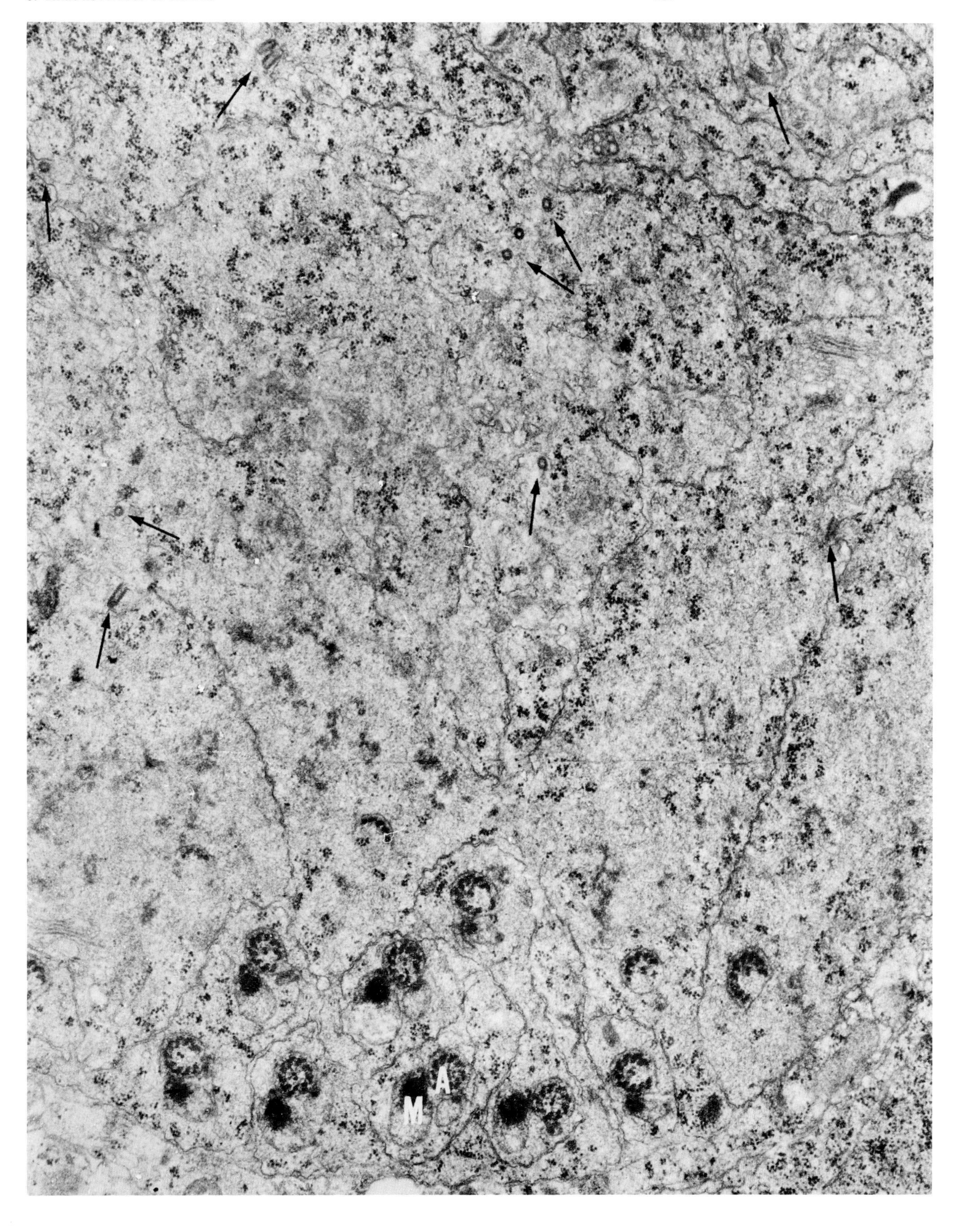
A
M

Fig. 3. An enlargement of one of the areas outlined in Fig. 1, showing sigma virions budding from the plasma membrane of *Drosophila melanogaster* spermatids. The arrows indicate filaments that may be virus nucleocapsids. a, Axonema; m, mitochondria. Micrograph and interpretation courtesy of Teninges (1972a), with permission of the Annales de L'Institut Pasteur, Paris. (×82,000.)

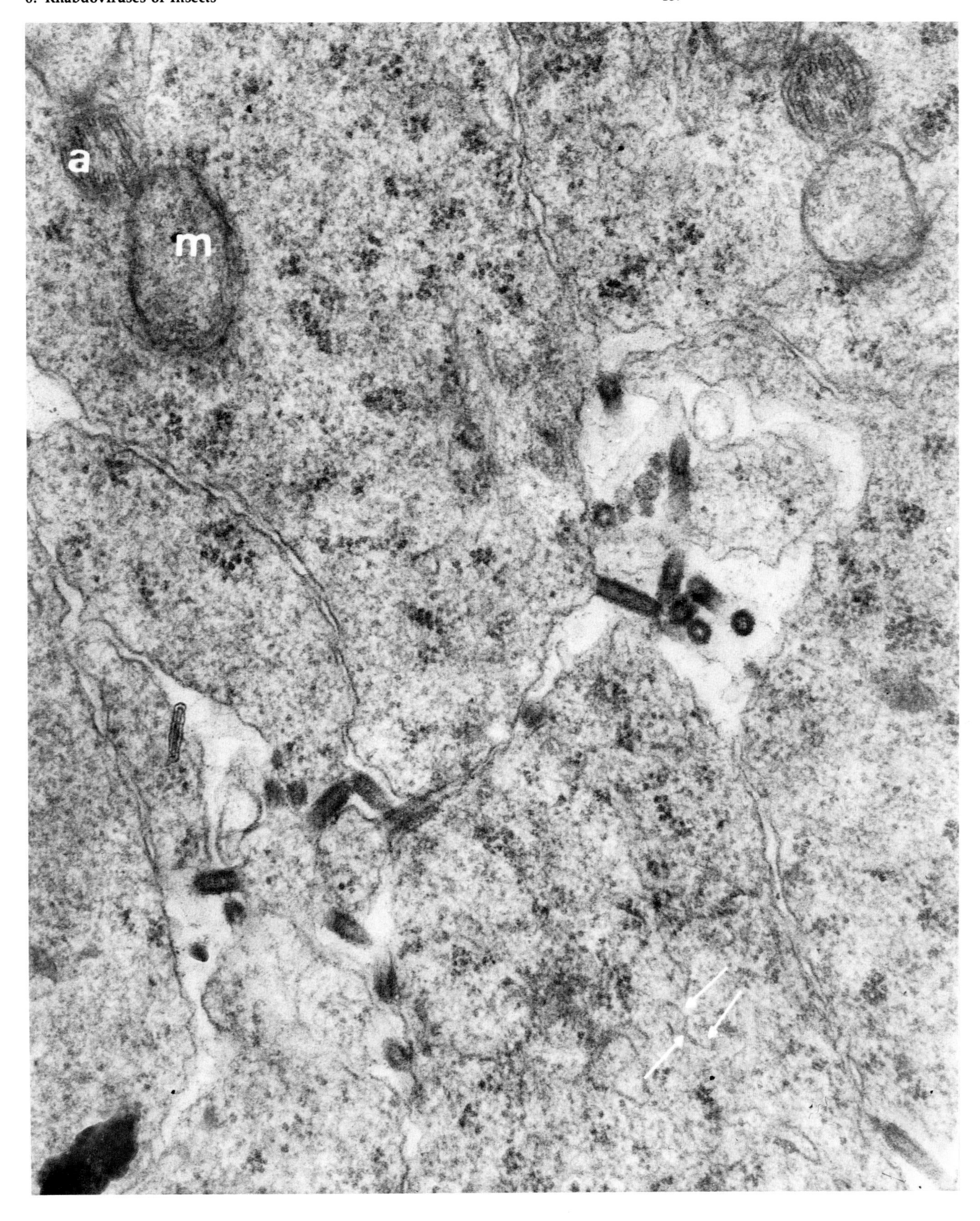
a
m

Chapter 7

Bee Viruses

L. BAILEY AND R. D. WOODS

I. Chronic Bee-Paralysis Virus 141
II. Chronic Bee-Paralysis Virus Associate 142
III. Sacbrood Virus 142
IV. Acute Bee-Paralysis Virus 143
V. Arkansas Bee Virus 143
VI. Bee Virus X 143
VII. Slow Bee-Paralysis Virus 144
VIII. Black Queen-Cell Virus 144
IX. Kashmir Bee Virus 144
X. *Apis* Iridescent Virus (L. Bailey, Brenda V. Ball, and R. D. Woods) 144
References 145

I. CHRONIC BEE-PARALYSIS VIRUS

Burnside (1945) reproduced the symptoms of "paralysis" of adult honeybees by spraying them with bacteria-free filtrates of extracts from paralytic bees from the field. Chronic bee-paralysis virus was shown to be the etiological agent by Bailey *et al.* (1963). The virus occurs in honeybees throughout the world and causes a variety of rather similar syndromes (Bailey, 1965, 1971). There is no other known host.

Chronic bee-paralysis virus has naked anisometric particles, mostly ellipsoidal in outline, and often with a small irregular proturberance at one end; otherwise they are featureless (Fig. 1a). They are serologically unrelated to any other known virus from bees. All have a modal width of about 22 nm but vary considerably in length. They resolve into three or four components during centrifugation, with modal lengths of about 30, 40, 55, and 65 nm and with sedimentation constants ($s_{20,w}$) of about 82, 97–106, 110–124, and 125–136 S (Bailey *et al.*, 1963). However, the particles, which are serologically indistinguishable, all migrate electrophoretically as a single component, all have the same buoyant density (1.33 gm/cm^3) in cesium chloride (Bailey, 1976), and all contain a single protein of 23,500 daltons (L. Bailey and J. M. Carpenter, unpublished data). The base ratio (A:C:G:U) of the nucleic acid is 24:28:20:28 (Bailey *et al.*, 1968).

Many empty particles, penetrated by negative stains, sometimes appear (Fig. 1b), and bizarre forms include rings, figures-of-8, stalked rings, branching rods, and lengths up to 640 nm. Partial flattening, causing central concavities, may well account for the apparent ring-shaped particles that are sometimes seen (Fig. 1c). The capsids are, nevertheless, very resistant, forming shells about 25 × 35 nm, with walls about 3–4 nm thick and containing no nucleic acid after the particles have been kept for many hours in cold 1 *N* HCl or cold KOH (Bailey *et al.*, 1968). The virus is insensitive to ether.

Chronic bee-paralysis virus multiplies most abundantly in honeybees kept in laboratory conditions at 30°; however, injected bees die the fastest, within a few days, when kept at 35° (Bailey and Milne, 1969). Between 10^{10} and 10^{11} particles can be extracted from a bee killed by the virus and about half this number are in the head (Bailey *et al.*, 1968). Numerous particles, morphologically similar to those of chronic bee-paralysis virus, can be seen in bees suffering from chronic paralysis. These particles appear in densely but randomly packed groups, either free or within vesicles, in the cytoplasm of thoracic and abdominal ganglia, gut, and mandibular and hypopharyngeal glands (Giauffret *et al.*, 1966, 1970) (Fig. 1d). However, they do not appear in the cytoplasm of fat or muscle tissue (Lee and Furgala, 1965).

II. CHRONIC BEE-PARALYSIS VIRUS ASSOCIATE

Small particles, 17 nm across (Fig. 2), are consistently associated with chronic bee-paralysis virus (Bailey, 1975). They have an absorption spectrum typical of a nucleoprotein, contain RNA, and have a sedimentation constant of 46 S. They are serologically unrelated to any known bee virus and do not multiply except in association with chronic bee-paralysis virus.

III. SACBROOD VIRUS

Sacbrood, a disease of honeybee larvae, was first described by White (1913) and was diagnosed by him as a virus disease (White, 1917). Sacbrood virus was identified by Bailey *et al.* (1964) and has naked, featureless, isometric particles 30 nm in diameter (Fig. 3). Similar particles from larvae, apparently with sacbrood, were described by Brčák and Kralik (1963). They have a sedimentation constant of 160 S, a buoyant density in CsCl of 1.33 gm/cm^3 (pH 7.0–9.0), are unstable below pH 4.0, and the base ratio (A : C : G : U) of their single-stranded nucleic acid is 32.1 : 17.9 : 19.1 : 30.9 (Newman *et al.*, 1973). They contain three major proteins of 25,000, 28,000, and 31,500 daltons (L. Bailey and J. M. Carpenter, unpublished data) and are serologically unrelated to any other known virus from bees. Some particles appear to be "empty" when negatively stained with neutral sodium phosphotungstate and examined in the electron microscope. These empty particles are fragile and are best seen by fixing virus preparations in 1% formaldehyde for 15 minutes before staining (Bailey *et al.*, 1964). Sacbrood virus is insensitive to ether. It has no known host other than honeybees.

Larvae killed by sacbrood virus each contain about 10^{13} particles, many of which occur in fluid between the "sac" (the final larval skin) and the body of the prepupa. Many particles, sometimes in crystalline array, have been seen in the cytoplasm of fat, muscle, and tracheal-end cells of larvae infected by feeding under laboratory conditions (Lee and Furgala, 1967a). Similar particles have also been seen in the cytoplasm of the fat bodies of apparently healthy adult bees that have been injected with the virus (Lee and Furgala, 1967b), and infection experiments have established that sacbrood virus does multiply in adult bees without causing obvious symptoms (Bailey, 1969), accumulating especially in the hypopharyngeal glands of worker bees and in the brains of drone (male) bees (Bailey and Fernando, 1972).

IV. ACUTE BEE-PARALYSIS VIRUS

Acute bee-paralysis virus commonly occurs in Britain as an inapparent infection of honeybees (Bailey *et al.*, 1963) and of bumblebees (Bailey and Gibbs, 1964). It has also been reported in Australia (Reinganum, 1969) and France (Giauffret *et al.*, 1969).

Acute bee-paralysis virus has naked, featureless particles 30 nm in diameter, some appearing empty when negatively stained in neutral sodium phosphotungstate (Fig. 4). They sediment in two fractions, corresponding to full and empty particles, with sedimentation constants of 160 and 80 S. The full particles are stable below pH 4.0, have a buoyant density in CsCl of 1.34, 1.36, and 1.42 gm/cm^3 at pH 7.0, 8.0, and 9.0, respectively, and their nucleic acid has a base ratio (A:C:G:U) of 30.3:20.5:18.8:30.4 (Newman *et al.*, 1973). They contain two major proteins of 23,500 and 31,500 daltons (L. Bailey and J. M. Carpenter, unpublished data), and are serologically unrelated to any other known virus from honeybees. The virus is insensitive to ether. Cytoplasmic crystalline arrays of particles have been seen in fat-body cells (Furgala and Lee, 1966) and brain cells (Bailey and Milne, 1969) of infected adult bees. Most virus accumulates in infected bees kept at 35°; however, infected bees die faster within a few days, when kept at 30°.

V. ARKANSAS BEE VIRUS

Arkansas bee virus was found as an inapparent infection of adult honeybees in Arkansas and substantially shortens the lives of honeybees when injected into them (Bailey and Woods, 1974). It also occurs in honeybees in Britain (L. Bailey, unpublished data).

Arkansas bee virus has naked isometric particles, 30 nm in diameter (Fig. 5). These contain RNA, have a sedimentation constant of 128 S, a buoyant density in CsCl of 1.37 gm/cm^3, and sometimes form dimers sedimenting at 170 S. They contain one major protein of 41,000 daltons (L. Bailey and J. C. Carpenter, unpublished data). They are serologically unrelated to any other virus from honeybees and are futher distinguished from these by their failure to react in immunodiffusion tests in agar containing 0.05 *M* ethylenediaminetetraacetic acid (EDTA).

VI. BEE VIRUS X

Bee Virus X commonly occurs in honeybees in Britain and is often abundant in adult bees found dead at the end of winter (Bailey and Woods, 1974; Bailey, 1975). Similar particles have been seen in extracts of honeybees in Australia (Reinganum, 1969), France (Giauffret *et al.*, 1969), and the United States (Kulinčevíc *et al.*, 1970).

Bee virus X has naked, featureless isometric particles, 35 nm across (Fig. 6). These contain RNA, have a sedimentation constant of 187 S, a buoyant density in CsCl of 1.37 gm/cm^3, contain one major protein of 54,500 daltons (L. Bailey and J. M. Carpenter, unpublished data), and are serologically unrelated to any other known virus from honey bees.

Bee virus X occurs in the abdomen of adult bees, mostly if not entirely in the gut. It does not multiply when injected into bees but multiplies very much when fed to young adult bees that are subsequently incubated at 30° for 3 to 5 weeks, and then it shortens their lives (L. Bailey, unpublished data). Very little bee virus X multiplies in similarly infected young bees when they are subsequently incubated at 35°.

VII. SLOW BEE-PARALYSIS VIRUS

Slow bee-paralysis virus has been found as an inapparent infection of adult honeybees in Britain and shortens their lives when injected into them (Bailey and Woods, 1974).

Slow bee-paralysis virus has naked, featureless, isometric particles, 30 nm in diameter (Fig. 7). These contain RNA, have a sedimentation constant of 173–178 S, a buoyant density in CsCl of 1.37 gm/cm^3, contain three major proteins of 27,000, 29,000, and 46,000 daltons (L. Bailey and J. M. Carpenter, unpublished data), and are serologically unrelated to any other known virus from honeybees. Purified particles aggregate spontaneously in agar of immunodiffusion plates when these contain 0.85% sodium or potassium chloride, but they aggregate much less readily in 0.01% KCl and are prevented from such aggregation when mixed with an equal volume of normal rabbit serum.

VIII. BLACK QUEEN-CELL VIRUS

Black queen-cell virus has been found in many dead prepupae of queen honeybees in Britain and some serological evidence indicates that it occurs also in the United States (Bailey *et al.*, 1976b).

The virus has isometric featureless particles, 30 nm across (Fig. 8), which contain RNA. They have a sedimentation constant of 153 S, a buoyant density in CsCl of about 1.34 gm/cm^3, one major protein of 30,000 daltons and are serologically unrelated to any other known bee virus. The virus also multiplies in adult honeybees and shortens their lives somewhat.

IX. KASHMIR BEE VIRUS

Kashmir bee virus has been isolated from the Eastern honeybee, *Apis cerana*, from Kashmir (Bailey *et al.*, 1976a).

Kashmir bee virus has featureless isometric particles, 30 nm across (Fig. 9), contains RNA, has a sedimentation constant of 172 S, and a buoyant density in CsCl of 1.37 gm/cm^3. It contains two major proteins of 27,000 and 39,000 daltons (L. Bailey and J. M. Carpenter, unpublished data). It multiplies abundantly when injected into adult European honeybees, *Apis mellifera*, and shortens their lives considerably.

It aggregates spontaneously into crystals in 0.85% sodium chloride.

X. *APIS* IRIDESCENT VIRUS*

Apis iridescent virus was isolated from sick Eastern hive-bees, *Apis cerana*, from Kashmir. It will multiply in the cytoplasm of various tissues of the Euro-

* This section was authored by L. Bailey, Brenda V. Ball, and R. D. Woods.

pean honeybee, *Apis mellifera*, forming particles in crystalline arrays (Fig. 10) visible to the naked eye as a blue iridescence. These particles contain DNA and closely resemble those of Tipula iridescent and similar viruses in physicochemical characters. For example, they have a sedimentation constant of 2216 S, a buoyant density in CsCl of about 1.31 gm/cm^3, and the outer icosahedral shell of the capsid is composed of isometric subunits, 7 nm across, that become detached from disintegrating capsids mostly as trisymmetrons, each composed of 55 subunits. However, *Apis* iridescent virus is only distantly related to the iridescent viruses of other insects (Bailey *et al.*, 1976a).

REFERENCES

Bailey, L. (1965). *J. Invertebr. Pathol.* **7,** 167–169.

Bailey, L. (1969). *Ann. Appl. Biol.* **63,** 483–491.

Bailey, L. (1971). *Rothamsted Exp. Stn. Rep. 1970* Part 2, pp. 171–183.

Bailey, L. (1975). *Rothamsted Exp. Stn. Rep. 1974* Part 1, pp. 115–116.

Bailey, L. (1976). *Adv. Virus Res.* **20,** 271–304.

Bailey, L., and Fernando, E. F. W. (1972). *Ann. Appl. Biol.* **72,** 27–35.

Bailey, L., and Gibbs, A. J. (1964). *J. Insect Pathol.* **6,** 395–407.

Bailey, L., and Milne, R. G. (1969). *J. Gen. Virol.* **4,** 9–14.

Bailey, L., and Woods, R. D. (1974). *J. Gen. Virol.* **25,** 175–186.

Bailey, L., Gibbs, A. J., and Woods, R. D. (1963). *Virology* **21,** 390–395.

Bailey, L., Gibbs, A. J., and Woods, R. D. (1964). *Virology* **23,** 425–429.

Bailey, L., Gibbs, A. J., and Woods, R. D. (1968). *J. Gen. Virol.* **2,** 251–260.

Bailey, L., Ball, B. V., and Woods, R. D. (1976a). *Rothamsted Exp. Stn. Rep. 1975* Part 1, 132–133.

Bailey, L., Ball, B. V., Carpenter, J. M., Simpson, J., and Woods, R. D. (1976b). *Rothamsted Exp. Stn. Rep. 1975* Part 1, 133.

Brčák, J., and Kralik, D. (1963). *J. Insect. Pathol.* **5,** 385–386.

Burnside, C. E. (1945). *Am. Bee J.* **85,** 354–355.

Furgala, B., and Lee, P. E. (1966). *Virology* **29,** 346–348.

Giauffret, A., Duthoit, J. L., and Tostain-Caucat, M. J., (1966). *Bull. Apic. Doc. Sci. Tech. Inf.* **9,** 221–228.

Giauffret, A., Duthoit, J. L., Poutiers, F., and Tostain-Caucat, M. J. (1969). *Bull. Apic. Doc. Sci. Tech. Inf.* **12,** 13–22.

Giauffret, A., Duthoit, J. L., and Tostain-Caucat, M. J. (1970). *Bull. Apic. Doc. Sci. Tech. Inf.* **13,** 115–126.

Kulinčevic, J. M., Stairs, G. R., and Rothenbuhler, W. C. (1970). *J. Invertebr. Pathol.* **16,** 423–426.

Lee, P. E., and Furgala, B. (1965). *J. Invertebr. Pathol.* **7,** 170–174.

Lee, P. E., and Furgala, B. (1967a). *J. Invertebr. Pathol.* **9,** 178–187.

Lee, P. E., and Furgala, B. (1967b). *Virology* **32,** 11–17.

Newman, J. F. E., Brown, F., Bailey, L., and Gibbs, A. J. (1973). *J. Gen. Virol.* **19,** 405–409.

Reinganum, C. (1969). *Victorian Plant Res. Inst., Rep.* No. 5, p. 28.

White, G. F. (1913). *U. S., Dep. Agric., Bur. Entomol., Circ.* **169,** 1–5.

White, G. F. (1917). *U. S., Dep. Agric., Bull.* **431,** 1–55.

Fig. 1. Chronic bee-paralysis virus. (a) and (c) Particles from Britain and (b) from Texas after purification and mounting in neutral sodium phosphotungstate. (d) Ultrathin section of part of the hind-gut tissue of an adult honey bee (*Apis mellifera*) suffering from chronic paralysis and showing characteristic aggregates of cytoplasmic particles. (Courtesy of Dr. A. Giauffret, Laboratorie Régional de Recherches Vétérinaries, Nice, France.)

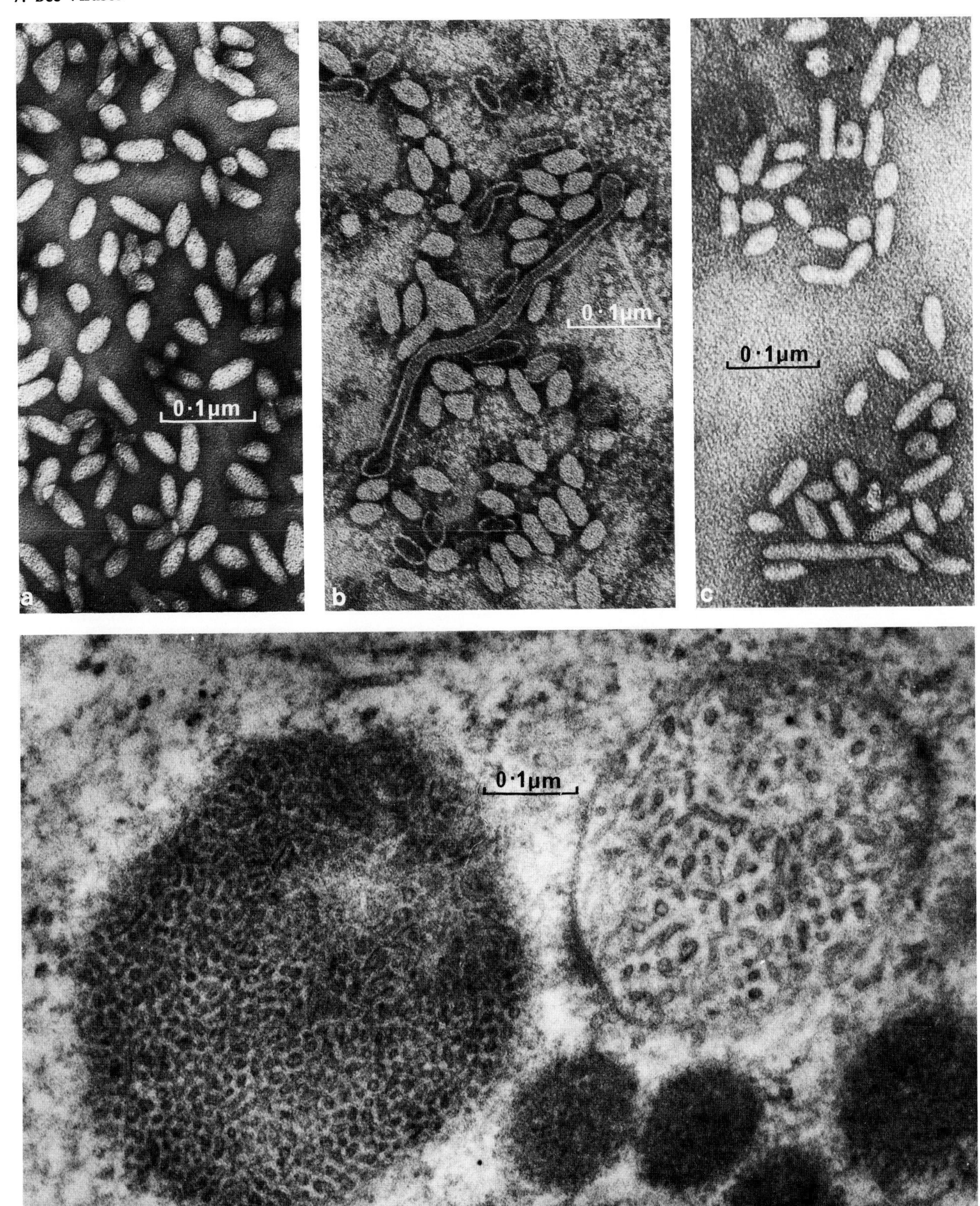
0·1μm
0·1μm
0·1μm
0·1μm
a
b
c
d

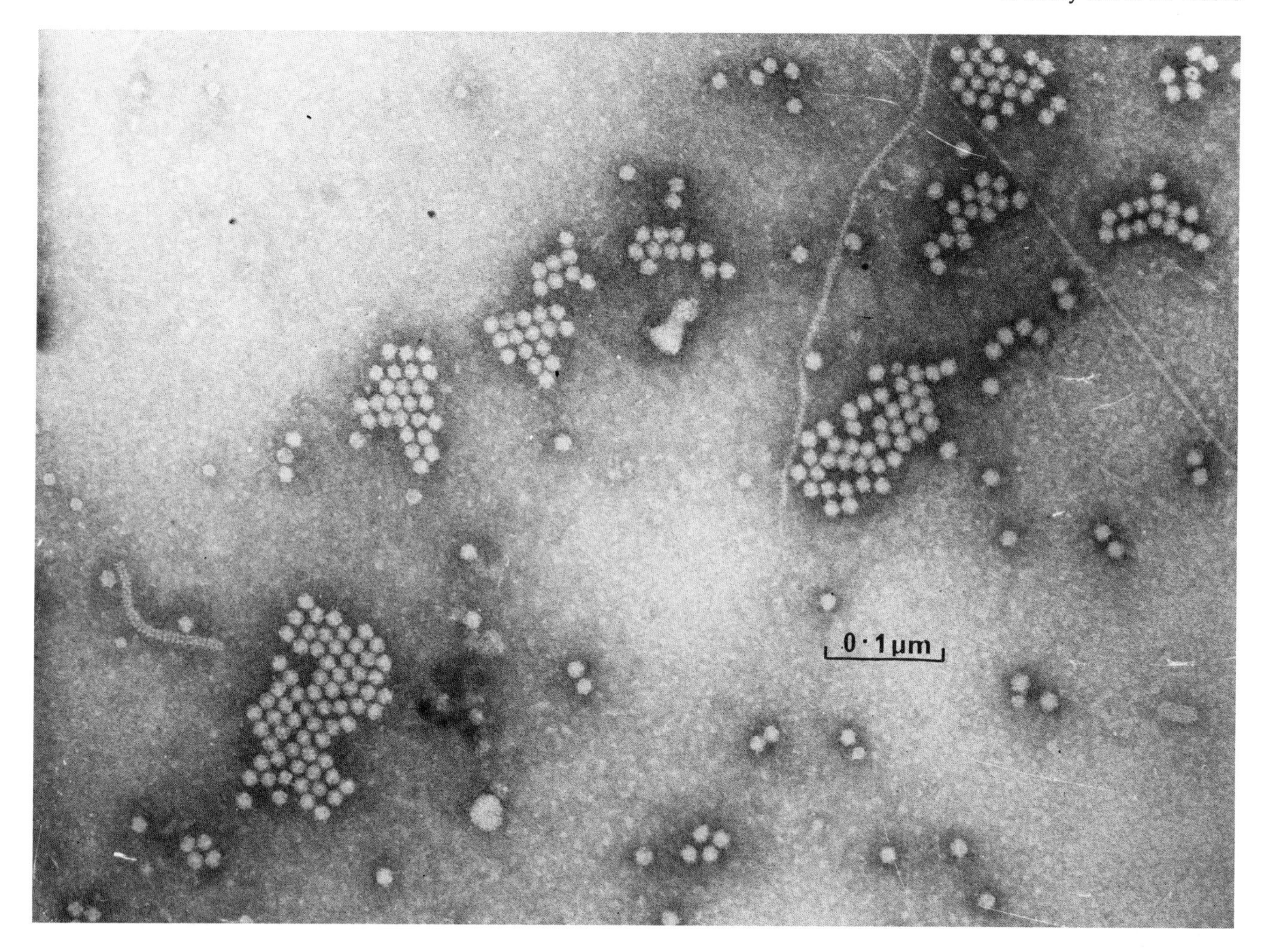

Fig. 2. Chronic bee-paralysis virus associate. Electron micrograph of particles after purification and mounting in neutral sodium phosphotungstate.

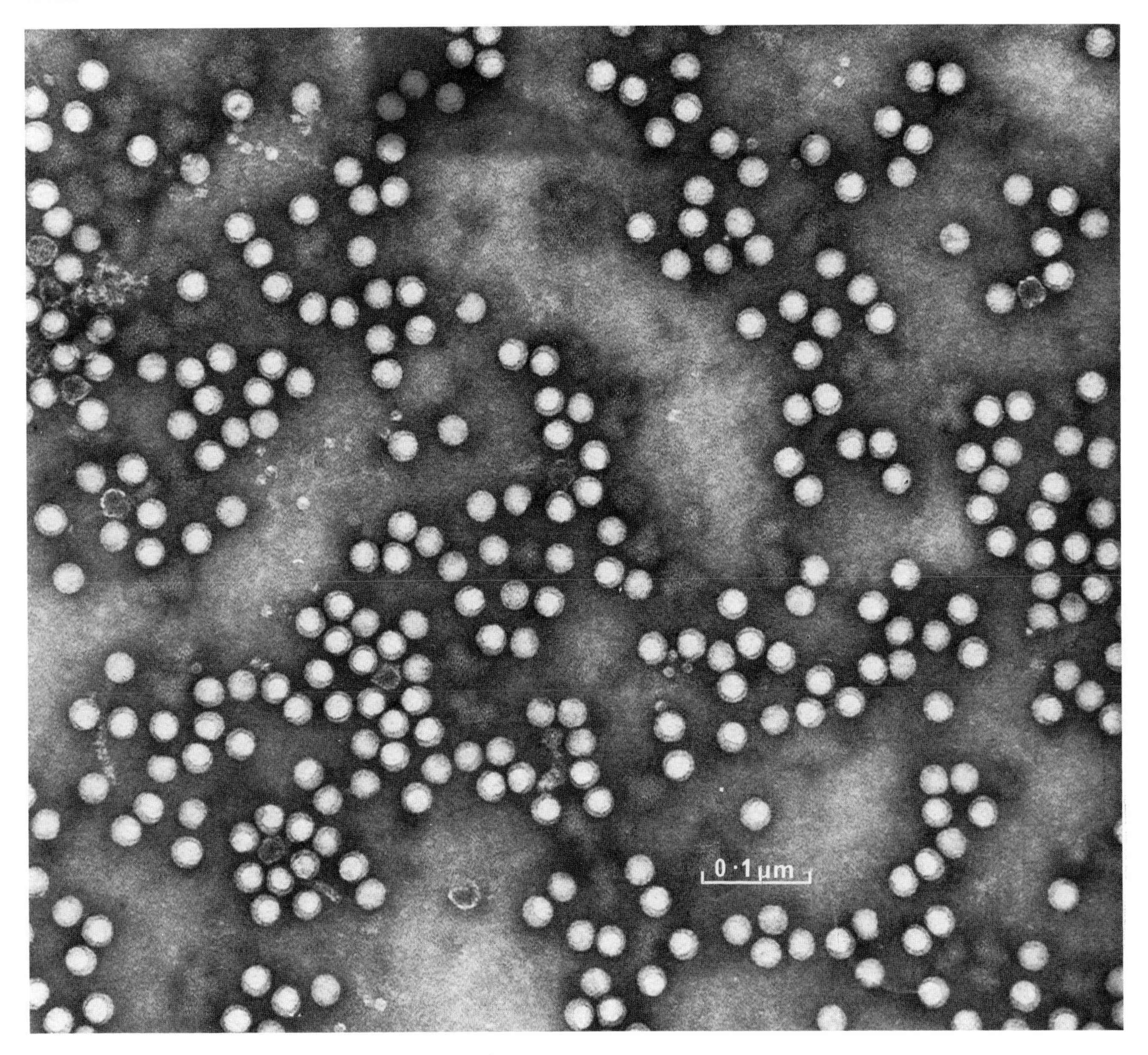

Fig. 3. Sacbrood virus. Electron micrograph of particles after purification and mounting in neutral sodium phosphotungstate.

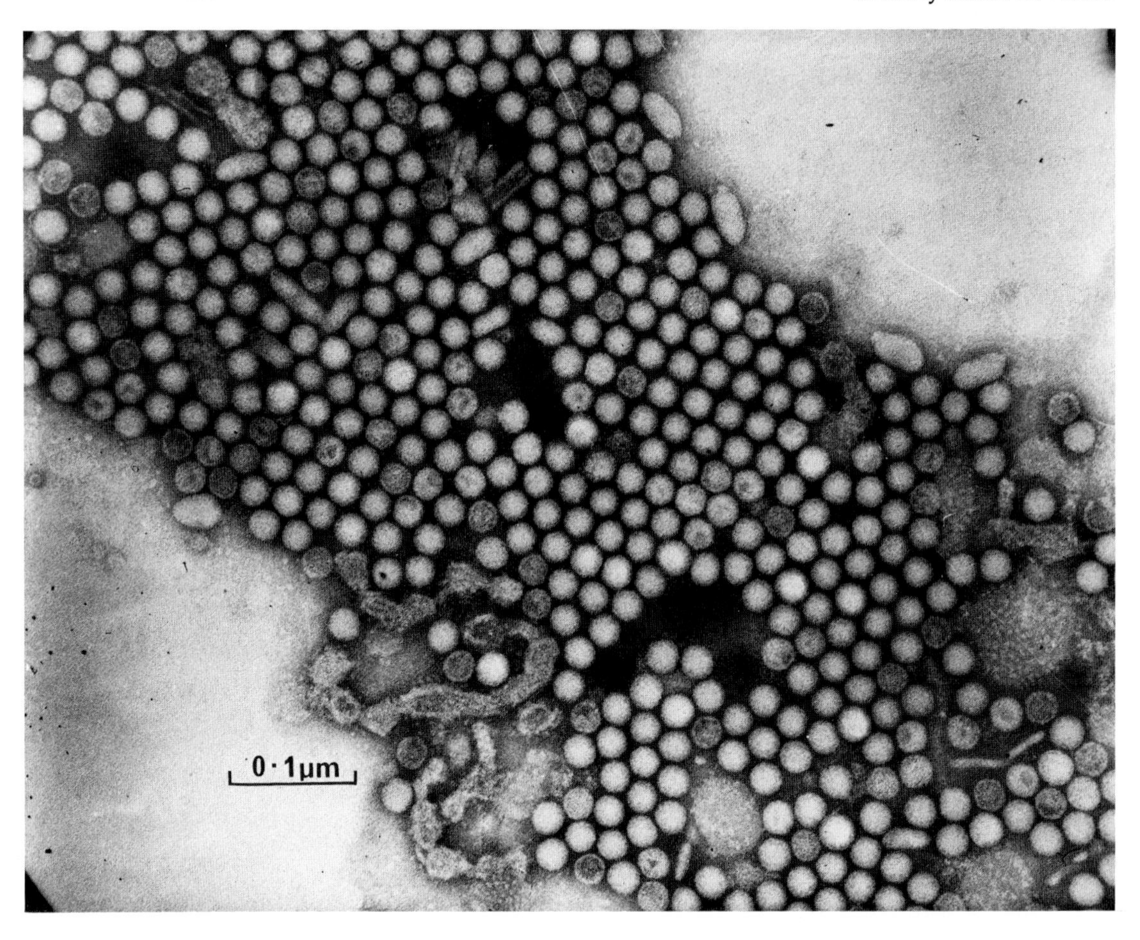

Fig. 4. Acute bee-paralysis virus. Electron micrograph of particles after purification and mounting in neutral sodium phosphotungstate.

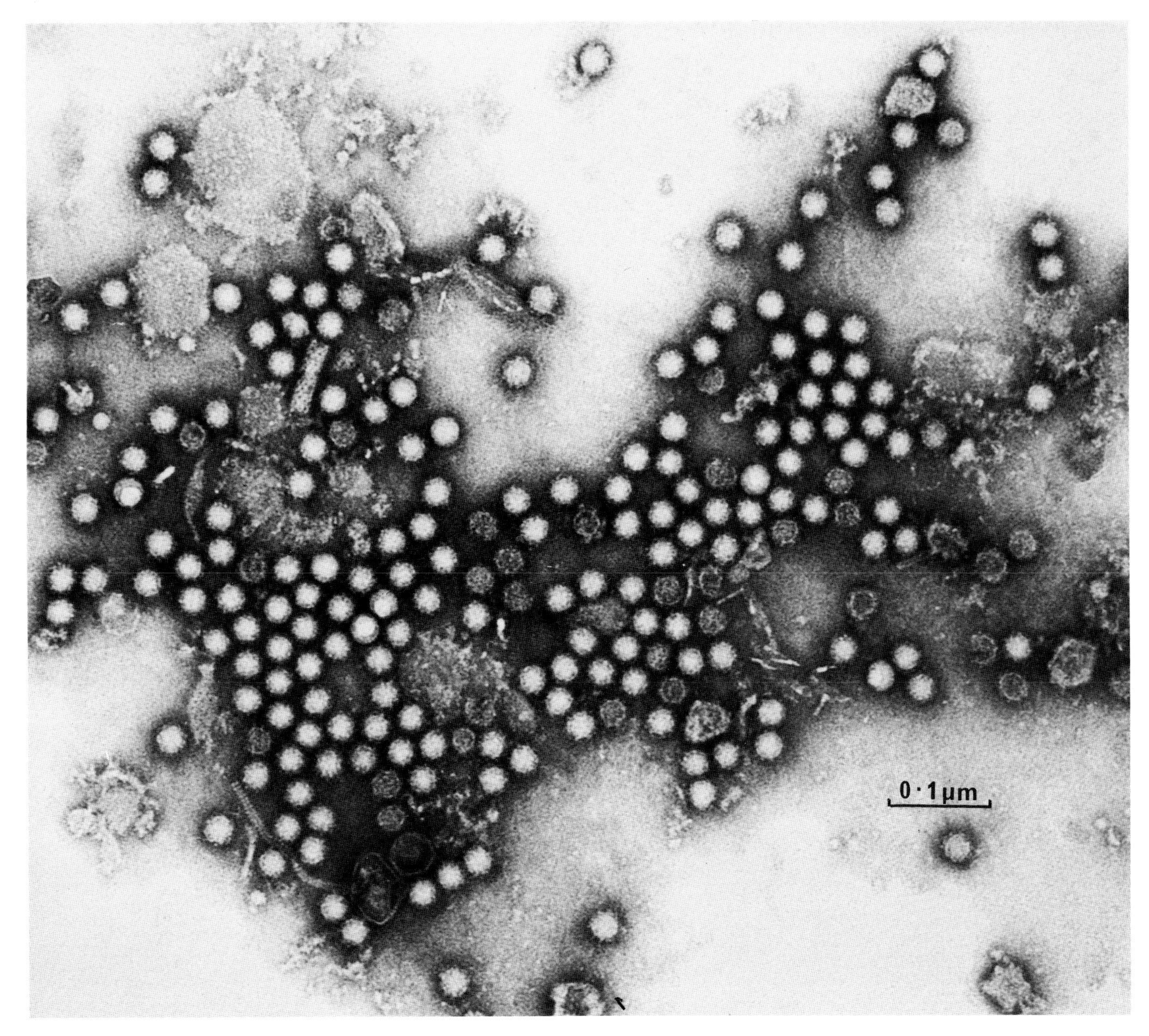

Fig. 5. Arkansas bee virus. Electron micrograph of particles after purification and mounting in neutral sodium phosphotungstate.

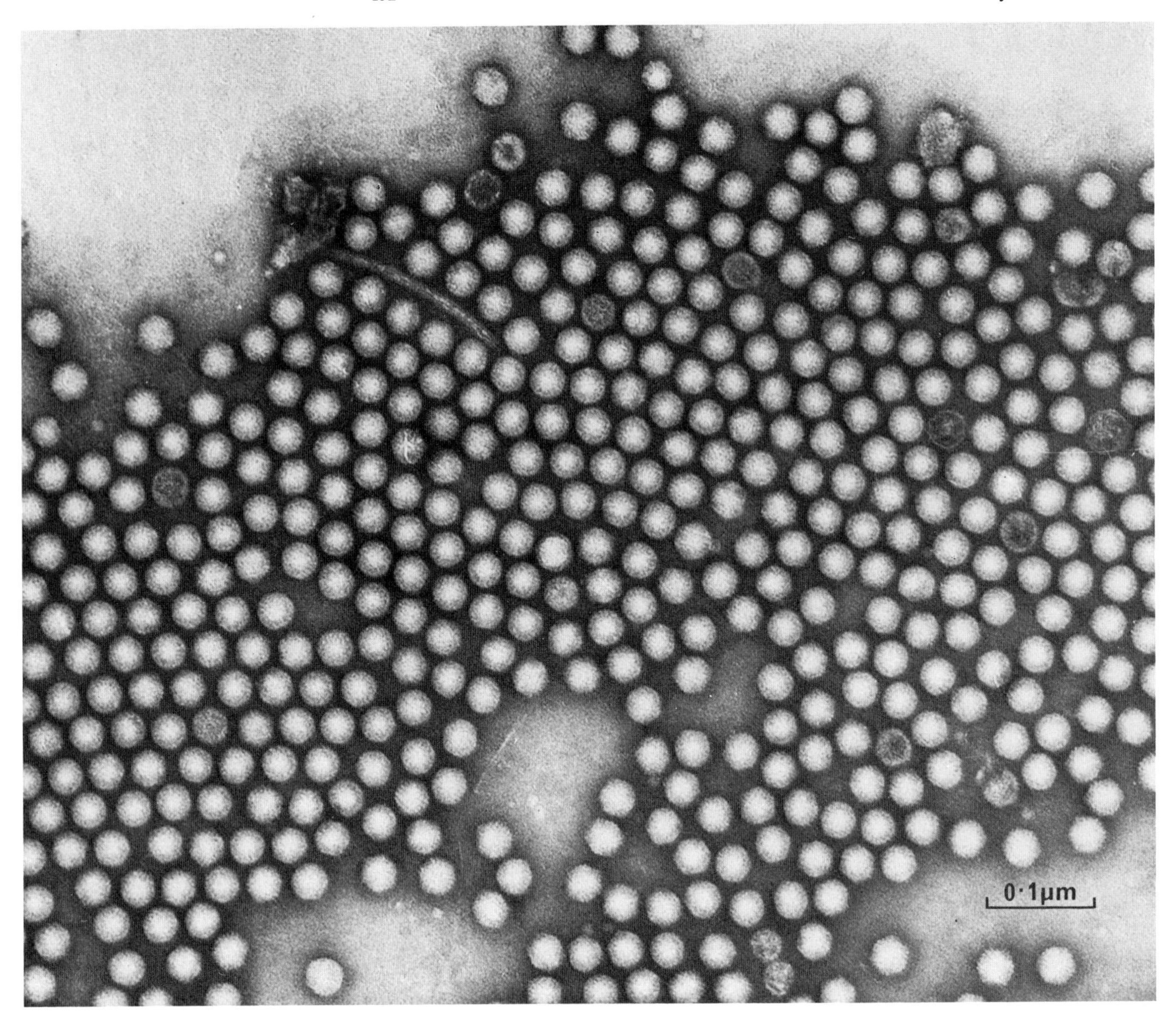

Fig. 6. Bee virus X. Electron micrograph of particles after purification and mounting in neutral sodium phosphotungstate.

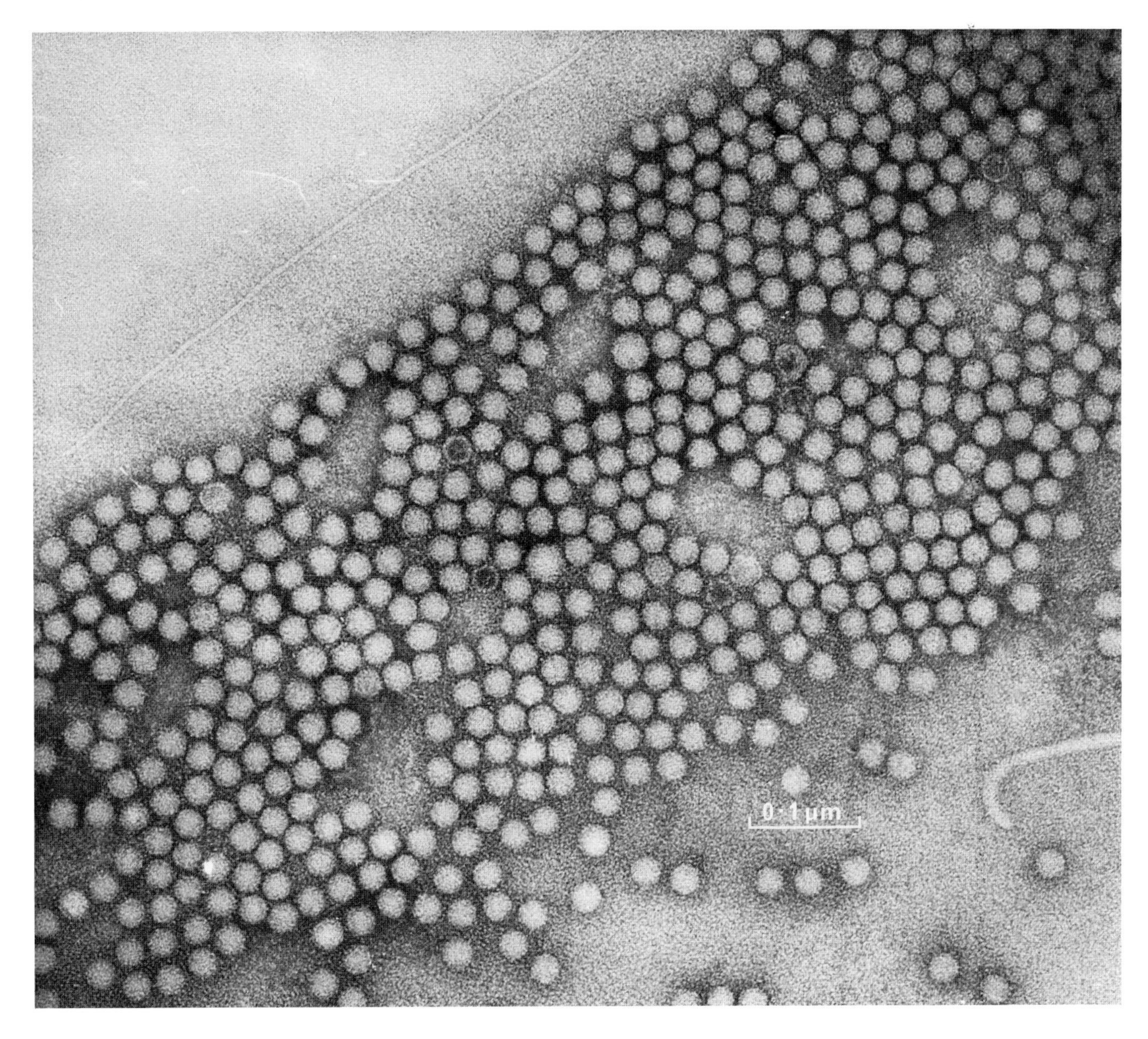

Fig. 7. Slow bee-paralysis virus. Electron micrograph of particles after purification and mounting in neutral sodium phosphotungstate.

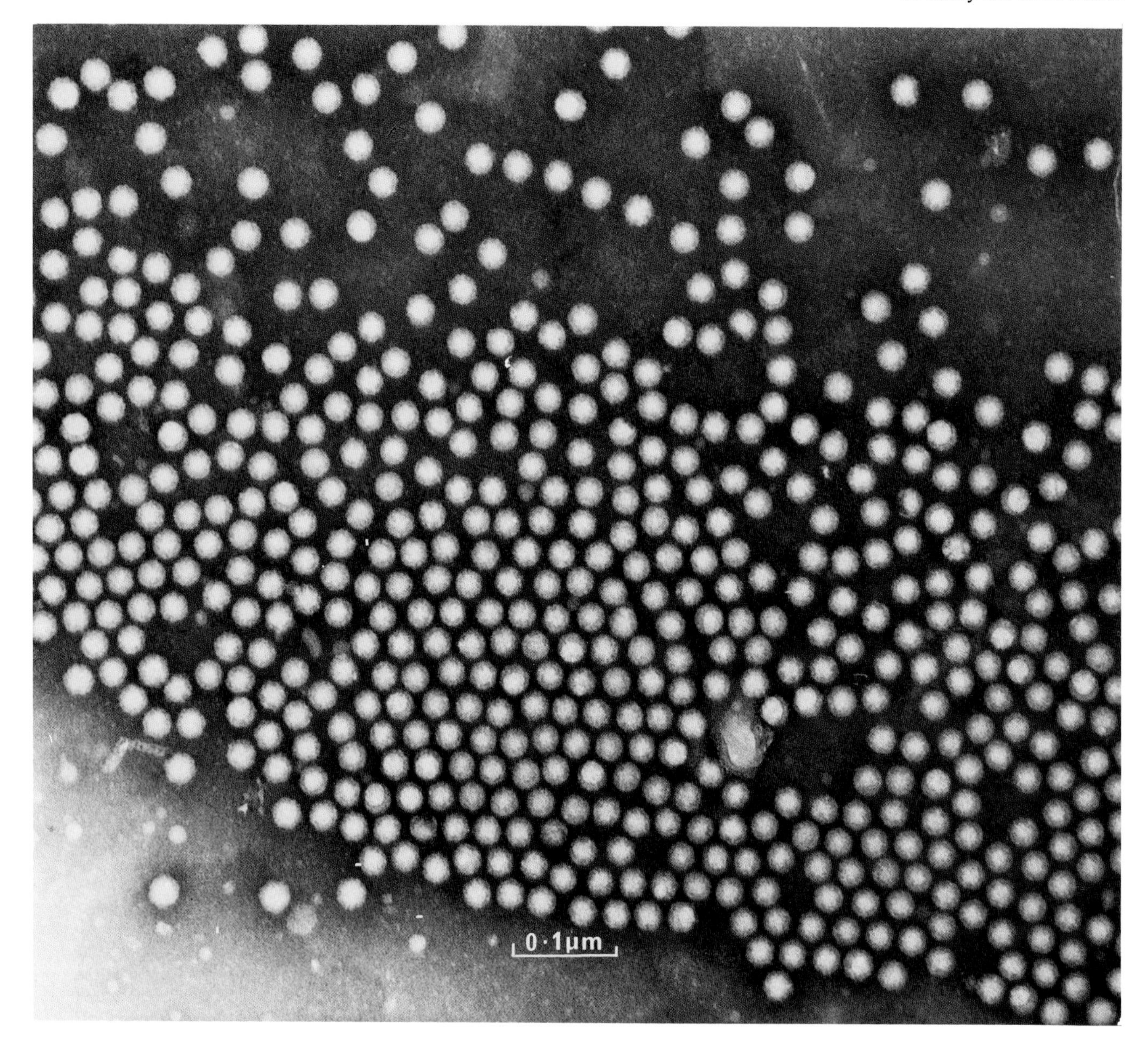

Fig. 8. Black queen cell virus. Electron micrograph of particles after purification and mounting in neutral sodium phosphotungstate.

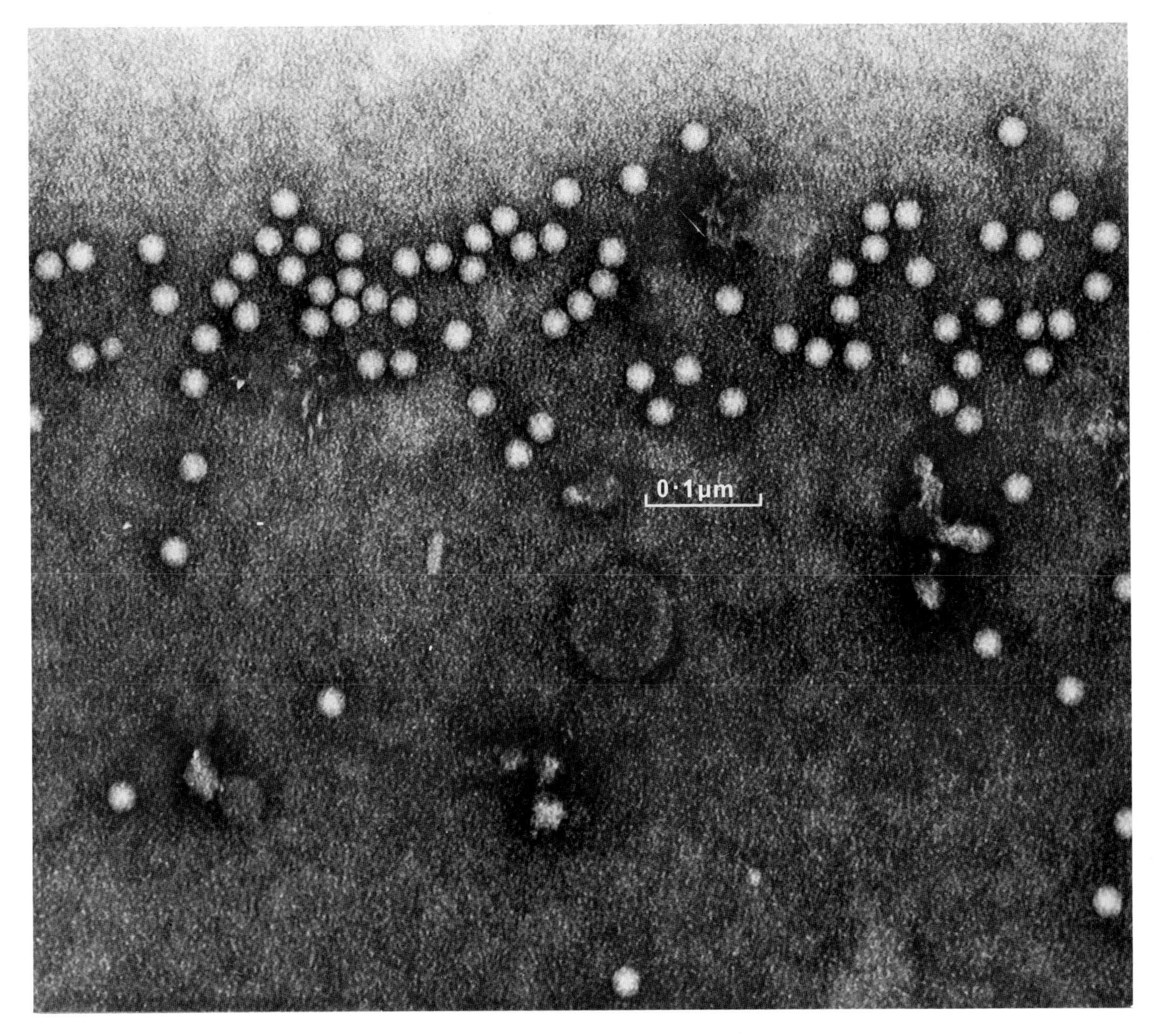

Fig. 9. Kashmir bee virus. Electron micrograph of particles after purification and mounting in neutral sodium phosphotungstate.

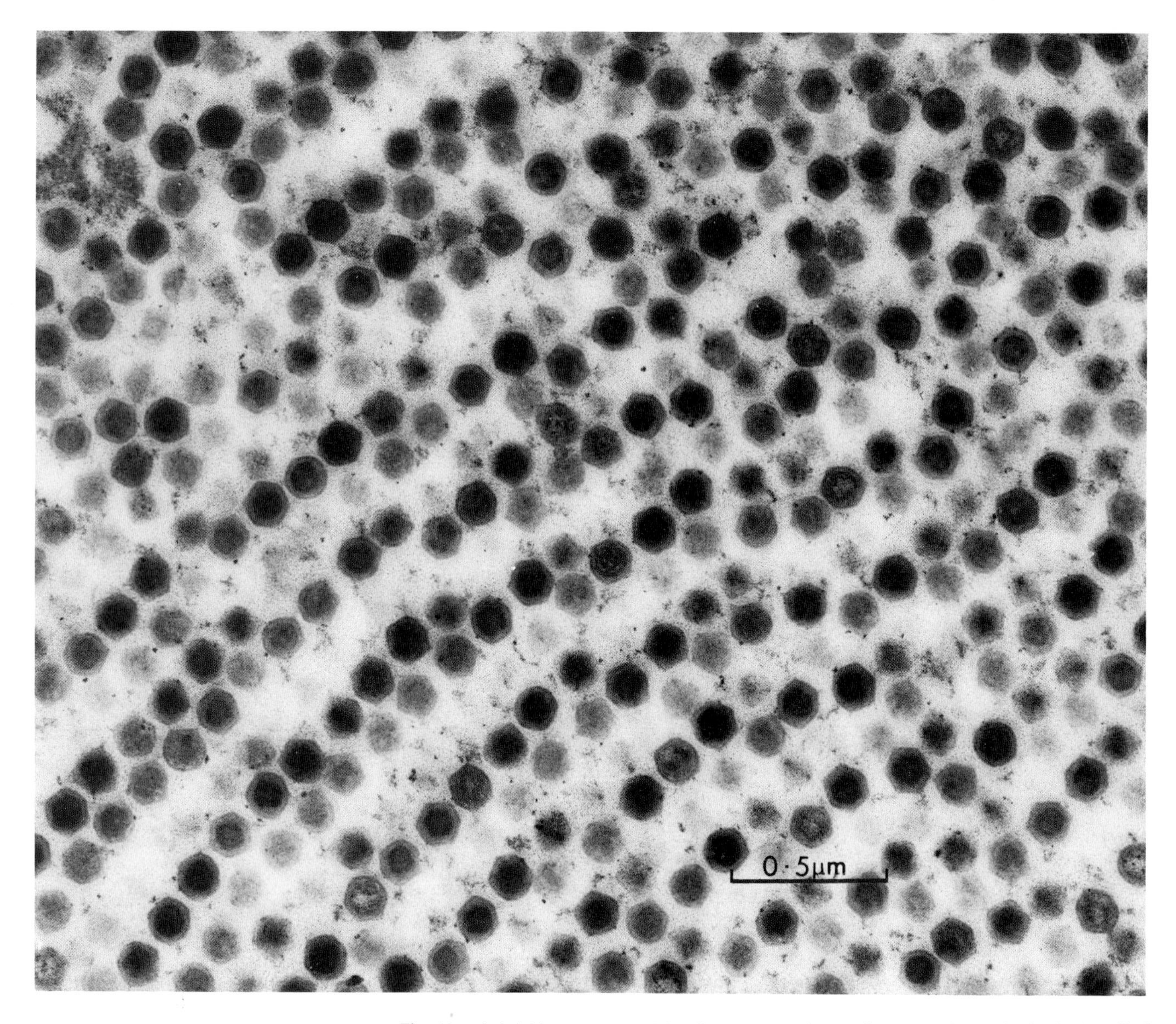

Fig. 10. *Apis* iridescent virus. Ultrathin section of part of the cytoplasm of a fat-body cell of *Apis mellifera* infected with *Apis* iridescent virus.

PART II

Plant Viruses

Chapter 8

Cauliflower Mosaic Virus (DNA Virus of Higher Plants)

ROBERT J. SHEPHERD

I. Introduction 159
II. Biological Properties and Transmissibility 159
III. Physical and Chemical Properties 161
IV. Cytopathological Effects in Infected Plants 162
V. Summary 163
References 163

I. INTRODUCTION

The cauliflower mosaic virus is the first virus of higher plants shown to contain DNA as its genomic material (Shepherd *et al.*, 1968). At least two other plant viruses are now recognized to contain DNA. These and other viruses share many attributes with cauliflower mosaic virus and can be placed in a discrete taxonomic group for which the name caulimovirus has been adopted.

Other viruses in the group are dahlia mosaic virus, carnation etched ring virus, mirabilis mosaic virus, and strawberry vein banding virus. The properties of these viruses, as a group, have been reviewed in considerable detail recently by the author (Shepherd, 1976). The DNA viruses are easily distinguished from other plant viruses because of their unique properties. The more salient features of caulimoviruses are described herein using cauliflower mosaic as a well characterized prototype of the group.

II. BIOLOGICAL PROPERTIES AND TRANSMISSIBILITY

The cauliflower mosaic virus (CaMV) and other caulimoviruses induce mosaic-mottle types of diseases, which are not readily distinguished from diseases caused by other types of plant viruses. Hence, the viruses are not easily distinguished from other viruses by symptoms alone; other means are required for their diagnosis.

CaMV and other caulimoviruses have restricted host ranges. For example, only members of a single family, the Cruciferae, are found infected by CaMV in nature. In addition, only 2 solanaceous species, other than cruciferous plants, *Nicotiana clevelandii* and *Datura stramonium*, have been infected experimentally. A list of hosts and illustrations of symptoms of CaMV on various plants is given by Broadbent (1957).

Caulimoviruses can be diagnosed by their restricted host ranges, high thermal inactivation points, aphid-transmissibility, morphology of the virion, and the unique nature of their inclusion bodies. Diagnostic hosts and other distinguishing biological properties of CaMV are given by Shepherd (1970).

A convenient way to diagnose CaMV in chronically infected plants is by light-microscopic observation of the distinctively appearing inclusion bodies in strips of epidermis stained in 1% phloxine. After thorough rinsing of excess stain with saline or water, only the inclusion bodies remain deeply stained (Fig. 1,A). These appear as well-defined spherical or somewhat irregularly lobed bodies in the cytoplasm.

The quick-dip method of preparing virus for electron microscopy directly from infected plants does not generally work well for CaMV because of the low concentration of the virus in cell exudates. It seems probable that very little virus exists free in the cell since most is occluded in the inclusion bodies which do not break down readily when cells are broken. For these reasons there is generally too little virus present to give positive serological reactions with crude extracts used in the conventional precipitin tests for identification of plant viruses (Pirone *et al.*, 1961). Consequently, partially purified preparations should be used for serological diagnostic tests. The virus is serologically closely related to dahlia mosaic virus and carnation etched ring virus.

Caulimoviruses are spread in nature by insects and by infected vegetatively propagated planting material. A single group of sucking insects, aphids (Aphididae, Hemiptera, Insecta), are the vectors in nature. These two forms of transmission appear to be the sole means of spread in nature as none of the viruses are transmitted by other types of arthropods or seed.

At least 27 different species of aphids have been reported as vectors of CaMV (Kennedy *et al.*, 1962), hence, these viruses exhibit relatively little specificity with their aphid vectors. Although CaMV is apparently a stylet-borne entity, its transmission has some unconventional characteristics when compared to other nonpersistent stylet-borne viruses. Pre- or postacquisition feeding for example, has little effect on acquisition and retention, and the virus can be retained for much longer periods than other stylet-borne viruses. CaMV has been reported to be retained by aphids for 3–20 hours, regardless of whether insects are feeding or fasting. It may be retained for periods as long as 36 hours by insects subjected to postacquisition fasting.

Several investigators have suggested an internal type of retention of CaMV in aphid vectors because of its relatively long retention, but no convincing evidence has been obtained for this. In fact, the evidence supports a stylet type of relationship with the most pertinent observation being the loss of transmission ability when the insects undergo molting (Day and Venables, 1961), which results in the insects casting off mouthparts along with the exoskeleton. Similarly, insects fail to transmit virus when it is injected into the hemolymph (Day and Venables, 1961). If the virus was internally borne one would expect transmission to result from hemolymph injections.

Recent experiments (Lung and Pirone, 1973, 1974) with some noninsect-transmissible isolates of CaMV have shown that an accessory factor in addition to the virus itself must be taken up by aphids in order for transmission to occur. Aphids probing through membranes into solutions of purified, normally aphid-transmissible isolates of CaMV do not transmit virus (Pirone and Megahed, 1966). However, if aphids are allowed to probe first into plants infected with an aphid-transmissible isolate and then through membranes into a solution of a different strain of purified virus (even nontransmissible isolates) they will acquire and transmit the latter (Lung and Pirone, 1973, 1974). Therefore, some virus-specified gene product, in addition to the virus itself, is required for acquisition and dissemination of the virus in nature. One can conjecture that the caulimoviruses have evolved this mechanism to ensure their dispersal by aphids in nature.

III. PHYSICAL AND CHEMICAL PROPERTIES

CaMV is the only member of the caulimovirus group that has been well characterized chemically and physically. Unfortunately, the other viruses are very difficult to prepare in quantities adequate for chemical and physical characterization.

The virions of CaMV consist of spherical particles 50 nm in diameter which distort easily when air-dried on electron microscope grids (Pirone *et al.*, 1961). The particles are penetrated by potassium phosphotungstate and the 50 nm particles appear to have a hollow center of about 20 nm diameter. None of the common electron-dense stains reveal any readily discernible external structure on the particles (Fig. 1B). In tissue sections stained with lead and osmium salts, particles appear 45–50 nm in diameter and both "empty" and "full" images are present. The empty particles are probably an artifact of staining as empty virus shells have not been encountered in purified preparations of the virus.

CaMV contains 16% DNA based on its phosphorus content (1.63%) and quantitative diphenylamine tests. The sedimentation coefficient is 208 S in 0.1 *M* NaCl, 0.01 *M* phosphate, pH 7.2, at infinite dilution. The measured partial specific volume is 0.704 gm/cm^3 and the diffusion coefficient is 0.75×10^{-11} m^2/sec. From these features the molecular weight equivalent for the particle can be calculated as 22.8×10^6. Hence, the molecular weight of the DNA is about 3.9×10^6 (Hull *et al.*, 1976).

The caulimoviruses have extremely stable virions that are not degraded by any of the more common procedures used for the dissociation of viruses. The virions of CaMV, for example, are not dissociated by 6 *M* guanidine hydrochloride or by emulsification with phenol in the presence of 1% sodium dodecyl sulfate (SDS). Boiling in 6 *M* guanidine hydrochloride or 2–3% SDS will dissociate the virions. Proteolysis with pronase or fungal protease K in the presence of 0.25% SDS have been used for the preparation of the DNA of CaMV (Shepherd *et al.*, 1970).

The DNA of CaMV has been determined to be double-stranded by a number of criteria (Shepherd *et al.*, 1970). The nucleic acid exhibits a cooperative-type melting curve with a T_m = 87.2 and 33–36% hyperchromicity in 0.15 *M* NaCl–0.015 *M* sodium citrate pH 7.0 (SSC). The DNA is nonreactive to 1–5% formaldehyde at room temperature and has a buoyant density in cesium chloride of 1.702 gm/ml. Nucleotide analyses after either acid or enzymatic hydrolysis show a molar equivalence between dAMP and TMP and between dGMP and dCMP. The GC content of the nucleic acid from the buoyant density, melting temperature, and nucleotide analyses is about 42%.

The DNA of CaMV consists of circular and linear forms with the same contour length of 2.31 μm, as revealed by electron microscopic observations (Fig. 2) (Shepherd and Wakeman, 1971). This length corresponds to a molecular weight of 4.4×10^6. Similarly, sedimentation experiments with fresh preparations show two types of molecules with sedimentation coefficients of 17.1 and 19.0 S. The faster can be converted to the slower form by mild DNase digestion, indicating that the faster component is the nicked-circular form and the slower component is the linear form. The $s_{20,w}$ value for the linear form suggests a molecular weight of about 4.0×10^6 for the DNA. In sucrose density gradients and gel electrophoresis experiments, the nicked-circular form is the infectious entity (Hull and Shepherd, 1975).

Buoyant density experiments with CaMV DNA in cesium chloride in the presence of ethidium bromide reveal only one isopycnic species (Shepherd *et al.*, 1970), indicating that the nucleic acid consists solely of linear and nicked-circular forms. No evidence has been found indicating that the DNA exists in a supercoiled condition as does that of the small DNA viruses of animals.

The DNA of CaMV after mild alkali treatment gives a somewhat heterogeneous peak of 8–10 S indicating a single-stranded molecular weight of 4×10^5, considerably lower than one might expect for the intact single strand (Hull and Shepherd, 1975). Treatment with other strand-separating procedures, such as heat and formaldehyde, gives products that correspond to the natural single strand. This and other evidence suggests that viral DNA is made up of smaller segments linked together by RNA.

The virions of CaMV consist of four proteins in addition to the DNA. Two major proteins of 37,000 and 64,000 daltons occur in a molar ratio of about 5 to 1 and make up more than 90% of the protein portion of the capsid. Two minor polypeptides of higher molecular weight, 88,000 and 96,000, give faint but indecisive staining for glycoprotein with periodate-Shiff reagent (Hull and Shepherd, 1976). One can calculate that about 420 copies of the 37,000 component, about 55 copies of the 64,000 component, and 12 or so copies of each of the larger polypeptides per virion, suggesting a $T = 7$ icosahedral structure with perhaps a $T = 1$ core (Hull and Shepherd, 1975).

Amino acid analyses of whole virus have revealed that the protein of CaMV is very basic in nature, with 18% lysine and 5% arginine (Brunt *et al.*, 1975). Although it is not known how the amino acids are distributed among the various proteins, it is apparent that the protein is extremely basic with a very strong electrostatic interaction between the protein and DNA. This may be one feature responsible for the unusual stability of the virion, in terms of chemical dissociation.

IV. CYTOPATHOLOGICAL EFFECTS IN INFECTED PLANTS

The caulimoviruses cause distinctive cytological changes accompanied by the production of characteristic inclusion bodies in infected cells. The inclusion bodies are unlike any produced by other types of plant viruses and several observations suggest they play an important role in replication of the viruses.

With the light microscope, cells infected with CaMV, or other caulimoviruses, are found to contain compact ovoid masses in the cytoplasm that are unique to infected tissues (Fig. 1,A). These bodies, the inclusion bodies, are very refractive and vary in size from a fraction of one to several μm in diameter. When such cells are suitably fixed, sectioned, and stained for electron microscopic observations, the inclusion bodies are seen to consist of an electron-dense, granular matrix in which the spherical virions are embedded (Figs. 1,C and D) (Fujisawa *et al.*, 1967). Throughout the matrix, in an apparent random distribution, are transparent vacuolelike areas of roughly circular outline. These may be portions of a reticulate system of lacunae that radiate throughout the inclusion body (Figs. 1,C and D). These vacuolated conglomerative type inclusion bodies are unique to the caulimoviruses and provide a reliable characteristic for distinguishing these viruses from other groups of plant viruses.

The inclusions of caulimoviruses are found predominately, if not solely, in the cytoplasm of infected cells and are frequently near the nucleus or dictyosomes. Generally only one, or occasionally more than one, appear per cell.

The caulimoviruses are remarkably restricted in their intracellular distribution in comparison with other plant viruses because of their almost total association with the inclusion body (Fujisawa *et al.*, 1967). Only occasionally have solitary or a few scattered virions been found in cells not in association with the inclusion bodies.

The matrix of the inclusion consists of finely granular, electron-dense material that may consist largely of protein, based on its staining properties and susceptibility to proteolytic enzymes (Martelli and Castellani, 1971; Conti *et al.*, 1972). This material appears to consist of a self-coherent mass which comprises the bulk of the inclusion. At high resolution the matrix material is resolved as parallel electron-dense fibers about 70 Å thick (Kitajima *et al.*, 1969).

The caulimoviruses cause a very characteristic morphological transformation of the plasmadesmata of infected cells (Kitajima and Lauritis, 1969). The most obvious change is a transformation to a structure of larger diameter, 60–80 nm in diameter rather than 25–35 nm for the normal type. Virions are frequently found in these transformed plasmodesmata. One may speculate that this is related to cell-to-cell movement of these viruses since virions are too large to move through unmodified plasmodesmata.

Incipient inclusions appear as minute patches of matrix material in the cytoplasm of infected cells. These can be distinguished by their affinity for electron-dense stains and the prominent clustering of ribosomes around their periphery. The latter feature suggests there may be intense protein synthesis at these foci. As the inclusions grow by accretion of newly synthesized matrix protein, mature virions appear embedded in the matrix. Hence, the inclusions are sites for virus assembly. In addition, autoradiographic evidence indicates that the inclusions are active sites for DNA synthesis (Kamei *et al.*, 1969; Fujisawa *et al.*, 1971, 1972).

V. SUMMARY

The most characteristic features of the caulimoviruses are their reproduction in the cytoplasm, an unusual trait for small DNA viruses, and the replication of DNA and assembly of virions in cytoplasmic foci within an electron-dense matrix. No other viruses of plants exhibit this feature. The caulimoviruses are the smallest independently replicating DNA's known at the present time in plant cells and provide the only available examples of simple DNA's whose gene products are known and easily recognized. Hence, these viruses will probably play an important role in molecular biological investigations of plant cells.

REFERENCES

Broadbent, L. (1957). "Investigations of Virus Diseases of Brassica Crops." Cambridge Univ. Press, London and New York.

Brunt, A. A., Barton, R. J., Tremaine, J. H., and Stace-Smith, R. (1975). *J. Gen. Virol.* **27,** 101–106.

Conti, G. G., Vegetti, G., Bassi, M., and Favali, M. A. (1972). *Virology* **47,** 694–700.

Day, M. F., and Venables, D. G. (1961). *Aust. J. Biol. Sci.* **14,** 187–197.

Fujisawa, I., Rubio-Huertos, M., Matsui, C., and Yamaguchi, A. (1967). *Phytopathology* **57,** 1130–1132.
Fujisawa, I., Rubio-Huertos, M., and Matsui, C. (1971). *Phytopathology* **61,** 681–684.
Fujisawa, I., Rubio-Huertos, M., and Matsui, C. (1972). *Phytopathology* **62,** 810–811.
Hull, R., and Shepherd, R. J. (1975). *Proc. Int. Congr. Virol., 3rd, 1975* Abstracts, p. 92.
Hull, R., and Shepherd, R. J. (1976). *Virology* **70,** 217–220.
Hull, R., Shepherd, R. J., and Harvey, J. D. (1976). *J. Gen. Virol.* **31,** 93–100.
Kamei, T., Rubio-Huertos, M., and Matsui, C. (1969). *Virology* **37,** 506–508.
Kennedy, J. S., Day, M. F., and Eastop, V. F. (1962). "A Conspectus of Aphids as Vectors of Plant Viruses." Commonwealth Institute of Entomology, London.
Kitajima, E. W., and Lauritis, J. A. (1969). *Virology* **37,** 681–685.
Kitajima, E. W., Lauritis, J. A., and Swift, H. (1969). *Virology* **39,** 240–249.
Lung, M. C. Y., and Pirone, T. P. (1973). *Phytopathology* **63,** 910–914.
Lung, M. C. Y., and Pirone, T. P. (1974). *Virology* **60,** 260–264.
Martelli, G. P., and Castellano, M. A. (1971). *J. Gen. Virol.* **13,** 133–140.
Pirone, T. P., and Megahed, E.-S. (1966). *Virology* **30,** 631–637.
Pirone, T. P., Pound, G. S., and Shepherd, R. J. (1961). *Phytopathology* **51,** 541–546.
Shepherd, R. J. (1970). *CMI/AAB Descriptions Plant Viruses* No. 24, pp. 1–4.
Shepherd, R. J. (1976). *Adv. Virus Res.* **20,** 305–339.
Shepherd, R. J., and Wakeman, R. J. (1971). *Phytopathology* **61,** 188–193.
Shepherd, R. J., Wakeman, R. J., and Romanko, R. (1968). *Virology* **36,** 150–152.
Shepherd, R. J., Bruening, G., and Wakeman, R. J. (1970). *Virology* **41,** 339–347.

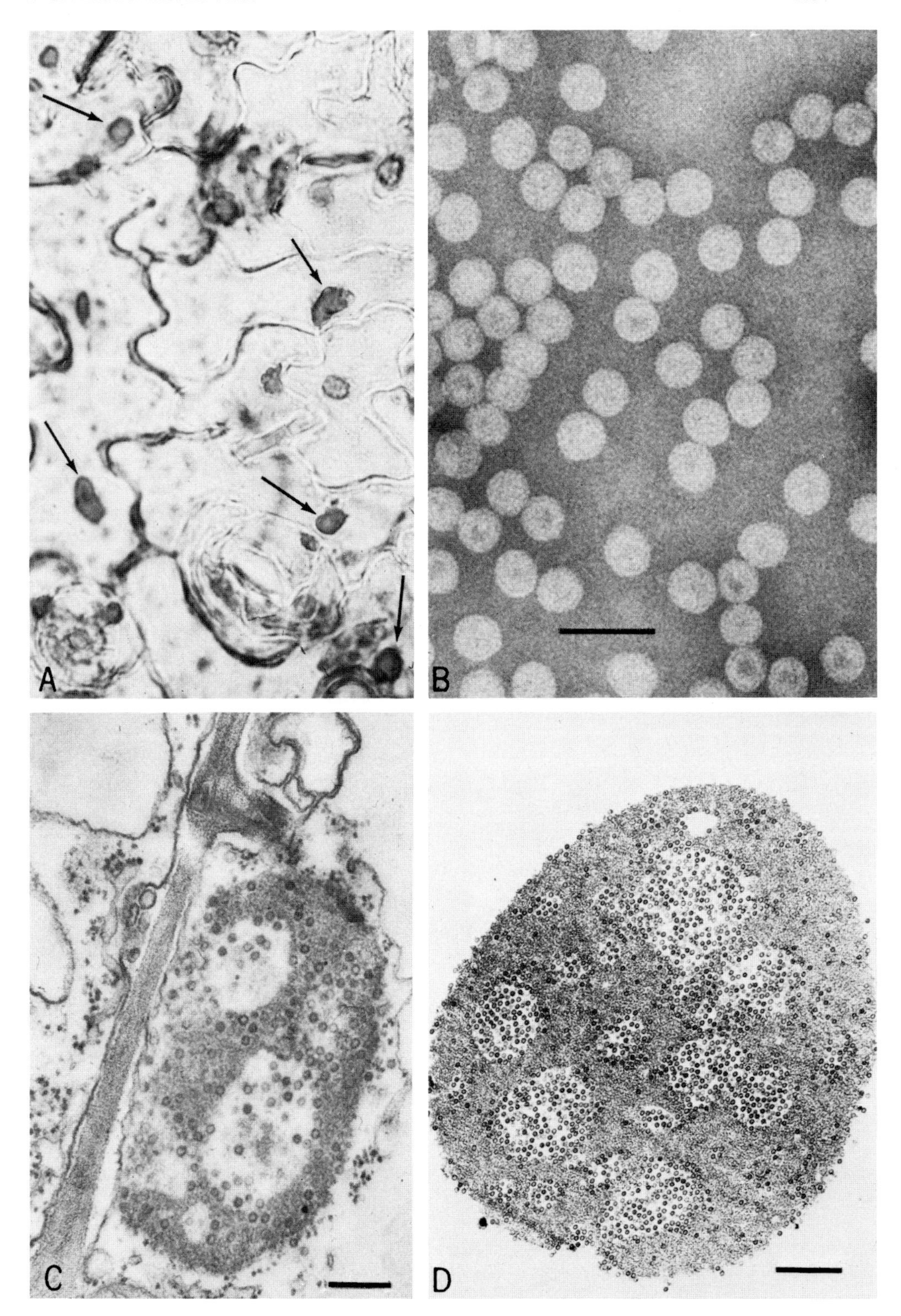

Fig. 1. (A) An epidermal strip from *Brassica campestris* infected with cauliflower mosaic virus showing viral inclusion bodies after staining with phloxine (arrows). (B) Purified cauliflower mosaic virus mounted in uranyl acetate. (Bar represents 100 nm.) (C) An inclusion body of cauliflower mosaic virus *in situ* in the cytoplasm of *B. campestris*. (Bar represents 500 nm.) (Photo courtesy of Dr. A. Allison.) (D) An inclusion body of cauliflower mosaic virus after isolation *in vitro*. (Bar represents about 1 μm.)

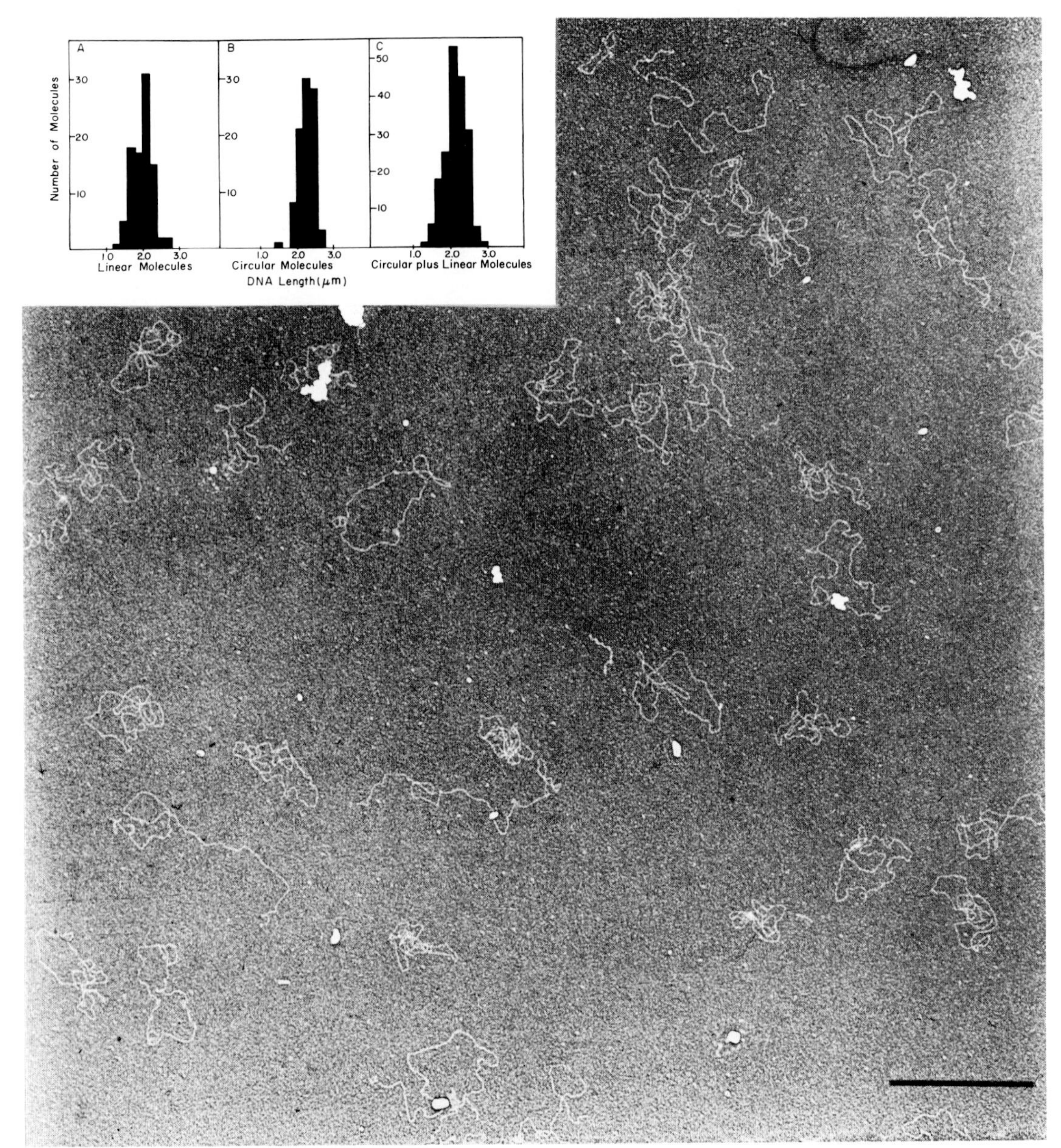

Fig. 2. DNA of cauliflower mosaic virus stained with uranyl acetate. The histogram in the insert shows the contour length distributions of linear and circular molecules. (Bar represents about 0.5 μm.)

Chapter 9

Comoviruses

A. VAN KAMMEN AND J. E. MELLEMA

I. Introduction 167
II. Viral Genome and Genetic Properties 168
III. Composition and Morphology of the Virus Capsid 169
IV. Cytopathic Structures in Infected Cells 170
V. Replication Process 171
References 173

I. INTRODUCTION

Comoviruses are a group of small, spherical plant viruses containing single-stranded RNA. A virus group is, according to the definition of the Plant Virus Subcommittee of the International Committee on Nomenclature of Viruses (Harrison *et al.*, 1971), "a collection of viruses and/or virus strains each of which shares with the type member all or nearly all the main characteristics of the group." The type member of the comovirus group is cowpea mosaic virus (CPMV). The main characteristics of CPMV are that it has three components of different buoyant density (centrifugal components), all isometric particles and about 24 nm in diameter, with capsids composed of two structural proteins. The sedimentation coefficients of the components are about 115, 95, and 58 S and the RNA contents, 36%, 25%, and 0%, respectively. Both components containing RNA are required for infection to occur. The virus particles occur in two electrophoretic forms, each composed of the three centrifugal components. The comoviruses, or the viruses of the cowpea mosaic virus group, represent, moreover, a separate structural class of icosohedral viruses since the structural proteins of the capsids of the virus particles are arranged in interpenetrating $T = 1$ lattices.

Members of the comovirus group are bean pod mottle virus (BPMV), broad bean stain virus (BBSV), broad bean true mosaic virus (BBTMV) (=echtes Ackerbohenenmosaic virus), cowpea mosaic virus (CPMV), radish mosaic virus (RaMV), red clover mottle virus (RCMV), and squash mosaic virus (SMV). The group is probably equivalent taxonomically to what vertebrate, invertebrate, and bacterial virologists regard as families, but further comparative studies are necessary to warrant such a classification.

The biological properties of the different comoviruses pertaining to diseases caused, host range, symptomatology, modes of transmission, and serology, which are of great importance for the identification of the viruses, have been summarized in the respective "Descriptions of Plant Viruses" distributed by the Commonwealth Mycological Institute and the Association of Applied Biologists in England, to which we refer the reader. Here we shall mainly review recent data on the molecular and cellular aspects of these viruses. Since CPMV has been studied most comprehensively, the emphasis in this chapter will be on this virus.

II. VIRAL GENOME AND GENETIC PROPERTIES

Comoviruses have a genome composed of two single-stranded RNA molecules, which are encapsidated separately in particles, that form M and B virions. Both virions or both RNA molecules are necessary for virus multiplication (for review, see Van Kammen, 1972; Jaspars, 1975). The two RNA's (M-RNA and B-RNA) of CPMV, sedimenting at 26 and 34 S, have molecular weights of 1.37×10^6 daltons and 2.02×10^6 daltons, respectively (Reijnders *et al.*, 1974). The base compositions of M-RNA of CPMV are: G, 20.7; A, 28.7; C, 19.7; and U, 31.6; and for B-RNA G, 22.9; A, 28.7; C, 17.2; and U, 31.4; indicating a high A + U content for both RNA's (Van Kammen and Van Griensven, 1970). The RNA's of the other comoviruses have similar sizes and base compositions. Both M-RNA and B-RNA of CPMV contain a sequence of polyadenylic acid, 150–200 residues long, at their 3′-end (El Manna and Bruening, 1973). Polyadenylic acid sequences were also detected in BPMV RNA's (Semancik, 1974).

Recently it was demonstrated that the two CPMV RNA's have no $m^7G(5')ppp(5')N$ . . . ("cap") at their 5′ end nor is (p)(p)pN . . . present at the 5′ terminal ends (Klootwijk *et al.*, 1977). In the same study it was shown that no 2′-*O*-methylated nucleotides, N^6-methyladenosines, or pseudouridines are present in CPMV RNA's indicating a very low degree, if any, of postsynthetic modification of the nucleotides of CPMV RNA. M-RNA and B-RNA of CPMV have, except for the poly(A) sequence at the 3′-end, no base sequences in common as was demonstrated by molecular hybridization studies (Van Kammen and Rezelman, 1972). Both RNA's appear to be a unique piece of the viral genome.

Genetic analysis has shown that the genetic information for CPMV is distributed among the M and B components (Bruening, 1969; De Jager and Van Kammen, 1970; Wood, 1972; De Jager, 1976). Heterologous mixtures of components of wild-type and mutant strains of CPMV were used to locate mutations in the components and showed that each component carries genetic characters for the phenotype of the virus. Attempts to locate the information for the virion proteins on either the M or B component have led to confusing results. Wood (1972) reported that one of the structural genes for the capsid polypeptides is located on M-RNA. Siler *et al.* (1976) obtained evidence that the small capsid polypeptide of CPMV might be encoded on B-RNA. Moore and Scott (1971) reported that the serological characteristics of BPMV can be inherited by the M component. Kassanis *et al.* (1973) presented results on the inheritance of the serological properties of two strains of RaMV which could be explained if on each viral RNA component there resided a structural gene for each of the capsid polypeptides. Clearly the genes for the capsid proteins cannot yet be assigned to either of the components.

Recently De Jager *et al.* (1976) characterized a temperature-sensitive mutant of CPMV with a defect in the synthesis, or the assembly, of the virus-RNA replicase, which was located in the M component.

III. COMPOSITION AND MORPHOLOGY OF THE VIRUS CAPSID

The structure of CPMV has been studied by image analysis of electron micrographs. Agrawal (1964) examined virus preparations, after negative staining, with potassium phosphotungstate at pH 5.1–7.0. Such preparations showed polyhedral particles. Most of the particles apparently contained nucleic acid, but some appeared empty, presumably top component particles. The particles measured 240–270 Å (side to side, between extreme points) for those containing RNA and 230–250 Å for those without nucleic acid. No morphological subunits (capsomers) were observed. By comparing the structure of particles in electron micrographs, using the rotation technique of Markham *et al.* (1963) to enhance image details, with a model viewed in different orientations Agrawal concluded that the particles were icosahedral with 5:3:2 axial symmetry and constructed of 60 subunits. That proposal implied that the protein subunits in the capsid were arranged in a $T = 1$ surface lattice (Caspar and Klug, 1972). At that time no data were available on the protein composition of the virus capsid.

A more detailed study of the structure of CPMV was made after it had been found that the capsids of CPMV consist of two different proteins in equimolar amounts (Wu and Bruening, 1971; Geelen *et al.*, 1972). The molecular weights of the two proteins, estimated by SDS gel electrophoresis, are 44,000 and 25,000 (Geelen, 1974) and, as the molecular weight of the RNA-free top component of CPMV is 3.8×10^6 as determined by sedimentation equilibrium, it could be calculated that the capsid contains about 60 molecules of each of the two proteins. Three dimensional image reconstruction from electron micrographs (Crowther *et al.*, 1970; Crowther, 1971) was then applied to preparations of CPMV negatively stained with uranyl acetate at pH 4.2 (Crowther *et al.*, 1974). In Fig. 1 a representative view of particles lying over a hole in the carbon film is shown. Such particles measured 200–240 Å. These values for the size of CPMV particles are probably the most accurate, since the calibration of the magnification with a grating was done very carefully in this case and the deformation of particles suspended in stain over holes in the carbon film is minimal. As shown in Fig. 1, the negatively stained particles show characteristic views suggesting icosahedral symmetry. By analyzing digital transforms of the projections for the presence of icosahedral symmetry it was confirmed that CPMV does have icosahedral symmetry, at least to a resolution of about 25 Å, allowed by the negative staining technique. A three-dimensional density map was calculated to this resolution by inversion of the data in three-dimensional Fourier space, which was constructed from several particles. The results of the three-dimensional image analysis and reconstruction showed that the surface of CPMV does not consist of clear capsomers such as occur, for example, in turnip yellow mosaci virus. At the five- and three-fold lattice positions, protruding bumps of protein were located, as can be visualized in an equatorial trace of the particle (Fig. 2). The radii of these protrusions were 120 Å and 150 Å, respectively. There was a good agreement in the comparison of the most striking computer-generated projections of CPMV and some observed views of particles in the electronmicrographs. Figure 3 represents a drawing of the virus capsid based upon the data of the three-dimensional reconstruction. This structure is consistent with the chemical data, assuming that the 60 molecules of the larger of the two protein subunits (MW about 44,000) are clustered to form 12 knobs at the five-fold positions and the 60 molecules of the smaller subunits (M.W. about 25,000) in 20 smaller lumps around the three-fold positions.

Unlike the icosahedral single-stranded RNA plant viruses, which have capsids constructed of 180 identical subunits in a $T = 3$ surface lattice (like turnip yellow mosaic virus, the bromovirus group, and others), CPMV represents a class of viruses with capsids consisting of 60 subunits composed of two structural proteins arranged in interpenetrating $T = 1$ lattices. It is probable, although independent evidence is lacking for the other members of the group, that all the comoviruses have this type of capsid structure.

The particle weights of the M and B components of CPMV, as determined by sedimentation-equilibrium centrifugation are 5.15×10^6 daltons and 5.87×10^6 daltons, in good agreement with the molecular weight of the empty capsid and the molecular weights of M-RNA and B-RNA and the RNA-content of the components given in sections I and II (Geelen, 1974).

A remarkable finding is that the coat protein of CPMV, but not of BPMV, contained about 1.9% carbohydrates covalently linked on the viral protein (Partridge *et al.*, 1974). This was related to the seed-transmissibility of CPMV, while BPMV is nonseed-transmitted.

Particles of CPMV, BPMV, RaMV, and BBTMV occur in two electrophoretic forms, the slow moving (S) and fast moving form (F). The S and F forms differ in the small capsid protein. Both S and F contain the large protein with a molecular weight of about 44,000, but the small protein of the S form has a molecular weight of about 25,000 and that of the F form a molecular weight of about 22,000. The ratio of the S and F forms was found to vary according to the age of infection. In CPMV and BBTMV the relative amount of the F form increased with time (Niblett and Semancik, 1969; Blevings and Stace-Smith, 1976), whereas in BPMV the relative amount of S form increased with time (Gillaspie and Bancroft, 1965). The changing S:F ratios appear to be the result of a conversion of one electrophoretic form to the other, rather than the appearance and disappearance of two distinct forms, by limited proteolysis in the plant. The conversion of the S form into F and CPMV and BBTMV and of the F form into S with BPMV can be achieved *in vitro* by incubation with proteolytic enzymes. The conversion *in vivo* and *in vitro* results in a loss of a peptide with a molecular weight of approximately 2500 from the smaller capsid protein. The molecular weight of that protein becomes, then, about 22,000. The conversion of one electrophoretic form into the other does not lead to an observable reorganization of the protein capsid.

IV. CYTOPATHIC STRUCTURES IN INFECTED CELLS

Shortly after infection, cytopathic structures that do not occur in healthy cells can be observed in CPMV-infected cells (Assink, 1974; De Zoeten *et al.*, 1974). Figure 4 presents an overall picture of a mesophyll cell from an infected primary cowpea leaf. Whereas healthy cells have characteristically a thin layer of parietal cytoplasm, the appearance of the cells after CPMV-infection is changed by morbid growth of membranes. As can be seen, especially at higher magnifications (Fig. 5), rays of many vesicles penetrate the cytopathic structure forming a kind of reticulum. The spaces between the vesicles of the reticulum are filled with electron dense material that does not seem to have a clear structure. CPMV virus particles can be seen embedded in this material. The cytopathic structures are often found adjacent to the nucleus; they are not surrounded by a membrane but occur in the cytoplasm; they contain osmiophilic globules; and many mitochondria are seen near them. The vesicles often contain fibrils (Fig. 6). Such fibrils in vesicles also occur with some other plant virus infections, like pea enation mosaic virus (De Zoeten *et al.*, 1972; Burgess *et al.*, 1974b), beet western yellow virus (Esau and Hoefert, 1972), and cowpea chlorotic mottle virus (Burgess *et al.*, 1974a). The fibrils are reminiscent of nucleic acid fibers as visualized in mitochondria and chloroplasts, but hard evidence of the nature of the fibrillar material in the vesicles is lacking.

The ontogeny of the cytopathic structure has not been studied and no data are available on the origin of the vesicles or the structure per se. The chemical nature of the electron dense material is also unresolved.

The cytopathic structures appear rather early after infection and they are found in virtually all mesophyll cells of cowpea leaves on day 4 after innoculation, when virus multiplication as measured by increase of infectivity enters its most rapid phase. If isolated cowpea mesophyll protoplasts are infected with CPMV, cytopathic structures similar to those in cells of CPMV-infected leaves, are found 24 hours after inoculation (Hibi *et al.*, 1975).

The involvement of the cytopathic structures with CPMV infection and multiplication is confirmed by the finding that also in tobacco protoplasts similar structures arise after inoculation with CPMV (Huber *et al.*, 1976). The cytopathic structures in CPMV infected tobacco protoplasts have the same features as those in cowpea cells. Evidently the origin of these structures and their characteristics are determined by CPMV and very little, if at all, by the host cell.

The cytopathic structures are so large that they can be seen by the light microscope (De Zoeten *et al.*, 1974). They probably correspond to the inclusion bodies observed in CPMV-infected leaf cells, after staining with phloxine, as a red amorphous mass near or surrounding the nucleus (Agrawal, 1964; Swaans and Van Kammen, 1973).

Similar cytopathic structures like those in CPMV-infected cells have been reported to arise in RaMV-infected turnip epidermal leaf tissue (Štefanac and Ljubešić, 1971), but for the other comoviruses there is no information on the occurrence of such structures.

Other aberrations of cellular structures following infection by CPMV have been described (Van der Scheer and Groenewegen, 1971) but these occur only later, after infection, and are rarely found. They appear therefore to be secondary effects of CPMV infection on the cell structure. This applies for the fine membrane structures in the nucleoplasm of mesophyll cells of primary cowpea leaves 13 days after inoculation and for single rows of virus-like particles in narrow tubules found not only in the cytoplasm but also in the vacuoles, in the nucleus, and often parallel to the surface against and within the cell wall of mesophyll cells (Fig. 8). Similar structures of rows of particles within membrane-bound tube-like structures have been reported for BPMV (Kim and Fulton, 1971) and RaMV (Honda and Matsui, 1972; Hooper *et al.*, 1972). The significance of these structures is not at all clear. Figure 8 shows that these tubules with particles might arise as protrusions from the plasmalemma penetrating into the cytoplasm.

V. REPLICATION PROCESS

For CPMV it has been demonstrated that the replication of viral RNA is closely associated with the membranes of the vesicles of the cytopathic structures. It is possible to isolate the cytopathic structures more or less intact by fractionation of a homogenate of CPMV-infected cowpea leaves on discontinuous sucrose gradients (Assink *et al.*, 1973; De Zoeten *et al.*, 1974; Van

Kammen, 1974). It was shown that the fractions carrying the fragments of the cytopathic structures contained the replicative form (RF) of CPMV, characterized by its resistance to ribonucleases, its ability to hybridize with ^{3}H-labeled CPMV RNA, its behavior on polyacrylamide gel electrophoresis, and by electron microscopy. When CPMV-infected leaf tissue was labeled with ^{3}H-uridine for 2 hours, the RF's in the fractions containing cytopathic structures became rapidly labeled. The majority of RNA, complementary to CPMV-RNA and presumably involved in viral RNA synthesis, was therefore found with the cytopathic structures. The association of this RNA with the vesicular membranes was demonstrated by autoradiography (De Zoeten *et al.*, 1974) performed on sections of intact CPMV-infected leaves and on sections of pellets of isolated cytopathic structures from tissue treated with ^{3}H-uridine for 2 hours. The autoradiograms showed that the ^{3}H-label was mainly in the vesicular membranes of the cytopathic structure (Fig. 7), rather than in the electron-dense material. In further studies it was shown that CPMV-infected cowpea leaves contain an RNA-dependent RNA polymerase closely bound with cytoplasmic membranes and which is able to synthesize *in vitro* both double-stranded and single-stranded viral RNA (Zabel *et al.*, 1974). Thus, the replication of CPMV RNA takes place on the cytopathic membrane vesicles and is therefore located in the cytoplasm of the infected cell. The membrane-bound replicase was solubilized from the membranes by washing with a Mg^{2+}-deficient buffer. Further purification of the solubilized replicase resulted in a stable enzyme preparation which was completely template dependent (Zabel *et al.*, 1976).

The synthesis of CPMV-specific proteins in cowpea plants was inhibited by low levels (0.5–1 μg/ml) of cycloheximide but was not affected by several inhibitors of the 70 S chloroplast ribosomes (Owens and Bruening, 1975). The CPMV-specific proteins are therefore probably synthesized on 80 S cytoplasmic ribosomes.

Recently it was found that the two CPMV RNA's with molecular weights of 1.37×10^6 daltons and 2.02×10^6 daltons were translated in a messenger-dependent *in vitro* protein synthesizing system from rabbit reticulocytes into products which include large polypeptides with apparent molecular weights of 130,000 and 220,000, respectively. The size of these polypeptides corresponds approximately to the coding capacity of the two RNAs' (Pelham and Stuik, 1976). Among the products of *in vitro* translation in this system no CPMV coat proteins were detected. This suggests that either both CPMV RNA's are translated in one large precursor polypeptide from which subsequently the functional proteins, among which are the coat proteins, arise by posttranslational cleavage or it suggests that the RNA's must first undergo some modification, such as a pretranslational cleavage or replication of smaller translation units.

There is no information on the assembly of virus RNA and proteins into virions. Newly synthesized virus particles accumulate in the cytoplasm but not in any of the cell organelles. Sometimes particles are aggregated into clusters, very rarely in crystalline arrangements. Aggregates of virus particles, either scattered or arranged regularly in microcrystals, embedded in a membranous matrix have been reported for RaMV in cells of infected turnip leaves (Štefanac and Ljubešić, 1971).

ACKNOWLEDGMENTS

The authors wish to express their thanks to J. Groenewegen, Department of Virology, Agricultural University, Wageningen, The Netherlands, for his help with the electron micrographs on cytopathic structures and to Dr. J. Davies for correcting the English.

Figures 4–8 were obtained from the Archives of the Laboratory of Virology, Wageningen, and were taken by Mrs. C. van der Scheer (Figs. 4 and 8) and Dr. G. A. de Zoeten (Figs. 5–7).

The senior author's work described here was supported by the Netherlands Foundation for Chemical Research (S.O.N.) with financial aid from the Netherlands Organization for the Advancement of Pure Research (Z.W.O.).

REFERENCES

Agrawal, H. O. (1964). *Meded. Landbouwhogesch. Wageningen* **64,** 1–53.
Assink, A. M. (1974). Doctoral Thesis, pp. 1–70. Agricultural University, Wageningen.
Assink, A. M., Swaans, H., and Van Kammen, A. (1973). *Virology* **53,** 384–391.
Blevings, S., and Stace-Smith, R. (1976). *J. Gen. Virol.* **31,** 199–210.
Bruening, G. (1969). *Virology* **37,** 577–584.
Burgess, J., Motoyoshi, F., and Fleming, E. N. (1974a). *Planta* **117,** 135–144.
Burgess, J., Motoyoshi, F., and Fleming, E. N. (1974b). *Planta* **119,** 247–256.
Caspar, D. L. D., and Klug, A. (1962). *Cold Spring Harbor Symp. Quant. Biol.* **27,** 1–24.
Crowther, R. A. (1971). *Philos. Trans. R. Soc. London, Ser. B* **261,** 221–230.
Crowther, R. A., De Rosier, D. J., and Klug, A. (1970). *Proc. R. Soc. London, Ser. A* **317,** 319–340.
Crowther, R. A., Geelen, J. L. M. C., and Mellema, J. E. (1974). *Virology* **57,** 20–27.
De Jager, C. P., Zabel, P., Van der Beek, C. P., and Van Kammen, A. (1977). *Virology* **76,** 164–172.
De Jager, C. P., and Van Kammen, A. (1970). *Virology* **41,** 281–287.
De Jager, C. P., Zabel, P., Van der Beek, C. P., and Van Kammen, A. (1976). *Virology* **76,** 164–172.
De Zoeten, G. A., Gaard, G., and Diez, F. B. (1972). *Virology* **48,** 638–647.
De Zoeten, G. A., Assink, A. M., and Van Kammen, A. (1974). *Virology* **59,** 341–355.
El Manna, M. M., and Bruening, G. (1973). *Virology* **56,** 198–206.
Esau, K., and Hoefert, L. L., (1972). *Virology* **48,** 724–738.
Geelen, J. L. M. C. (1974). Doctoral Thesis, pp. 1–85. Agricultural University, Wageningen.
Geelen, J. L. M. C., Van Kammen, A., and Verduin, B. J. M. (1972). *Virology* **49,** 205–213.
Gillaspie, A. G., and Bancroft, J. B. (1965). *Phytopathology* **55,** 906–908.
Harrison, B. D., Finch, J. T., Gibbs, A. J., Hollings, M., Shepherd, R. J., Valenta, V., and Wetter, C. (1971). *Virology* **45,** 356–363.
Hibi, T., Rezelman, G., and Van Kammen, A. (1975). *Virology* **64,** 308–318.
Honda, Y., and Matsui, C. (1972). *Phytopathology* **62,** 448–452.
Hooper, G. R., Spink, G. C., and Myers, R. L. (1972). *Virology* **47,** 833–837.
Huber, R., Rezelman, G., Hibi, T., and Van Kammen, A. (1977). *J. Gen. Virol.* **34,** 315–323.
Jaspars, E. M. J. (1975). *Adv. Virus Res.* **19,** 37–149.
Kassanis, B., White, R. F., and Woods, R. D. (1973). *J. Gen. Virol.* **20,** 277–285.
Kim, K. S., and Fulton, J. P. (1971). *Virology* **43,** 329–337.
Klootwijk, J., Klein, I., Zabel, P., and Van Kammen, A. (1977). *Cell* (in press).
Markham, R., Frey, S., and Hills, G. J. (1963). *Virology* **20,** 88–102.
Moore, B. J., and Scott, H. A. (1971). *Phytopathology* **61,** 831–835.
Niblett, C. L., and Semancik, J. S. (1969). *Virology* **38,** 685–693.
Owens, R. A., and Bruening, G. (1975). *Virology* **64,** 520–530.
Partridge, J. E., Shannon, L. M., Gumpf, D. J., and Colbaugh, P. (1974). *Nature (London)* **247,** 391–392.
Pelham, H. R. B., and Stuik, E. J. (1976). *Collogues Intern. C. N. R. S. No. 261, Acides Nucléiques et Synthèse des Protéines chez les Végétaux.* 691–695.
Reijnders, L., Aalbers, A. M. J., and Van Kammen, A. (1974). *Virology* **60,** 515–521.
Semancik, J. S. (1974). *Virology* **62,** 288–291.
Siler, D. J., Babcock, J., and Bruening, G. (1976). *Virology* **71,** 560–567.
Štefanac, Z., and Ljubešić, N. (1971). *J. Gen. Virol.* **13,** 51–57.
Swaans, H., and Van Kammen, A. (1973). *Neth. J. Plant Pathol.* **79,** 257–265.
Van der Scheer, C., and Groenewegen, J. (1971). *Virology* **46,** 493–497.
Van Kammen, A. (1972). *Annu. Rev. Phytopathol.* **10,** 125–150.
Van Kammen, A. (1974). *Proc. Intersect. Congr. IAMS, 1st, 1974* Vol. 3, pp. 93–100.
Van Kammen, A., and Rezelman, G. (1972). *Proc. Int. Congr. Virol., 2nd, 1971* p. 235.
Van Kammen, A., and Van Griensven, L. J. L. D. (1970). *Virology* **41,** 274–280.
Wood, H. A. (1972). *Virology* **49,** 592–598.
Wu, G-j., and Bruening, G. (1971). *Virology* **46,** 596–612.
Zabel, P., Weenen-Swaans, H., and Van Kammen, A. (1974). *J. Virol.* **14,** 1049–1055.
Zabel, P., Jongen-Neven, I., and Van Kammen, A. (1976). *J. Virol.* **17,** 679–685.

Fig. 1–3. Structure of cowpea mosaic virus as determined by image analysis of electron micrographs of virus particles.

Fig. 1. Cowpea mosaic virus negatively stained with uranyl acetate at pH 4.2, lying over a hole in the carbon film. Full and empty particles can be seen. Particles 2, 3, and 5 indicate particles with features of a two-fold, three-fold, and five-fold axis of symmetry, respectively. The bar represents approximately 500 Å.

Fig. 2. The outer contour of an image-reconstructed virus particle around an equator normal to a two-fold axis. The positions of the two-fold, three-fold, and five-fold axes lying in the plane are indicated.

Fig. 3. Model for the structure of the capsid of CPMV based on the three-dimensional image reconstruction and on the protein composition of the capsid.

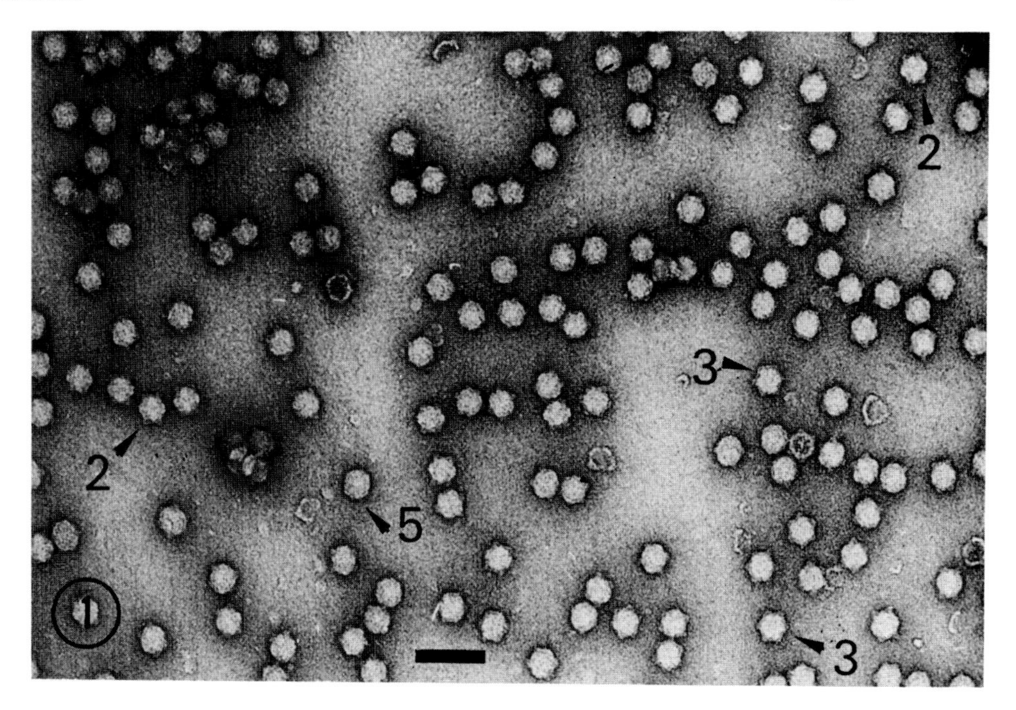
2
3
2
5
3
1

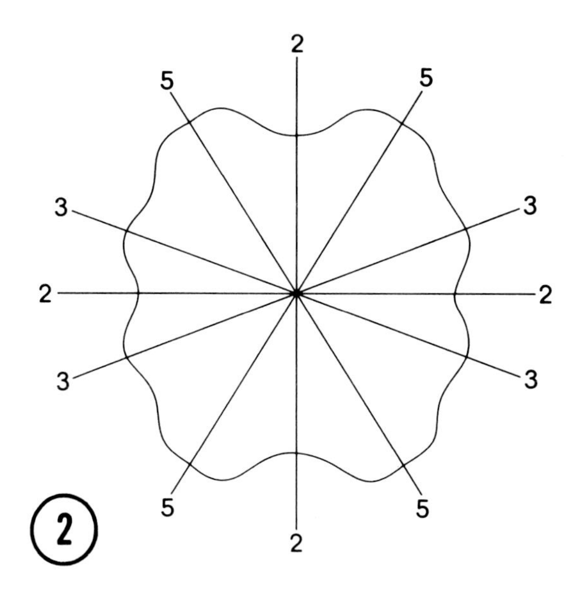
2
5
5
3
3
2
2
3
3
5
5
2
2

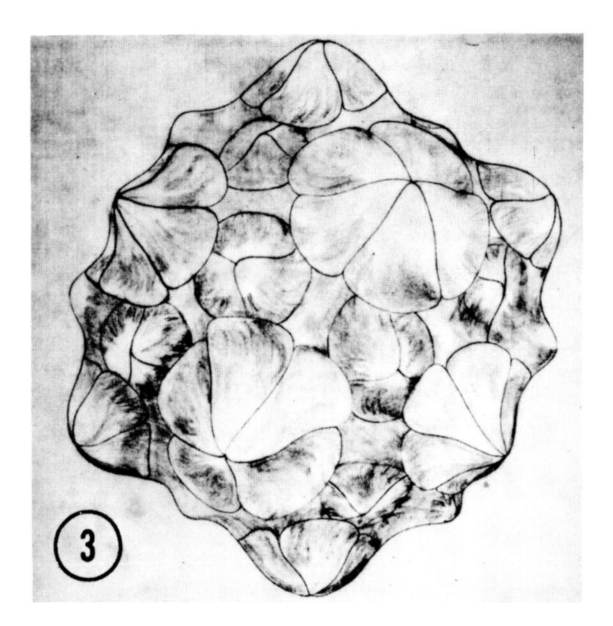
3

Fig. 4–8. Cytopathic structures in CPMV-infected cowpea mesophyll cells.

Fig. 4. Section through a CPMV-infected mesophyll cell from primary cowpea leaves, 4 days after inoculation of the leaves. Note the extensiveness of the cytopathic structure consisting of numerous membrane vesicles and electron dense material. Note the osmiophylic globules. Virus particles occur embedded in the electron dense material and scattered through the cytoplasm but are difficult to distinguish from ribosomes because of their size. (×24,000.)

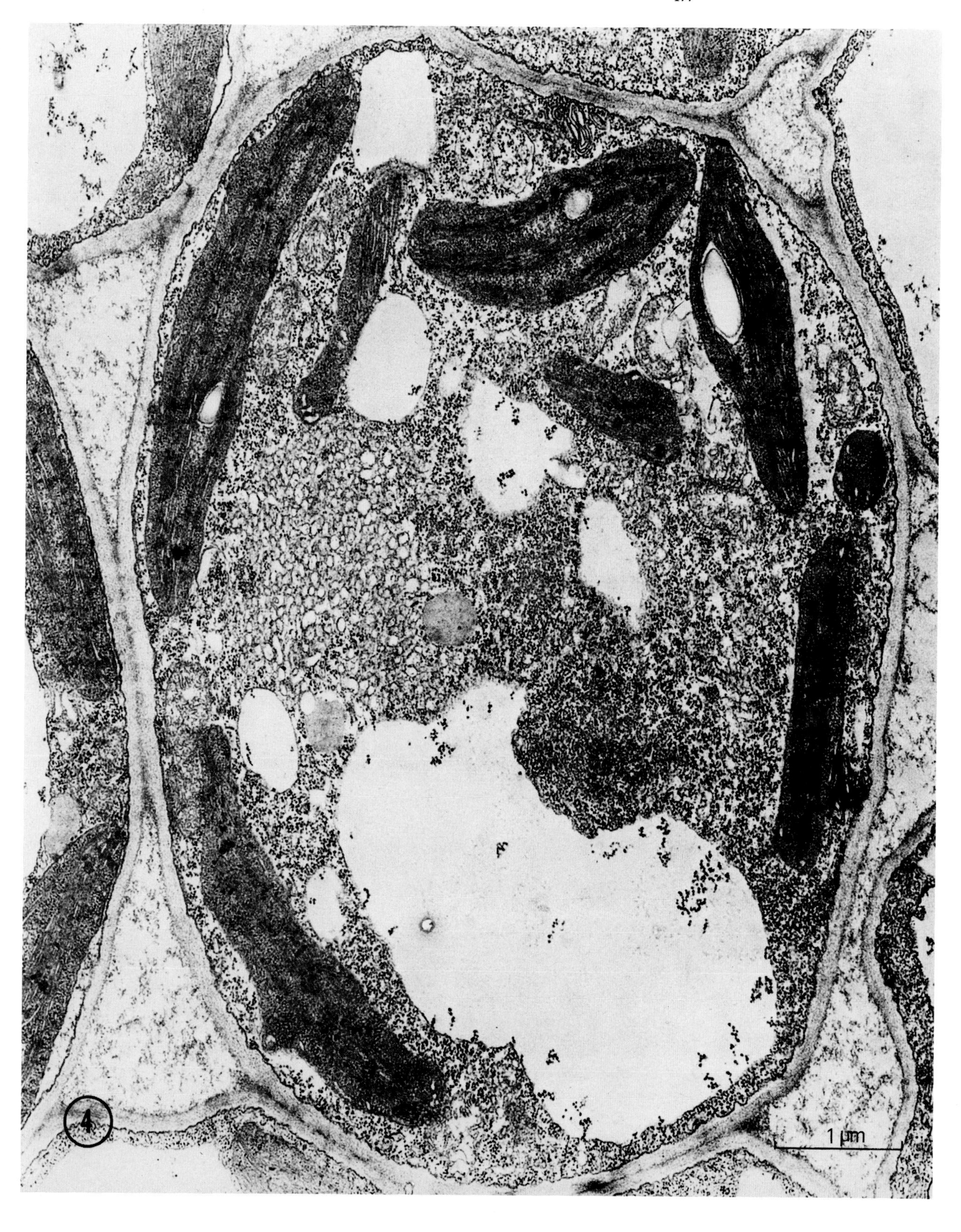
4
1 μm

Fig. 5. Part of a cytopathic structure showing the salient features. Rays of vesicles penetrate the cytopathic structure forming a kind of reticulum and the spaces between the constituting parts are filled with electron dense material that does not seem to have a clear structure. Some virus particles occur embedded in this material. Some vesicles contain fibrillar material. (×20,000.)

Fig. 6. Detail of cytopathic structure at higher magnification. Note the vesicles containing fibrillar material characteristic of this structure. (×80,000.)

Fig. 7. Part of an electron microscopic autoradiogram of a section from CPMV-infected leaf tissue 4 days after inoculation and incubated with ^{3}H-uridine in the presence of actinomycin D and thymidine. The grains from ^{3}H-uridine incorporated in newly synthesized RNA are mainly associated with the membrane vesicles of the cytopathic structures and not with the electron-dense material or the virus particles. (×80,000.)

Fig. 8. In CPMV-infected cells at longer periods after inoculation, tubules with virus-like particles are sometimes found. (×13,000.) The insert is a higher magnification (×26,000) to show the virus-like particles in the tubules more clearly. The tubules might arise from the plasmalemma protruding into the cytoplasm.

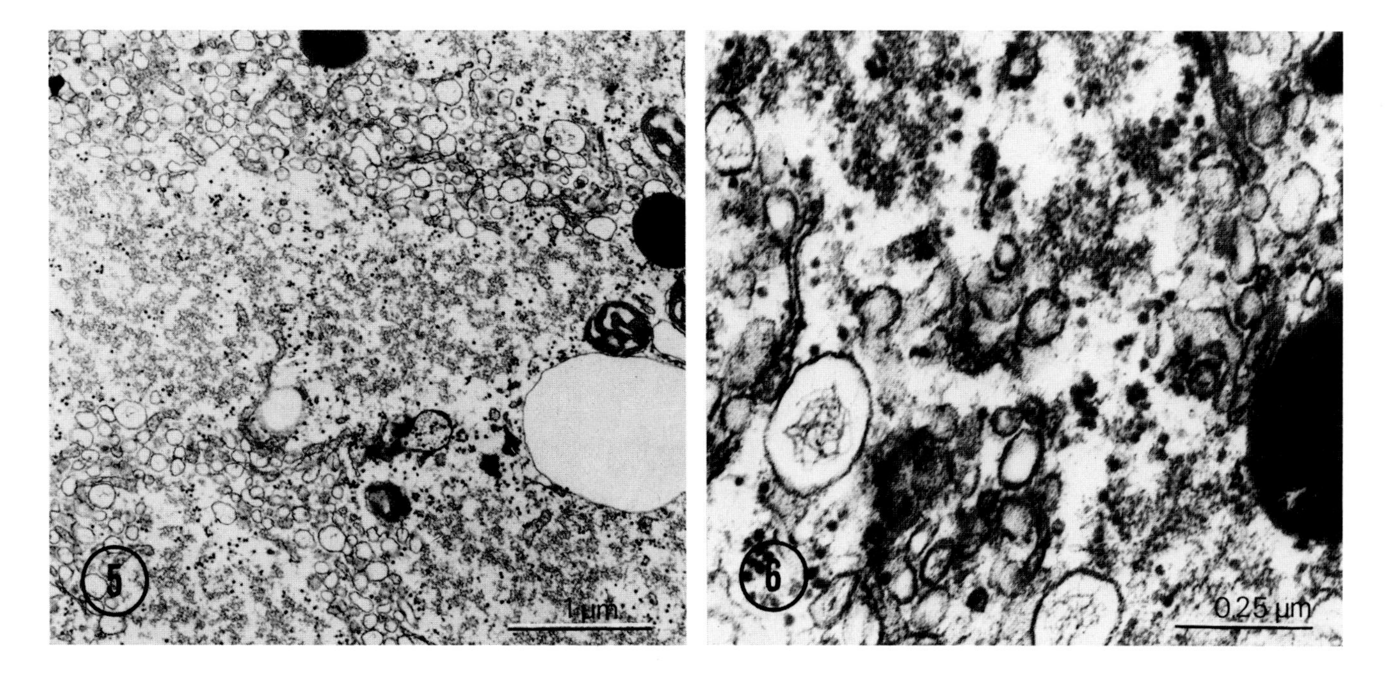
5
1 µm
6
0.25 µm

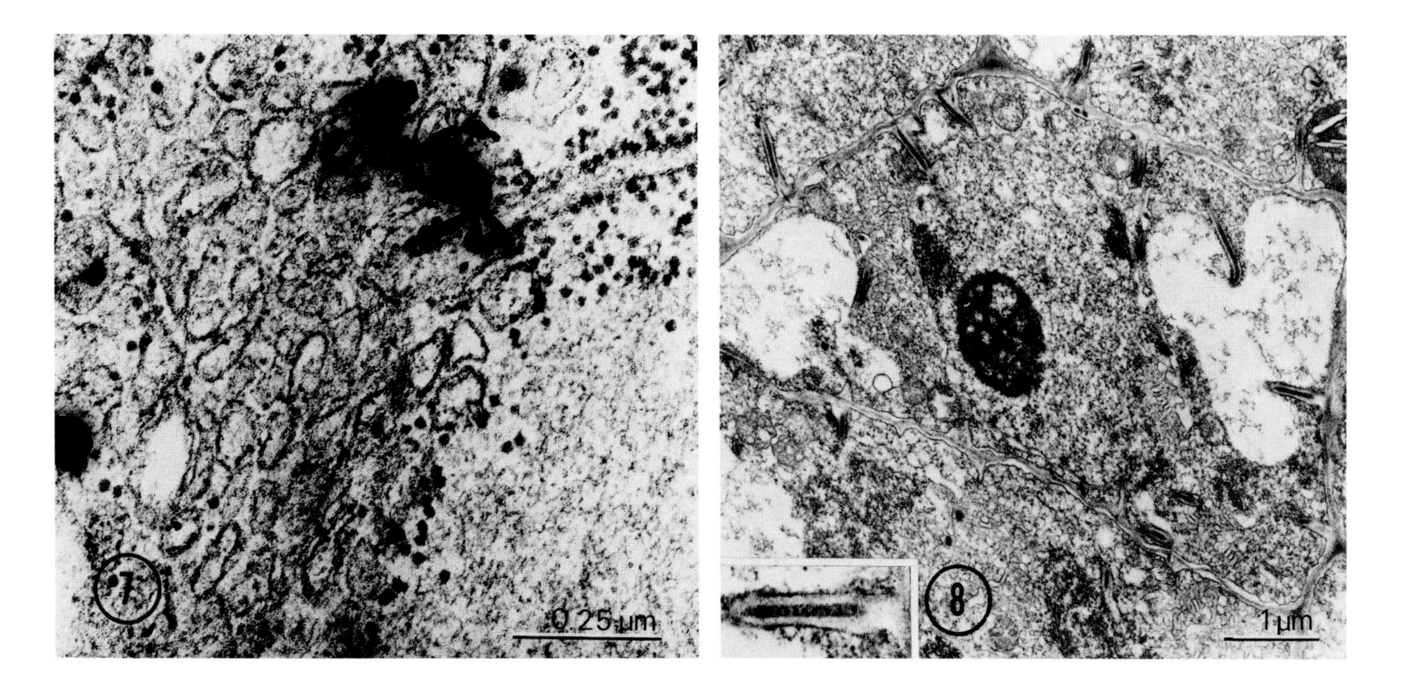
7
0.25 µm
8
1 µm

Chapter 10

Rhabdoviruses of Plants

G. P. MARTELLI AND M. RUSSO

I. Introduction 181
II. Morphological and Physicochemical Properties 186
A. Morphology and Structure of the Virion 186
B. The Surface Projections 187
C. The Viral Envelope 188
D. The Nucleocapsid 188
E. Hydrodynamic Properties 190
III. Serological Properties 190
IV. Interactions with the Host Cells 191
A. Intracellular Appearance and Distribution of Virus Particles 191
B. Morphogenesis 191
V. Relationships with Other Viruses 192
References 194

I. INTRODUCTION

Enveloped plant viruses with bacilliform or bullet-shaped particles and animal viruses having a comparable gross morphology are included, under the name of rhabdoviruses (Melnick and McCombs, 1966), in the same taxonomic group (Wildy, 1971). Plant rhabdoviruses, therefore, constitute a cluster of large RNA-containing entities characteristically endowed with a host-derived enveloping membrane. This feature differentiates them from the rest of known viruses infecting plants, except for tomato spotted wilt, which also possesses membrane-bound particles. Its morphology, however, is distinctively different (Best, 1968).

These viruses have been treated by Hull (1970) and more recently by Francki (1973) in an exhaustive review also containing detailed information on aspects which are not covered in this chapter. Additional articles reviewing, comparatively, animal and plant rhabdoviruses, although with a stronger emphasis on the former, have been written by Howatson (1970), Hummeler (1971), Knudson (1973), and Wagner (1975).

Since the first record of a rhabdovirus in a plant host (Herold *et al.*, 1960), a number of additional putative members of the group have been discovered, to

TABLE I

The Plant Rhabdovirus Group

Virus and host plants	Vector	Sap transmission	Site of maturation[a]	Reference (first morphologic observation)
Monocotyledons				
1. Maize mosaic (MMV)	Leafhopper	–	Nucleus	Herold *et al.* (1960)
2. American wheat striate mosaic (WSMV)	Leafhopper	–	Nucleus and cytoplasm	Lee (1964)
3. Northern cereal mosaic (NCMV)	Leafhopper	–	Cytoplasm ?	Shikata and Lu (1967)
4. Russian winter wheat mosaic (RWWMV)	Leafhopper	–	Cytoplasm	Razvyazkina and Polyakova (1967)
5. Rice transitory yellowing (RTYV)	Leafhopper	–	Nucleus	Shikata and Chen (1969)
6. Wheat chlorotic streak (WCSV)	Unknown	–	Cytoplasm	Signoret *et al.* (1972)
7. Barley yellow striate mosaic (BYSMV)	Leafhopper	–	Cytoplasm (viroplasm)	Conti (1969)
8. Ryegrass bacilliform (RBV)	Unknown	–	Nucleus	Plumb and James (1975)
9. *Dendrobium* × *Phalaenopsis* hybrid	Unknown	–	Nucleus (viroplasm) and cytoplasm ?	Ali *et al.* (1974)
10. Orchids (LORV)	Unknown	–	Cytoplasm	Lesemann and Doraiswamy (1975b)
11. *Colocasia esculenta*	Unknown	–	Cytoplasm (viroplasm) and nucleus ?	James *et al.* (1973)
12. Pineapple chlorotic leaf streak (PCLSV)	Unknown	–	Nucleus	Kitajima *et al.* (1975)
Dicotyledons				
1. Lettuce necrotic yellows (LNYV)	Aphid	+	Nucleus and cytoplasm (viroplasm)	Harrison and Crowley (1965)
2. Broccoli necrotic yellows (BNYV)	Aphid	+	Cytoplasm	Hills and Campbell (1968)
3. Potato yellow dwarf (PYDV)	Leafhopper	+	Nucleus	McLeod *et al.* (1966)

4. Sowthistle yellow vein (SYVV)	Aphid	–	Nucleus	Richardson and Sylvester (1968)
5. Sowthistle yellow net (SYNV)	Aphid	+	Nucleus	Christie *et al.* (1974)
6. Eggplant mottled dwarf (EMDV)	Unknown	+	Nucleus	Martelli (1969)
7. Strawberry crinkle (SCV)	Aphid	–	Cytoplasm	Richardson *et al.* (1972)
8. *Melilotus* latent (MLV)	Unknown	–	Nucleus	Kitajima *et al.* (1969)
9. Clover enation (CEV)	Unknown	–	Nucleus	Rubio-Huertos and Bos (1969)
10. Sugarbeet leaf curl (BLCV)	Tingid bug	–	Unknown	Eisbein (1970)
11. Lucerne enation	Unknown	–	Nucleus	Alliot *et al.* (1972)
12. *Gomphrena* (GV)	Unknown	+	Nucleus	Kitajima and Costa (1966)
13. *Atropa belladonna*	Unknown	–	Unknown	Lesemann (1972)
14. Red clover mosaic (RCMV)	Unknown	–	Nucleus	Vela and Rubio-Huertos (1974)
15. Melon variegation (MVV)	Unknown	–	Cytoplasm (viroplasm)	Rubio-Huertos and Peña-Iglesias (1973)
16. Stunting of *Chondrilla juncea*	Unknown	–	Nucleus	Hasan *et al.* (1973)
17. Plantain	Unknown	–	Unknown	Hitchborn *et al.* (1966)
18. Saintpaulia leaf necrosis	Unknown	–	Unknown	Ciampor and Dokoupil (1974)
19. *Euonymus* fasciations	Unknown	–	Nucleus	Codaccioni and Jonsson (1972)
20. *Cynara* rhabdovirus (CyRV)	Unknown	+	Cytoplasm	Peña-Iglesias *et al.* (1972)
21. Raspberry vein chlorosis (RVCV)	Aphid	–	Cytoplasm	Putz and Meignoz (1972)
22. *Laburnum* yellow vein (LYVV)	Unknown	–	Nucleus	Schultz and Harrap (1975)
23. *Sarracenia purpurea*	Unknown	–	Cytoplasm	Barckhaus and Weinert (1975)
24. Thyme virus (TV)	Unknown	–	Nucleus	Schultz *et al.* (1975)
25. Parsley rhabdovirus	Aphid	+	Unknown	Tomlinson and Webb (1974); Tomlinson (personal communication, 1976)

(*continued*)

TABLE I *(continued)*

Virus and host plants	Vector	Sap transmission	Site of maturation[a]	Reference (first morphologic observation)
Dicolyledons *(continued)*				
26. Pigeon pea proliferation	Leafhopper	–	Nucleus	Maramorosch *et al.* (1974)
27. *Sonchus* rhabdovirus	Unknown	+	Cytoplasm	Vega *et al.* (1976)
28. Carnation bacilliform virus	Aphid	–	Cytoplasm ?	K. S. Milne (personal communication, 1976)
29. *Pittosporum* vein yellowing	Unknown	–	Nucleus	Plavšić-Banjac *et al.* (1976)

[a] Question marks indicate that the virus may mature at that site, but better evidence would be desirable.

total about 40 (Table I). For some of them the inclusion in this grouping may appear questionable being based on scanty information (i.e., particle morphology only). Thus, for instance, the bacilliform virus associated with sugarbeet leafcurl (Eisbein, 1970) might qualify as a rhabdovirus, although better electron micrographic documentation would be desirable to substantiate this likelihood. Possible rhabdoviruses found in lupin (Tomlinson *et al.*, 1972), *Cajanus cajan* (Hirumi *et al.*, 1973; Maramorosch *et al.*, 1974), and tomato (Peña-Iglesias *et al.*, 1975) are too little known for proper designation. Others require additional studies such as the bacillary entity associated with enation of lucerne in Spain (Rodriguez-Sardiña and Torres-Barros, 1975) which should be compared with a similar virus found in France (Alliot *et al.*, 1972), to establish possible similarities or identities. Also, more accurate investigations are required to establish the extent of the relationship, if any, between the mechanically transmissible rhabdovirus isolated from *Sonchus* in Argentina (Vega *et al.*, 1976) and the agents of broccoli necrotic and lettuce necrotic yellows.

Finally, there is a whole group of smaller-sized bacilliform viruses observed in orchids and other plants (Petzold, 1971; Kitajima and Costa, 1972; Kitajima *et al.*, 1972) whose rhabdovirus nature is not yet unequivocally established, as will be discussed later.

Rhabdovirus particles are structurally complex and very unstable both in crude extracts of plants or insects and in clarified plant sap. Their thermal inactivation point is about 50° C and the longevity *in vitro* at room temperature lasts a few hours. They do not withstand treatment with organic solvents and need to be stabilized in various ways during extraction and purification for retaining infectivity (Francki, 1973).

In nature, rhabdoviruses infect many different plant species belonging to widely separated botanical families (Table I). Among dicotyledons, potato yellow dwarf (PYDV) and lettuce necrotic yellows (LNYV) viruses cause diseases of economic importance (Black, 1937; Stubbs and Grogan, 1963), whereas eggplant mottled dwarf (EMDV) and lucerne enation (LEV) viruses are potentially destructive pathogens owing to the severity of the syndromes they elicit (Martelli and Cirulli, 1969; Alliot *et al.*, 1972). In monocotyledonous plants, at least six of the eight known cereal rhabdoviruses are responsible for severe field diseases (Francki, 1973).

The symptomatological response of the hosts is extremely varied ranging from various patterns of chromatic discoloration of the foliage to flower color break, localized to extensive necrosis, stunting, and unfruitfulness. A few rhabdoviruses infect their hosts symptomlessly. In any case, the plants are always invaded systemically and virions can be found in all tissues (epidermis to conducting elements) and different organs, where they reach fairly high concentrations.

Most of the rhabdoviruses have no known vector, seven are transmitted by leafhoppers, eight by aphids, and, oddly enough, one by a tingid bug (Table I). The transmission is of the persistent type and, in several instances, the viruses have been reported to multiply in the vector. Transovarial passage has been recorded for the *Agallia constricta* form of PYDV (Black, 1953), sowthistle yellow vein virus (SYVV) (Sylvester, 1969), and LNYV (Francki, 1973). Some isolates of SYVV are actually pathogenic to the vectors whose longevity is reduced (Sylvester, 1973; Sylvester and Richardson, 1971). A small minority of plant rhabdoviruses (8 out of 41) can be transmitted by inoculation of sap to a rather restricted range of herbaceous plants. Comparisons of the host range responses are not thoroughly indicative of close similarities between these viruses (Francki, 1973), although some of them infect the same test plants inducing somewhat similar symptoms.

II. MORPHOLOGICAL AND PHYSICOCHEMICAL PROPERTIES

A. Morphology and Structure of the Virion

Although two morphologically distinct types of particles, i.e., bacilliform with both ends rounded and bullet-shaped with one rounded and one planar end, are normally found with most plant rhabdoviruses, both *in situ* (thin sections) and *in vitro* (leaf dips or purified preparations), it is generally agreed, with a few exceptions (Wagner, 1975), that the mature forms are bacilliform in shape (Howatson, 1970; Hummeler, 1971; Knudson, 1973; Francki, 1973). Hence, bullet-shaped elements are either immature virions in various stages of development, or they originate from the bacilliform ones as a consequence of preparative artifacts. Rice transitory yellowing virus (RTYV) represents a noteworthy exception, for its particles are reported to be prevalently of the bullet type both in the host cells and in free preparations. Only occasional bacilliform virions can be observed (Chen and Shikata, 1971). In this respect RTYV comes close to animal rhabdoviruses whose particle shape is thought to be bullet-like (Wagner, 1975). However, evidence has recently been obtained that an animal rhabdovirus (i.e., vesicular stomatitis virus) also has true bacilliform particles (Orenstein *et al.*, 1976).

Aberrant particles may easily arise during manipulation (i.e., negative stain mounts) for electron microscopic observations. Excellent examples of particle variability due to preparative artifacts have been reported for LNYV, and the explanation of their possible origin given (Harrison and Crowley, 1965; Wolanski *et al.*, 1967; Wolanski and Francki, 1969; Francki, 1973). Fixation with glutaraldehyde prior to negative staining helps retain the true particle shape (McLeod, 1968; Peters and Kitajima, 1970; Ahmed *et al.*, 1970; Russo and Martelli, 1973).

The size of the particles varies a great deal not only among different members of the group but also for the same virus depending on method of measurement: virions in thin sections or in negatively stained preparations. Usually, dehydrated particles have a diameter lower (up to 30%) than those of comparable material in PTA mounts. The reported range of variation in breadth is from 45 (BYSMV) to 100 nm (SYVV). Particle length seems much less affected by the mode of preparation. Thus, most of the recorded differences are intrinsic to the diverse viruses, which range in length from the 130 nm of the bullet RTYV (Chen and Shikata, 1971) and the 150 nm of the bacilliform *Chondrilla juncea* stunt virus (Hasan *et al.*, 1973) to the 430–500 nm of raspberry vein chlorosis virus (Stace-Smith and Lo, 1973; Jones *et al.*, 1974).

Negatively stained rhabdoviruses may exhibit different properties, depending on the extent of stain penetration into the virion. This is evidently related to the degree of damage suffered by the outer envelope. Thus, superficial penetration of PTA shows the hexamer layer of the particles (Fig. 3), whereas a deeper penetration resolves the inner component (coiled nucleocapsid), which appears as a regularly cross-banded structure (Figs. 1 and 2) with striations approximately 45–55 Å apart (Howatson, 1970; Francki, 1973).

When transversely sectioned in the host cells, the particles appear as multilayered circular bodies. The number of layers in each particle (i.e., 2 or 3) does not seem to reflect intrinsic structural differences between diverse members of the group, but, rather, depends on whether the envelope is resolved as a single or a double-layer entity. This interpretation, originally given for LNYV (Wolanski and Chambers, 1972), was supported by subsequent studies on EMDV, demonstrating how the outer look of cross-sectioned virions could be drastically affected by fixing, embedding and staining procedures so as to appear single, double, or triple-layered (Martelli and Castellano, 1970; Russo and Martelli, 1973). The nucleocapsid, however, always appears as a single inner ring irrespective of the mode of tissue processing (Figs. 4, 5, and 6).

Based on electron microscopic evidence derived from profiles of negatively stained or sectioned particles and on available physicochemical data, it seems generally agreed that a same architectural model applies to both animal and plant rhabdoviruses (Howatson, 1970; Knudson, 1973; Francki, 1973; Wagner, 1975; Hull, 1976). Schematically, this model consists of a helically wound nucleoprotein filament enveloped by a lipoprotein membrane studded with an array of short spikes protruding from its surface (Fig. 7).

As yet, it is unclear if the central canal, i.e., the area internal to the nucleocapsid helix is hollow or contains materials mimicking an inner core. Francki (1973) maintains that there is no evidence for the existance of organized structures in the center of the virions. Therefore, he suggests that the ring-like constituent, 18–20 nm in diameter, shown by cross-sectioned particles of many different rhabdoviruses, derives from fixation or staining artifacts. Although this may also be true, it seems interesting that in EMDV the presence of an inner axial core and its level of resolution were shown to be dependent on the fixing and dehydrating schedule used, as though certain treatment prevented removal of materials from the interior of the particles whereas others did not (Russo and Martelli, 1973). In any case, if the central core is an artifact, it is indeed a very consistent one, as it shows up regularly in micrographs produced from different laboratories, even in those produced recently (Jones *et al.*, 1974; Christie *et al.*, 1974; Plumb and James, 1975; Lesemann and Doraiswamy, 1975b; Russo *et al.*, 1975; Martelli *et al.*, 1975; Lawson and Ali, 1975).

B. The Surface Projections

Electron microscopy of negatively stained (Figs. 1 and 2) or sectioned (Figs. 5 and 6) rhabdovirions reveals the presence of surface projections lining the particle contour and protruding 6–10 nm from the envelope. Recent evidence obtained with vesicular stomatitis virus (VSV), an animal rhabdovirus, indicates that the projections are not superficial structures but are associated with the nucleocapsid, thus penetrating the whole thickness of the enveloping membrane (Brown *et al.*, 1974). The spikes cover the whole surface of the virion and are hexagonally arranged (Francki, 1973; Hull, 1976). Actually, the lattices found in optical diffractions of different rhabdoviruses are believed to arise from the hexagonal disposition of surface projections (Hull, 1976). These are composed of glycoproteins (protein G), each projection possibly being made up by two glycopolypeptides (Hull, 1976). In VSV the projections seem to represent the only source of glycosylated proteins (Wagner, 1975), whereas this may not apply to SYVV, where L protein is also glycosylated (Ziemiecki and Peters, 1976). However, in plant as in animal rhabdoviruses, it is conceivable that protein G is the primary source of glycopolypeptides.

A protein species with molecular weight similar to that of protein G of animal rhabdoviruses was identified in PYDV, LNYV, and SYVV, the only plant rhabdoviruses whose protein composition has been analyzed so far (Table II). Furthermore, the occurrence of neuraminic acid, a sugar contained also in the glycopolypeptides of VSV (McSharry and Wagner, 1971) was ascertained in SYVV (Lee *et al.*, 1972). In American wheat striate mosaic virus (WSMV), about 3% of the particle weight is suspected to be carbohydrates (Sinha and Becki, 1972) and sugars were also detected in maize mosaic virus (MMV) (Lastra and Acosta, 1975).

TABLE II

Chemical Composition of Some Plant Rhabdoviruses

Virus	Lipids	RNA MW × 10^6	Proteins MW × 10^3						Reference
			L	G	N	NS	M_1	M_2	
Potato yellow dwarf	20%	4.6	Present	78	56	45	33	22	Knudson and McLeod (1972)
Lettuce necrotic yellows	Present	4.0	>100	75	55	—	—	22	Francki and Randles (1975)
Sowthistle yellow vein	Present	—	150	83	60	—	44	36	Ziemiecki and Peters (1976)

The biological role of surface projections in plant rhabdoviruses is unknown. In VSV they take part in the initiation of infection, for G protein seems to be required for particle attachment to the plasma membrane of the host cells (Wagner, 1975). A similar function was postulated for plant rhabdoviruses with regard to their attachment to the plasma membrane of insect cells where they multiply (Francki, 1973). In this connection, it is worth mentioning that hemagglutination activity, for which spike glycoproteins are responsible, was detected in PYDV (Knudson, 1973).

C. The Viral Envelope

The envelope of plant rhabdoviruses constitutes a bilayered membrane about 10 nm thick. There is plenty of circumstantial and experimental evidence that it contains proteins and lipids (Francki, 1973). Chemical analysis of PYDV, WSMV, and MMV has indicated the lipid content to be above 20, 24, and 40% of the particle weight, respectively (Ahmed *et al.*, 1964; Sinha and Becki, 1972; Lastra and Acosta, 1975). The protein moiety consists of the matrix or M protein which may occur as two species (M_1 and M_2) with a different molecular weight (Table II). The matrix protein is believed to form a tubular structure (hexamer layer) surrounding the nucleocapsid, on which it exerts a stabilizing function (Knudson, 1973; Hull, 1976). It is made up of subunits which, as discussed by Hull (1976), seem to be arranged in individual hexamers. M proteins are structurally interrelated with G proteins which, as mentioned previously, are also disposed in a hexagonal lattice. Since available quantitative data on the protein composition of animal and plant rhabdoviruses indicate that M and G proteins occur in a 3:1 ratio (Hull, 1976), a structural model fitting the above ratio is that proposed for VSV, in which M and G proteins form separate superimposed hexamers with each G protein subunit surrounded by three M protein subunits (Cartwright *et al.*, 1972) (Fig. 8). In such a model the end capping would have icosahedral simmetry as in other tubular viruses (Hull, 1976).

D. The Nucleocapsid

A nucleoprotein helix forming a hollow cylinder constitutes the internal component of rhabdoviruses. The nucleic acid moiety is a molecule of single-stranded RNA with molecular weight equal to or higher than 4 × 10^6 daltons

(Table II). Estimates of RNA percentage vary a great deal and were reported to be 0.6% in PYDV (Knudson, 1973) and 5% in WSMV (Sinha and Becki, 1972). Interestingly, the viral RNA is a "negative strand" which is not infectious by itself (i.e., when deprived of proteins) and does not serve as a messenger. Therefore, it requires transcription for translation to occur (Baltimore, 1971). In VSV, transcriptase activity proved to be associated with the nucleocapsid residing in the L and, as a cofactor, in NS proteins (Wagner, 1975). A similar situation is likely to occur in plant rhabdoviruses. In fact, an endogenous nucleocapsid-associated RNA-dependent RNA polymerase (transcriptase) was detected also in LNYV (Francki and Randles, 1972, 1973, 1975).

The bulk of the protein moiety of the nucleocapsid is the N protein which is a structural constituent tightly bound to the RNA filament. The consensus is that the subunits of N protein are helically arranged along the cylindrical portion of the nucleocapsid. However, different views exist on the conformation of the emispherical end of the bullet. This, according to some authors (Nakai and Howatson, 1968; Vernon *et al.*, 1972; Wagner, 1975), is made up by several turns of the helix of diminishing diameter, whereas others propose that it is constructed from protein subunits arranged in cubic symmetry (Peters and Schultz, 1975).

Subunits of N proteins are likely to be interrelated with the adjacent G proteins but the way in which this is achieved in not yet clearly understood (Hull, 1976).

Nucleocapsids of animal rhabdoviruses are evidently bullet-shaped (Howatson, 1970; Wagner, 1975). Claims have been made that plant rhabdoviruses, irrespective of the bacilliform shape of the particles, also contain a bullet-like internal component. Convincing evidence of this was provided for SYVV, whose particles, after glutaraldehyde fixation, show an inner component hemispherical at one end and planar at the other (Peters and Kitajima, 1970). Furthermore, removal of the viral envelope by detergent or enzymatic treatment again reveals a bullet-shaped nucleocapsid (Figs. 9, 10, and 12) that uncoils starting from the blunt end (Fig. 11). With several other rhabdoviruses, indications of the presence of bullet-like internal components, either in a unenveloped state (Fig. 14) or enclosed within bacilliform particles, can be clearly obtained from profiles of longitudinally cut virions in thin sectioned cells (Rubio-Huertos and Bos, 1969; Kitajima *et al.*, 1969; Russo and Martelli, 1972; Lawson and Ali, 1975).

Despite these findings, which support Peters and Kitajima's (1970) hypothesis, the notion that bacilliform rhabdovirions contain a nonbacilliform nucleocapsid does not meet with general favor. Thus, Francki (1973) stated that the bacillary shape of plant rhabdoviruses arises from an end-to-end fusion of two bullets, specifying that each LNYV or PYDV particle may enclose two viral nucleocapsids in a back-to-back position. Similar views are shared by Wagner (1975).

Although the above suggestion is intriguing, we feel that it is not sufficiently substantiated by experimental evidence. For example, the occurrence in many rhabdoviruses of particles with a median cleavage (i.e., two bullets abutted to each other), which was advocated as possible supporting proof of this hypothesis, seems to be against rather than in favor of it. In fact, such particles are often referred to as "long bacilliform," having about twice the length of the supposedly mature virions (Kitajima *et al.*, 1969; Martelli and Castellano, 1970; Sinha, 1971; Lee and Peters, 1972; Hasan *et al.*, 1973; Vela and Rubio-Huertos, 1974; Lawson and Ali, 1975). Hence, they are indeed true doublets formed by two nucleocapsids encased in the same enveloping membrane but should not be regarded as ordinary infectious units. The infectious units are identifiable with the shorter bacilliform particles which, as mentioned above, are liable to possess a bullet-shaped helical component.

E. Hydrodynamic Properties

The difficulty of obtaining satisfactorily purified preparations has greatly hampered studies for thorough characterization of plant rhabdoviruses. As yet, no more than six or seven members of the group have been successfully purified, and incomplete biophysical data are available for only four of these. Thus, the following are the reported values of sedimentation coefficient and buoyant density, respectively: (a) LNYV = 945 ± 16 S; 1.19–1.20 gm/cm^3 (Harrison and Crowley, 1965); (b) PYDV = 810–950 S; 1.17 gm/cm^3 (Brakke, 1958); (c) BNYV = 847 ± 41 S; 1.18–1.19 gm/cm^3 (Lin and Campbell, 1972); (d) WSMV = 900 S (Sinha and Becki, 1972).

Molecular weight estimates for LNYV and PYDV range from 450 to 1400 × 10^6 daltons, depending on particle size (Knudson, 1973).

III. SEROLOGICAL PROPERTIES

Information on the antigenic constitution of plant rhabdoviruses is scanty, for very few of them have been investigated serologically.

Since intact virions, because of their size, do not diffuse through agar media, gel diffusion tests are of limited value unless virus particles are previously degraded. Soluble antigens, however, can easily be picked up by this method (Thottapilly and Sinha, 1973). With WSMV, techniques such as precipitin ring and precipitin tube tests proved to be more sensitive than agar gel tests for detecting and titrating antigens and relative antibodies (Sinha and Thottapilly, 1974).

Although rhabdoviruses in general appear to be rather poor immunogens, as judged from the low titer of their homologous antisera, a serum anti-SYVV was prepared with a titer of 1/10,000 (Peters and Black, 1970). The use of immune sera free from antibodies to normal plant constituents has revealed the absence of healthy host-specific antigens in rhabdovirions, thus indicating that viral infection modifies profoundly the chemical composition of the host membrane from which the particle envelope originates. This is an established fact with some animal rhabdoviruses where viral G and M proteins synthesized in infected cells are inserted into cytoplasmic membranes prior to particle budding (Wagner, 1975).

However, it is conceivable that virions with such a complex structure have a comparably complex antigenic nature. In fact, serological studies on LNYV (McLean *et al.*, 1971), BNYV (Lin and Campbell, 1972), and WSMV (Thottapilly and Sinha, 1973) have provided evidence that all these viruses have indeed three or more antigens which are presumably associated with different subviral particulates. So far, a direct correlation between antigenic and structural components has been worked out only for the two varieties of PYDV, in which four different antigens with diverse serotype specificity were identified with G, N, M_1, and M_2 proteins (Knudson and McLeod, 1972).

To our knowledge, serological relationships between different plant rhabdoviruses have not been investigated. No cross-reaction was found in the only reported attempt to look for a possible serological correlation between an animal (VSV) and a plant (SYVV) rhabdovirus (Hackett *et al.*, 1968).

IV. INTERACTIONS WITH THE HOST CELLS

A. Intracellular Appearance and Distribution of Virus Particles

As mentioned previously, the *in situ* appearance of plant rhabdoviruses may be bullet-shaped, bacilliform, or long bacilliform. These correspond, respectively, to profiles of (a) immature particles in the process of being formed, (b) mature virions, and (c) dimeric particles with two nucleocapsids enveloped together. Particles having both basic shapes (i.e., bullet and bacillary) are also encountered in cells of viruliferous vectors (Herold and Munz, 1965; O'Loughlin and Chambers, 1967; Sylvester and Richardson, 1970; Chen and Shikata, 1972) and in insect cell culture (Chiu *et al.*, 1970). Rhabdovirions are easily identified in infected cells owing to their large size and to their peculiar outward appearance which differentiates them from ordinary cell constituents. They are seldom isolated, tending to accumulate in large numbers, sometimes in a paracrystalline array (Fig. 4) near the nucleus or in the cytoplasm. Whatever their localization, the particles are never free in the surrounding medium but occur in membrane-bound enclaves, i.e., areas between the lamellae of the nuclear envelope (Figs. 13, 14, 15, and 17) or cytoplasmic vesicles (Fig. 16). These vesicles may originate either from a dilation of preexisting endoplasmic reticulum cisternae, from *de novo* synthesis of intracytoplasmic membranes, or from membranous elements pinched off the outer lamella of the nuclear envelope, carrying virus particles in the cytoplasm away from the nucleus.

Unenveloped nucleocapsids have also been observed in the nucleus (Fig. 14) and cytoplasm (Fig. 19) of infected plant and insect cells, occasionally in massive aggregates (Kitajima and Costa, 1966; O'Loughlin and Chambers, 1967; Richardson and Sylvester, 1968; Rubio-Huertos and Bos, 1969; Kitajima *et al.*, 1969).

B. Morphogenesis

Cell membranes are undoubtedly involved in rhabdovirus morphogenesis, as they represent the site of particle assembly. It is generally agreed that animal rhabdovirions mature by budding at intracytoplasmic or marginal membranes, whereas those infecting plants acquire their coating at the inner lamella of the nuclear envelope (Figs. 14 and 15) or from the endoplasmic reticulum (Fig. 20) (Howatson, 1970; Hull, 1970; Hummeler, 1971; Knudson, 1973; Francki, 1973; Wagner, 1975). Interestingly, plant rhabdoviruses were never seen budding from plasma membranes nor when multiplying in the vector cells. They seem to retain the same site of maturation in insect and plant hosts. This was particularly well illustrated by Chiu *et al.* (1970) with PYDV in leafhopper cell culture.

Although some of the plant rhabdoviruses appear to develop in close association with the nucleus and others are formed in the cytoplasm (Table I); it is presently unsafe to draw a sharp separation between the two groups. Such a subdivision seems to be supported by the existing differences in the particle diameters of nuclear and cytoplasmic viruses (Hull, 1970), but still we agree with Knudson (1973) and Francki (1973) that much more work needs to be done for a better understanding of where and how many of the reported rhabdoviruses are assembled. Nevertheless, with a certain number of them it is possible to be reasonably confident that the nucleus is the primary, if not the only, site of particle production. In these instances the direct involvement of such an organelle in virus replication is also supported by the occurrence of various types of intranuclear inclusions in infected cells (Kitajima and Costa, 1966; Kitajima *et al.*, 1969; Rubio-Huertos and Bos, 1969; Vela and Lee, 1974; Lawson and Ali, 1975) or by the "washed out" appearance of the nucleoplasm (Figs. 13 and 17), accompanied by the nonstaining condition of the central area (Martelli and Castellano, 1970; Chen and Shikata, 1971; Christie *et al.*, 1974; Kitajima *et al.*, 1975; Martelli *et al.*, 1975) caused by the depletion of nucleohistones (Russo and Martelli, 1975). Often, such nuclear alterations appear to be the only cytopathic effects of any consequence arising from viral infection.

True cytoplasmic rhabdoviruses are more difficult to identify, since the occurrence of an early nuclear phase in multiplication cannot be excluded when one considers the findings relative to LNYV (Wolanski and Chamber, 1971). It is evident, however, that in certain stages of infection, which might not necessarily be late ones, virus production is not associated with the nucleus. A primary example of this is provided by particle formation in cytoplasmic "viroplasms" (Figs. 21 and 22) as in BYSMV (Conti and Appiano, 1973). Moreover, there are indications that supposedly cytoplasmic rhabdoviruses, when maturing in close proximity to the nucleus, acquire their coat from the outer rather than from the inner lamella of the nuclear envelope (Fig. 18) (Russo *et al.*, 1975).

As for the process leading to particle development, there is increasing agreement that it might be similar for plant and animal rhabdoviruses. Recently, Peters and Schultz (1975) proposed a unitarian model of particle morphogenesis for both groups of viruses based on two main concepts: (1) ribonucleoprotein strands are coiled into nucleocapsids starting from the hemispherical end and (2) coiling is concurrent with envelopment.

Such an ingenious model, however, does not take into account the possibility of nucleocapsid formation independent of coating nor of lateral envelopment of preassempled nucleocapsids. For these reasons, Peters and Schultz (1975), contrary to current views, interpret aspects such as those illustrated in Fig. 14 as steps in a refusion process by which the viral envelope is being reincorporated in the inner nuclear lamella. This would then be followed by release of the stripped off nucleocapsids into the karyoplasm where they accumulate.

Lack of adequate time–course studies makes it practically impossible to verify the validity of the above series of events. However, it seems difficult to reconcile this hypothesis with cases, such as the *Gomphrena* virus, in which nucleocapsids appear to develop from a modified nucleolar matrix to which many of them are still attached (Kitajima and Costa, 1966) or with the occurrence of naked nucleocapsids within and around nonmembrane-bound viroplasms (Conti and Appiano, 1973; Lawson and Ali, 1975). Moreover, lateral ensheathment of two preassembled bullet nucleocapsids could explain the formation of dimeric particles, perhaps more readily than the mechanisms proposed by Peters and Schultz (1975), i.e., joining of the envelopes of two particles at the moment of detachment or assembly before particle detachment of a second nucleocapsid starting at the blunt end of the first one. In this regard, it is intriguing that a very high proportion of dimeric virions were found in the perinuclear spaces of cells infected with red clover mosaic virus (RCMV), a situation in which, according to published micrographs, nucleocapsids being apparently side-coated are plentiful but particles budding "head on" at the nuclear membrane are hardly visible (Vela and Rubio-Huertos, 1974).

V. RELATIONSHIPS WITH OTHER VIRUSES

A few years ago, the occurrence of bacillary or bullet-shaped particles in orchid plants with necrotic ringspots were reported from several European laboratories (Düvel and Peters, 1971; Lesemann and Begtrup, 1971; Petzold, 1971). Similar particles, which for practical purposes will be referred to as KORV ("kurze Orchideen-Rhabdoviren", Lesemann and Doraiswamy, 1975b), were subsequently observed in diseased coffee (Kitajima and Costa, 1972) and citrus (Kitajima *et al.*, 1972) plants, as well as in numerous orchids from Brasil, Japan, and the United States (Kitajima *et al.*, 1974; Cho *et al.*, 1973; Lesemann and Doraiswamy, 1975a). Presently, such particles have been recorded in over 40 orchid species of at least 20 different genera (Kitajima *et al.*, 1974; Lesemann and Doraiswamy, 1975a).

Strictly speaking, these structures should only be regarded as virus-like particles, for their infectivity has not yet been experimentally established, except for a single report of successful manual transmission (Cho *et al.*, 1973). Nevertheless, the morphological and structural features of KORV particles and the relationship they establish with the host cells leave little doubt of their viral nature. In addition, in citrus these particles are associated with a disease (citrus leprosis) which is apparently transmitted by a mite (Knorr, 1968; Rossetti *et al.*, 1969).

The question then arises of whether these viruses are rhabdoviruses or not. Although several authors favor the rhabdovirus hypothesis (Petzold, 1971; Düvel and Peters, 1971; Lesemann and Begtrup, 1971; Kitajima *et al.*, 1972) others regard them as a sort of *species inquirendae* (Knudson, 1973; Francki, 1973; Peters and Schultz, 1975).

Indeed, at first sight, some of these entities recall baculoviruses of insects, but, on closer examination, their differences become more apparent, owing to the basic differences existing in the nucleocapsid structure and mode of envelope acquisition (for a review, see David, 1975).

Remarkable differences also exist with rhabdoviruses: (1) KORV particles are smaller. Their nucleocapsids in negative stain mounts measure 40–50 × 105–120 nm, and in thin sections 30–38 × 100–128 nm (see discussion in Lesemann and Doraiswamy, 1975a). Enveloped mature particles are rarely seen and, when present, measure 55–65 × 120 nm (monomers) and 55–65 × 250 (dimers) (Lesemann and Doraiswamy, 1975a). (2) Particles are consistently present in an uncoated form within nuclei of infected cells but never accumulate perinuclearly. (3) Infected nuclei have a sinous outline because of the many blebs of variable size which protrude outward into the cytoplasm (Fig. 23). These blebs are local evaginations of the nuclear envelope (both lamellae), on the inner side of which viral nucleocapsids are perpendicularly attached (Fig. 24). It is conceivable that the vesicles, upon detachment from the nucleus, give rise to the globular "spokewheel-like" structures (Fig. 25) which are highly characteristic of KORV (Petzold, 1971; Lesemann and Begtrup, 1971; Kitajima *et al.*, 1972, 1974; Lesemann and Doraiswamy, 1975a). (4) The mechanism of particle envelopment is obscure. Nucleocapsids egrees from the nucleus either by means of "spokewheels" or directly through the envelope without acquiring necessarily the outer membrane (Lesemann and Doraiswamy, 1975a).

KORV similarities to rhabdoviruses are as follows: KORV have bullet-shaped nucleocapsids helically wound (Begtrup, 1972); an outer envelope with surface projections; and, most probably, a hexamer layer (Lesemann and Doraiswamy, 1975a).

Because of these structural characteristics Lesemann and Doraiswamy (1975a,b) have reaffirmed the eligibility of KORV as a member of the rhabdovirus group. However, although the ultrastructural evidence seems to make this idea tenable, an unequivocal classification of these viruses will have to await the support of physicochemical data. In any case, should these entities prove to be true rhabdoviruses, obviously they will constitute a subgroup distinct from that which includes the members presently discussed.

Finally, virus-like bacilliform particles averaging 24 × 70 nm were observed in *Mycoplasma*-infected aster cells (Allen, 1972). Although there are indications that these particles may possess an envelope, they hardly resemble rhabdovirions. In fact, it is not even known whether they are plant or *Mycoplasma* pathogens (Allen, 1972).

ACKNOWLEDGMENTS

Grateful thanks are due to Drs. A. Appiano, R. Hull, E. W. Kitajima, D. Lesemann, D. Peters, M. Rubio-Huertos, E. Shikata, K. S. Milne, and J. A. Tomlinson for kindly providing illustrations, unpublished data, or manuscripts prior to publication. The work carried out in our laboratory was supported by the Consiglio Nazionale delle Ricerche, Rome.

REFERENCES

Ahmed, M. E., Black, L. M., Perkins, E. G., Walker, B. L., and Kummerov, F. A. (1964). *Biochem. Biophys. Res. Commun.* **17,** 103.
Ahmed, M. E., Sinha, R. C., and Hochster, R. M. (1970). *Virology* **41,** 768.
Ali, S., Lawson, R H., and Ishii, M. (1974). *Am. Orchid Soc. Bull.* **43,** 529.
Allen, T. C. (1972). *Virology* **47,** 491.
Alliot, B., Giannotti, J., and Signoret, P. A. (1972). *C. R. Hebd. Seances Acad. Sci.* **274,** 1974.
Baltimore, D. (1971). *Bacteriol. Rev.* **35,** 235.
Barckhaus, R. H., and Weinert, H. (1975). *Protoplasma* **84,** 101.
Begtrup, J. (1972). *Phytopathol. Z.* **75,** 268.
Best, R. J. (1968). *Adv. Virus Res.* **13,** 65.
Black, L. M. (1937). *N. Y. Agric. Exp. Sta., Ithaca, Mem.* **209.**
Black, L. M. (1953). *Phytopathology* **43,** 9.
Brakke, M. K. (1958). *Virology* **6,** 96.
Brown, F., Smale, C. J., and Horzinek, M. C. (1974). *J. Gen. Virol.* **22,** 455.
Cartwright, B., Smale, C. J., Brown, F., and Hull, R. (1972). *J. Virol.* **10,** 256.
Chen, M.-J., and Shikata, E. (1971). *Virology* **46,** 786.
Chen, M.-J., and Shikata, E. (1972). *Virology* **47,** 483.
Chiu, R. J., Lin, H. Y., McLeod, R., and Black, L. M. (1970). *Virology* **40,** 387.
Cho, S., Arai, K., Doi, Y., and Yora, K. (1973). *Ann. Phytopathol. Soc. Jpn.* **39,** 171.
Christie, S. R., Christie, R. G., and Edwardson, J. R. (1974). *Phytopathology* **64,** 840.
Ciampor, F., and Dokoupil, M. (1974). *Acta Virol. (Engl. Ed.)* **18,** 355.
Codaccioni, M., and Jonsson, G. (1972). *Bull. Soc. Bot. Fr.* **119,** 51.
Conti, M. (1969). *Phytopathol. Z.* **66,** 275.
Conti, M., and Appiano, A. (1973). *J. Gen. Virol.* **21,** 315.
David, W. A. L. (1975). *Annu. Rev. Entomol.* **20,** 97.
Düvel, D., and Peters, K. R. (1971). *Gartenwelt* **71,** 52.
Eisbein, K. (1970). *Zentralbl. Bakteriol., Parasitenkd., Infectionskr. Hyg.* **125,** 515.
Francki, R. I. B. (1973). *Adv. Virus Res.* **18,** 257.
Francki, R. I. B., and Randles, J. W. (1972). *Virology* **47,** 270.
Francki, R. I. B., and Randles, J. W. (1973). *Virology* **54,** 359.
Francki, R. I. B., and Randles, J. W. (1975.). *In* "Negative Strand Viruses" (B. W. J. Mahy and R. D. Barry, eds.), Vol. 1, p. 185. Academic Press, New York.
Hackett, A. J., Sylvester, E. S., Richardson, J., and Wood, P. (1968). *Virology* **36,** 693.
Harrison, B. D., and Crowley, N. C. (1965). *Virology* **26,** 297.
Hasan, S., Giannotti, J., and Vago, C. (1973). *Phytopathology* **63,** 791.
Herold, F., and Munz, K. (1965). *Virology* **25,** 412.
Herold, F., Bergold, G. H., and Weibel, J. (1960). *Virology* **12,** 335.
Hills, G. J., and Campbell, R. N. (1968). *J. Ultrastruct. Res.* **24,** 134.
Hirumi, H., Maramorosch, K., and Hichez, E. (1973). *Phytopathology* **63,** 202.
Hitchborn, J. H., Hills, G. J., and Hull, R. (1966). *Virology* **28,** 768.
Howatson, A. F. (1970). *Adv. Virus Res.* **16,** 195.
Hull, R. (1970). *In* "The Biology of Large RNA Viruses" (R. D. Barry and B. W. J. Mahy, eds.), p. 153. Academic Press, New York.

Hull, R. (1976). *Adv. Virus Res.* **20,** 1.
Hummeler, K. (1971). *In* "Comparative Virology" (K. Maramorosch and E. Kurstak, eds.), p. 361. Academic Press, New York.
James, M., Kenten, R. H., and Woods, R. D. (1973). *J. Gen. Virol.* **21,** 145.
Jones, A. T., Roberts, I. M., and Murant, A. F. (1974). *Ann. Appl. Biol.* **77,** 283.
Kitajima, E. W., and Costa, A. S. (1966). *Virology* **29,** 523.
Kitajima, E. W., and Costa, A. S. (1972). *Cienc. Cult. (Sao Paulo)* **24,** 542.
Kitajima, E. W., Lauritis, J. A., and Swift, H. (1969). *J. Ultrastruct. Res.* **29,** 141.
Kitajima, E. W., Müller, G. W., Costa, A. S., and Yuki, W. (1972). *Virology* **50,** 245.
Kitajima, E. W., Blumenschein, A., and Costa, A. S. (1974). *Phytopathol. Z.* **81,** 280.
Kitajima, E. W., Giacomelli, E. J., Costa, A. S., Costa, C. L., and Cupertino, F. P. (1975). *Phytopathol. Z.* **82,** 83.
Knorr, L. C. (1968). *Proc. Conf. Int. Organ. Citrus Virol., 4th, 1966* p. 332.
Knudson, D. L. (1973). *J. Gen. Virol.* **20,** 105.
Knudson, D. L., and McLeod, R. (1972). *Virology* **47,** 285.
Lastra, J. R., and Acosta, J. M. (1975). *Int. Virol.* **3,** 244 (abstr.).
Lawson, R. H., and Ali, S. (1975). *J. Ultrastruct. Res.* **53,** 345.
Lee, P. E. (1964). *Virology* **23,** 145.
Lee, P. E., and Peters, D. (1972). *Virology* **48,** 739.
Lee, P. E., Boerjan, M., and Peters, D. (1972). *Virology* **50,** 309.
Lesemann, D. (1972). *Phytopathol. Z.* **73,** 83.
Lesemann, D., and Begtrup, J. (1971). *Phytopathol. Z.* **71,** 257.
Lesemann, D., and Doraiswamy, S. (1975a). *Phytopathol. Z.* **83,** 27.
Lesemann, and Doraiswamy, S. (1975b). *Phytopathol. Z.* **84,** 201.
Lin, M. T., and Campbell, R. N. (1972). *Virology* **48,** 30.
McLean, G. D., Wolanski, B. S., and Francki, R. I. B. (1971). *Virology* **43,** 480.
McLeod, R. (1968). *Virology* **34,** 771.
McLeod, R., Black, L. M., and Moyer, F. H. (1966). *Virology* **29,** 540.
McSharry, J. J., and Wagner, R. R. (1971). *J. Virol.* **7,** 412.
Maramorosch, K., Hirumi, H., Kimura, M., Bird, J., and Vakili, N. G. (1974). *FAO Pland Prot. Bull.* **22,** 32.
Martelli, G. P. (1969). *J. Gen. Virol.* **5,** 319.
Martelli, G. P., and Castellano, M. A. (1970). *Phytopathol. Mediterr.* **9,** 39.
Martelli, G. P., and Cirulli, M. (1969). *Ann. Phytopathol.* **1,** 393.
Martelli, G. P., ad Russo, M. (1973). *CMI/AAB Descriptions Plant Viruses* No. 115.
Martelli, G. P., Russo, M., and Malaguti, G. (1975). *Phytopathol. Mediterr.* **14,** 138.
Melnick, J. L., and McCombs, R. M. (1966). *Prog. Med. Virol.* **8,** 400.
Nakai, T., and Howatson, A. F. (1968). *Virology* **35,** 268.
O'Loughlin, G. T., and Chambers, T. C. (1967). *Virology* **33,** 262.
Orenstein, J., Johnson, L., Shelton, E., and Lazzarini, R. A. (1976). *Virology* **71,** 291.
Peña-Iglesias, A., Rubio-Huertos, M., and Moreno-San Martin, R. (1972). *An. Inst. Nac. Invest. Agrar. (Spain), Ser.: Prot. Veg.* **2,** 123.
Peña-Iglesias, A., Ayuso-Gonzales, P., Moreno-San Martin, R., and Rubio-Huertos, M. (1975). *Int. Virol.* **3,** 273 (abstr.).
Peters, D., and Black, L. M. (1970). *Virology* **40,** 847.
Peters, D., and Kitajima, E. W. (1970). *Virology* **41,** 135.
Peters, D., and Schultz, M. G. (1975). *Proc. K. Ned. Akad. Wet., Ser. C* **78,** 172.
Petzold, H. (1971). *Phytopathol. Z.* **70,** 43.
Plavšić-Banjac, B., Miličić, D., and Erić Ž. (1976) *Phytopathol. Z.* **86,** 225.
Plumb, R. T., and James, M. (1975). *Ann. Appl. Biol.* **80,** 181.
Putz, C., and Meignoz, R. (1972). *Phytopathology* **62,** 1477.
Razvazkina, G. M., and Polyakova, G. P. (1967). *Dokl. Akad. Nauk SSSR* **174,** 1435.
Richardson, J., and Sylvester, E. S. (1968). *Virology* **35,** 347.
Richardson, J., Frazier, N. W., and Sylvester, E. S. (1972). *Phytopathology* **62,** 491.
Rodriguez-Sardiña, J., and Torres-Barros, J. (1975). *Int. Virol.* **3,** 158 (abstr.).
Rossetti, V., Lasca, C. C., and Negretti, S. (1969). *Proc. 1st Int. Symp. Citrus* **3,** 1453.
Rubio-Huertos, M., and Bos. L. (1969). *Neth. J. Plant Pathol.* **75,** 329.
Rubio-Huertos, M., and Peña-Iglesias, A. (1973). *Plant Dis. Rep.* **75,** 329.
Russo, M., and Martelli, G. P. (1972). *Phytopathol. Mediterr.* **11,** 136.
Russo, M., and Martelli, G. P. (1973). *Virology* **52,** 39.
Russo, M., and Martelli, G. P. (1975). *Phytopathol. Z.* **83,** 97.
Russo, M., Martelli, G. P., and Rana, G. L. (1975). *Phytopathol. Z.* **83,** 223.
Schultz, M. G., and Harrap, K. A. (1975). *Ann. Appl. Biol.* **79,** 247.
Schultz, M. G., Harrap, K. A., and Land, J. B. (1975). *Ann. Appl. Biol.* **80,** 251.
Shikata, E., and Chen, M.-J. (1969). *J. Virol.* **3,** 261.

Shikata, E., and Lu, Y. T. (1967). *Proc. Jpn. Acad.* **43,** 918.
Signoret, P.-A., Giannotti, J., and Alliot, B. (1972). *Ann. Phytopathol.* **4,** 45.
Sinha, R. C. (1971). *Virology* **44,** 342.
Sinha, R. C., and Becki, R. M. (1972). *CMI/AAB Descriptions Plant Viruses* No. 99.
Sinha, R. C., and Thottapilly, G. (1974). *Phytopathol. Z.* **81,** 124.
Stace-Smith, R., and Lo, R. (1973). *Can. J. Bot.* **51,** 1343.
Stubbs, L. L., and Grogan, R. G. (1963). *Aust. J. Agric. Res.* **14,** 439.
Sylvester, E. S. (1969). *Virology* **38,** 440.
Sylvester, E. S. (1973). *Virology* **56,** 632.
Sylvester, E. S., and Richardson, J. (1970). *Virology* **42,** 1023.
Sylvester, E. S., and Richardson, J. (1971). *Virology* **46,** 310.
Thottapilly, G., and Sinha, R. C. (1973). *Virology* **53,** 312.
Tomlinson, J. A., and Webb, M. J. W. (1974). *Report Nat. Veg. Stn. Wellesbourne, 1973,* p. 100.
Tomlinson, J. A., Webb, M. J. W., and Faithfull, E. M. (1972). *Ann. Appl. Biol.* **71,** 127.
Vega, J., Gracia, O., Rubio-Huertos, M., and Feldman, J. M. (1976). *Phytopathol. Z.* **85,** 7.
Vela, A., and Lee, P. E. (1974). *J. Ultrastruct. Res.* **47,** 169.
Vela, A., and Rubio-Huertos, M. (1974). *Phytopathol. Z.* **79,** 343.
Vernon, S. K., Neurath, A. R., and Rubin, B. A. (1972). *J. Ultrastruct. Res.* **41,** 29.
Wagner, R. R. (1975). *Compr. Virol.* **4,** 1.
Wildy, P. (1971). *Virol. Monogr.* **5,** 51.
Wolanski, B. S., and Chambers, T. C. (1971). *Virology* **44,** 582.
Wolanski, B. S., and Chambers, T. C. (1972). *Virology* **47,** 656.
Wolanski, B. S., and Francki, R. I. B. (1969). *Virology* **37,** 437.
Wolanski, B. S., Francki, R. I. B., and Chambers, T. C. (1967). *Virology* **33,** 287.
Ziemiecki, A., and Peters, D. (1976). *J. Gen. Virol.* **32,** 369.

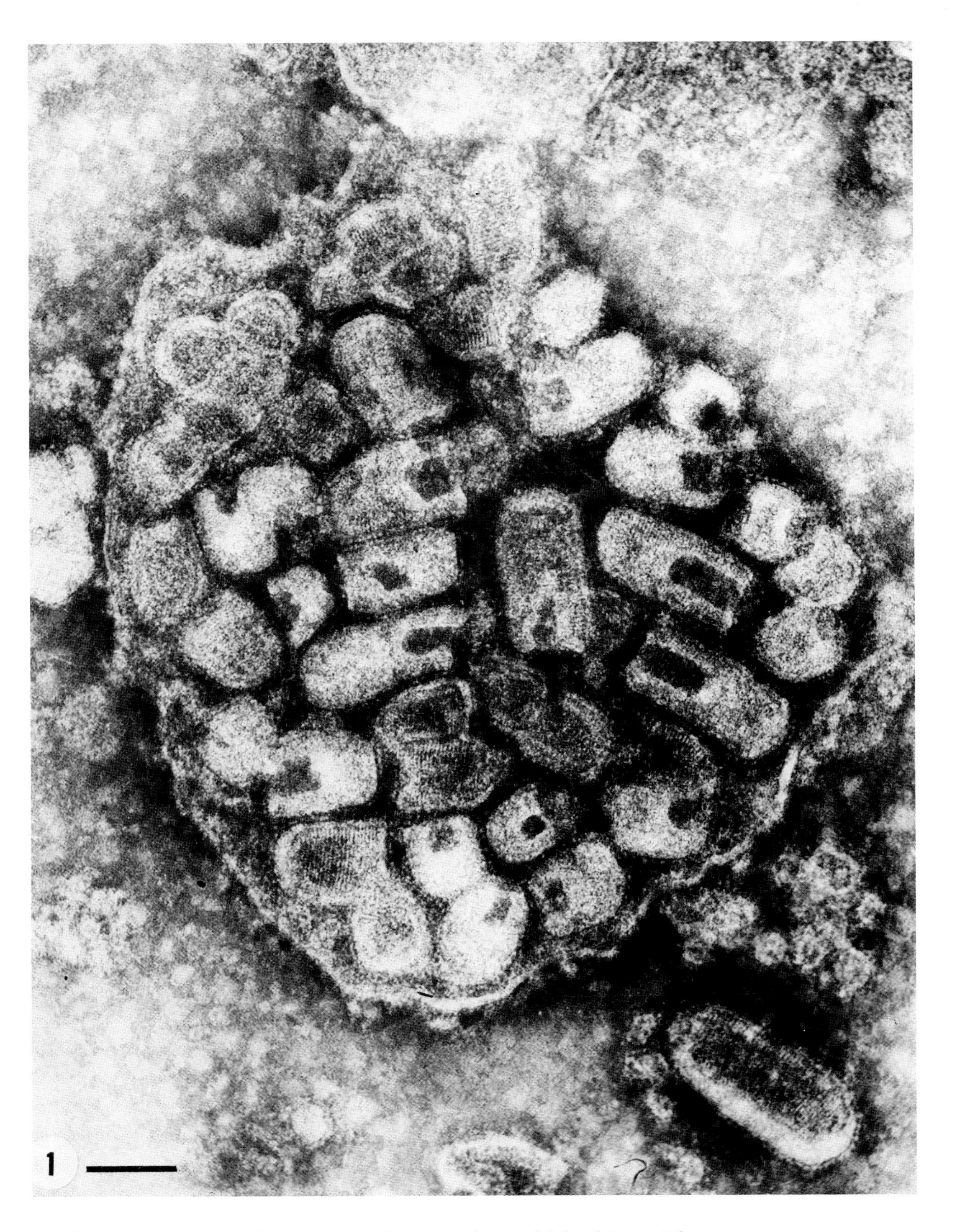

Fig. 1. A negatively stained dip preparation showing eggplant mottled dwarf virus particles within a membranous structure. Most of the virions are bullet-shaped, except for a bacilliform one, (lower right corner) and exhibit surface projections and transverse striations. Bar = 100 nm. (From Martelli and Russo, 1973, courtesy of the copyright holders Commonwealth Agricultural Bureaux and Association of Applied Biologists.)

Figs. 2 and 3. Purified preparations of sowthistle yellow vein virus in phosphotungstic acid (Fig. 2) and uranyl acetate (Fig. 3) mounts. The particles were fixed in glutaraldehyde prior to staining. Both sets of particles are bacilliform, some showing the inner helical component (Fig. 2) and others the patterns of the hexamer layer (Fig. 3). Bars = 100 nm. (From Peters and Kitajima, 1970, courtesy of the authors.)

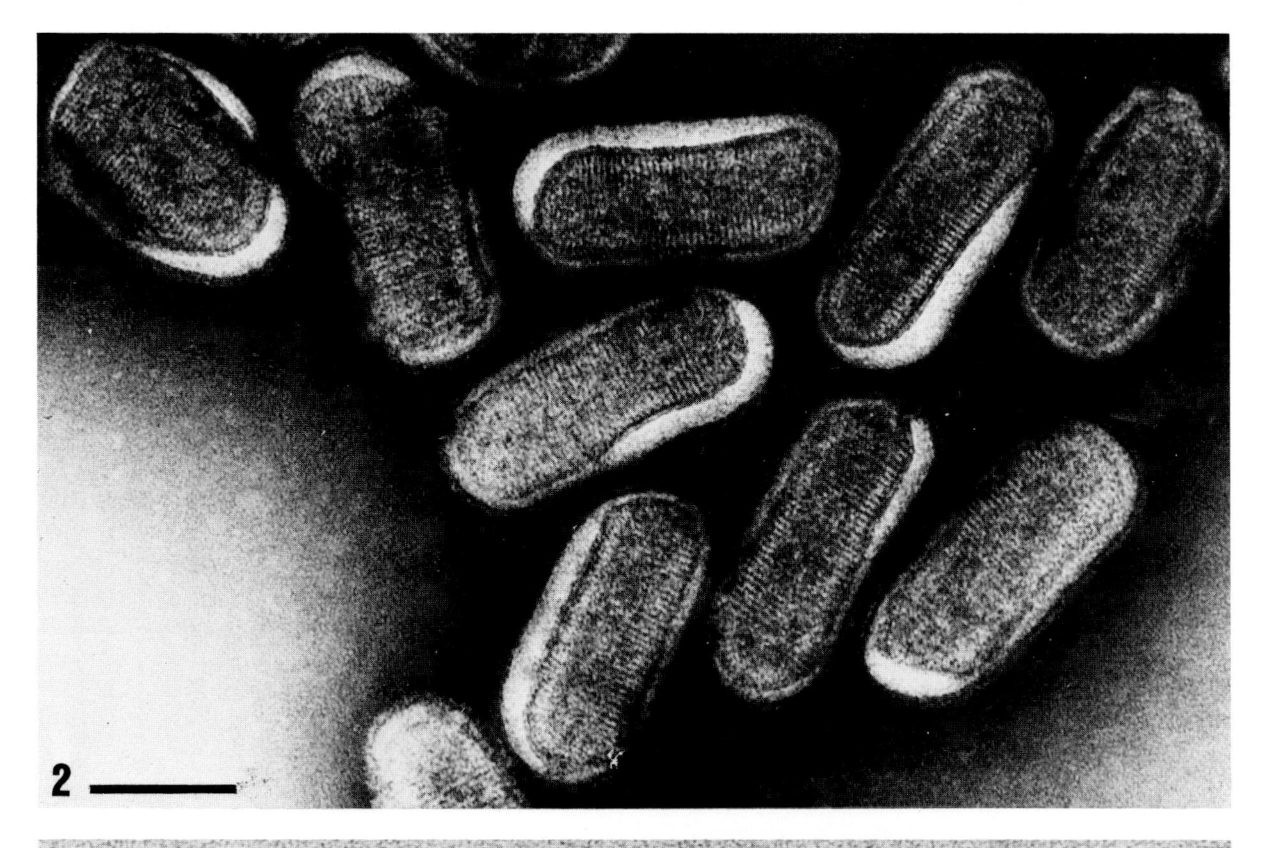
2

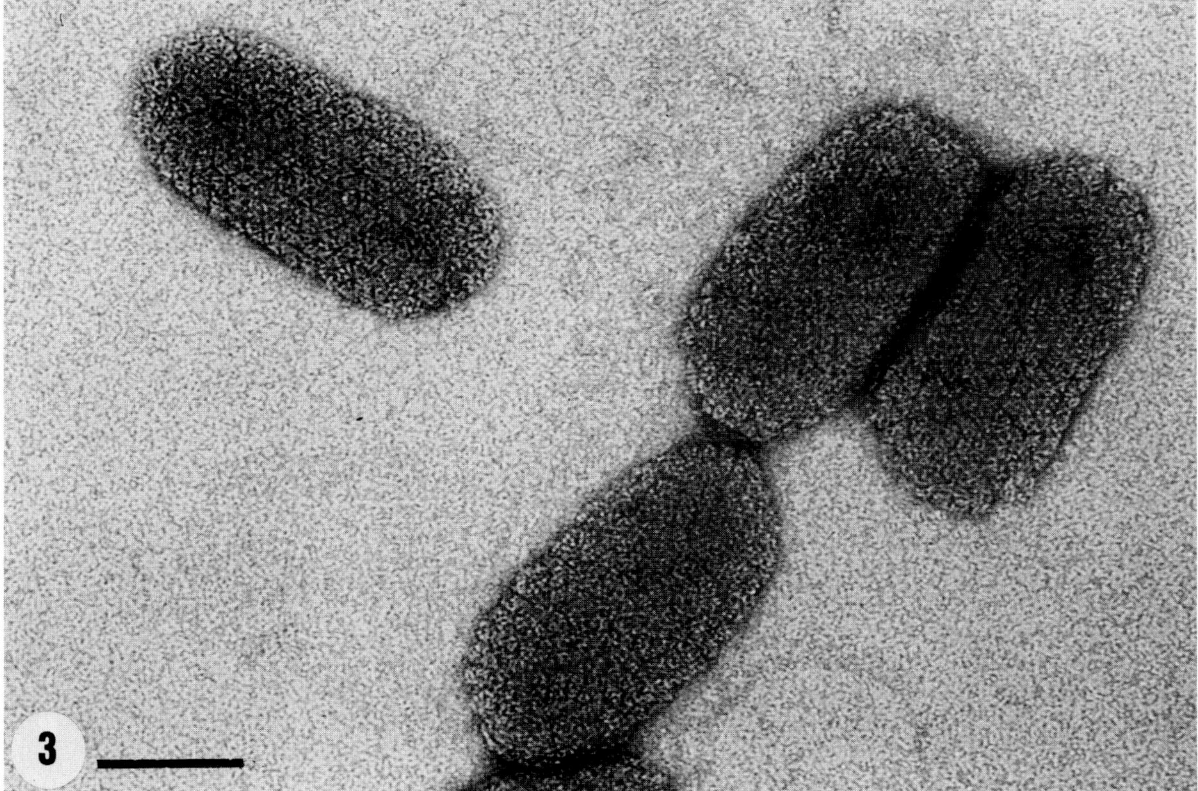
3

Figs. 4–6. Different aspects of plant rhabdovirus particles seen in transverse sections in tissues treated in various ways. Eggplant mottled dwarf virus particles fixed in aldehydes only (uranyl soak method), embedding in Epon–Araldite (Fig. 4); prolonged $KMnO_4$ fixation, embedding in styrene butyl methacrylate (Fig. 5). Particles of maize mosaic virus fixed in aldehydes and osmium, embedding in Araldite (Fig. 6). Bars = 100 nm.

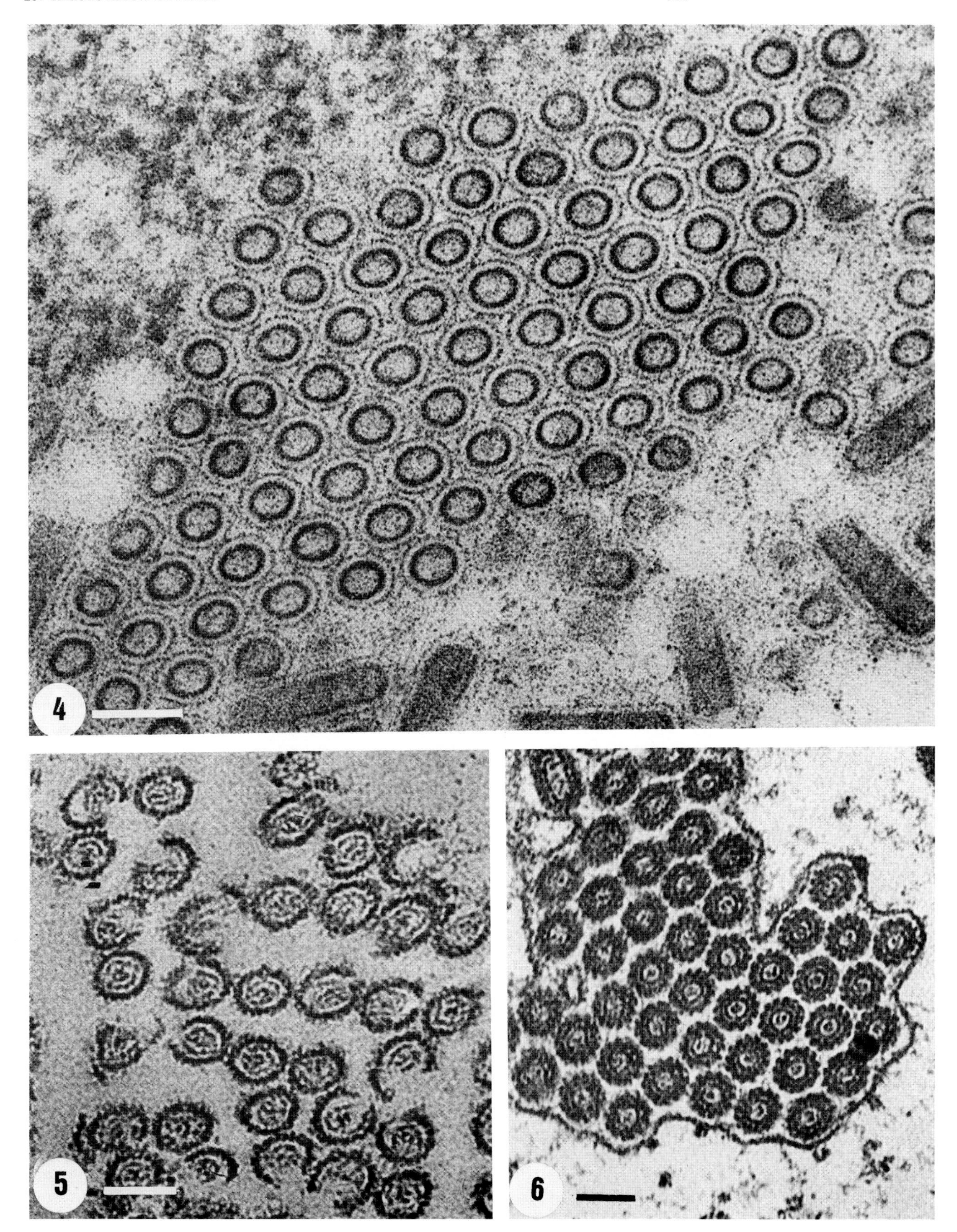
4
5
6

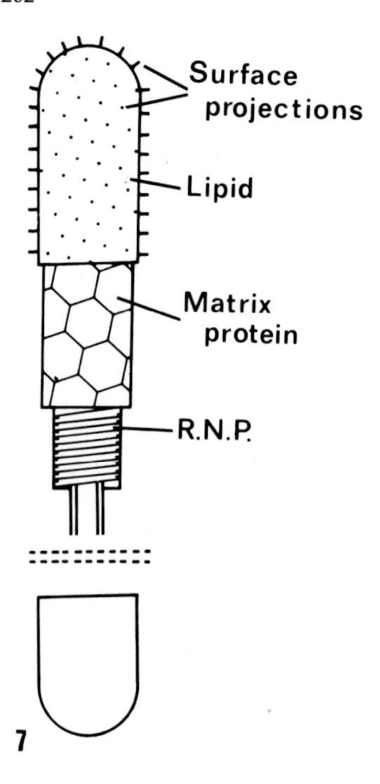

Fig. 7. A diagrammatic reconstruction of broccoli necrotic yellows virus showing the relative positions of the various layers constituting the virion. (Courtesy of R. Hull.)

Fig. 8. Diagrammatic representation of the interrelationship between M and G proteins. In the hexagonal lattice, the matrix protein subunits and the surface projections are arranged as separate hexamers. Each surface projection (★) is surrounded by three matrix protein subunits (●). (Redrawn from Cartwright *et al.*, 1972, courtesy of the authors.)

Figs. 9–12. Aspects of nucleocapsids of variously treated sowthistle yellow vein virus particles. Two bullet-shaped nucleocapsids enveloped by a single membrane in a glutaraldehyde fixed negatively stained particle (Fig. 9). Nucleocapsids obtained after treatment with 0.2% Nonidet P40 (Figs. 10 and 12). The same after treatment with 1% saponin showing uncoiling starting from the planar end (Fig. 11). Bar = 50 nm (Fig. 10); 100 nm (Figs. 9, 11, and 12). (Fig. 9 from Lee and Peters, 1972. Figs. 10–12 courtesy of A. Ziemiecki and D. Peters.)

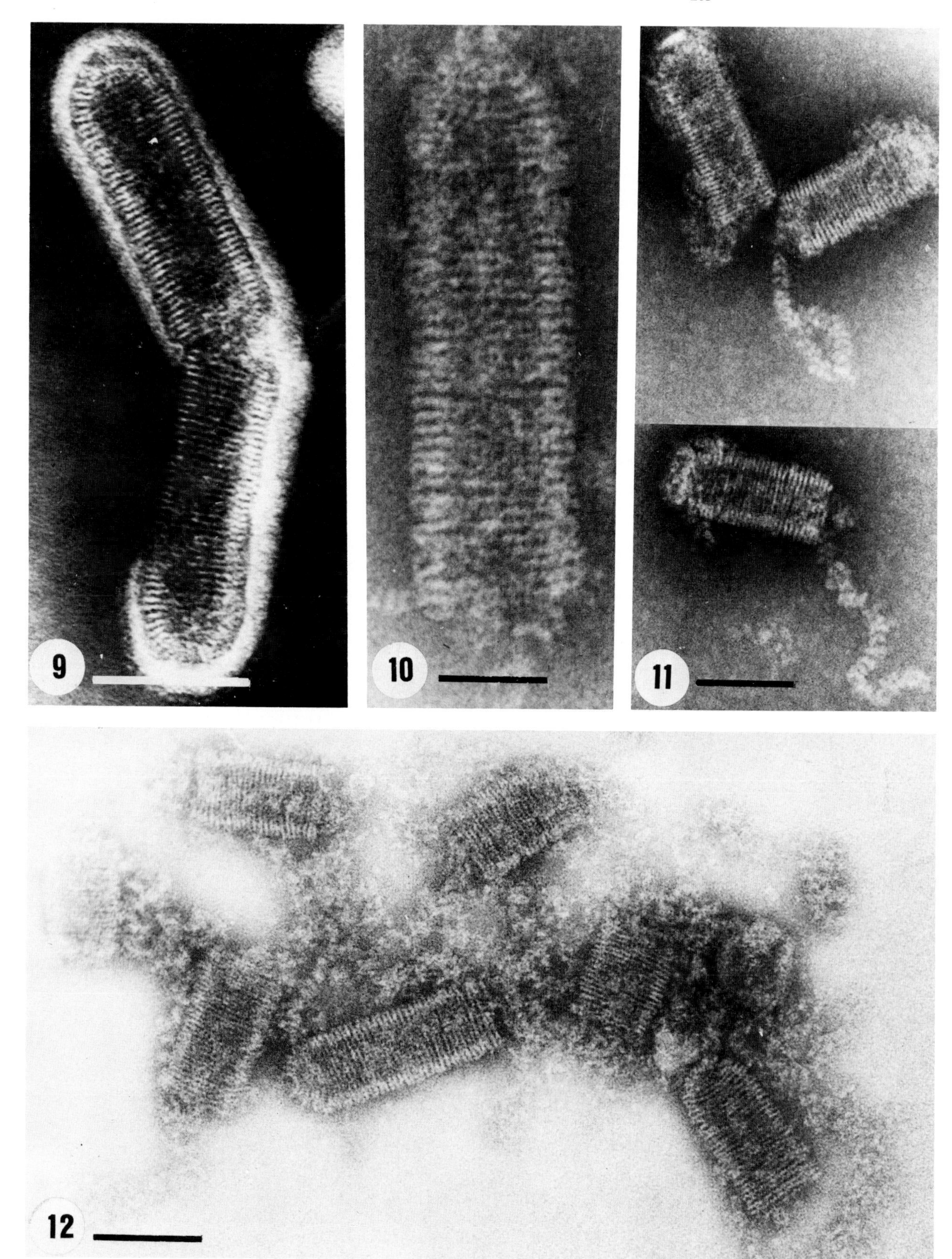
9
10
11
12

Fig. 13. Transverse section through an eggplant cell showing a nucleus (N) surrounded by aggregates of virus particles in perinuclear position. Bar = 1 μm. (From Martelli, 1969, courtesy of the copyright holder, Cambridge University Press.)

Fig. 14. Perinuclear aggregates of *Melilotus* latent virus particles (V). Naked nucleocapsids (t) in longitudinal and cross section are visible in the nucleus (N). Some are being enveloped at the inner lamella (im) of the nuclear membrane (arrows). Bar = 250 nm. (From Kitajima *et al.*, 1969, courtesy of the authors.)

Fig. 15. Eggplant mottled dwarf virus particles budding at the inner lamella (IL) of the nuclear envelope. Note continuity of the membrane with the outer contour of the particles. N, Nucleus; Np, nuclear pore; OL, outer lamella. Bar = 250 nm.

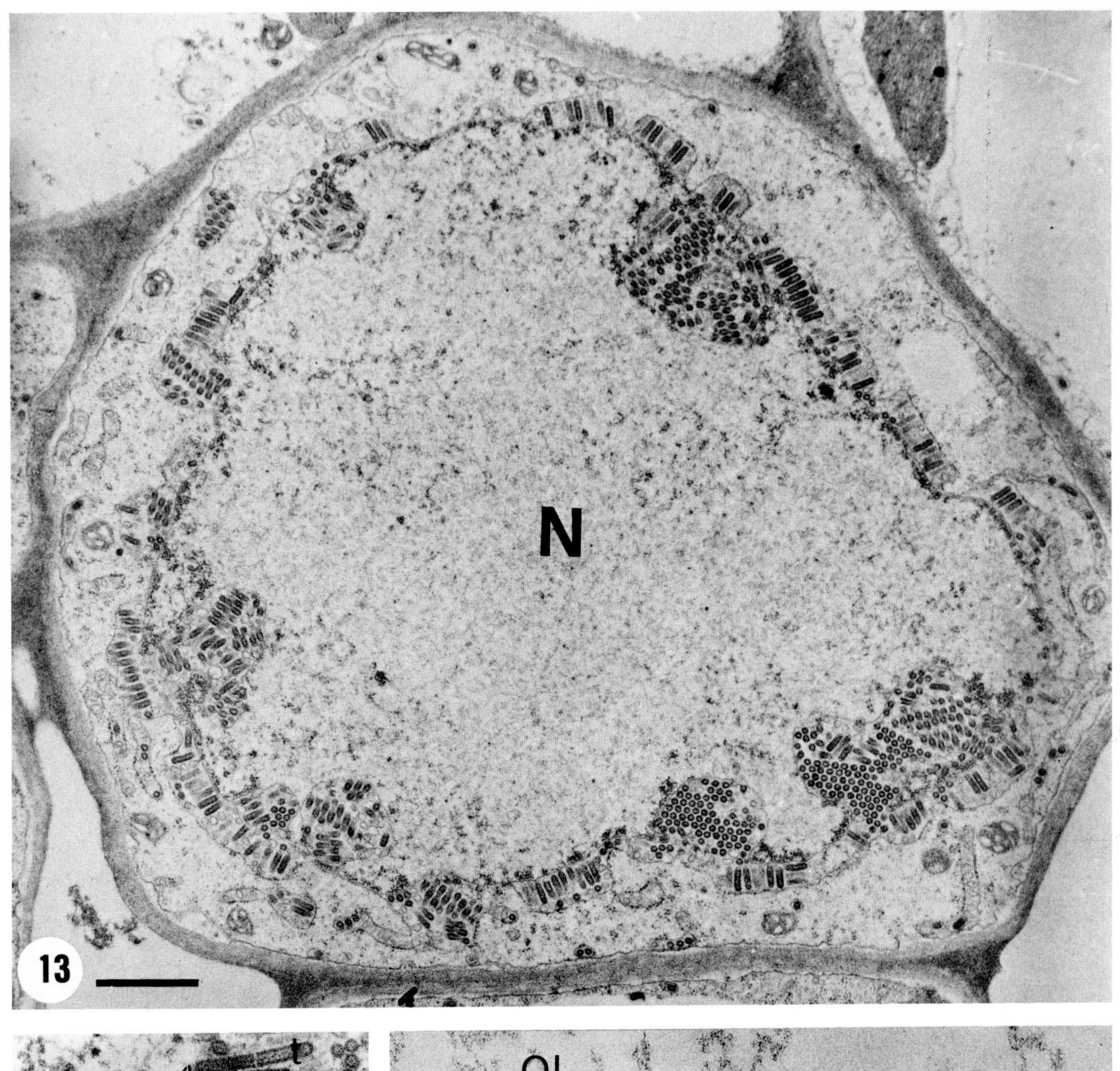
N
13

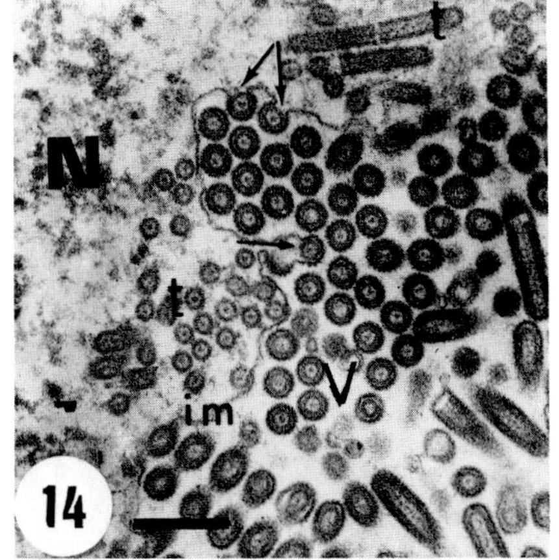
t
N
t
im
V
14

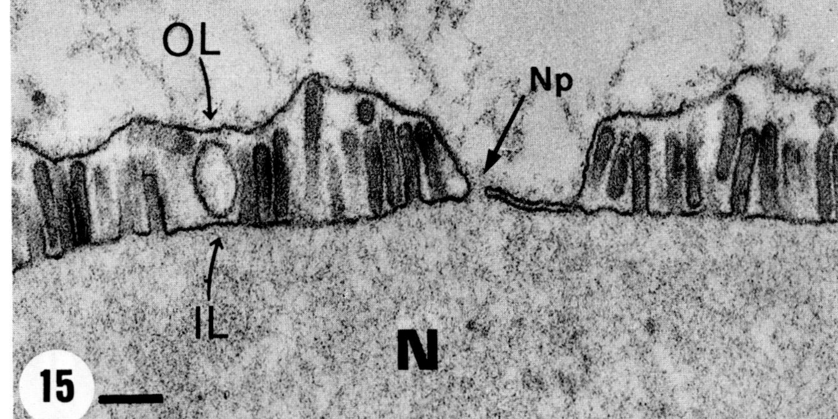
OL
Np
IL
N
15

Fig. 16. Cytoplasmic membrane-bound pockets containing maize mosaic virus particles. Continuity of the membrane of the cytoplasmic enclaves with the nuclear envelope is visible (arrow); N, nucleus. Bar = 300 nm.

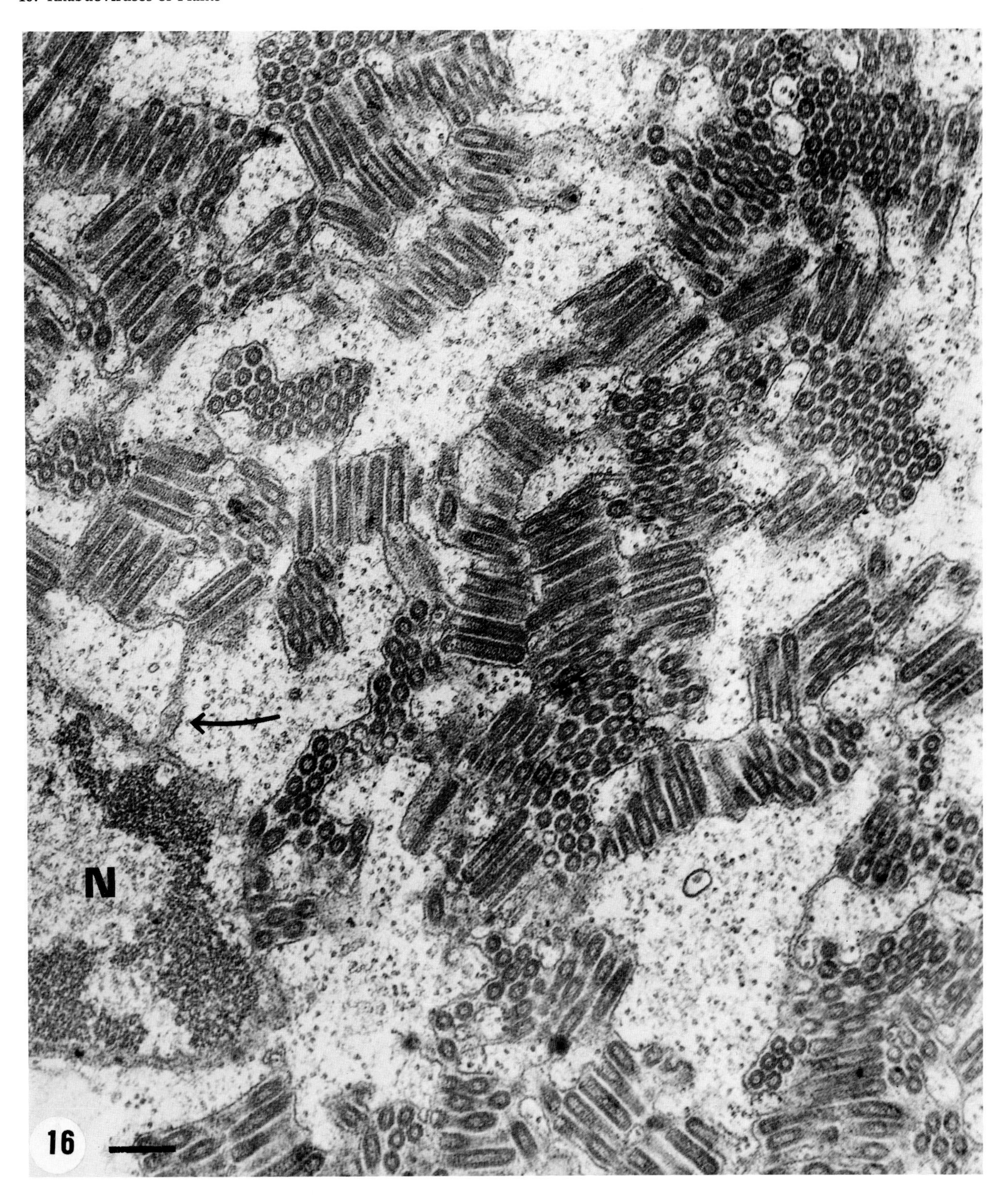
N
16

Fig. 17. Transverse section through a nucleus (N) of a rice cell infected with rice transitory yellowing virus. Bullet-shaped particles accumulate perinuclearly. Bar = 300 nm. (From Chen and Shikata, 1971, courtesy of the authors.)

Fig. 18. *Cynara* rhabdovirus particles contained within membranous structures deriving from the nuclear membrane. The external coat of a particle is continuous with a membrane originating from the outer lamella (OL) of the nuclear envelope. N, Nucleus; IL, inner lamella. Bar = 200 nm. (From Russo *et al.*, 1975, courtesy of the copyright holder, Paul Parey Verlag, Berlin.)

Figs. 19 and 20. Melon variegation virus. Rod-shaped structures interpreted as unenveloped nucleocapsids in the cytoplasm of an infected melon cell (Fig. 19). Particles budding through endoplasmic reticulum membranes in a forward (upper Fig. 20) and lateral (lower Fig. 20) position. Bar = 200 nm (Fig. 19); 100 nm (Fig. 20). (From Rubio-Huertos and Peña-Iglesias, 1973, in *Plant Disease Reporter*, Washington, courtesy of the authors.)

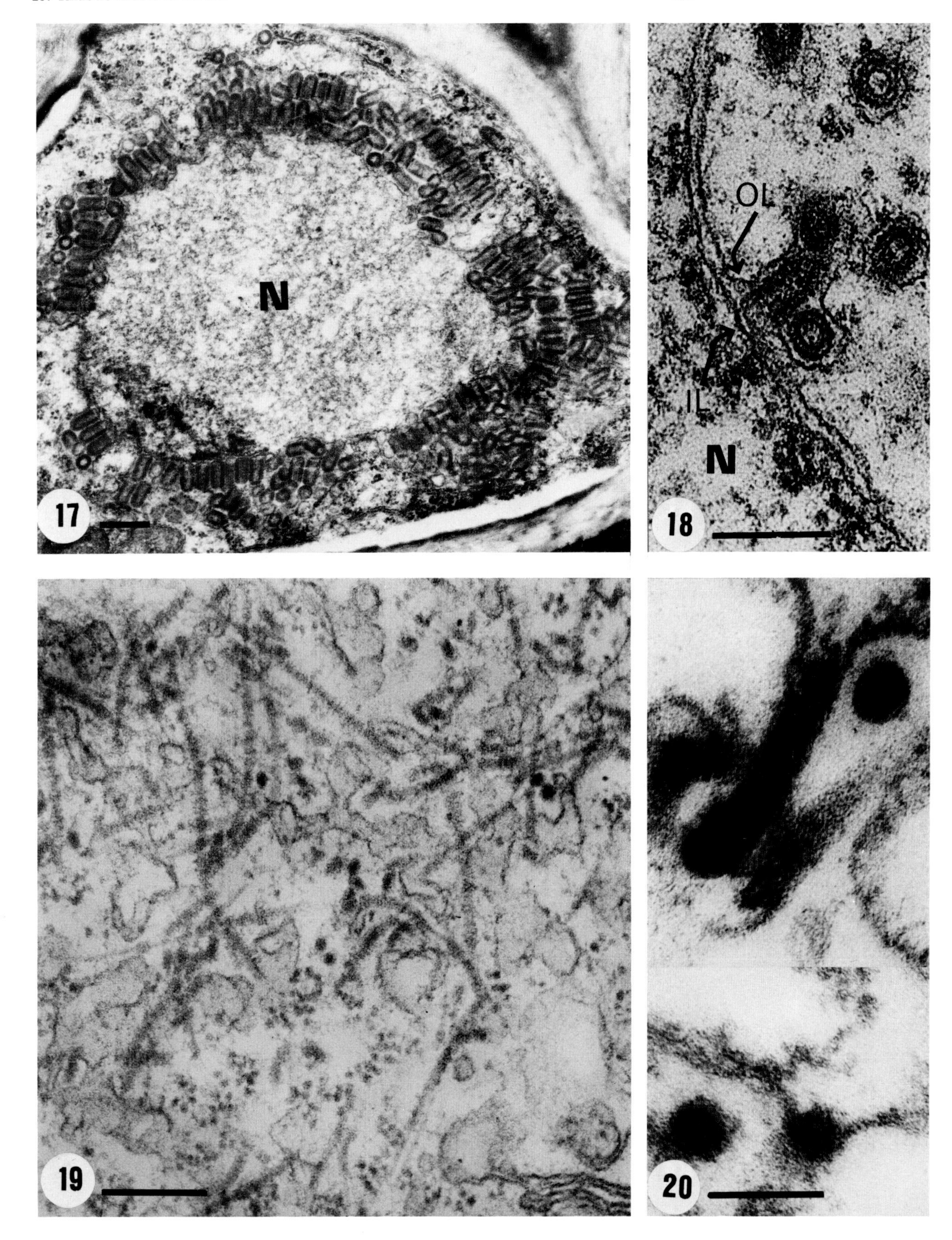
N
17
OL
IL
N
18
19
20

Figs. 21 and 22. Viroplasm body (Vp) surrounded by virus particles in a cell infected by barley yellow striate mosaic virus (Fig. 21) and details of particles budding from the viroplasm matrix (Fig. 22). N, Nucleus. Bar = 500 nm (Fig. 21); 200 nm (Fig. 22). (Fig. 21 from Conti and Appiano, 1973, courtesy of the authors and the copyright holder Cambridge University Press.)

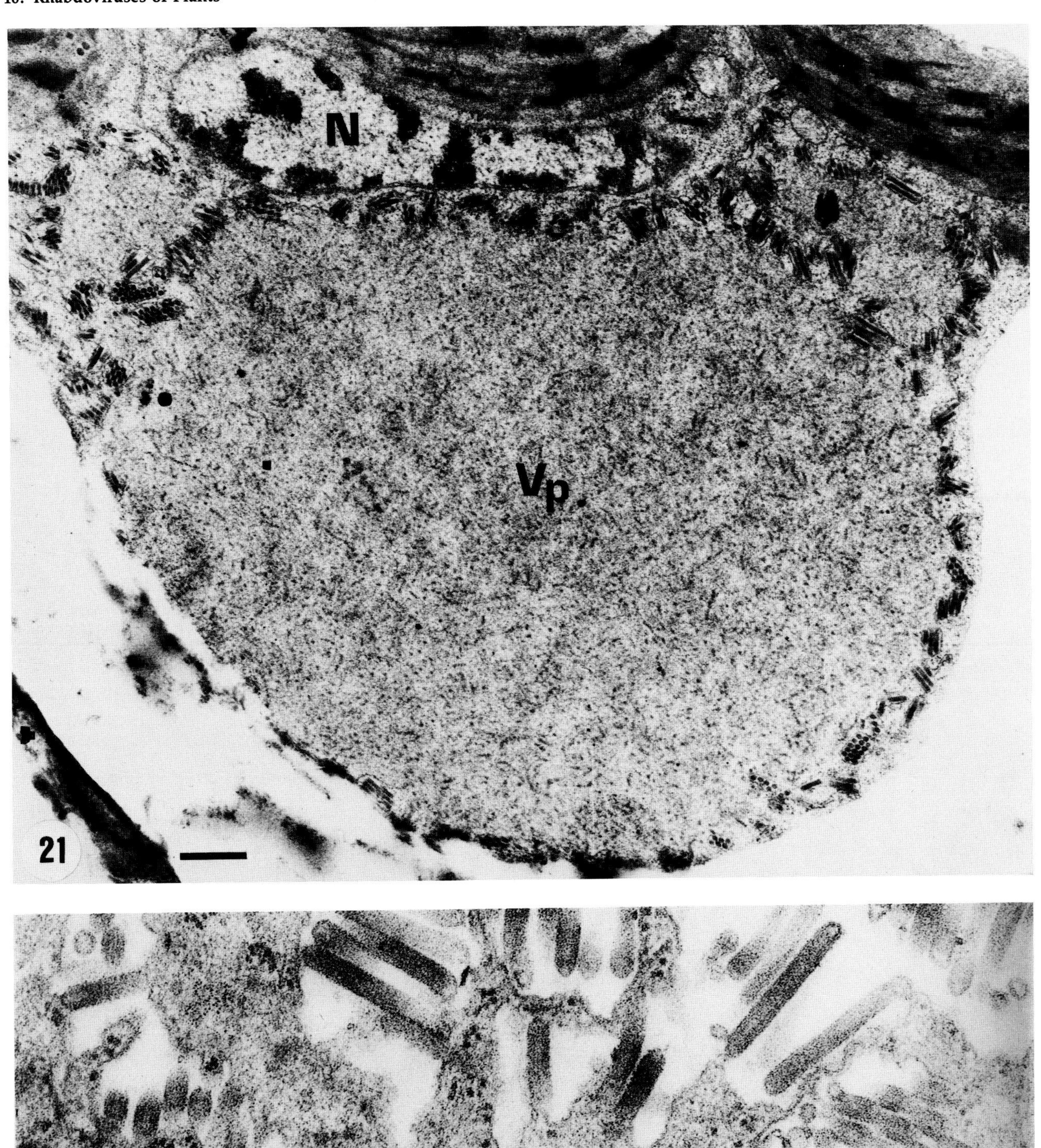
N
Vp
21
Vp
22

Fig. 23. A nucleus (N) of the orchid *Miltonia spectabilis* infected by KORV. Many blebs containing viral nucleocapsids are visible at the nuclear margins. Bar = 1 μm.

Fig. 24. A close-up of nuclear evaginations showing attachment of nucleocapsids perpendicularly to the nuclear envelope. N, nucleus. Bar = 250 nm.

Fig. 25. A "spokewheel" (arrow head) in the cytoplasm of an infected *M. spectabilis* cell. Bar = 250 nm. (From Lesemann and Doraiswamy, 1975a, courtesy of the authors and the copyright holder Paul Parey Verlag, Berlin.)

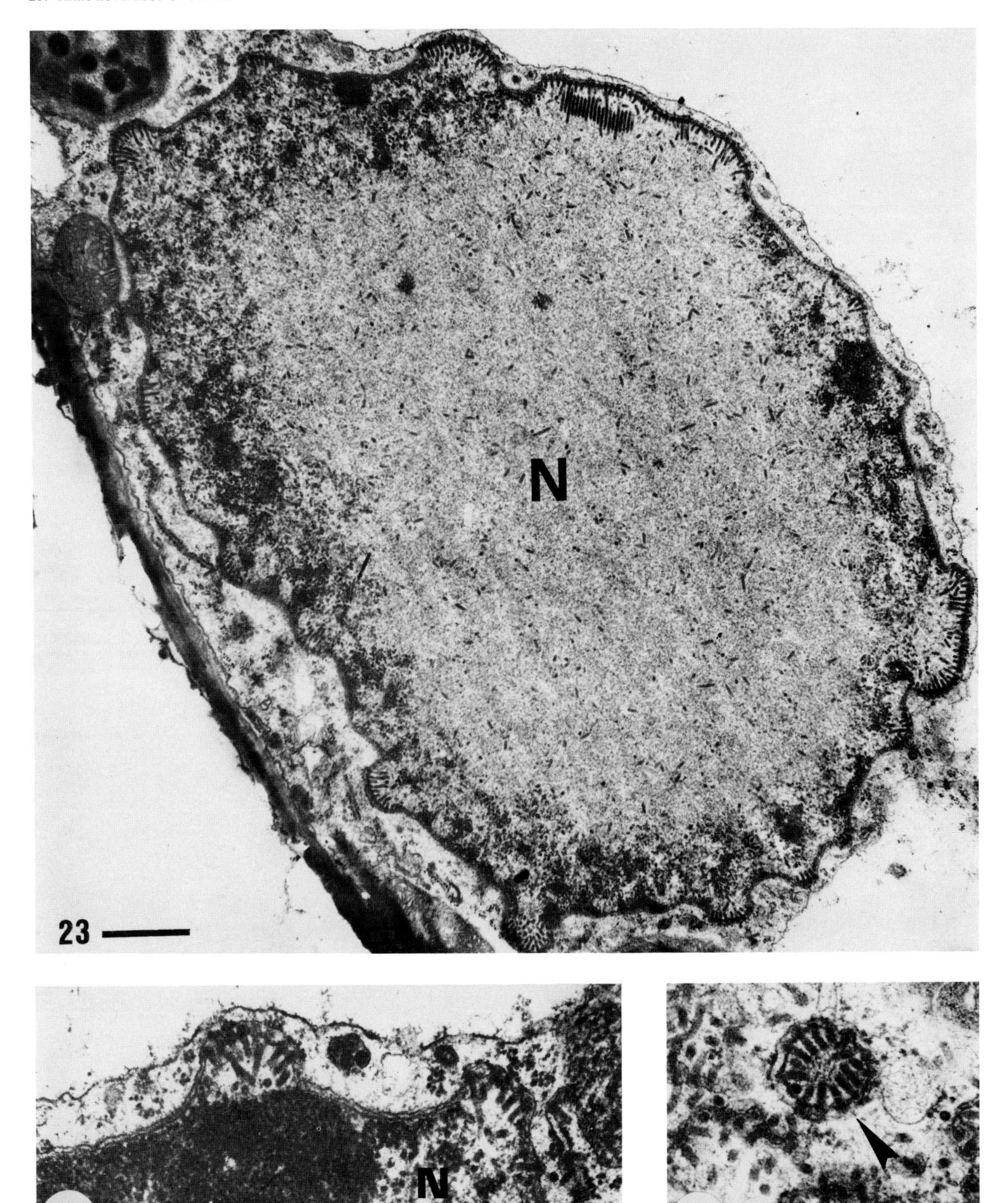
N
23
24
N
25

Chapter 11

Tobravirus (Tobacco Rattle Virus) Group

E. M. J. JASPARS

I. Introduction 215
II. Particle Structure and Composition 215
III. Genome Constitution and Replication 216
References 216

I. INTRODUCTION

The viruses related to tobacco rattle virus (TRV) are called tobraviruses (Harrison *et al.*, 1971). They are rigid rod-shaped, nematode-transmitted, positive-stranded RNA viruses of higher plants with a bipartite genome. Structurally they are related to tobacco mosaic and barley stripe mosaic viruses, which also have rigid rod-shaped particles with helically arranged protein subunits, but the width of tobravirus particles is 25 nm as compared to 18 nm for tobacco mosaic virus and 20 nm for barley stripe mosaic virus. The cryptogram of the type member of the tobraviruses, the PRN strain of TRV is R/1:2.5/5 + 1.0/5:E/E:S/Ne (Cooper and Mayo, 1972). This cryptogram applies to all strains of tobacco rattle and pea early-browning viruses except for the molecular weight of the small RNA species which is variable (Harrison, 1970, 1973). Tobraviruses are geographically widespread and occur in many host plants. Pea early-browning virus infects peas systemically whereas TRV does not. The viruses have been reviewed by Lister (1969). Kaper (1975) summarizes structural and chemical properties and compares the dissociation and reassembly behavior with that of tobacco mosaic virus.

II. PARTICLE STRUCTURE AND COMPOSITION

Two major length classes of particles are found in all strains (Fig. 1). The larger class is about 200 nm and has a particle weight of about 50×10^6 daltons.

The smaller class is variable in length. Depending on the strain, particles of about 40 to 110 nm are found. The RNAs of the long and short particles have molecular weights of 2.5 and 0.6 to 1.3 × 10^6 daltons, respectively. The coat of the particles is built from a single species of protein with a molecular weight of 22,000 daltons (Mayo and Robinson, 1975). Amino acid compositions have been reported by Semancik (1966). In the particles the subunits are arranged in a helix with a pitch of 25.5 Å (Offord, 1966).

Conditions for polymerization of TRV protein and for reassembly of TRV particles have been reported (Fritsch *et al.*, 1973; Morris and Semancik, 1973; Abou Haidar *et al.*, 1973).

III. GENOME CONSTITUTION AND REPLICATION

The RNA species of long and short particles represent two different parts of the viral genome. However, by competitive hybridization studies it has been shown that they have sequences of 600 nucleotides in common (Darby and Minson, 1973). The homologous sequences are not located in the 3′-terminal region of the molecules (Minson *et al.*, 1976). They may represent a common cistron or common recognition sites for ribosomes, coat protein, or replicase. Base ratio U:C:A:G is 29:17:29:25 for both RNA species (Semancik and Kajiyama, 1967). Their 3′-terminal sequence is —$GCCC_{OH}$ (Minson and Darby, 1973).

Serological and fingerprint analysis of the coat protein of the progeny of particle combinations derived from different strains showed that the coat protein cistron is in the RNA of the small particles (Sänger, 1968; Ghabrial and Lister, 1973). This has been confirmed also by *in vitro* translation studies (Mayo *et al.*, 1976). For infectivity the small particles are not necessary. When small particles are absent in the inoculum the long particle RNA replicates independently. This has been found both in intact plants (Lister, 1966) and in protoplasts (Kubo *et al.*, 1975). No particles are found then and the infection is difficultly transmissible with sap. These so-called labile infections sometimes occur spontaneously in nature. Single symptoms caused by dilute inocula often contain infections of the labile type. The fact that long-particle RNA can replicate autonomously demonstrates that it has all the genetic information for the replicase. By characterizing hybrid progenies, several symptom markers have been assigned to either RNA species. Not all strains are compatible with regard to hybrid formation. This may depend on whether the replicase recognizes the heterologous short particle RNA. In one case a hybrid combination was compatible, whereas the reverse combination was not (Ghabrial and Lister, 1973).

In infected leaf cells of *Nicotiana clevelandii*, X-bodies have been found largely composed of abnormal mitochondria (Harrison *et al.*, 1970). Electron microscopy of thin sections also revealed that with some strains the mitochondria of the infected cells were surrounded by long particles (Harrison and Roberts, 1968) (Fig. 2).

ACKNOWLEDGMENT

The author wishes to express his thanks to Dr. B. D. Harrison for kindly providing him with copies of the electron micrographs reproduced in this article.

REFERENCES

Abou Haidar, M., Pfeiffer, P., Fritsch, C., and Hirth, L. (1973). *J. Gen. Virol.* **21,** 83.
Cooper, J. I., and Mayo, M. A. (1972) *J. Gen. Virol.* **16,** 285.

Darby, G., and Minson, A. C. (1973). *J. Gen. Virol.* **21,** 285.
Fritsch, C., Witz, J., Abou Haidar, M., and Hirth, L. (1973). *FEBS Lett.* **29,** 211.
Frost, R. R., and Harrison, B. D., and Woods, R. D. (1967). *J. Gen. Virol.* **1,** 57.
Ghabrial, S. A., and Lister, R. M. (1973). *Virology* **52,** 1.
Harrison, B. D., (1970). *CMI/ABB Descriptions Plant Viruses* No. 12.
Harrison, B. D. (1973). *CMI/ABB Descriptions Plant Viruses* No. 120.
Harrison, B. D., and Roberts, I. M. (1968). *J. Gen. Virol.* **3,** 121.
Harrison, B. D., Štefanac, Z., and Roberts, I. M. (1970). *J. Gen. Virol.* **6,** 127.
Harrison, B. D., Finch, J. T., Gibbs, A. J., Hollings, M., Shepherd, R. J., Valenta, V., and Wetter, C. (1971). *Virology* **45,** 356.
Kaper, J. M. (1975). "The Chemical Basis of Virus Structure, Dissociation and Reassembly." North-Holland Publ., Amsterdam.
Kubo, S., Harrison, B. D., Robinson, D. J., and Mayo, M. A. (1975). *J. Gen. Virol.* **27,** 293.
Lister, R. M. (1966). *Virology* **28,** 350.
Lister, R. M. (1969). *Fed. Proc., Fed. Am. Soc. Exp. Biol.* **28,** 1875.
Mayo, M. A., and Robinson, D. J. (1975). *Intervirology* **5,** 313.
Mayo, M. A., Fritsch, C., and Hirth, L. (1976). *Virology* **69,** 408.
Minson, A. C., and Darby, G. (1973). *J. Mol. Biol.* **77,** 337.
Minson, A. C., Darby, G., and Gugerli, P. (1976). *Ann. Microbiol. Paris* **127a,** 91 (abstr.).
Morris, T. J., and Semancik, J. S. (1973). *Virology* **53,** 215.
Offord, R. E. (1966). *J. Mol. Biol.* **17,** 370.
Sänger, H. L. (1968). *Mol. Gen. Genet.* **101,** 346.
Semancik, J. S. (1966). *Phytopathology* **56,** 1190.
Semancik, J. S., and Kajiyama, M. R. (1967). *Virology* **33,** 523.

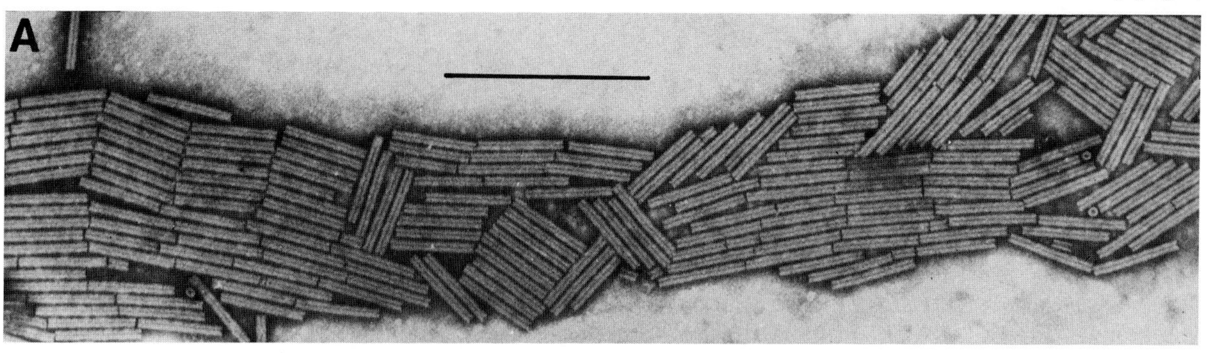

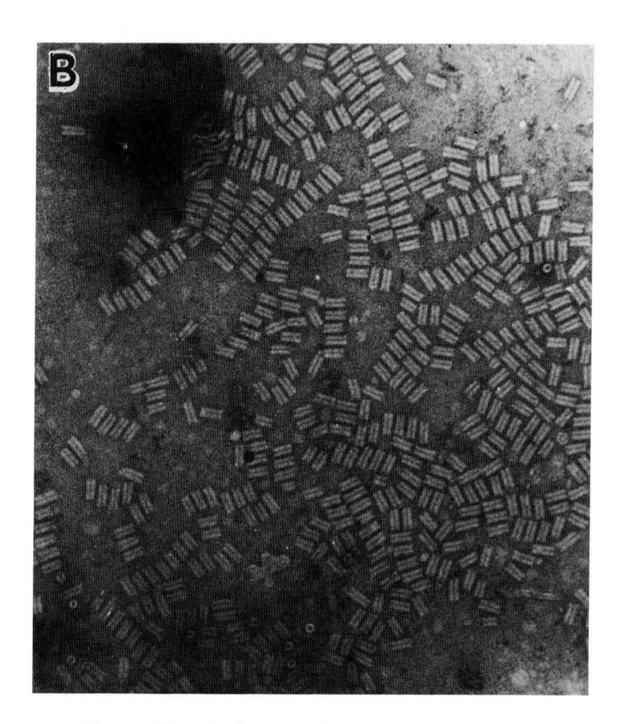

Fig. 1. Negatively stained (phosphotungstic acid) long (A) and short (B) particles of the CAM strain of tobacco rattle virus. These very pure preparations were obtained by two cycles of velocity gradient centrifugation. Bar = 500 nm. From Frost *et al.* (1967), reproduced with the permission of the Cambridge University Press.

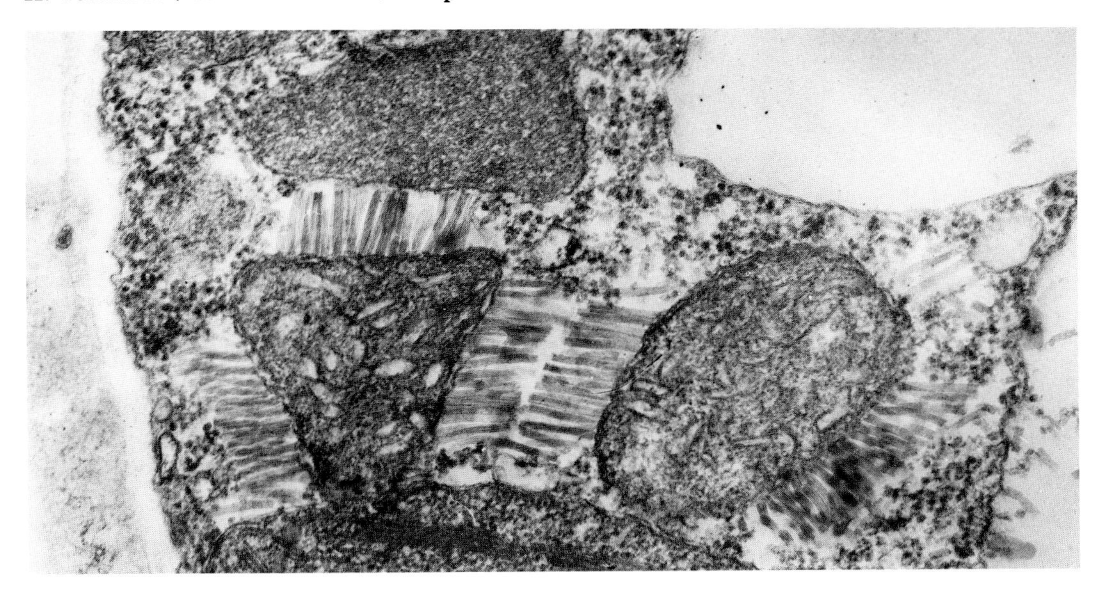

Fig. 2. Association of long particles of the CAM strain of tobacco rattle virus with mitochondria in leaf cells of *Nicotiana clevelandii*. Sections were fixed in glutaraldehyde and postfixed in OsO_4. (×48,000.) From Harrison and Roberts (1968), reproduced with the permission of the Cambridge University Press.

Chapter 12

Nepovirus (Tobacco Ringspot Virus) Group

R. I. B. FRANCKI AND T. HATTA

I. Virus Composition and Structure 222
II. Cytology of Virus-Infected Plants 224
References 225

Nepo is the acronym for *ne*matode-*po*lyhedral and refers to two of the main characteristics of these viruses that distinguish them from those of other groups (Cadman, 1963). Nepoviruses are transmitted by nematode vectors belonging to the genera *Xiphinema* and *Longidorus* (Martelli, 1975) and their particles are small polyhedra about 30 nm in diameter (Wildy, 1971). Other distinguishing properties of these viruses are that they can be mechanically transmitted as well as being seedborne; they have wide host ranges inducing ringspot or mottle symptoms and they are relatively stable *in vitro* (Wildy, 1971). The nomenclature used in the naming of nepoviruses is not altogether satisfactory. Some serologically related members have been described as strains of one virus as, for example, a number of tobacco ringspot virus (TRSV) variants (Gooding, 1970), whereas other similarly related ones such as cherry leafroll virus (CLRV), elm mosiac virus (EMV), and golden elderberry virus (GEV) are described under distinct names (Jones and Murant, 1971). In Table I we list TRSV, the nepovirus type member, with the other six viruses of the group; there is no demonstrable antigenic relationship between any viruses of the group. In addition, we list with each of the seven viruses, other viruses which we regard as serotypes or synonyms. Should strains of nepoviruses be found which are antigenically related to more than one of the seven nepoviruses considered as distinct then this system of nomenclature will need revision.

TABLE I

Members of the Nepovirus Group

Virus (and "strains")[a]	Cryptogram[b]	References
Tobacco ringspot virus (TRSV)	R/1:2.3/40+1.4/28 (or 2 × 1.4/44)S/S:S/Ne [57]	Stace-Smith (1970a); Mayo *et al.* (1971); Harrison *et al.* (1972)
Arabis mosaic virus (AMV) Raspberry yellow dwarf virus Rhubarb mosaic virus Grapevine fanleaf virus Grapevine yellow mosaic virus Grapevine vein banding virus Hop line pattern virus	R/1:2.4/41+1.4/29 (or 2×1.4/44):S/S:S/Ne [54]	Murant (1970b); Hewitt *et al.* (1970); Mayo *et al.* (1971); G. P. Martelli (private communication; H. F. Dias (private communication)
Raspberry ringspot virus (RRV) Raspberry Scottish leafcurl virus Red currant ringspot virus	R/1:2.4/43+1.4/30 (or 2×1.4/46):S/S:S/Ne [54]	Murant (1970a); Mayo *et al.* (1971); Harrison *et al.* (1972)
Tomato black ring virus (TBRV) Grapevine chrome mosaic virus Cocoa necrosis virus Artichoke Italian latent virus Chicory chlorotic ringspot virus Potato bouquet virus Beet ringspot virus Potato pseudo-aucuba virus Lettuce ringspot virus Celery yellow vein virus Tomato top necrosis	R/1:2.5/38 × 1.5/28:S/S:S/Ne [60]	Murant (1970c); Martelli and Quacquarelli (1972); Kenten (1972); Murant *et al.* (1973)
Tomato ringspot virus (TomRSV) Grape yellow vein virus Peach yellow bud mosaic virus Peach rosette mosaic virus	R/1:2.3/40+2.3/40:S/S:S/Ne [*]	Stace-Smith (1970b); Schneider *et al.* (1974)
Cherry leafroll virus (CLRV) Elm mosaic virus Golden elderberry virus	R/1:2.4/43+2.1/40:S/S:S/Ne [54]	Cropley and Tomlinson (1971); Jones and Mayo (1972)
Strawberry latent ringspot virus (SLRV)	R/1:2.6/37+(2×1.6/42):S/S:S/Ne [29+44]	Murant (1974); Mayo *et al.* (1974)

[a] Seven antigenically distinct nepoviruses are recognized here. Other names refer either to serotypes or synonyms mentioned in the literature.

[b] Cryptograms are presented in the conventional form (Wildy, 1971) and in addition the molecular weight of the viral polypeptide(s) is given in brackets ($\times 10^3$ daltons).

I. VIRUS COMPOSITION AND STRUCTURE

Purified preparations of most of the nepoviruses contain three types of polyhedral particles, top (T), middle (M), and bottom (B) components with sedimentation coefficients of 50–55, 91–119, and 121–130 S, respectively. The only exceptions to the presence of three components appear to be an absence of T in CLRV and an absence of M in strawberry latent ringspot virus (SLRV). Particles of T, M, and B appear to be indistinguishable in size (about 28–30 nm) and protein composition, but T consists of only protein whereas M and B contain 28–30 and 38–43% RNA, respectively. Each particle of M component always contains a single molecule of single-stranded RNA with a molecular weight of approximately 1.4×10^6 daltons. B-component particles of all the nepoviruses contain a single molecule of single-stranded RNA with a molecular weight $2.3–2.6 \times 10^6$ daltons. However, in the case of TRSV, arabis mosaic virus (AMV), raspberry ringspot virus (RRV), and SLRV, some B-component particles contain two molecules of RNA similar in size to those in M-component particles. Thus, all nepoviruses contain two molecules of RNA that differ in size, and the method of encapsidation varies. The two RNA types have been referred to as RNA_1 and RNA_2 but this is confusing since some authors refer to the larger molecule as RNA_1 and the smaller as RNA_2 whereas others refer to the larger as RNA_2 and the smaller as RNA_1. To avoid further confusion we will refer to the RNA with the shorter or smaller molecule as RNA-S and the longer or larger one as RNA-L. It should be emphasized here that there may be significant errors in the reported molecular weights of nepovirus RNA's presented in Table I. Most of the values have been determined by polyacrylamide-gel electrophoretic techniques using marker molecules under nondenaturing conditions. These methods may well yield spurious results and it has already been shown that a discrepancy exists in the molecular weight of RNA-L of TRSV when determined by polyacrylamide-gel electrophoresis and the formylation-sedimentation technique (Kaper and Waterworth, 1973). Significant differences in values for the molecular weight of the RNA's of tomato black ring virus (TBRV) were obtained when determined in aqueous polyacrylamide gels and in gels containing formamide to denature the polyribonucleotides (J. W. Randles, private communication). It would appear that the molecular weights of nepovirus RNA's require further research.

In addition to RNA-L and RNA-S of TRSV and TBRV, satellite RNA's with molecular weights of about 1.2×10^5 and 5×10^5 daltons, respectively have been detected (Sogo *et al.*, 1974; Murant *et al.*, 1973). The satellite RNA's are encapsidated in protein shells indistinguishable from those of the normal viruses and they are dependent for their replication on the RNA's of their helper viruses. The biological significance of the satellite RNA's is not understood.

The significance of RNA-L and RNA-S of nepoviruses has been of considerable interest since their first detection in preparations of TRSV by Diener and Schneider (1966). These workers demonstrated that infectivity was correlated with RNA-L and suggested that TRSV RNA was synthesized in the form of two noninfectious pieces (RNA-S) which were later joined to form infectious RNA-L. Recently, Rezaian and Francki (1974) have demonstrated by RNA–RNA annealing experiments that the nucleotide sequences of RNA-L are distinct from those of RNA-S although the RNA's may have sequences of up to 900 nucleotides in common. Similar experiments also demonstrated that the RNA-S isolated from M-component particles have nucleotide sequences indistinguishable from those of RNA-S from B-component particles (Rezaian and Francki, 1974). Harrison *et al.* (1972) have reported that highly purified preparations of RNA-L and RNA-S from RRV and TRSV were poorly infective but that infectivity was greatly stimulated when homologous, but not heterologous, RNA-L and RNA-S preparations were mixed. Furthermore, by constructing pseudorecombinants from RNA-L and RNA-S of RRV strains with appropriate genetic markers, it was demonstrated that each RNA carried certain determinants (Harrison *et al.*, 1974). These observations indicate that the nepovirus genome is bipartite. However, it remains to be explained why it has not been possible to completely eliminate infectivity from preparations of purified RNA-L. It also remains obscure that when RRV RNA (Murant *et al.*, 1972) or

TRSV RNA (Francki, unpublished data) preparations are fractionated by sucrose density-gradient centrifugation, the distribution of infectivity in the gradients is correlated with the presence of RNA-L and not a mixture of RNA-L and RNA-S.

Electron micrographs of negatively stained TRSV preparations such as that in Fig. 1 indicate the polyhedral nature of the capsids. The appearance of other nepovirus particles is very similar. In such micrographs, some particles are penetrated by the stain whereas others are not and the presence of penetrated particles has sometimes been taken as evidence for the presence of T particles. However, it was demonstrated by Davison and Francki (1969) that both penetrated and unpenetrated particles can occur in preparations of isolated T, M, or B components and that the conditions of staining can affect the proportion of the particles which become penetrated.

With the aid of the rotation technique on micrographs of negatively stained TRSV particles and model building, Chambers *et al.* (1965) concluded that the capsid has 5:3:2 symmetry and is most probably an icosahedron with 42 capsomeres (Fig. 2). Similar studies by Agrawal (1967) indicate that the arabis mosaic virus (AMV) capsid has a similar structure. An icosahedral structure with 42 capsomeres will require 240 polypeptide molecules (Caspar and Klug, 1962). With the exception of SLRV, results of polyacrylamide-gel electrophoretic determinations indicate that nepovirus capsids consist of only one type of polypeptide with molecular weights between 5.4 and 6.0×10^4 daltons (Table I). Construction of a capsid with 42 capsomeres from such polypeptides would yield a structure with molecular weight of at least 13×10^6 daltons. This is about four times that obtained by using Svedberg's equation or the proportion of RNA in capsids containing RNA. Thus, Mayo *et al.* (1971) have suggested that the nepovirus capsid is built of 60 structural subunits, each consisting of a single polypeptide. It would seem that more research is needed to resolve the question of the fine structure of nepovirus particles.

II. CYTOLOGY OF VIRUS-INFECTED PLANTS

The direct visualization of nepovirus particles in thin sections of plant cells is difficult because of their very similar size and appearance to cytoplasmic ribosomes (Crowley *et al.*, 1969). However, the presence of virus-like particles within tubules in the cytoplasm and in association with plasmodesmata (Figs. 3–5 and 7–9) are commonly observed in infected plants indicating that they are virus particles. Furthermore, large numbers of virus-like particles are often seen in cell vacuoles of infected plants, again suggesting that they must be virus particles (Figs. 3 and 6). In a few instances, small crystal-like aggregates of virus-like particles have been seen associated with cells of infected plants (Gerola *et al.*, 1965, 1966; De Zoeten and Gaard, 1969; Roberts *et al.*, 1970); but generally, crystalline or paracrystalline cellular aggregates of nepovirus particles are uncommon.

Cytoplasmic tubules enclosing virus-like particles were first observed, by Walkey and Webb (1968), in apical meristem squashes of *Nicotiana rustica* infected with CLRV and later in thin sections of virus invaded cells. It seems unlikely that these tubules are involved in the replication of virus. Atchison and Francki (1972) presented evidence that there is very little, if any, replication of TRSV in bean root tip cells, and yet tubules containing virus-like particles are particularly numerous in such tissue (Davison, 1969; Crowley *et al.*, 1969). It would seem more likely that the tubules are involved in the translocation of virus particles from cell to cell because of their common association with plasmodesmata (Fig. 4).

The significance of virus-like particle accumulation in cell vacuoles is not clear. However, the apparently intact tonoplast and absence of any cytoplasmic abnormalities of many cells whose vacuoles contain particles (Crowley *et al.*, 1969; Roberts *et al.*, 1970) indicate that the cells are still metabolically functional. Some micrographs are highly suggestive of a mechanism that exists by which particles are extruded from the cytoplasm into the vacuole without damage to the tonoplast (Fig. 6).

Nepoviruses spread rapidly throughout the tissues of an infected plant. For example, CLRV was readily detected in all the tested vegetative and reproductive tissues of infected plants (Walkey and Webb, 1970; Jones *et al.*, 1973). The ability to invade meristematic plant tissues appears to be a general feature of nepoviruses and the cytology of both meristematic and mesophyll cells of plants infected with several nepoviruses has been studied. Other than the presence of virus-like particles in cytoplasmic tubules, plasmodesmata, and vacuoles, no apparent changes have been reported in the cellular organization of meristematic cells. However, in cells of the mesophyll, extensive changes have been observed, including severe modifications to the chloroplast structure (Gerola *et al.*, 1965, 1966) and the induction of cell-wall protrusions (Jones *et al.*, 1973; Halk and McGuire, 1973). These consequences of virus infection probably occur in cells of plants at advanced stages of infection. At the early stages, cells of infected plants develop what has been termed "inclusion bodies," usually in the proximity of the nucleus (Roberts and Harrison, 1970). These inclusions are composed of endoplasmic reticulum, complex membrane structures, and ribosomes and have been observed in cells of plants infected with AMV (Gerola *et al.*, 1966), SLRV (Roberts and Harrison, 1970), and CLRV (Jones *et al.*, 1973). In the case of cells from cucumber leaves systemically infected with TRSV, we have also observed extensive cytoplasmic membrane development. However, these changes tend to be distributed diffusely through the cytoplasm without forming distinct inclusions; they are also not always in proximity to the nucleus. The changes involve the appearance of numerous golgi bodies (Figs. 7, 8, 10, and 11) which can be detected in cells of recently infected tissues which show no nuclear, chloroplast, or mitochondrial abnormalities. Recent biochemical studies indicate that TRSV protein and RNA synthesis takes place in the cytoplasm (Rezaian *et al.*, 1976).

It is of considerable interest that comoviruses, viruses similar in some respects to nepoviruses in that they have similar polyhedral particles and a bipartite genome, induce similar cytological abnormalities. These include similar proliferation of cytoplasmic membranes, presence of tubules containing virus-like particles in the cytoplasm and in association with plasmodesmata, and the presence of cell-wall protrusions as observed in cells of bean pod mottle virus-infected plants (Kim and Fulton, 1971, 1972).

ACKNOWLEDGMENTS

This work was supported by grants from The Rural Credits Development Fund of the Reserve Bank of Australia and the Australian Research Grants Committee. We thank Professor T. C. Chambers for the micrograph used in Fig. 1, Dr. E. M. Davison for those used in Figs. 3–6, and Drs. G. Martelli and H. Dias for information regarding their unpublished results.

REFERENCES

Agrawal, H. O. (1967). *J. Ultrastruct. Res.* **17,** 84–90.
Atchison, B. A., and Francki, R. I. B. (1972). *Physiol. Plant Pathol.* **2,** 105–111.
Cadman, C. H. (1963). *Annu. Rev. Phytopathol.* **1,** 143–172.

Caspar, D. L. D., and Klug, A. (1962). *Cold Spring Harbor Symp. Quant. Biol.* **27,** 1–24.
Chambers, T. C., Francki, R. I. B., and Randles, J. W. (1965). *Virology* **25,** 15–21.
Cropley, R., and Tomlinson, J. A. (1971). *CMI/AAB Descriptions Plant Viruses* No. 80.
Crowley, N. C., Davison, E. M., Francki, R. I. B., and Owusu, G. K. (1969). *Virology* **39,** 322–330.
Davison, E. M. (1969). *Virology* **37,** 694–695.
Davison, E. M., and Francki, R. I. B. (1969). *Virology* **39,** 235–239.
De Zoeten, G. A., and Gaard, G. (1969). *J. Cell Biol.* **40,** 814–823.
Diener, T. O., and Schneider, I. R. (1966). *Virology* **29,** 100–105.
Gerola, F. M., Bassi, M., and Betto, E. (1965). *Caryologia* **18,** 353–375.
Gerola, F. M., Bassi, M., and Belli, G. G. (1966). *Caryologia* **19,** 481–491.
Gooding, G. V. (1970). *Phytopathology* **60,** 708–713.
Halk, E. I., and McGuire, J. M. (1973). *Phytopathology* **63,** 1291–1300.
Harrison, B. D., Murant, A. F., and Mayo, M. A. (1972). *J. Gen. Virol.* **16,** 339–348.
Harrison, B. D., Murant, A. F., Mayo, M. A., and Roberts, I. M. (1974). *J. Gen. Virol.* **22,** 233–247.
Hewitt, W. B., Martelli, G., Dias, H. F., and Taylor, R. H. (1970). *CMI/AAB Descriptions Plant Viruses* No. 28.
Jones, A. T., and Mayo, M. A. (1972). *J. Gen. Virol.* **16,** 349–358.
Jones, A. T., and Murant, A. F. (1971). *Ann. Appl. Biol.* **69,** 11–15.
Jones, A. T., Kinninmonth, A. M., and Roberts, I. M. (1973). *J. Gen. Virol.* **18,** 61–64.
Kaper, J. M., and Waterworth, H. E. (1973). *Virology* **51,** 183–190.
Kenten, R. H. (1972). *Ann. Appl. Biol.* **71,** 119–126.
Kim, K. S., and Fulton, J. P. (1971). *Virology* **43,** 329–337.
Kim, K. S., and Fulton, J. P. (1972). *Virology* **49,** 112–121.
Martelli, G. P. (1975). *In* "Nematode Vectors of Plant Viruses" (F. Lamberti, C. E. Taylor, and J. W. Seinhorst, eds.), pp. 233–252. Plenum, New York.
Martelli, G. P., and Quacquarelli, A. (1972). *CMI/AAB Descriptions Plant Viruses* No. 103.
Mayo, M. A., Murant, A. F., and Harrison, B. D. (1971). *J. Gen. Virol.* **12,** 175–178.
Mayo, M. A., Murant, A. F., Harrison, B. D., and Goold, R. A. (1974). *J. Gen. Virol.* **24,** 29–37.
Murant, A. F. (1970a). *CMI/AAB Descriptions Plant Viruses* No. 6.
Murant, A. F. (1970b). *CMI/AAB Descriptions Plant Viruses* No. 16.
Murant, A. F. (1970c). *CMI/AAB Descriptions Plant Viruses* No. 38.
Murant, A. F. (1974). *CMI/AAB Descriptions Plant Viruses* No. 126.
Murant, A. F., Mayo, M. A., Harrison, B. D., and Goold, R. A. (1972). *J. Gen. Virol.* **16,** 327–338.
Murant, A. F., Mayo, M. A., Harrison, B. D., and Goold, R. A. (1973). *J. Gen. Virol.* **19,** 275–278.
Rezaian, M. A., and Francki, R. I. B. (1974). *Virology* **59,** 275–280.
Rezaian, M. A., Francki, R. I. B., Chu, P. W. G., and Hatta, T. (1976). *Virology* **74,** 481–488.
Roberts, D. A., Christie, R. G., and Archer, M. C. (1970). *Virology* **42,** 217–220.
Roberts, I. M., and Harrison, B. D. (1970). *J. Gen. Virol.* **7,** 47–54.
Schneider, I. R., White, R. M., and Civerolo, E. L. (1974). *Virology* **57,** 139–146.
Sogo, J. M., Schneider, I. R., and Koller, T. (1974). *Virology* **57,** 459–466.
Stace-Smith, R. (1970a). *CMI/AAB Descriptions Plant Viruses* No. 17.
Stace-Smith, R. (1970b). *CMI/AAB Descriptions Plant Viruses* No. 18.
Walkey, D. G. A., and Webb, M. J. W. (1968). *J. Gen. Virol.* **3,** 311–313.
Walkey, D. G. A., and Webb, M. J. W. (1970). *J. Gen. Virol.* **7,** 159–166.
Wildy, P. (1971). *Monogr. Virol.* **5,** 59.

Figs. 1 and 2. Structure of nepovirus particles.

Fig. 1. A preparation of tobacco ringspot virus (TRSV) negatively stained with phosphotungstic acid neutralized with KOH. Note that some of the particles have been penetrated by the stain whereas others have not (see text for discussion of this feature). The particles of other nepoviruses appear very similar to those of TRSV when examined by this technique.

Fig. 2. Model of TRSV proposed by Chambers *et al.* (1965) deduced from micrographs of negatively stained virus preparations like that in Fig. 1. Each particle consists of an icosahedron built from 42 subunits (see text for discussion of this model).

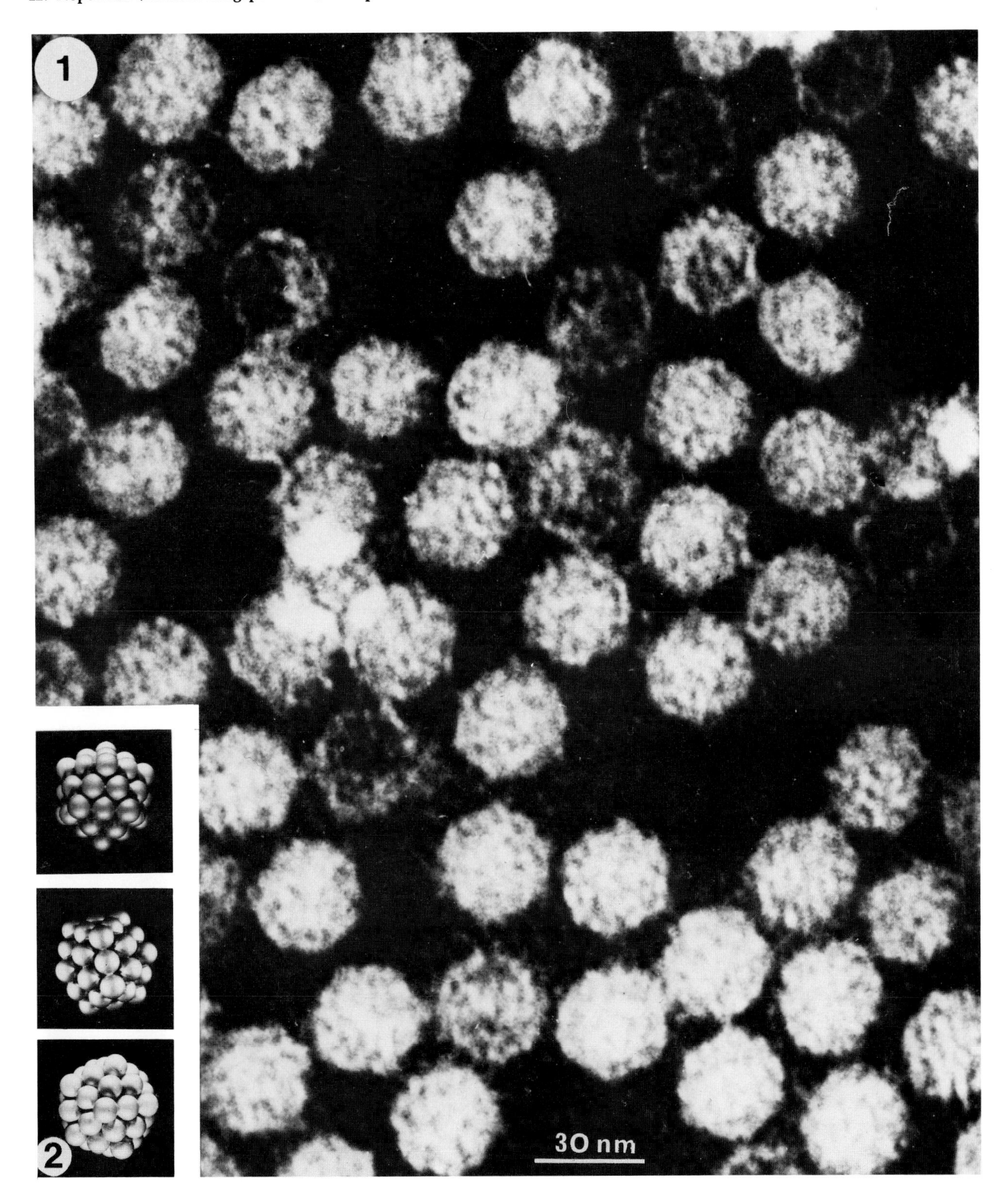
1
2
30 nm

Figs. 3 and 4. Meristematic French bean root-tip cells invaded by TRSV.

Fig. 3. Cell wall (W) dividing two young root-tip cells joined by plasmodesmata, one of which contains densely staining virus-like particles (arrow). Virus-like particles can also be seen in the vacuole (Vac). The presence of virus-like particles in plasmodesmata and vacuoles is a general feature of meristematic cells invaded by nepoviruses.

Fig. 4. A tubule containing virus-like particles (arrow) in the cytoplasm of a cell from a virus-invaded root tip. Such tubules are often seen in cells of plants infected by nepoviruses.

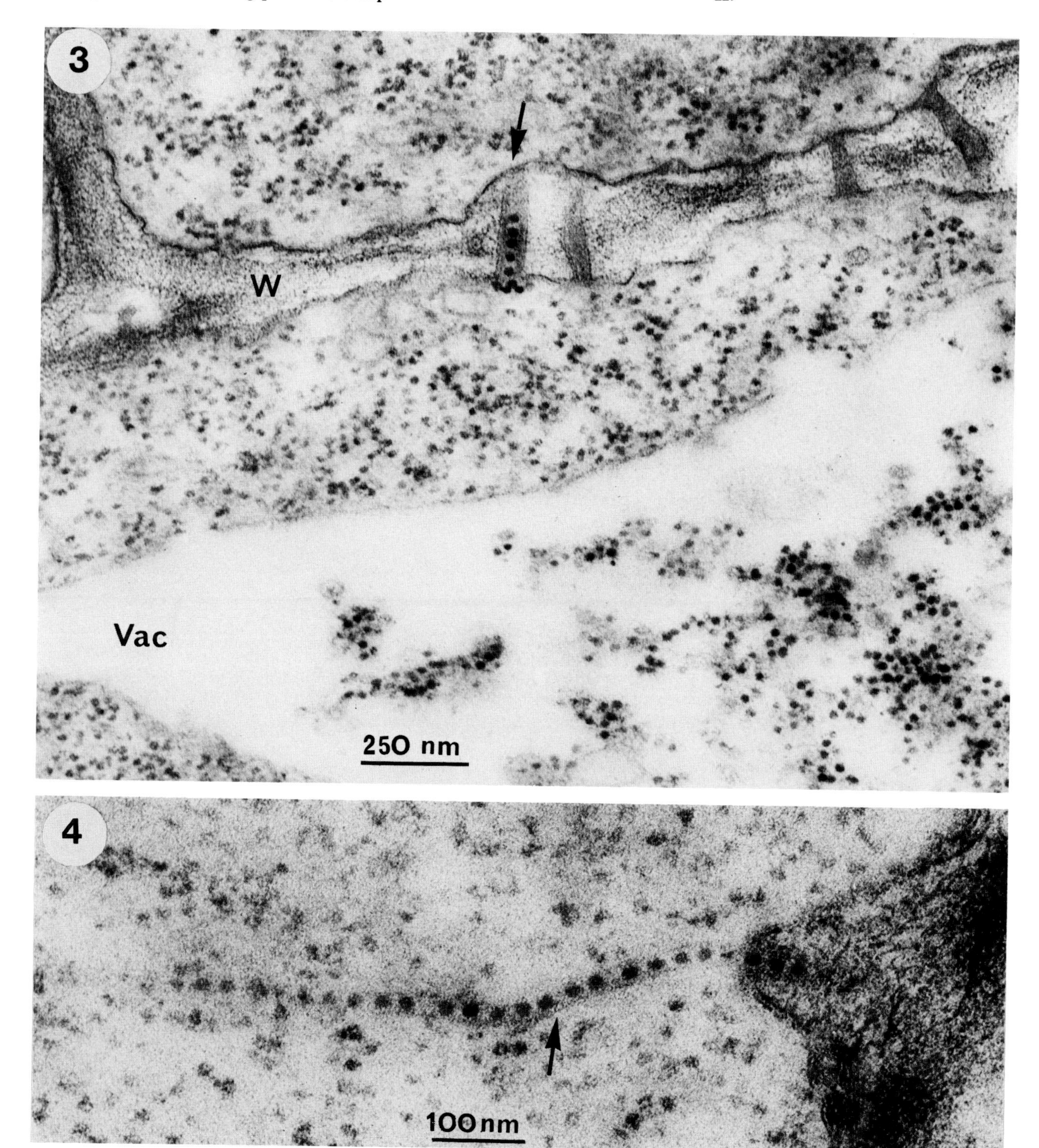
3
W
Vac
250 nm
4
100nm

Figs. 5 and 6. Meristematic French bean root-tip cells invaded by TRSV.

Fig. 5. Virus-like particles inside a plasmodesma.

Fig. 6. Part of a root-tip cell still in the process of division in which virus-like particles (arrow) appear to be passing into the vacuole (Vac). Parts of the cell wall (W), nucleus (N), and a mitochondrion (M) can also be seen.

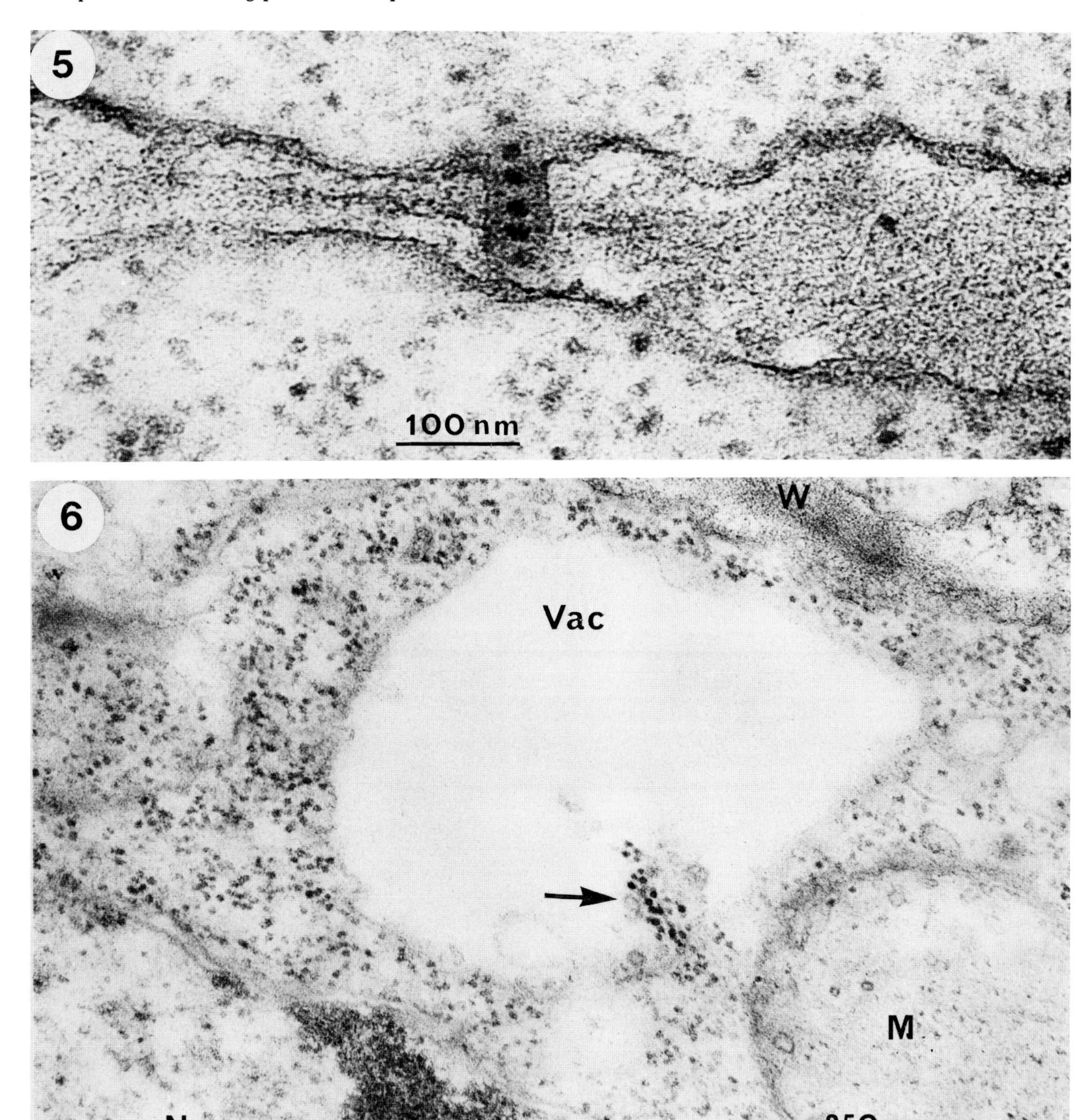
5
100 nm
6
W
Vac
M
N
250 nm

Figs. 7–9. Cells of young cucumber leaf at early stages of systemic infection by TRSV.

Fig. 7. The cytoplasm of an infected cell in proximity of the nucleus (N). Several golgi bodies (G) can be seen in association with proliferation of cellular membranes. Tubules containing virus-like particles (arrow) can also be seen in the cytoplasm.

Fig. 8. Area of the cytoplasm in Fig. 7 with prominent golgi bodies (G) and tubules containing virus-like particles (Tu). Particles loosely embedded in the cytoplasm between G and Tu may be virus particles by virtue of their dense staining. However, it is not possible to distinguish virus particles from ribosomes on the basis of size (see text for discussion on the problem associated with distinguishing ribosomes and nepovirus particles).

Fig. 9. Cytoplasm of virus-infected cell showing tubules containing virus-like particles (arrow). Some of the tubules appear to be associated with a developing vacuole (upper right corner) whereas others appear to be embedded in the body of the cytoplasm (lower left corner).

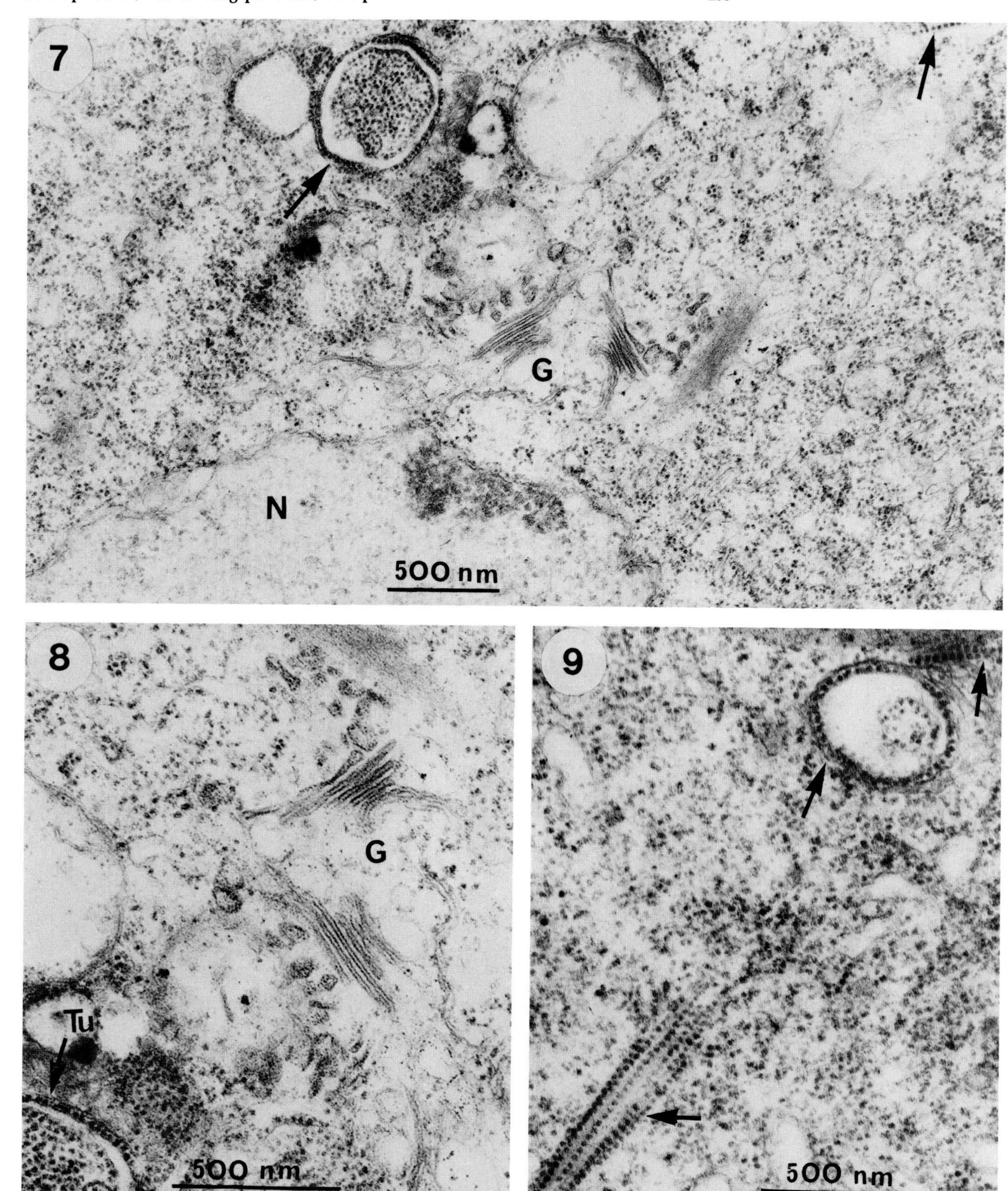
7
G
N
500 nm
8
G
Tu
500 nm
9
500 nm

Figs. 10 and 11. Cells of young cucumber leaf at early stages of systemic infection by TRSV.

Fig. 10. Cytoplasm in proximity of a chloroplast (C) showing typical changes associated with virus infection. This involves the appearance of groups of golgi bodies (G) and vesiculation of membranes in their proximity and also development of tubules containing virus-like particles (arrow). The tonoplast (T) lining the vacuole (Vac) appears to be intact.

Fig. 11. Cytoplasm in proximity of the nucleus (N) showing typical changes associated with virus infection. This involves the appearance of groups of golgi bodies (G) and areas in the cytoplasm with intricate membrane systems (arrow). In this micrograph the characteristic association of four golgi bodies can be seen together with prominent vesiculation of the cytoplasm between the golgi bodies.

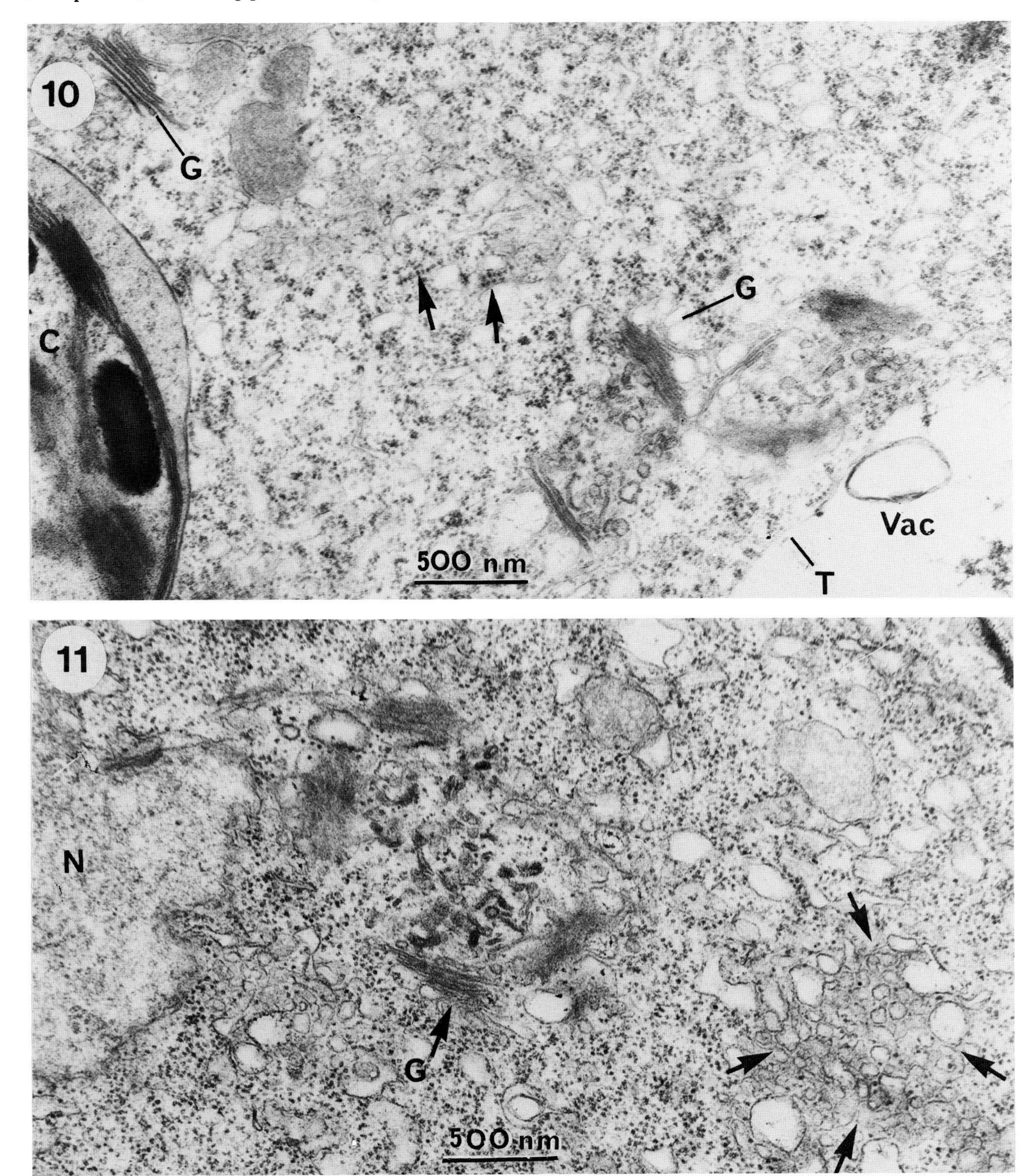
10
G
G
C
Vac
T
500 nm
11
N
G
500 nm

Chapter 13

Tabamovirus (Tobacco Mosaic Virus) Group

EISHIRO SHIKATA

I. General Description (R/1 : 2/5 : E/E : S/O) 237
II. Composition of Virions 237
III. Localization of the Virus Particles *in Situ* 239
References 241

I. GENERAL DESCRIPTION (R/1:2/5:E/E:S/O)

The representative virus in this group is the tobacco mosaic virus, which is the most extensively studied plant virus. This group includes tobacco mosaic virus (TMV) (Zaitlin and Israel, 1975), cucumber green mottle mosaic virus (CGMMV) (Hollings *et al.*, 1975), *Odontoglossum* ringspot virus (ORSV) (Paul, 1975), sunn hemp mosaic virus (SHMV) (Kassanis and Varma, 1975), ribgrass mosaic virus (RMV) (Oshima and Harrison, 1975), and possibly Sammon's *Opuntia* virus (SOV) (Brandes and Chessin, 1965).

General morphological characteristics of this group are (1) the rigid rod shape of the virions, about 300 nm long, 18 nm wide; (2) the helical structure of the nucleocapsids with a pitch of 2.3 nm; (3) no envelope-enclosing virions; (4) single-stranded RNA as a viral genome. Most viruses belonging to this group are present in high concentration in the host plant, about 1–3 mg/liter. They are easily transmitted by sap inoculation, but not by insects, and are tolerant to high temperatures, about 90° C for 10 minutes.

II. COMPOSITION OF VIRIONS

Properties of the virions belonging to this group are summarized in Table I. Approximate molecular weights (Table I) of the virions are 39–40 $\times$ 10^6, and their sedimentation coefficients range between 187–212 S. The particles consist

TABLE I

Properties of TMV Virions

Viruses	Molecular weight of virion	Sedimentation coefficient	Number of protein subunits	Molecular weight of protein subunits	Number of amino acid residues	Percent of RNA content	Molecular weight of RNA	Base composition of RNA (A:G:C:U)
TMV	39×10^6 (Knight, 1964)	190 S (Lauffer and Stevens, 1968)	2200 (Knight, 1964)	17,538 (Knight, 1964) 17,493 (Knight, 1975)	158 (Knight, 1964; Kado *et al.*, 1968)	5% (Knight and Woody, 1958; Knight, 1975)	2×10^6 (Knight, 1975)	28:24:20:28 (Knight, 1964)
CGMMV	39×10^6 (Knight, 1964)	195 S (Brčák *et al.*, 1962)	2100 (Knight, 1964)	16,102 16,940 (Japan) (Knight, 1975)	158 (van Regenmortel, 1967; Nozu *et al.*, 1971)	5% (Knight and Woody, 1958; Knight, 1975)		25.7:25.7:18.7:30.2 (Knight, 1964)
ORV		212 S (Paul, 1975)		17,598 (Paul, 1975)	158 (Kado *et al.*, 1968) 157 (Paul, 1975)	5% (Paul, 1975)		
SHMV		187 S (Kassanis and MacCarthy, 1967; Kassanis and Varma, 1975)		18,062 (Kassanis and Varma, 1975)	161 (Kassanis and Varma, 1975)			
RMV	40×10^6 (Oshima and Harrison, 1975)		2000 (Oshima and Harrison, 1975)	17,500 (Oshima and Harrison, 1975)	156 (Oshima and Harrison, 1975)	5% (Oshima and Harrison, 1975)	2×10^6 (Oshima and Harrison, 1975)	29.3:25.8:18.0:27.0 (Oshima and Harrison, 1975)

of 2000–2200 protein subunits, with molecular weights of about 16,000–18,000. About 156–161 amino acid residues are detected in a protein subunit. The virions contain 5% RNA genome with a molecular weight of 2×10^6.

Serological relationships between TMV, CGMMV, ORSV, SHMV, RMV, and SDV have been reported (Kado *et al.*, 1968; Paul, 1975).

The length of TMV particles, determined by electron microscopy, has been reported to be 280 nm or 300 nm since 1939 (Shikata, 1966). In 1946, Rawlins *et al.* concluded that the average length of TMV was 300 nm. This value was confirmed more than 10 years later by precise electron microscopic measurements (Williams, 1957, 1959; Hall, 1958).

Electron microscopic studies of morphological changes on TMV inactivation ascertained that the infectivity depended on rods of average length and that serological reaction was associated with shortened rods until denaturation (Shikata, 1966). On the other hand, after the reconstitution studies of TMV particles (Fraenkel-Conrat and Williams, 1955), the first demonstration of viral-RNA infectivity was shown by TMV (Gierer and Schramm, 1956; Fraenkel-Conrat, 1956).

The fine structure of TMV particles has been established by X-ray diffraction by Franklin *et al.* (1959). Their model indicated that 2200 of the small identical protein subunits were helically arranged with a pitch 2.3 nm around the particle axis. Maximum diameter of the particles is 18 nm and mean radius is 7.5 nm. The subunits repeat 6.9 nm in the axial direction, and the repeat consists of 49 subunits distributed over three turns. Visual confirmation of this ultrastructure and its morphology by means of an electron microscope was published in 1964 (Finch, 1964), much later than X-ray studies. Finch (1972) also analyzed the electron microscopic images of TMV particles, which were tilted through small angles, and showed that the basic helix is right-handed. Following the development of high resolution electron microscopes, attempts were made to visualize the helix structures directly (Figs. 1 and 2) (Williams and Fischer, 1970; Nonomura and Ohno, 1974; Hart and Yoshiyama, 1975).

According to Franklin's model, RNA in TMV particles is deeply embedded in the protein subunit array. It is in the form of a single long-chain molecule which follows the line of the flat helix, binding the protein subunits at the core, which is 4 nm in diameter. Recently, high resolution electron micrographs taken by Nonomura and Ohno (1974) clearly revealed the RNA site in TMV and CGMMV protein disks (Figs. 3 and 4). A narrow ring was observed at a radius of 4 nm corresponding to the site of RNA obtained by X-ray diffraction studies of TMV. The rings were observed in the stacked protein disks and in the end view of the helical rods prepared in acidic solution from viral protein without RNA, while no such doughnut ring could be seen in the end view of the intact particles.

A possible assembly mechanism of the TMV particle from its RNA and protein has been proposed by Okada and Ohno (1972). The results indicated that reconstitution takes place in the formation of the initial complex with the 5′-end of TMV-RNA and the 20–30 S protein aggregate. The elongation of the rod from the initial complex proceeds by stepwise addition of protein subunits or 4 S aggregates.

III. LOCALIZATION OF THE VIRUS PARTICLES *IN SITU*

Localization of TMV particles in infected cells has long been studied by electron microscopy. As summarized by Esau (1968), virus accumulations have been found in nuclei, plastids, and chloroplasts, as well as in the cytoplasm. Usually, large masses of TMV particles, aggregated side by side and end to end, appeared within the cytoplasm (Figs. 5–11). Large crystalline aggregates are frequently formed in the epidermal hair cells. According to Esau (1968), the virus is found in the cells of parenchyma, mesophyll, sieve elements and phloem, tracheary elements, and in guard cells of stoma.

Host cells infected with TMV usually contain two types of inclusions. One is a crystalline form appearing as hexagonal plates, needle shaped, or fibrous structures. The other is an amorphous and vacuolate body, generally called X body (Fig. 12, inset).

Electron microscopic studies by Steere and Williams (1953) revealed that the crystalline inclusions were composed of TMV particles only. Ultrathin sections of the infected cells revealed the internal structure of the crystalline inclusions. Three dimensional hexagonal inclusions consist of layers of single particles with parallel alignment, corresponding in length to a TMV particle (Warmke and Edwardson, 1966) (Figs. 6 and 7). Needle inclusions consist of end to end aggregations, and their cross sections reveal a hexagonal, regular arrangement (Fig. 8).

It has been known that X bodies are rich in virus related protein, but its nature and fine structure were determined only recently. As a result of the progress in electron microscopic techniques, such as fixation and embedding, accumulations of filamentous or tubular structures, about 28 nm wide, associated with the TMV infected leaves, became apparent (Shalla, 1964; Kolehmainen *et al.*, 1965; Milne, 1966b). Esau and Cronshaw (1967) demonstrated by comparative light and electron microscope studies that the amorphous X bodies are the regions of aggregation of tubular structures associated with virus particles, ribosomes, endoplasmic reticulum, and some vacuoles without any surrounding membranes (Figs. 12 and 13). They also indicated that the earliest stages of tubular structure which appear as aggregates of granules consisted of a mixture of granules and flexuous tubules (Fig. 5). When the tubules increase in number, cell components can be diffentiated and the complexes can be detected with the light microscope (Fig. 12). Virus particles are frequently observed in the X bodies.

What the actual role or nature of the X bodies is still remains to be solved. Some workers assumed that the X bodies were involved in virus coat protein synthesis or aggregation (Kolehmainen *et al.*, 1965; Esau and Cronshaw, 1967), but others did not agree with these interpretations (Milne, 1966a). In any event, X bodies are closely associated with TMV infection and are involved in the ultrastructure cytopathic changes resulting from virus multiplication (Matsui and Yamaguchi, 1966).

Protoplasts isolated from tobacco mesophyll cells provided a tool for *in vitro* investigation of TMV entry and subsequent multiplication in host cells. Electron microscopic studies performed by Otsuki *et al.* (1972) resulted in the first visualization of the initial infection of plant cells by TMV rods. It was shown that one end of the TMV rod was absorbed by the plasmalemma (Fig. 14). Then the plasmalemma invagination at the site of virus attachment took place, and the vesicles containing the virus particles were seen (Fig. 15). This indicated the pinocytotic uptake of TMV particles into the plant host cells, similar to the process occuring in animal virus uptake (Dales and Gomatos, 1965). According to Otsuki *et al.* (1972), this process proceeds as early as 10 minutes after the addition of the virus. During the 4 hour "eclipse period" following inoculation, no virus particles were detected in ultrathin sections of inoculated protoplasts. Single progeny virus particles were first detected 6 hours after inoculation, sometimes accompanied by small groups of particles in the cytoplasm (Fig. 16).

The progeny virus accumulations grew rapidly in size and increased in number (Fig. 17). They mentioned that at later stages of infection, endoplasmic reticulum and electron dense substances of unknown nature were often present in the vicinity of the virus aggregations, but no such structures as tubules or electron opaque filaments were seen. According to our observations, the electron dense substances of unknown nature found in the infected protoplasts (Otsuki *et al.*, 1972) are identical with the granular X materials appearing during the earliest stages of infection in tobacco plants (Esau, 1968).

TMV particles were shown in the cells around the periphery of local lesions in *Nicotians glutinosa* (Hayashi and Matsui, 1963, 1966; Milne, 1966b; Shikata, 1966). The filamentous or tubular structures identical with those found in cells systemically infected with TMV were also observed in the cells of local lesions of *Chenopodium amaranticolor*. It was shown that the lesions occur after the virus has been synthesized (Milne, 1966a).

ACKNOWLEDGMENTS

The author is deeply indebted to Dr. Y. Nonomura, Department of Pharmacology, Faculty of Medicine, Tokyo University, Tokyo; Dr. C. Matsui and Dr. Y. Honda, Department of Plant Pathology, Faculty of Agriculture, Nagoya University, Nagoya, Dr. I. Takebe and Dr. Y. Otsuki, Institute for Plant Virus Research, Chiba, for their kind help in supplying original prints of electron micrographs. He also wishes to express his appreciation to Dr. Katherine Esau, Department of Biological Sciences, University of California, Santa Barbara, for permission to use her electron micrographs from the book "Viruses in Plant Hosts," The University of Wisconsin Press, 1968, and to Dr. K. Maramorosch who read the manuscript.

REFERENCES

Brandes, J., and Chessin, M. (1965). *Virology* **25,** 673–674.
Brčák, J., Ulrychová, M., and Čech, M. (1962). *Virology* **16,** 105–114.
Dales, S., and Gomatos, P. (1965). *Virology* **25,** 193–211.
Esau, K. (1968). "Viruses in Plant Hosts." Univ. of Wisconsin Press, Madison.
Esau, K., and Cronshaw, J. (1967). *Virology* **33,** 26–35.
Finch, J. T. (1964). *J. Mol. Biol.* **8,** 872–874.
Finch, J. T. (1972). *J. Mol. Biol.* **66,** 291–294.
Fraenkel-Conrat, H. (1956). *J. Am. Chem. Soc.* **78,** 882–883.
Fraenkel-Conrat, H., and Williams, R. C. (1955). *Proc. Natl. Acad. Sci. U.S.A.* **41,** 690–698.
Franklin, R. E., Caspar, D. L. D., and Klug, A. (1959). *In* "Plant Pathology: Problems and Progress, 1908–1958" (C. S. Holton *et al.*, eds.), pp. 447–461. Univ. of Wisconsin Press, Madison.
Gierer, A., and Schramm, G. (1956). *Nature (London)* **177,** 702–703.
Hall, C. E. (1958). *J. Am. Chem. Soc.* **80,** 2556–2557.
Hart, R. G., and Yoshiyama, J. M. (1975). *J. Ultrastruct. Res.* **51,** 40–45.
Hayashi, T., and Matsui, C. (1963). *Virology* **21,** 525–527.
Hayashi, T., and Matsui, C. (1966). *Phytopathology* **56,** 192–196.
Hollings, M., Komuro, Y., and Tochihara, H. (1975). *CMI/AAB, Descriptions Plant Viruses* No. 154.
Kado, C. I., van Regenmortel, M. H. V., and Knight, C. A. (1968). *Virology* **34,** 17–24.
Kassanis, B., and MacCarthy, D. (1967). *J. Gen. Virol.* **1,** 425–440.
Kassanis, B., and Varma, A. (1975). *CMI/AAB, Descriptions Plant Viruses* No. 152.
Knight, C. A. (1964). *In* "Plant Virology" (M. K. Corbett and H. D. Sisler, eds.), pp. 292–314. Univ. of Florida Press, Gainesville.
Knight, C. A. (1975). "Chemistry of Viruses," 2nd ed. Springer-Verlag, Berlin and New York.
Knight, C. A., and Woody, B. R. (1958). *Arch. Biochem.* **78,** 460–467.
Kolehmainen, L., Zech, H., and von Wettstein, D. (1965). *J. Cell Biol.* **25,** 77–97.
Lauffer, M. A., and Stevens, C. L. (1968). *Adv. Virus Res.* **13,** 1–63.
Matsui, C., and Yamaguchi, A. (1966). *Adv. Virus Res.* **12,** 127–174.
Milne, R. G. (1966a). *Virology* **28,** 79–89.
Milne, R. G. (1966b). *Virology* **28,** 520–526.
Milne, R. G. (1966c). *Virology* **28,** 527–532.

Nonomura, Y., and Ohno, T. (1974). *J. Mol. Biol.* **90,** 523–527.
Nozu, Y., Tochihara, H., Komuro, Y., and Okada, Y. (1971). *Virology* **45,** 577–585.
Okada, Y., and Ohno, T. (1972). *Mol. Gen. Genet.* **114,** 205–213.
Oshima, N., and Harrison, B. D. (1975). *CMI/AAB, Descriptions Plant Viruses* No. 152.
Otsuki, Y., Takebe, I., Honda, Y., and Matsui, C. (1972). *Virology* **49,** 188–194.
Paul, H. L. (1975). *CMI/AAB, Descriptions Plant Viruses* No. 155.
Rawlins, T. E., Roberts, C., and Nedra, M. U. (1946). *Am. J. Bot.* **33,** 356–363.
Shalla, T. A. (1964). *J. Cell Biol.* **21,** 253–264.
Shikata, E. (1966). *J. Fac. Agric., Hokkaido Univ.* **55,** 1–110.
Steere, R. L., and Williams, R. C. (1953). *Am. J. Bot.* **40,** 81–84.
van Regenmortel, M. H. V. (1967). *Virology* **31,** 467–480.
Warmke, H. E., and Edwardson, J. R. (1966). *Virology* **30,** 45–57.
Williams, R. C. (1957). *Spec. Publ., N. Y. Acad. Sci.* **5,** 207–215.
Williams, R. C. (1959). *In* "Plant Pathology: Problems and Progress, 1908–1958" (C. S. Holton *et al.*, eds.), pp. 437–445. Univ. of Wisconsin Press, Madison.
Williams, R. C., and Fisher, H. W. (1970). *J. Mol. Biol.* **52,** 121–123.
Zaitlin, M., and Israel, H. W. (1975). *CMI/AAB, Descriptions Plant Viruses* No. 151.

Fig. 1. TMV particles, negatively stained with uranyl formate. Note that the helical structure of the protein subunits in each particle is clearly shown. (Courtesy of Dr. Y. Nonomura.)

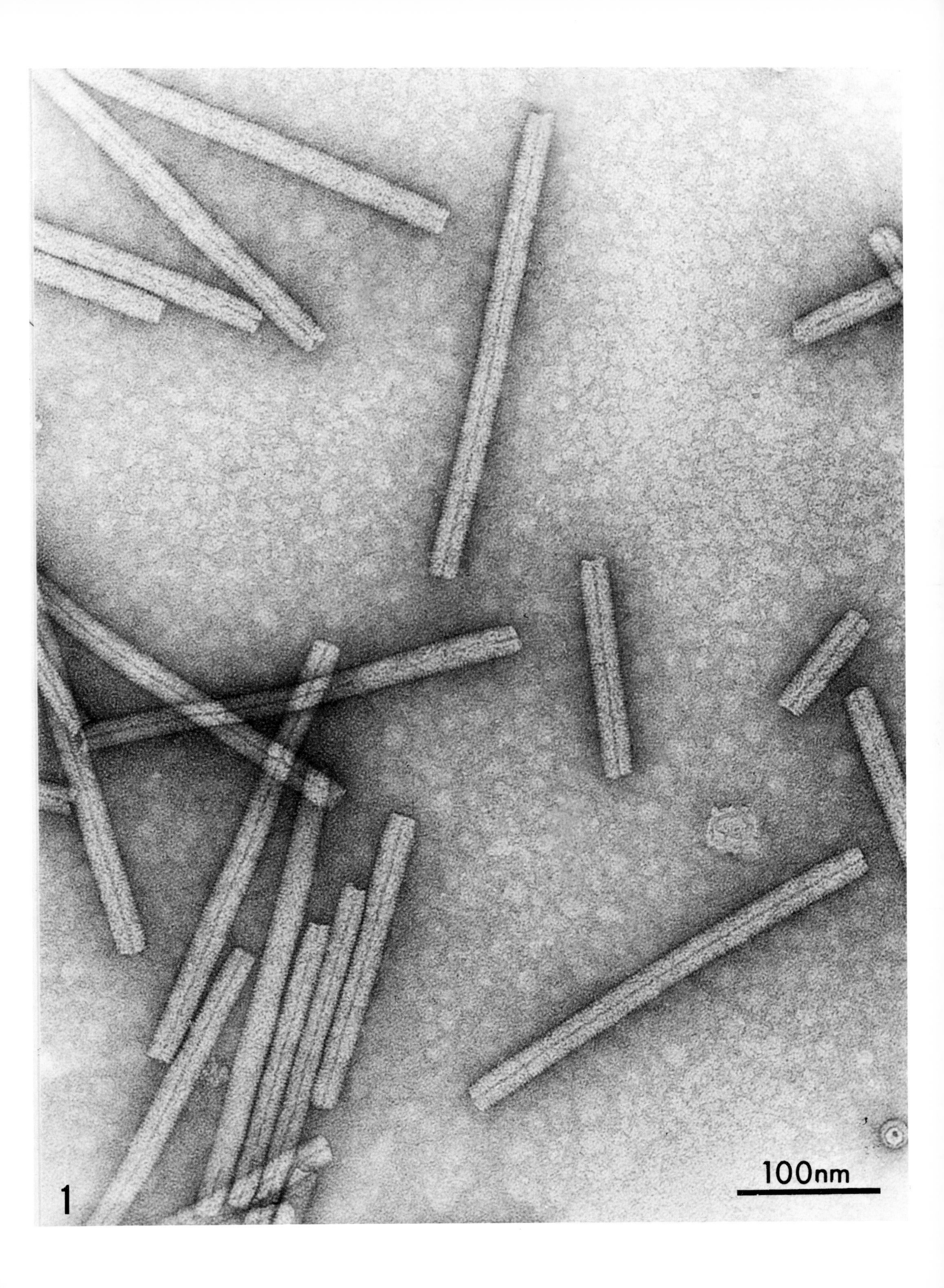
100nm
1

Fig. 2. CGMMV particles, negatively stained with uranyl formate. Note that the morphology of CGMMV is identical with TMV and the helical structure of the protein subunits is clearly indicated. (Courtesy of Dr. Y. Nonomura.)

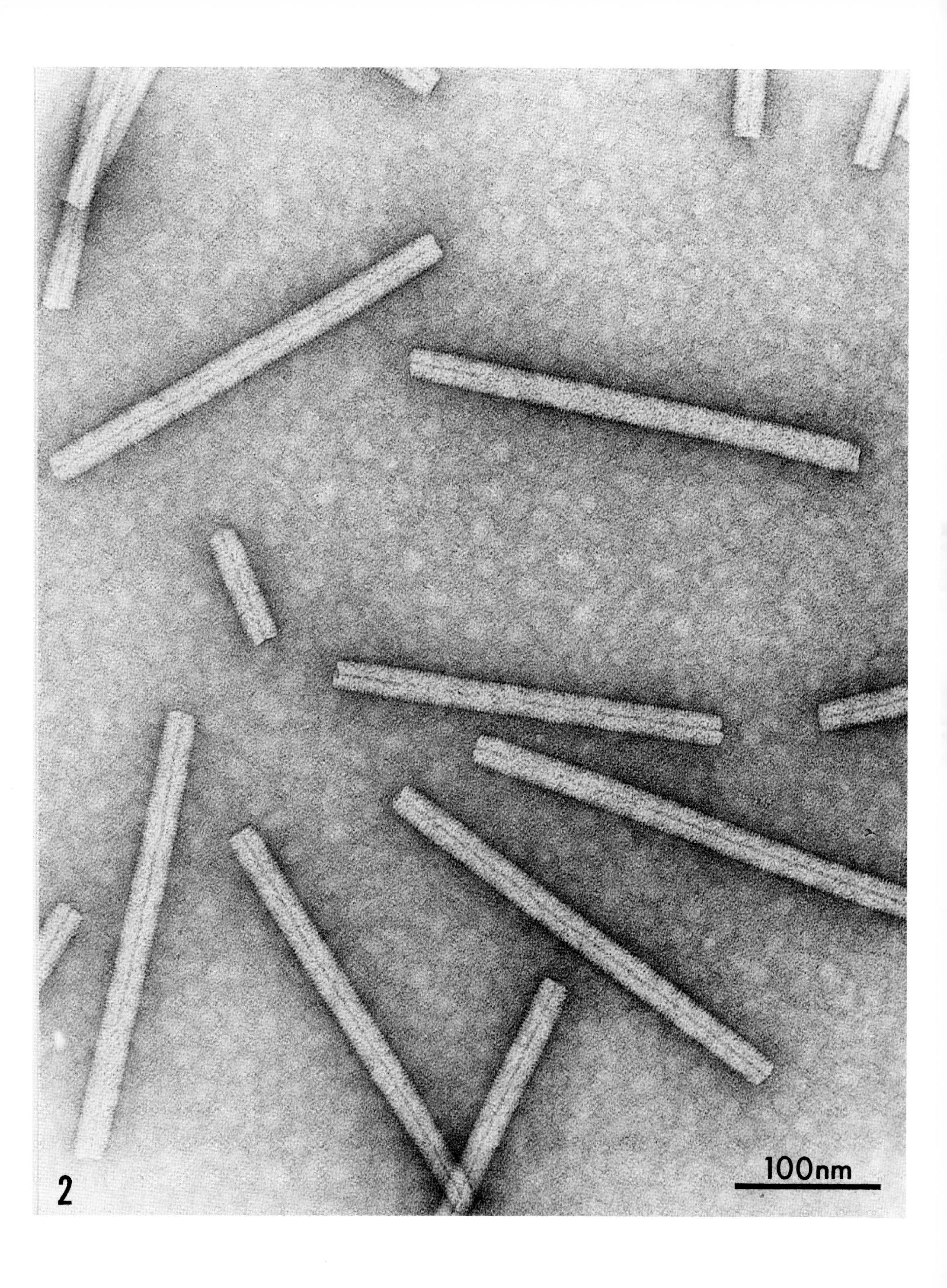
100nm
2

Fig. 3. Partially stacked TMV protein disks. Most of the disks show a doughnut-shaped structure with a central hole and a ring at a distance of 4 nm from the center, indicating the RNA site. (Courtesy of Dr. Y. Nonomura.)

Fig. 4. Partially stacked CGMMV protein disks. Most of the disks are doughnut-shaped, with a central hole and a ring at a distance of 4 nm from the center, indicating the RNA site. (Courtesy of Dr. Y. Nonomura.)

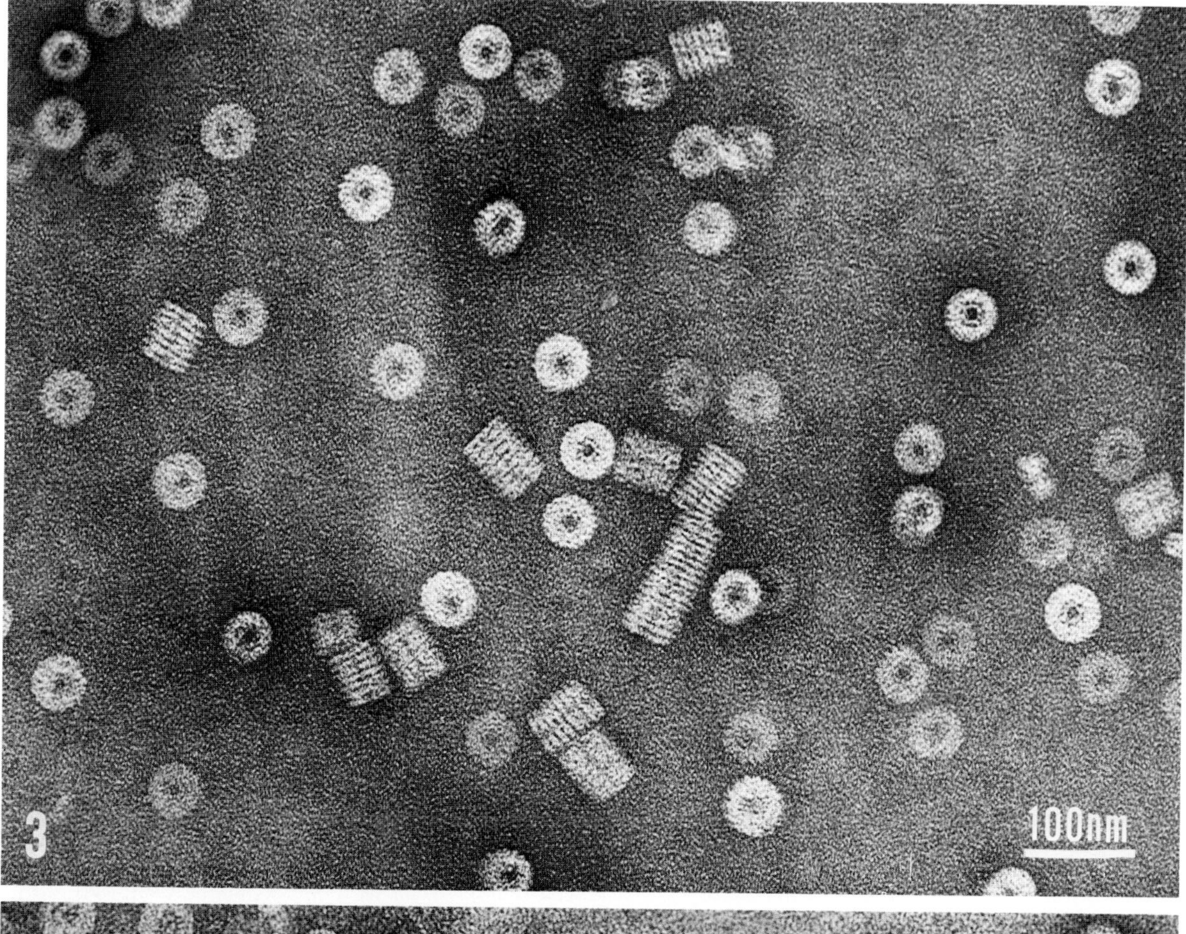
3
100nm

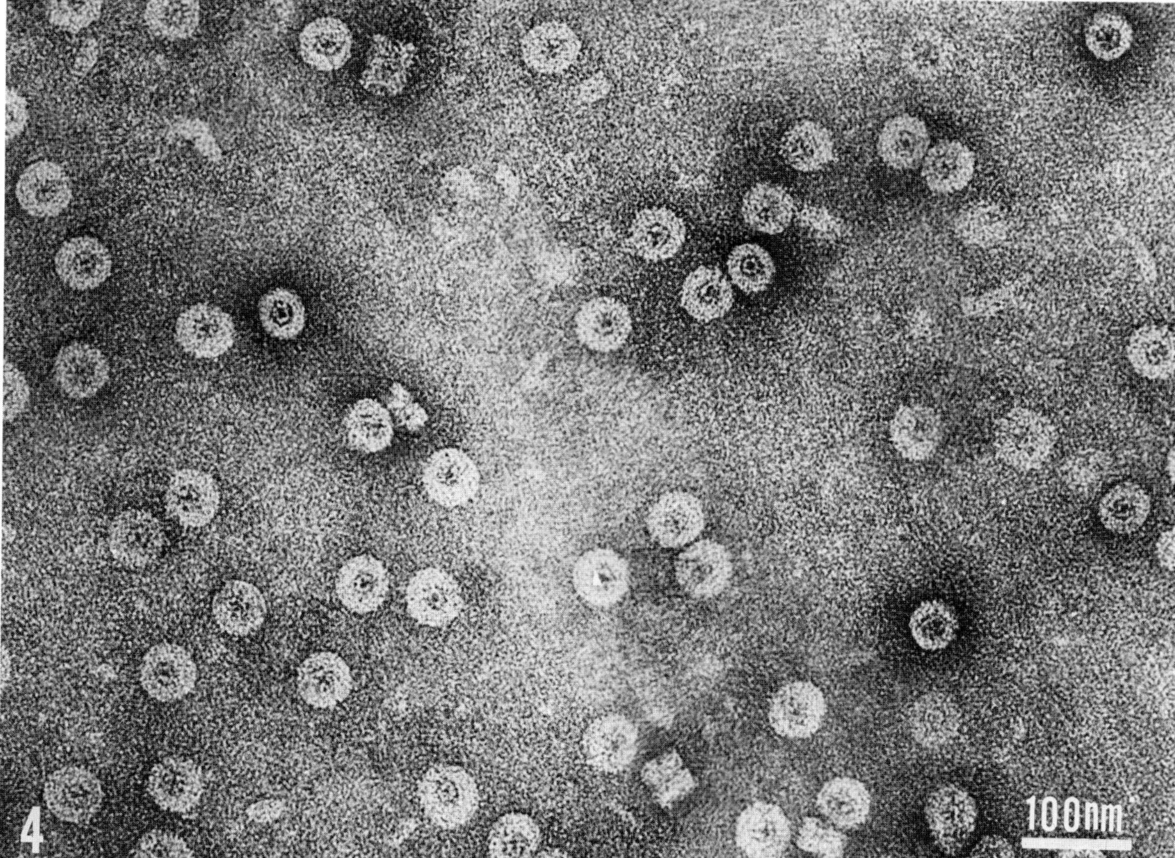
4
100nm

Fig. 5. An ultrathin section of a mesophyll cell from a leaf of *Nicotiana tabacum* infected with TMV. Accumulation of virus particles (V) in the cytoplasm and X body (X) are observed. VA, Vacuole; IS, intracellular space; CH, chloroplast; ST, starch. [From Katherine Esau, "Viruses in Plant Hosts" (Madison: The University of Wisconsin Press; © 1968 by the Regents of the University of Wisconsin), p. 206, Fig. 128.]

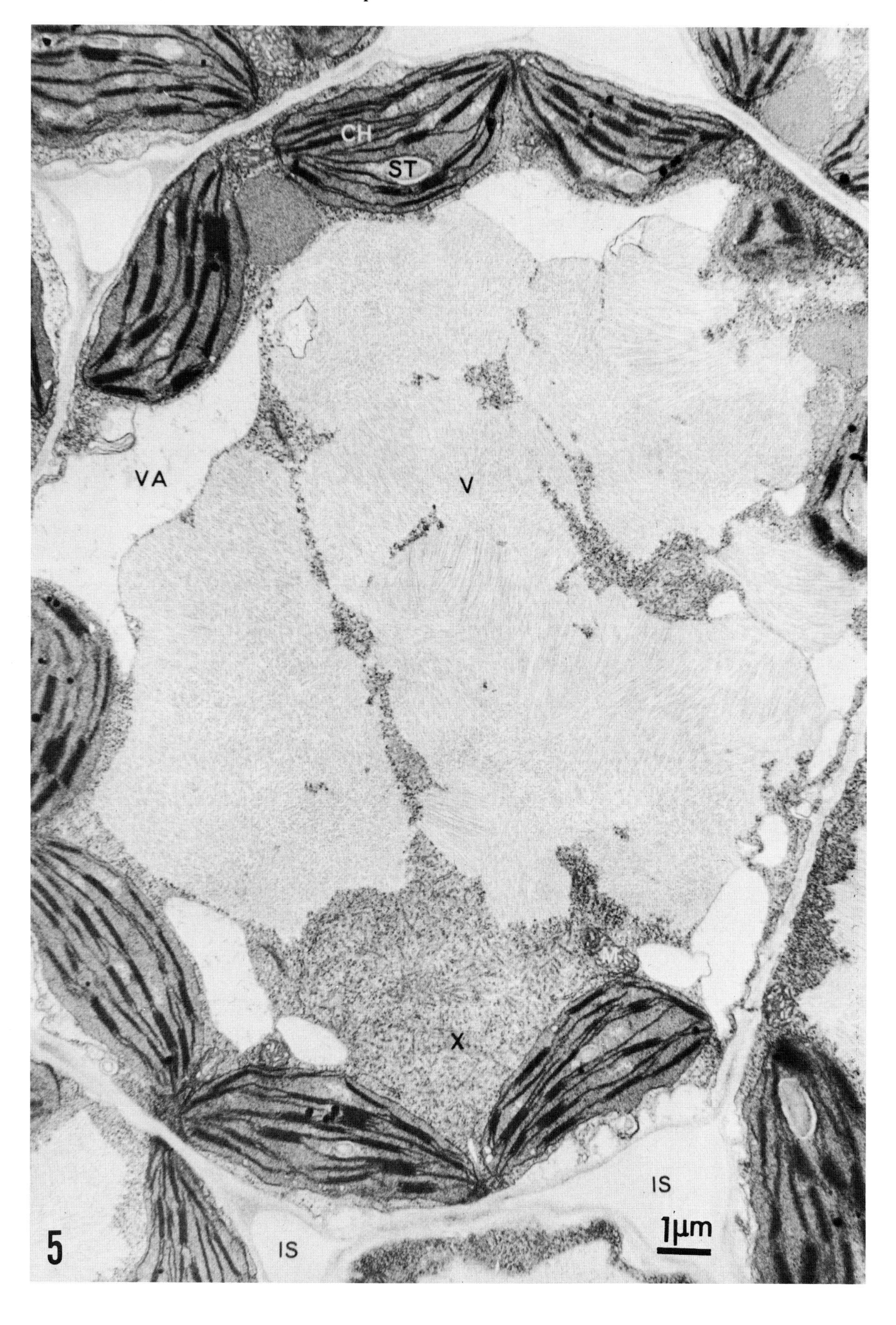
CH
ST
VA
V
X
IS
IS
1µm
5

Figs. 6–8. Ultrathin sections of parenchyma cells of *N. tabacum* infected with TMV, fixed with potassium permanganate. W, Cell wall; VA, vacuole; V, virus particles. [From Katherine Esau, "Viruses in Plant Hosts" (Madison: The University of Wisconsin Press; © 1968 by the Regents of the University of Wisconsin); p. 92, Figs. 1–3.]

Fig. 6. A section of TMV crystal. Single TMV rods arranged parallel and formed layers of the parallel band.

Fig. 7. A single layer of virus particles.

Fig. 8. A cross section of TMV crystal, showing hexagonal arrangement.

Figs. 9–11. High magnification of TMV particles in ultrathin sections of parenchyma cells of *N. tabacum*. [From Katherine Esau, "Viruses in Plant Hosts" (Madison: The University of Wisconsin Press; © 1968 by the Regents of the University of Wisconsin), p. 98, Figs. 10–12.]

Fig. 9. TMV particles in a transverse section showing a central electron lucent hole, electron dense RNA location, and outer protein coat of less dense area.

Fig. 10. A longitudinal section of TMV particles.

Fig. 11. Aggregation of TMV particles of usual appearance in an ultrathin section. At right, some helically arranged components are seen.

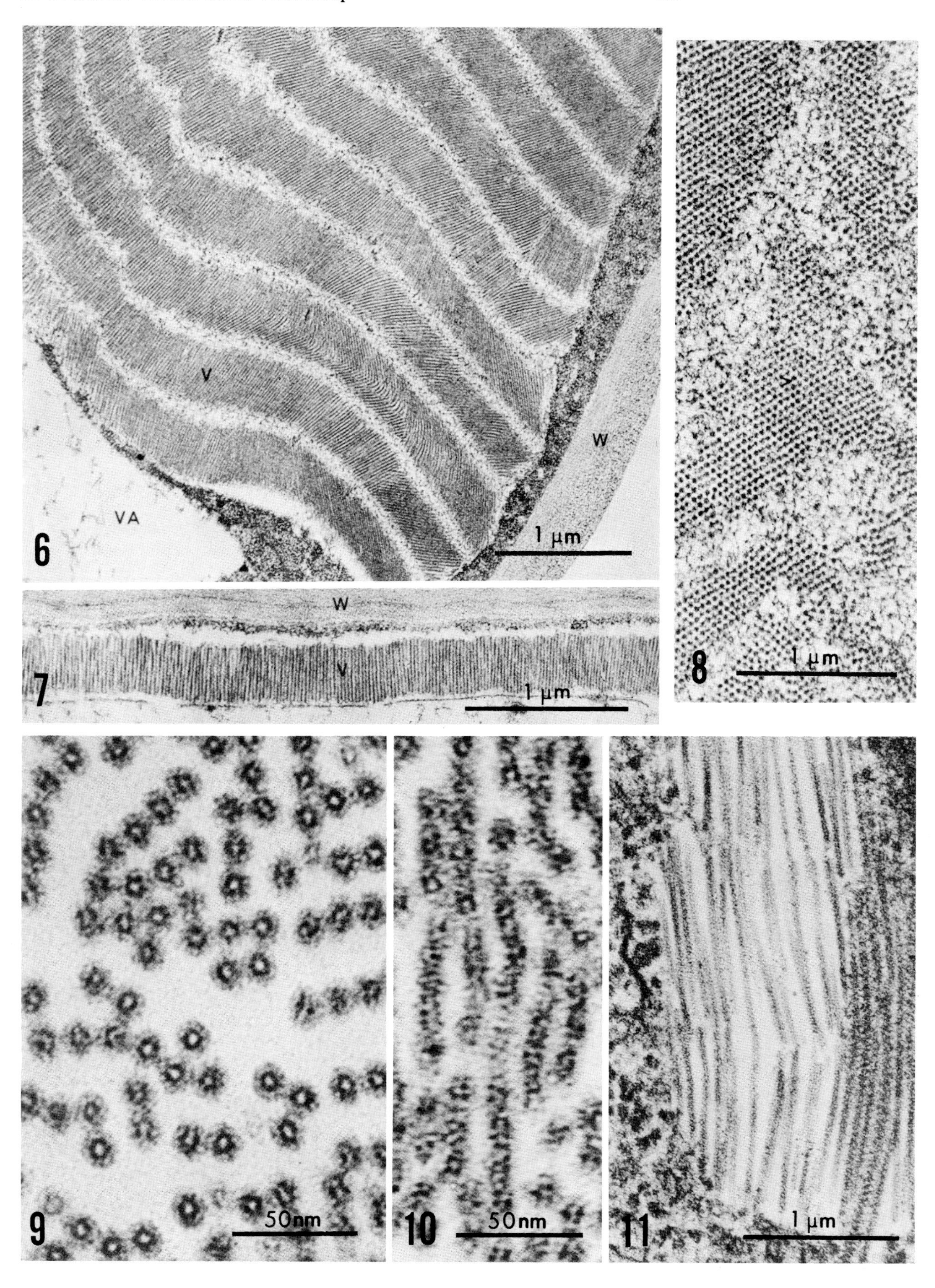
V
W
VA
1 μm
6
W
V
7
1 μm
1 μm
8
50nm
9
50nm
10
1 μm
11

Figs. 12–13. Ultrathin sections of parenchyma cells of *N. tabacum* infected with TMV. VA, Vacuole; W, cell wall; V, virus particles. [From Katherine Esau, "Viruses in Plant Hosts" (Madison: The University of Wisconsin Press; © 1968 by the Regents of the University of Wisconsin), p. 102, Figs. 14–16.]

Fig. 12. X body containing virus aggregates in a vacuole and accumulation of the bands or tubular structures (arrows) intermingled with ribosomes. A vacuolate X body by light microscope is shown in insert.

Fig. 13. High magnification of a part of an X body, showing tubular appearance of the bands (arrows). ER, Endoplasmic reticulum; RB, ribosome.

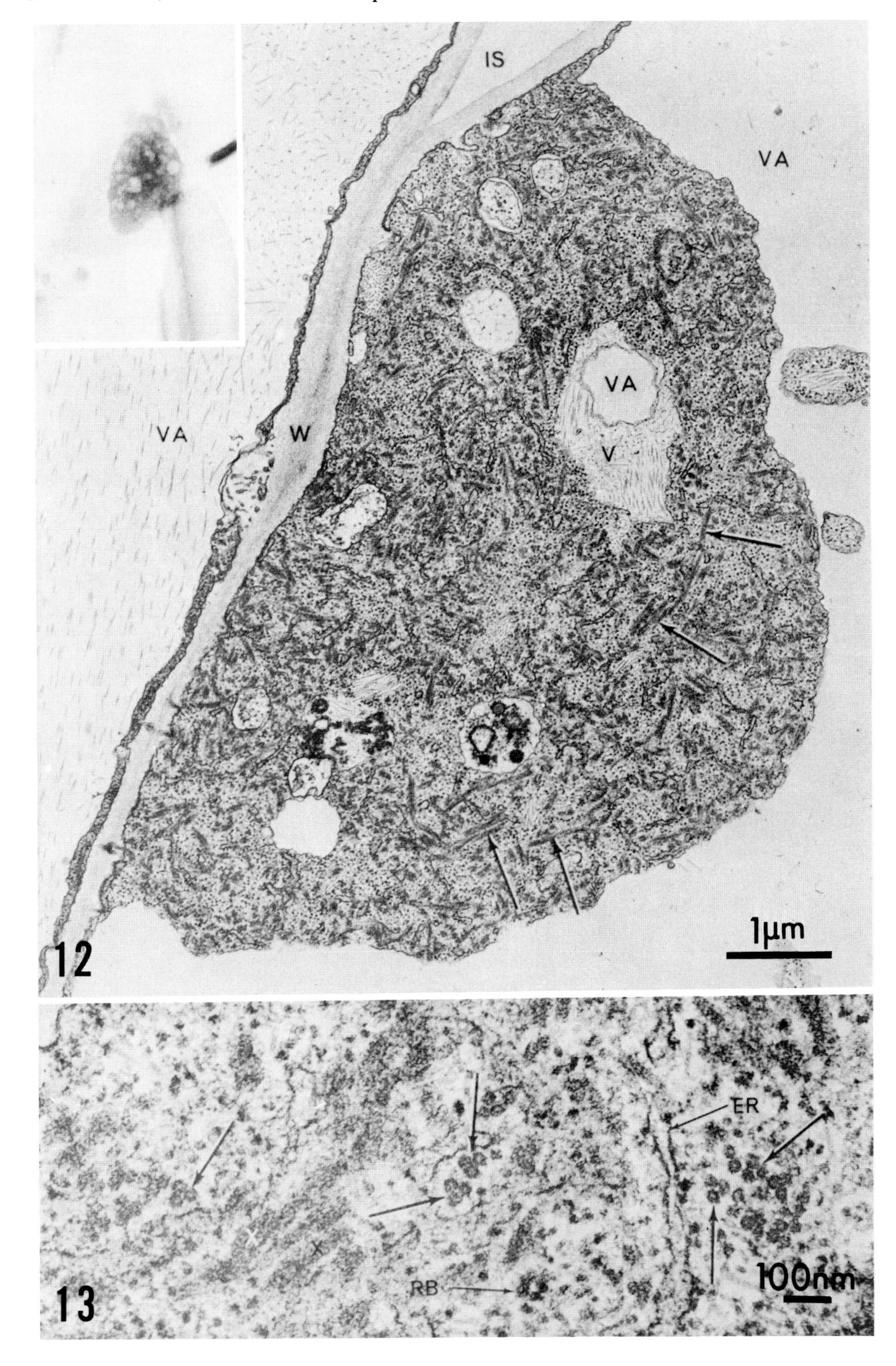
IS
VA
VA
V
VA
W
1μm
12
ER
RB
X
100nm
13

Figs. 14–17. TMV infection in tobacco (*N. tabacum* cv. *xanthi*) mesophyll protoplasts. (Courtesy of Drs. Y. Otsuki, I. Takebe, C. Matsui and Y. Honda.)

Fig. 14. Surface of tobacco mesophyll protoplast showing extensive invagination of the plasmalemma with adsorbed TMV particles. Protoplasts were fixed immediately after inoculation with 10 μg/ml TMV in the presence of 1 μg/ml poly-L-ornithine.

Fig. 15. TMV particles taken up into intracellular vesicles formed by invagination of the plasmalemma. Protoplasts were fixed immediately after inoculation. (From Otsuki *et al.*, 1972.)

Fig. 16. First progeny TMV particles appearing in protoplasts 6 hours after inoculation.

Fig. 17. Progeny TMV particles accumulated in protoplasts 47 hours after inoculation.

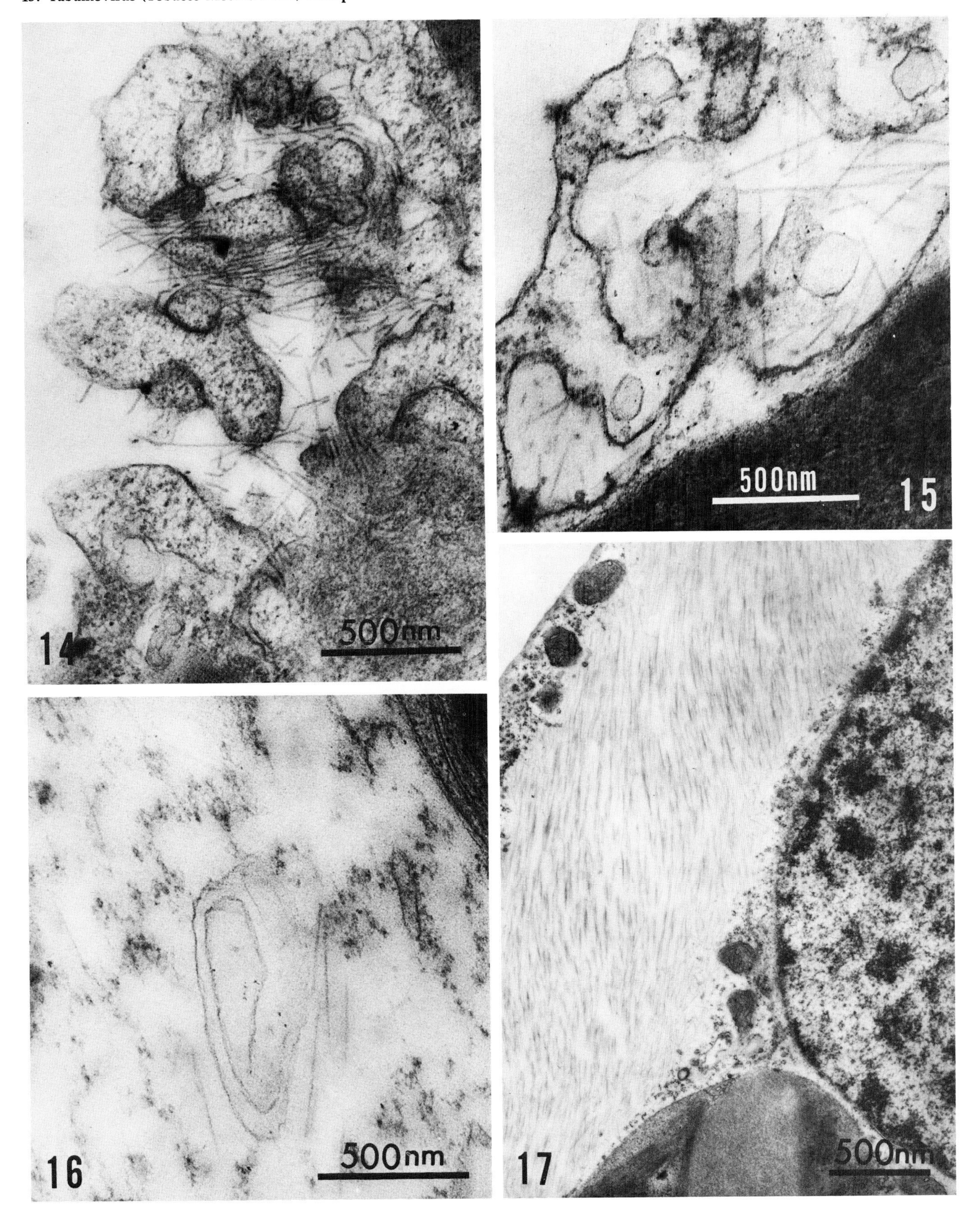
500nm
14
500nm
15
500nm
16
500nm
17

Chapter 14

Tombusvirus (Tomato Bushy Stunt Virus) Group

G. P. MARTELLI, M. RUSSO, AND A. QUACQUARELLI

I. Definition of the Group 257
II. General Characteristics 258
III. Morphological and Physicochemical Properties 259
  A. Morphology and Structure of the Virions 259
  B. The Protein Shell 259
  C. The Nucleic Acid 260
  D. Stabilizing Interactions 260
  E. Hydrodynamic Properties 260
IV. Serological Properties 261
V. Ultrastructure and Interaction with the Host Cell 262
  A. Intracellular Appearance and Distribution of Virus Particles 262
  B. Virus and Mitochondria 262
  C. Virus and Chloroplasts 263
  D. Virus and Nuclei 263
  E. Intracellular Inclusions 263
VI. Relationships with Other Viruses 264
VII. Note Added in Proof 265
  References 265

I. DEFINITION OF THE GROUP

Tombusviruses derive their name from tomato bushy stunt virus (TBSV, cryptogram : R/1 : 1.5/17 : S/S : S/*), the type member of the group (Harrison *et al.*, 1971). In the present formulation, the grouping includes viruses that are serologically related to one another, with the sole exception of the turnip crinkle virus whose inclusion, however, has not yet been unequivocally established. Hence, tombusviruses could easily represent a monotypic cluster if the members were grouped together as strains or serotypes of TBSV rather than being considered separate entities. Indeed, the strain approach was followed in the C.M.I./A.A.B. Descriptions of Plant Viruses, and consequently, a single description was issued covering all strains under tomato bushy stunt virus (Martelli *et al.*, 1971). However, the original definition of the group recognized five different members (Harrison *et al.*, 1971). Such a classification, although not entirely satisfactory, is substantiated, at least in part, by serological evidence indicating that large differences exist between some of the members (Hollings and Stone, 1975). For this reason, in this review, we follow the original classification scheme, comprised of the five viruses listed below (Section II), and discuss the membership of additional entities that are serologically unrelated to one another and to TBSV.

II. GENERAL CHARACTERISTICS

Tombusviruses are small RNA containing isometric entities about 30 nm in diameter. Their capsids show cubic symmetry, with protein subunits arranged in a dimer clustering pattern. The virions are remarkably stable as they withstand temperatures up to 90° and are not affected by organic solvents. Virus particles may reach extremely high concentrations in the host tissues (dilution end point of infectivity up to 10^{-6}), while in crude sap at 20° the infectivity is retained for 4–5 weeks. All members of the group are easily extracted from infected tissue and obtained in purified form in mg quantities (up to 100 mg per kg leaf tissue) by conventional methods, i.e., clarification with organic solvents and/or magnesium-bentonite and alternate cycles of differential centrifugation. Purified virus preparations contain a single centrifugal component sedimenting at about 135 S.

In nature, tombusviruses have been isolated from a relatively small number of plant species belonging to widely separated botanical families whereas, experimentally, they can infect an ample range of hosts.

Interestingly, infections very often remain localized in the experimental hosts whereas most naturally infected plants are invaded systemically. In diseased plants virus particles are easily visualized in different tissues, including epidermis, parenchyma, and the conducting elements. A great deal of circumstantial and experimental evidence exists that some tombusviruses spread in nature through the soil, but the vector has not yet been identified. Recent studies (Campbell *et al.*, 1975) suggest that the vector might be a root-infecting lower fungus that produces zoospores, which, in any case, is not *Olpidium brassicae* Wor.

The following members of the tombusvirus group are known:

1. *Tomato bushy stunt virus* (*TBSV*). Isolated from tomato in England (Smith, 1935) and Argentina (Pontis *et al.*, 1968), from cherry and apple in Canada (Allen and Davidson, 1967; Allen, 1969), from tulip in Scotland (Mowat, 1972), and from soil (TBSV-GCRI) in Britain (Hollings and Stone, 1975).
2. *Pelargonium leaf curl virus* (*PLCV*). Widespread in Europe and in the Mediterranean area. Occurs also in the United States (Hollings, 1962).
3. *Petunia asteroid mosaic virus* (*PAMV*). Reported from Italy (Lovisolo, 1957) and on grapevine from Germany (Bercks, 1967).
4. *Carnation Italian ringspot virus* (CIRV). Isolated in Britain from carnation plants grown in Italy and the United States (Hollings *et al.*, 1970).
5. *Artichoke mottled crinkle virus* (*AMCV*). Reported only from Mediterranean countries: Italy (Martelli, 1965), Morocco (Fischer and Lockhart, 1974), and Malta (Martelli *et al.*, 1976).

III. MORPHOLOGICAL AND PHYSICOCHEMICAL PROPERTIES

A. Morphology and Structure of the Virions

Tombusviruses have isodiametric particles with icosahedral symmetry (Caspar, 1956) and a diameter of 30 nm. When seen in the electron microscope, the particles exhibit a rounded or, more often, angular outline, but the details of the surface structure are not easily resolved (Fig. 1). In phosphotungstate mounts a few particles may be penetrated by the stain perhaps because they are damaged. Preparations of some members of the group must undergo formaldehyde fixation prior to mounting for electron microscopy (Martelli *et al.*, 1971). The chemical constitution of the virions is simple for they are composed of protein and nucleic acid (RNA) only.

An early structural model of TBSV (Klug *et al.*, 1966; Harrison, 1971) consisted of a central protein core (30 morphological units) and an outer protein shell (90 morphological units) with a layer of RNA in between. However, recent chemical, electron microscopic, and X-ray diffraction evidence (Michelin-Lausarrot *et al.*, 1970; Butler, 1970; Weber *et al.*, 1970; Ziegler *et al.*, 1974; Harrison and Jack, 1975; Jack *et al.*, 1975) has disproved the "spherical sandwich" concept in favor of an architectural organization based on dimer clustering of structural subunits located on the two-fold positions of a $T = 3$ icosahedral surface lattice (Finch *et al.*, 1970).

B. The Protein Shell

Chemical analysis has revealed that the coat protein of PAMV and of several strains of TBSV (De Fremery and Knight, 1954; Michelin-Lausarrot *et al.*, 1970; Tremaine, 1970a) differ very little in amino acid composition.

The capsid of TBSV and, conceivably, also that of other members of the group is made up of 90 morphological units each composed of two protein molecules, thus totalling 180 (Finch *et al.*, 1970). These morphological units form rings, of five units, about the five-fold axes and rings, of six units, about the three-fold axes of a rhombic triacontahedron (Fig. 2). Each dimer unit has roughly the shape of a V with two side arms (Fig. 3), the vertex being at the surface of the particle and the arms extending toward the five-fold and three-fold axes (Jack *et al.*, 1975).

In SDS polyacrylamide gel electrophoresis a major protein with molecular weight (MW) 41,000 (range of estimates, 38,000–42,000: Weber *et al.*, 1970; Butler, 1970; Michelin-Lausarrot *et al.*, 1970; Lesnaw and Reichmann, 1970; Carpenter *et al.*, 1971; Hull, 1971; Mayo and Jones, 1973) is resolved, along with two less represented protein species with molecular weights of 28,000 and 87,000, respectively (Butler, 1970; Hill and Shepherd, 1972; Ziegler *et al.*, 1974).

The smaller proteins were originally identified as specific subunits localized at the five-fold positions (Fig. 2) where an extra density was visible in the electron micrographs (Butler, 1970; Finch *et al.*, 1970; Crowther and Amos, 1971). Recently, however, it was found that the 28,000 MW polypeptides are cleavage products of the 41,000 MW major coat protein (Ziegler *et al.*, 1974) and that the filled appearance of the five-fold axis cavities is an artifact due either to lack of penetration of uranyl acetate in these positions or to withdrawing of the stain from the cavities under the influence of the electron beam (Jack *et al.*, 1975). Hence, TBSV prossesses protein subunits of one species, each folded into two compact portions of different size, connected by a protease-sentive bridge (Ziegler *et al.*, 1974). Proteolysis can remove about 25% of the molecule, i.e., the fraction which likely corresponds to the protruding tips of the subunits (Harrison and Jack, 1975).

The largest polypeptide (MW, 87,000) occurs as a single molecule per virion and is located inside the virus particle, bound to the RNA (Ziegler *et al.*, 1974). Its function is unknown but it was suggested that it may represent viral RNA-dependent RNA polymerase (Butler, 1970).

Based on the above data the molecular weight of the protein shell is about 7.5×10^6 daltons.

C. The Nucleic Acid

TBSV contains single-stranded RNA with the following molar percentages of nucleotides: G, 28.0; A, 24.9; C, 21.9; U, 25.3 (Markham and Smith, 1951). These values tally with those found for other TBSV isolates (Dorner and Knight, 1953; De Fremery and Knight, 1954) for PAMV (Ambrosino *et al.*, 1967) and AMCV (Quacquarelli *et al.*, 1966). Each virion has a single RNA molecule constituting the whole genome, with molecular weights ranging from 1.55 to 1.8×10^6 in TBSV according to different estimates (De Fremery and Knight, 1954; Tremaine, 1970a; Dorne and Pinck, 1971; Mayo and Jones, 1973) and 1.5×10^6 in AMCV (A. Quacquarelli and D. Gallitelli, unpublished information).

Infective RNA can be obtained by phenol extraction from TBSV and AMCV, but the yield is only 20–30% of the theoretical maximum (Rushizky and Knight, 1959; Quacquarelli *et al.*, 1966). AMCV RNA is also released by prolonged exposure to 1% SDS and subsequent polyacrylamide gel electrophoresis but not by freezing at −25° or by heating (A. Quacquarelli and P. Piazzolla, unpublished information).

Crystallographic evidence from X-ray diffraction analysis at 30 and 16 Å resolution, indicates that most of the RNA is tightly bound to protein subunits or subunit groups (15–20 nucleotides per subunit) and only 20–25% of the molecule resides in the center of the particle in a hydrated free form (Harrison, 1971; Harrison and Jack, 1975). It is possible, although not proved, that certain portions of the RNA molecule possess a secondary structure on the quasi-three-fold axes of the lattice (Harrison and Jack, 1975).

D. Stabilizing Interactions

TBSV interacts with organic mercurials that cause the particles to swell, but it is not affected by mercury chloride (Dorne and Hirth, 1968a,b, 1970, 1971). Such behavior suggests that the sulfhydryl groups are not located in structurally sensitive positions as in viruses where the dominating stabilizing interactions are protein–protein linkages (Kaper, 1975). On the other hand, AMCV particles are totally insensitive to freezing and thawing and resist heating in low salt. At the thermal denaturation midpoint (82°) virus particles precipitate without separating into the RNA and protein moieties (A. Quacquarelli and P. Piazzolla, unpublished information). These data provide additional evidence that the tombusvirus structure is primarily stabilized by RNA–protein interactions (Kaper, 1973, 1975).

E. Hydrodynamic Properties

Particles of all tombusviruses contain presumably the same amount of RNA (about 17% of particle weight) and therefore, constitute a homogeneous population sedimenting as a single component. The multiple light-scattering bands obtained in sucrose density gradient with PLCV, TBSV, and CIRV were either due to particle damage or to aggregation (Hollings and Stone, 1965).

Although no accessory particles are known to occur, small particles approximately 20 nm in diameter and sedimenting at about 40 S were obtained in degradation experiments of TBSV (Leberman and Finch, 1970). These consist entirely of proteins, have the same dimer clustering of the subunits of the virions (Fig. 2), and are thought to be aggregation products of the protein shell (Leberman and Finch, 1970).

Reported sedimentation coefficients of tombusviruses range between 131 and 140 S (Martelli *et al.*, 1971). Buoyant density values were found to be 1.347 gm/cm^3 for PAMV (Ambrosino *et al.*, 1967), 1.348–1.350 gm/cm^3 for TBSV (Brown and Hull, 1973; Mayo and Jones, 1973), and 1.349 gm/cm^3 for AMCV (A. Quacquarelli and P. Piazzolla, unpublished information). The molecular weight of whole particles is 9×10^6 daltons for TBSV (Weber *et al.*, 1970) and 8.8×10^6 for AMCV (A. Quacquarelli and P. Piazzolla, unpublished information). More recent molecular weight estimates for TBSV vary from 8.6×10^6 to $10 \pm 0.4 \times 10^6$ according to the method used for the determination (Mayo and Jones, 1973).

Other properties of TBSV virions are the following: diffusion coefficient ($D_{20} \times 10^{-7}$ cm^2/sec), 1.26 (Weber *et al.*, 1970); isoelectric point, pH 4.1 (McFarlane and Kekwick, 1938); partial specific volume; 0.71 cm^3/gm (Schachman and Williams, 1959); electrophoretic mobility, 4.96 and 4.65×10^5 sec^{-1} volt^{-1} for PAMV and TBSV, respectively (Ambrosino *et al.*, 1967); extinction coefficient, 4.5 (Ambrosino *et al.*, 1967).

IV. SEROLOGICAL PROPERTIES

Tombusviruses display remarkably complex serological behavior. Although all members of the group are serologically related to one another, the degree of relationship varies not only among different viruses but also among isolates of a same virus (Hollings and Stone, 1965, 1975). The interpretation of the results of serological tests may be further complicated by the type of antisera (i.e., whether they are broad-spectrum or strain-specific or originate from early or late bleedings after injection) or by the techniques (tube precipitin, gel-diffusion) used for estimating the relationship. Thus, it has been demonstrated that tube precipitin tests, being greatly influenced by the characteristics of individual antisera, are more useful for assessing antigen and antibody concentration than for quantitative differentiation between viruses. This is better achieved by gel-diffusion and intragel absorption tests (Wetter and Luisoni, 1969; Hollings and Stone, 1975).

These reasons explain why serological investigations carried out in different laboratories (Hollings and Stone, 1965; Hollings *et al.*, 1970; Bercks and Lovisolo, 1965; Martelli and Quacquarelli, 1966; Allen, 1968; Wetter and Luisoni, 1969) using different isolates of various tombusviruses yielded results which were sometimes conflicting and seldom comparable. Recently, Hollings and Stone (1975), in an extensive comparison of tombusviruses, identified two serological clusters of closely related strains, one of which is composed of PLCV isolates (except for 1 out of 10) and the other comprises AMCV, PAMV, TBSV-cherry, and TBSV-GCRI. TBSV type strain, CIRV and TBSV-tulip, seem to stand on their own, being related to one another and with any of the other strains, but to a lesser degree than the strains within each cluster.

Hence, as already pointed out by Wetter and Luisoni (1969), it appears that the current naming of tombusvirus strains does not correspond entirely to the serological typing.

The immunoelectrophoretic behavior of tombusviruses is consistent, yielding reproducible results (Bercks and Lovisolo, 1965; Martelli and Quacquarelli, 1966; Hollings and Stone, 1975) which, in general, support the grouping suggested by gel-diffusion tests (Hollings and Stone, 1975).

V. ULTRASTRUCTURE AND INTERACTION WITH THE HOST CELL

A. Intracellular Appearance and Distribution of Virus Particles

The majority of the cells of infected plants, irrespective of the host and virus strain, contain discernible virus particles. These appear as electron-dense rounded bodies with a smooth and regular outline, sometimes with hexagonal contour, and a diameter of about 26 nm. The virions often occur in huge accumulations in the ground cytoplasm and the central vacuole. The particles are either loosely scattered or appressed in a disorderly fashion to form rather compact accumulations or are regularly arranged in a lattice to constitute well defined crystals (Fig. 4).

Simple or geminate crystalline bodies having an apparently cubic close-packed structure (Russo *et al.*, 1968) are frequent in hosts infected by AMCV (Fig. 5) but are also encountered, although more rarely, in PCLV (Martelli and Russo, 1972) and TBSV (Russo and Martelli, 1972a) infections. Whether or not the ability to form crystals is a virus or a host-dependent characteristic has not been established.

Most frequently virus particles are scattered throughout the cytoplasm or accumulate within extensions of the tonoplast or of the plasma membrane which appear as bubble-like vesicles protruding into the central vacuole (Fig. 6) or in gaps formed between the cell wall and plasmalemma (Russo *et al.*, 1968). These vesicles become detached from the tonoplast through the pinching off of the bounding membrane and move away from the cytoplasm into the vacuole, where they release virions. Therefore they represent structures by which an active intracellular mechanism operates for driving virions into areas where plenty of room for accumulation is available. Such a mechanism represents a highly consistent and specific feature of tombusviruses as it was detected in all members of the group including PAMV and CIRV (Russo *et al.*, 1968; Martelli and Russo, 1972, 1973; Russo and Martelli, 1972a,b).

B. Virus and Mitochondria

In general, no particular relationship of any of the tombusviruses with mitochondria has been noticed except for AMCV whose particles were visualized within a few such organelles in two hosts (Russo *et al.*, 1968; Martelli and Russo, 1973). However, this localization is accidental, as in both cases the outer membrane of the mitochondria had breaches through which the virions could have entered from the cytoplasm. The modifications suffered by mitochondria following virus infections vary a great deal according to the virus–host combination. Thus, for instance, the relatively unaffected condition of mitochondria of AMCV-invaded plants (Russo *et al.*, 1968; Martelli and Russo, 1973) contrasts strikingly with the severe degeneration observed in PLCV infections (Martelli and Russo, 1972). Rather serious modifications of the matrix and cristae were occasionally observed in cells infected by TBSV, PAMV, and CIRV (Russo and Martelli, 1972a,b).

C. Virus and Chloroplasts

Usually, the chloroplasts of infected cells show clear signs of degeneration such as swelling, vesiculation, reduction, or abnormal development of the lamellar system. Disruption or complete disintegration may ensue, as with PLCV (Martelli and Castellano, 1969; Martelli and Russo, 1972). Particles of this virus, however, have been observed in the stroma of plastids with an apparently intact bounding membrane. These organelles exhibited a highly proliferating activity, releasing in the cytoplasm small, rounded or ovoid buds containing virus particles (Martelli and Russo, 1972) (Fig. 8). Virions were also found within seemingly intact chloroplasts of cells infected with CIRV and more rarely, with AMCV (Martelli and Russo, 1973).

Therefore, tombusviruses seem to establish with chloroplasts a type of relationship which differs from that hitherto recorded for other small isometric plant viruses, but whose significance remains to be elucidated.

D. Virus and Nuclei

As a general trend, tombusviruses show a close association with the nucleus of infected cells. In many instances virus particles have been visualized inside the nuclei either in the nucleolus, or in the nucleoplasm, or in both. Different situations, however, occur depending on the virus and the host. For instance, in artichoke cells infected with AMCV, no apparent alteration of the nuclei was detected nor were virus particles seen within them (Russo *et al.*, 1968). Conversely, the nuclei of *Chenopodium quinoa* Willd, infected by the same virus, exhibited most peculiar membranous inclusions (Fig. 9) and a degenerative condition leading to death of the organelle. Virions were present in the nucleoplasm (Martelli and Russo, 1973). Small groups of PLCV particles were seen within the nucleolar matrix and in the karyoplasm of severely altered nuclei of the host plant (Martelli and Castellano, 1969; Martelli and Russo, 1972). Likewise, TBSV particles accumulated in great abundance, sometimes in crystalline aggregates, in the nuclei of the invaded cells (Fig. 10), but in no instance did the organelles appear to suffer detectable injuries (Russo and Martelli, 1972a). CIRV and PAMV were also found in essentially unaffected nuclei of the respective hosts, although their particles were never abundant.

An unequivocal explanation of the occurrence of tombusvirses in the nucleus has not yet been obtained. However, there is much circumstantial evidence that this organelle does not represent a mere site of virus accumulation but, rather, it is actively involved in the synthesis and/or assembly of virus particles.

E. Intracellular Inclusions

A very consistent ultrastructural feature of tombusvirus infections is the occurrence of odd-looking cytoplasmic inclusions having the size of a small plastid and made up of three major components: (1) a peripherial enveloping membrane, (2) an electron-opaque finely granular of fibrillar matrix, and (3) many globose to ovoid vesicles containing a network of stranded material resembling nucleic acid (Fig. 7). The matrical matter may be very abundant and is sometimes arranged in a form resembling protein crystalloids, thus conferring a considerable electron density upon the inclusions. The vesicles line the internal boundary of the envelope and, when present, the central vacular area (Russo and Martelli, 1972a; Martelli and Russo, 1972, 1973).

The origin of these abnormalities is still obscure although it seems unlikely that they arise from degenerative changes of ordinary cell organelles. The participation of mitochondria and microbodies in the formation of the inclusions has been envisaged (Russo and Martelli, 1972a) but further findings suggest that the most probable origin of the bodies lies in the reorganization of preexsisting cellular components (e.g., endoplasmic reticulum strands, dyctiosomal vesicles) into a pseudo-organellar form. The function of these inclusions is unknown, although their formation as an early event of the infection process may indicate a direct relationship with virus multiplication.

VI. RELATIONSHIPS WITH OTHER VIRUSES

It has been proposed that tomato bushy stunt, sowbane mosaic, cucumber necrosis, turnip crinkle (TCV), and carnation mottle (CaMV) viruses are all members of a same group (Haselkorn, 1966; Tremaine, 1970b). Similarly, a possible grouping of TCV, CaMV, and pelargonium flower-break virus (PFBV) was recently envisaged, based on similarities in some biochemical and hydrodynamic properties (Stone and Hollings, 1973). On the same basis, carnation ringspot virus could perhaps be added to the list.

Indeed, all the above viruses are single-component entities containing single-stranded RNA, which accounts for 16 to 20% of the particle weight, are endowed with remarkable heat resistance (thermal inactivation point about or above 80°), and reach very high concentration in the host tissues. For some of them, however, this is as far as the similarity goes. Sowbane mosaic virus, for example, has biological physicochemical properties and intracellular behavior differentiating it clearly from TBSV (Milne, 1967; Kado, 1971). For other viruses such as cucumber necrosis (Dias and McKeen, 1972) and PFBV (Hollings and Stone, 1974), more information is needed for substantiating an unequivocal relationship with tombusviruses. However, PFBV has protein subunits of a size (MW, 41,000) that may be indicative of a structural organization similar to that of TBSV.

Carnation ringspot virus also has protein subunits close to the right dimensions (MW, 38,000) and physicochemical and hydrodynamic properties fitting broadly those of tombusviruses (Kalmakoff and Tremaine, 1967), but its epidemiological behavior (transmission by nematodes) is doubtful and the architectural organization is not known. Similarly, the structural details of CaMV are uncertain, but the capsid is reported to contain 234 identical subunits, each comprising 243 amino acids residues and with a molecular weight of 26,300. Based on the above data, two particle models have been proposed: a protein shell with 240 subunits in a T = 4 icosahedral symmetry or, alternately, an outer shell of 180 subunits (T = 3) and an internal core of 60 subunits (T = 1) with RNA sandwiched between (Tremaine, 1970b). Neither of these models, fits the architectural pattern of TBSV. Furthermore, the sedimentation coefficient (118 S) and molecular weight (7.7×10^6 daltons) of CaMV particles are lower than those found for tombusviruses (Tremaine, 1970b). However, a point of similarity exists in the intracellular behavior of the virus with particular reference to its association with the nucleus of infected cells (Castro-Robleda, 1973).

Elderberry latent (ELV) is another virus showing similarities to TBSV in particle size, including the occurrence of small particles 17 nm in diameter and sedimenting at 48 S (Jones, 1972), physical properties, and the sizes of its RNA and protein subunits. However, the two viruses are serologically unrelated, have different hydrodynamic properties and may differ in the number and arrangement of the protein shell subunits, so that ELV is presently regarded as a distinct entity (Mayo and Jones, 1973).

Based on current knowledge, only TCV appears to share enough properties with tombusviruses for qualifying as a possible member of the group. This virus has the same structure as TBSV (Finch *et al.*, 1970) and a capsid made up of 180 protein molecules (MW, 41,000) in dimeric clustering, plus a polypeptide with a molecular weight of 84,000 (Butler, 1970; Ziegler *et al.*, 1974). Moreover, sedimentation coefficient, RNA percentage and composition, and estimation of the molecular weight of whole particles tally with the figures known for tombusviruses (Hollings and Stone, 1972). If similarities rather than differences were also known to occur in the epidemiological (reportedly TCV is insect-borne) and intracellular (Stefanac, 1969) behavior, the relationship with TBSV and related strains would be thoroughly convincing.

VII. NOTE ADDED IN PROOF

Hollings *et al.* [*Ann. Appl. Biol.* **85,** 233 (1977)] have recently described cymbidium ringspot virus, a new putative member of of the tombusvirus group. This virus is not serologically related to any of the known tombusviruses but shares many biological and physicochemical properties with them.

REFERENCES

Allen, W. R. (1968). *Can. J. Bot.* **46,** 229.

Allen, W. R. (1969). *Can. J. Plant Sci.* **49,** 797.

Allen, W. R., and Davidson, T. R. (1967). *Can. J. Bot.* **45,** 2375.

Ambrosino, C., Appiano, A., Rialdi, G., Papa, G., Redolfi, P., and Carrara, M. (1967). *Atti Accad. Sci. Torino, Cl. Sci. Fis., Mat. Nat.* **101,** 301.

Bercks, R. (1967). *Phytopathol. Z.* **60,** 273.

Bercks, R., and Lovisolo, O. (1965). *Phytopathol. Z.* **52,** 96.

Brown, F., and Hull, R. (1973). *J. Gen. Virol.* **20,** 43.

Butler, P. J. G. (1970). *J. Mol. Biol.* **52,** 589.

Campbell, R. M., Lovisolo, O., and Lisa, V. (1975). *Phytopathol. Mediterr.* **14,** 82.

Carpenter, J. M., Cook, S. M., and Gibbs, A. J. (1971). *Rep., Rothamsted Exp. Stn., Harpenden, Engl.* p. 122.

Caspar, D. L. D. (1956). *Nature (London)* **177,** 475.

Castro-Robleda, S. (1973). *Phytopathol. Z.* **78,** 134.

Crowther, R. A., and Amos, L. (1971). *Cold Spring Harbor Symp. Quant. Biol.* **36,** 489.

De Fremery, D., and Knight, C. A. (1954). *J. Biol. Chem.* **214,** 559.

Dias, H. F., and McKeen, C. D. (1972). *CMI/AAB Descriptions Plant Viruses* No. 82.

Dorne, B., and Hirth, L. (1968a). *C. R. Hebd. Seances Acad. Sci., Ser. D* **267,** 127.

Dorne, B., and Hirth, L. (1968b). *C. R. Hebd. Seances Acad. Sci., Ser. D* **267,** 1027.

Dorne, B., and Hirth, L. (1970). *Biochemistry* **9,** 119.

Dorne, B., and Hirth, L. (1971).. *Biochemie* **53,** 469.

Dorne, B., and Pinck, L. (1971). *FEBS Lett.* **12,** 241.

Dorner, R. W., and Knight, C. A. (1953). *J. Biol. Chem.* **205,** 959.

Finch, J. T., Klug, A., and Leberman, R. (1970). *J. Mol. Biol.* **50,** 215.

Fischer, H. U., and Lockhart, B. E. (1974). *Plant Dis. Rep.* **58,** 1117.

Harrison, B. D., Finch, J. I., Gibbs, A. J., Hollings, M., Shepherd, R. J., Valenta, V., and Wetter, C. (1971). *Virology* **45,** 356.

Harrison, S. C. (1969). *J. Mol. Biol.* **42,** 457.

Harrison, S. C. (1971). *Cold Spring Harbor Symp. Quant. Biol.* **36,** 495.

Harrison, S. C., and Jack, A. (1975). *J. Mol. Biol.* **97,** 173.

Haselkorn, R. (1966). *Annu. Rev. Plant Physiol.* **17,** 137.

Hill, J. H., and Shepherd, R. J. (1972). *Virology* **47,** 817.

Hollings, M. (1962). *Ann. Appl. Biol.* **50,** 180.

Hollings, M., and Stone, O. M. (1965). *Ann. Appl. Biol.* **56,** 87.

Hollings, M., and Stone, O. M. (1970). *CMI/AAB Descriptions Plant Viruses* No. 7.
Hollings, M., and Stone, O. M. (1972). *CMI/AAB Descriptions Plant Viruses* No. 109.
Hollings, M., and Stone, O. M. (1974). *CMI/AAB Descriptions Plant Viruses* No. 130.
Hollings, M., and Stone, O. M. (1975). *Ann. Appl. Biol.* **80,** 37.
Hollings, M., Stone, O. M., and Bouttell, G. C. (1970). *Ann. Appl. Biol.* **65,** 299.
Hull, R. (1971). *Virology* **45,** 767.
Jack, A., Harrison, S. C., and Crowther, R. A. (1975). *J. Mol. Biol.* **97,** 163.
Jones, A. T. (1972). *Ann. Appl. Biol.* **70,** 49.
Kado, C. I. (1971). *CMI/AAB Descriptions Plant Viruses* No. 64.
Kalmakoff, J., and Tremaine, J. H. (1967). *Virology* **33,** 10.
Kaper, J. M. (1973). *Virology* **55,** 299.
Kaper, J. M. (1975). "The Chemical Basis of Virus Structure, Dissociation and Reassembly." North-Holland Publ., Amsterdam.
Klug, A., Finch, J. T., Longley, W., and Leberman, R. (1966). "Principles of Biomolecular Organisation." Little, Brown, Boston, Massachusetts.
Leberman, R., and Finch, J. T. (1970). *J. Mol. Biol.* **50,** 209.
Lesnaw, J. A., and Reichmann, M. E. (1970). *Virology* **42,** 724.
Lovisolo, O. (1957). *Boll. Stn. Patol. Veg., Roma* [3] **14,** 103.
McFarlane, A. S., and Kekwick, R. A. (1938). *Biochem. J.* **32,** 1607.
Markham, R., and Smith, J. D. (1951). *Biochem. J.* **49,** 401.
Martelli, G. P. (1965). *Phytopathol. Mediterr.* **4,** 58.
Martelli, G. P., and Castellano, M. A. (1969). *Virology* **39,** 610.
Martelli, G. P., and Quacquarelli, A. (1966). *Atti Congr. Un. Fitopatol. Mediterr., 1st, 1966* p. 195.
Martelli, G. P., and Russo, M. (1972). *J. Gen. Virol.* **15,** 193.
Martelli, G. P., and Russo, M. (1973). *J. Ultrastruct. Res.* **42,** 93.
Martelli, G. P., Quacquarelli, A., and Russo, M. (1971). *CMI/AAB Descriptions Plants Viruses* No. 69.
Martelli, G. P., Russo, M., and Rana, G. L. (1976). *Plant Dis. Rep.* **60,** 130.
Mayo, M. A., and Jones, A. T. (1973). *J. Gen. Virol.* **19,** 245.
Michelin-Lausarrot, P., Ambrosino, C., Steere, R. L., and Reichmann, M. E. (1970). *Virology* **41,** 160.
Milne, R. G. (1967). *Virology* **32,** 589.
Mowat, W. P. (1972). *Plant. Pathol.* **21,** 171.
Pontis, R. E., Gracia, O., and Feldman, J. M. (1968). *Plant. Dis. Rep.* **52,** 676.
Quacquarelli, A., Martelli, G. P., and Russo, M. (1966). *Atti Congr. Un. Fitopatol. Mediterr., 1st, 1966* p. 178.
Rushizky, G. W., and Knight, C. A. (1959). *Virology* **8,** 448.
Russo, M., and Martelli, G. P. (1972a). *Virology* **49,** 122.
Russo, M., and Martelli, G. P. (1972b). *Mikrobiologija* **9,** 177.
Russo, M., Martelli, G. P., and Quacquarelli, A. (1968). *Virology* **34,** 679.
Schachman, H. K., and Williams, R. C. (1959). *In* "The Viruses" (F. M. Burnet and W. M. Stanley, eds.), Vol. 1, p. 223. Academic Press, New York.
Smith, K. M. (1935). *Ann. Appl. Biol.* **22,** 731.
Štefanac, Z. (1969). *Acta Biol. Iugosl., Ser. B* **6,** 27.
Stone, O. M., and Hollings, M. (1973). *Ann. Appl. Biol.* **75,** 15.
Tremaine, J. H. (1970a). *Phytopathology* **60,** 454.
Tremaine, J. H. (1970b). *Virology* **42,** 611.
Weber, K., Rosenbusch, J., and Harrison, S. C. (1970). *Virology* **41,** 763.
Wetter, C., and Luisoni, E. (1969). *Phytopathol. Z.* **65,** 231.
Ziegler, A., Harrison, S. C., and Leberman, R. (1974). *Virology* **59,** 509.

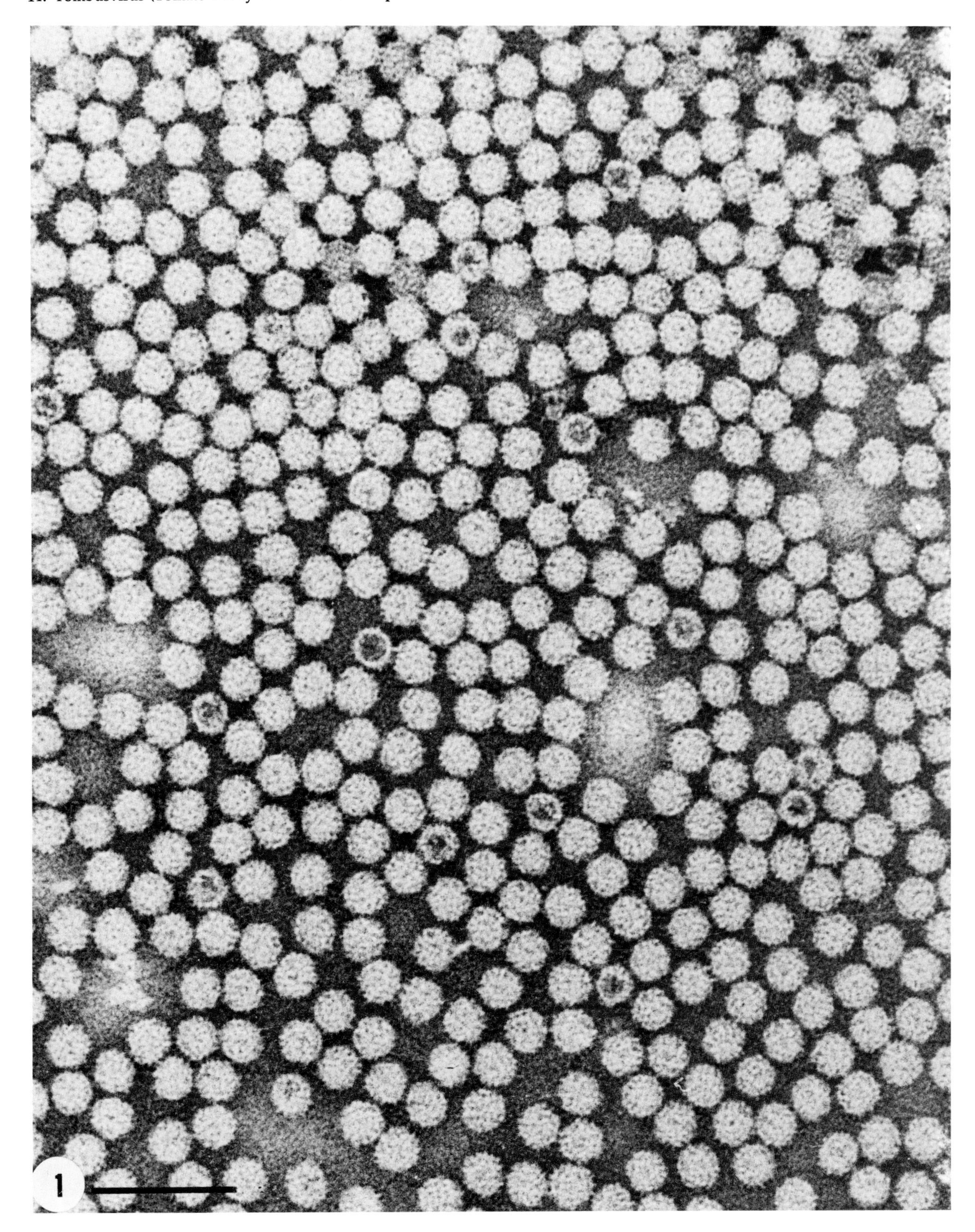

Fig. 1. Purified preparation of artichoke mottled crinkle virus negatively stained with potassium phosphotungstate. A few damaged particles are penetrated by the stain. Bar = 100 nm.

Fig. 2. Models of the large (upper row) and small (lower row) particles of tomato bushy stunt and turnip crinkle viruses. Each ball represents a dimer morphological unit. The models from left to right show the two-fold, three-fold, and five-fold view of the particles. The balls inserted in the pentagons represent the "extra density" seen in the electron micrographs of particles in these positions. These structures, formerly interpreted as the small-sized polypeptides (molecular weight, 28,000), have been recently shown to be artifacts. (From Finch *et al.*, 1970, by kind permission of the authors.)

Fig. 3. Diagramatic representation of subunit arrangement in tomato bushy stunt virus. (a) Rhombic triacontahedron, a solid which fits the subunit packing of the virus. (b) Disposition of dimeric subunits on the strict (2) and local (q.2) two-fold axes of half-facet of the triacontahedron. (c) Appearance of the protein subunits around different symmetry axes in an equatorial section of the virus particle. (From Jack *et al.*, 1975, by kind permission of the authors.)

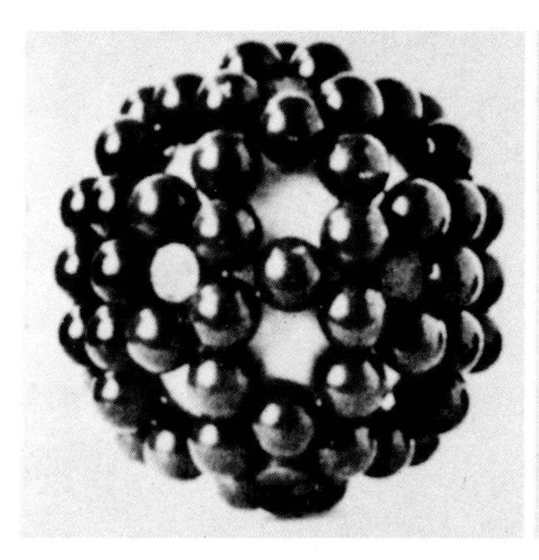
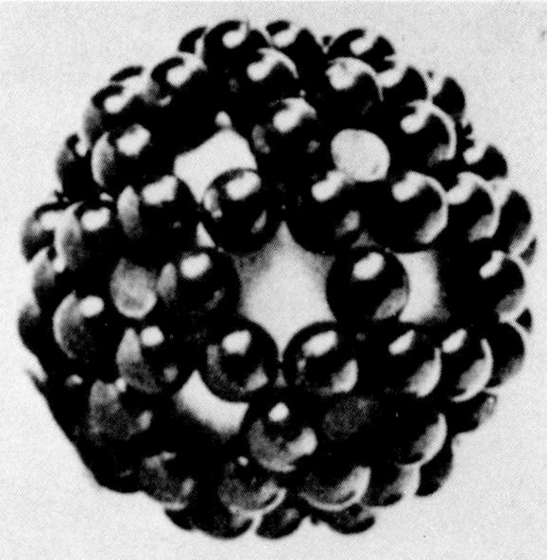
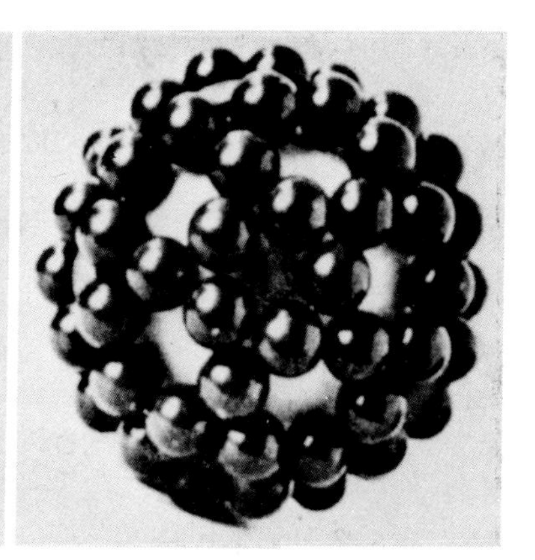

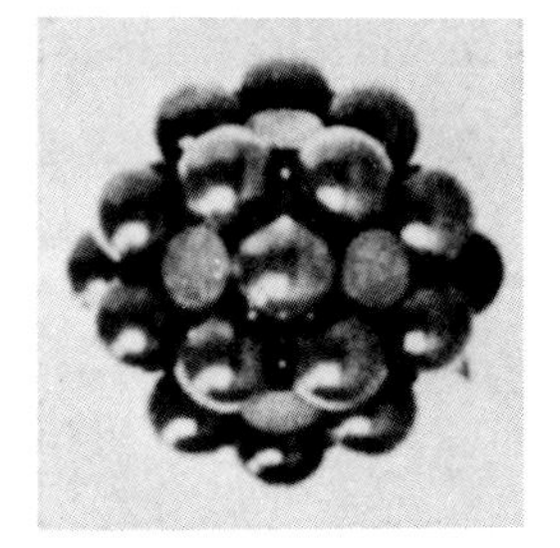
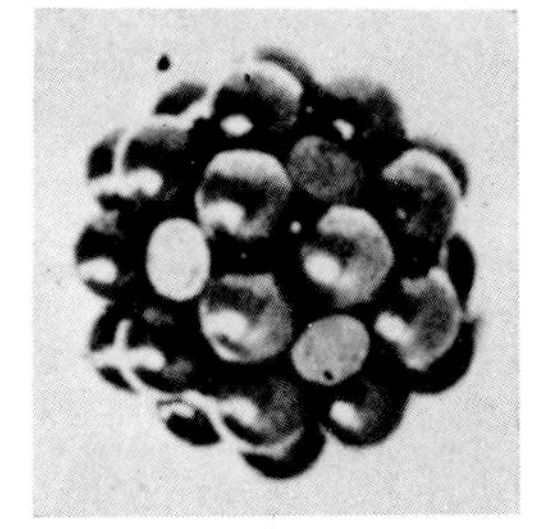
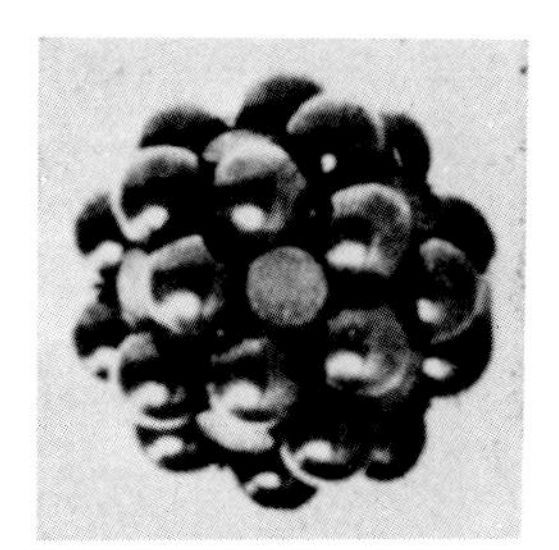

2

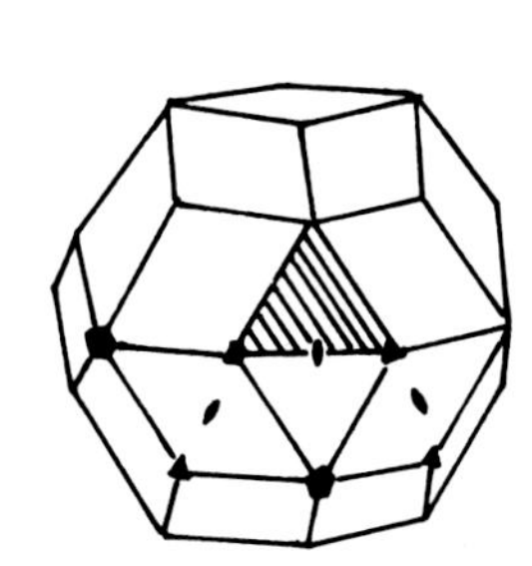

(a)

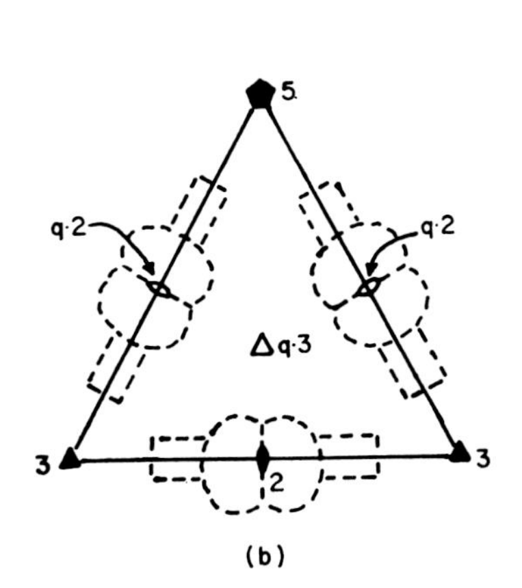

(b)

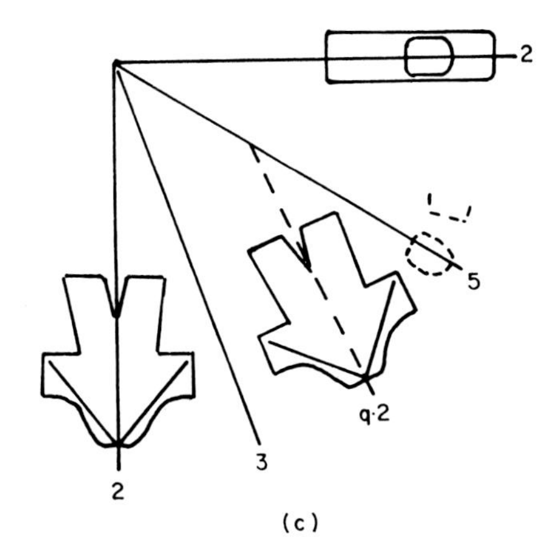

(c)

3

Fig. 4. Transverse section through epidermal cells of an artichoke seedling infected with mottled crinkle virus. Cytoplasm and vacuole (Vac) harbor large accumulations of virus particles in a disorderly fashion or in crystalline arrays. Bubble-like vesicles (Ve) containing virus particles, sometimes in a crystalline packing, are numerous. CW, Cell wall; bar = 500 nm.

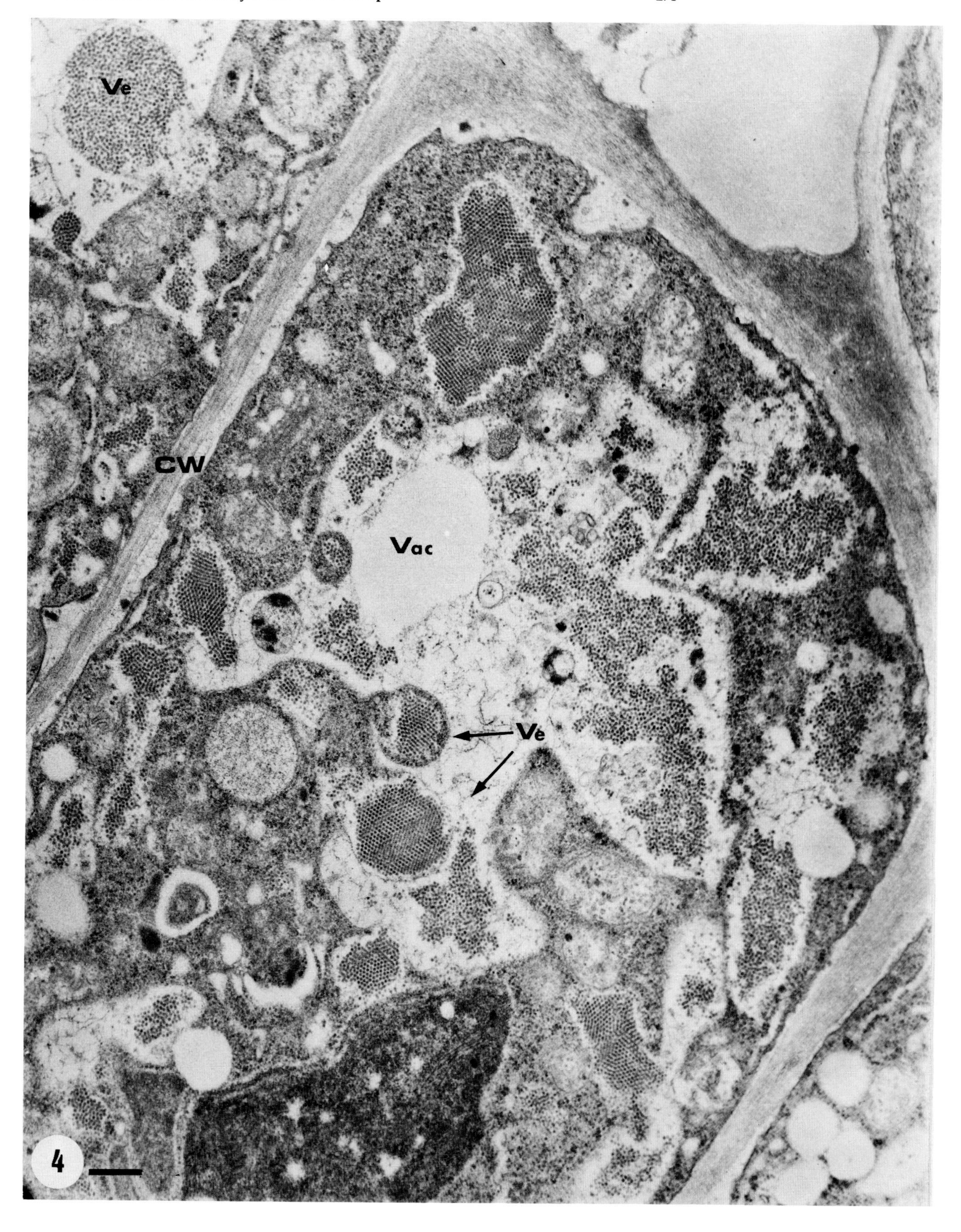
Ve
CW
Vac
Ve
4

Fig. 5. A large crystal of artichoke mottled crinkle virus in the vacuole of an infected cell. Bar = 250 nm.

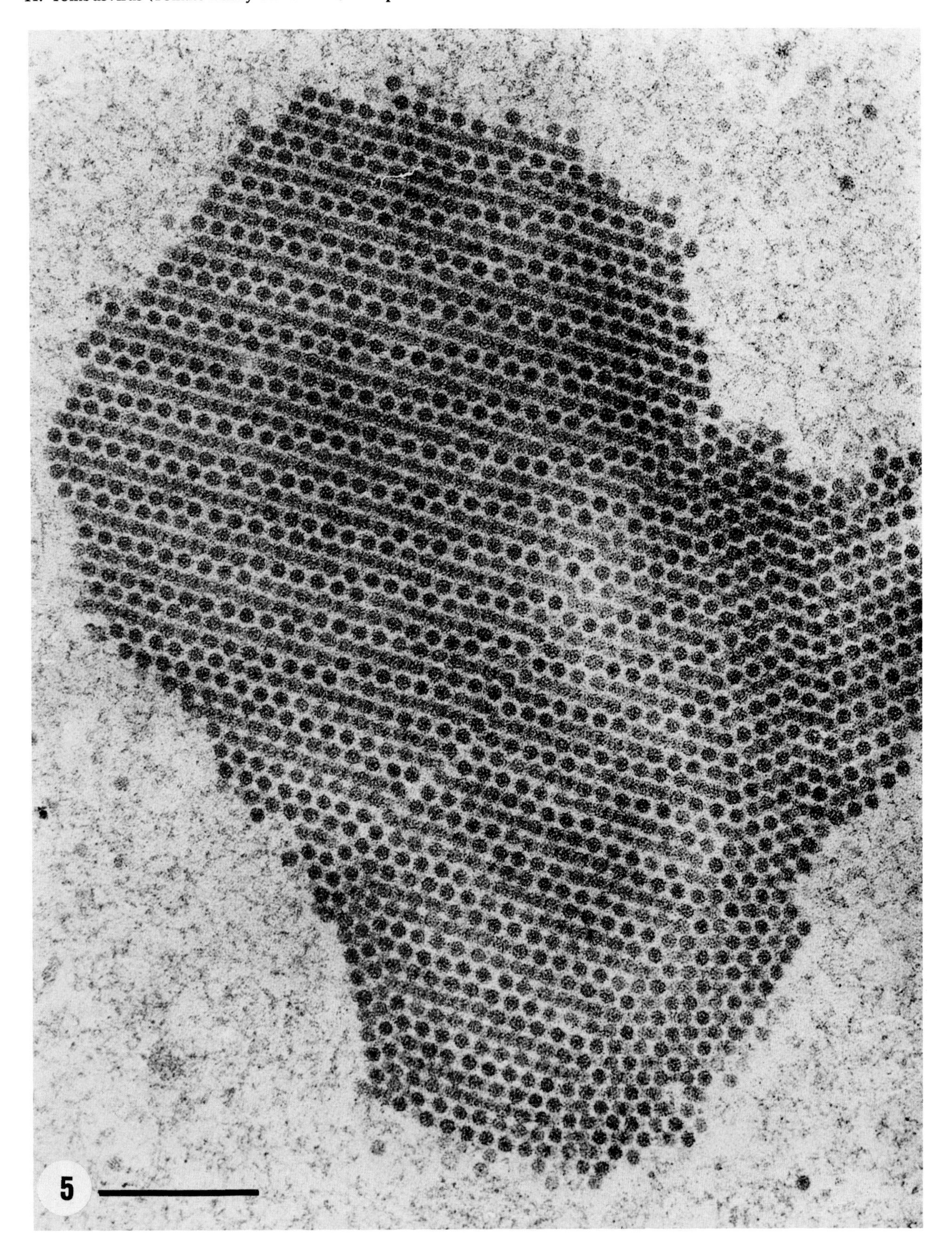
5

Fig. 6. Virus-containing vesicles protruding in the central vacuole (Vac) of a cell infected with artichoke mottled crinkle virus. Note the apparent streaming of virions from the cytoplasm into the vesicles. Bar = 250 nm. (Inset) A similar structure in a cell infected with tomato bushy stunt virus. Bar = 500 nm.

Fig. 7. Multivesicular inclusions (i) induced by petunia asteriod mosaic and artichoke mottled crinkle viruses (inset). Such pseudo-organellar bodies consist of a scanty fibrillar matrix and many small vesicles surrounded by an enveloping membrane. m, mitochondria; bar = 500 nm.

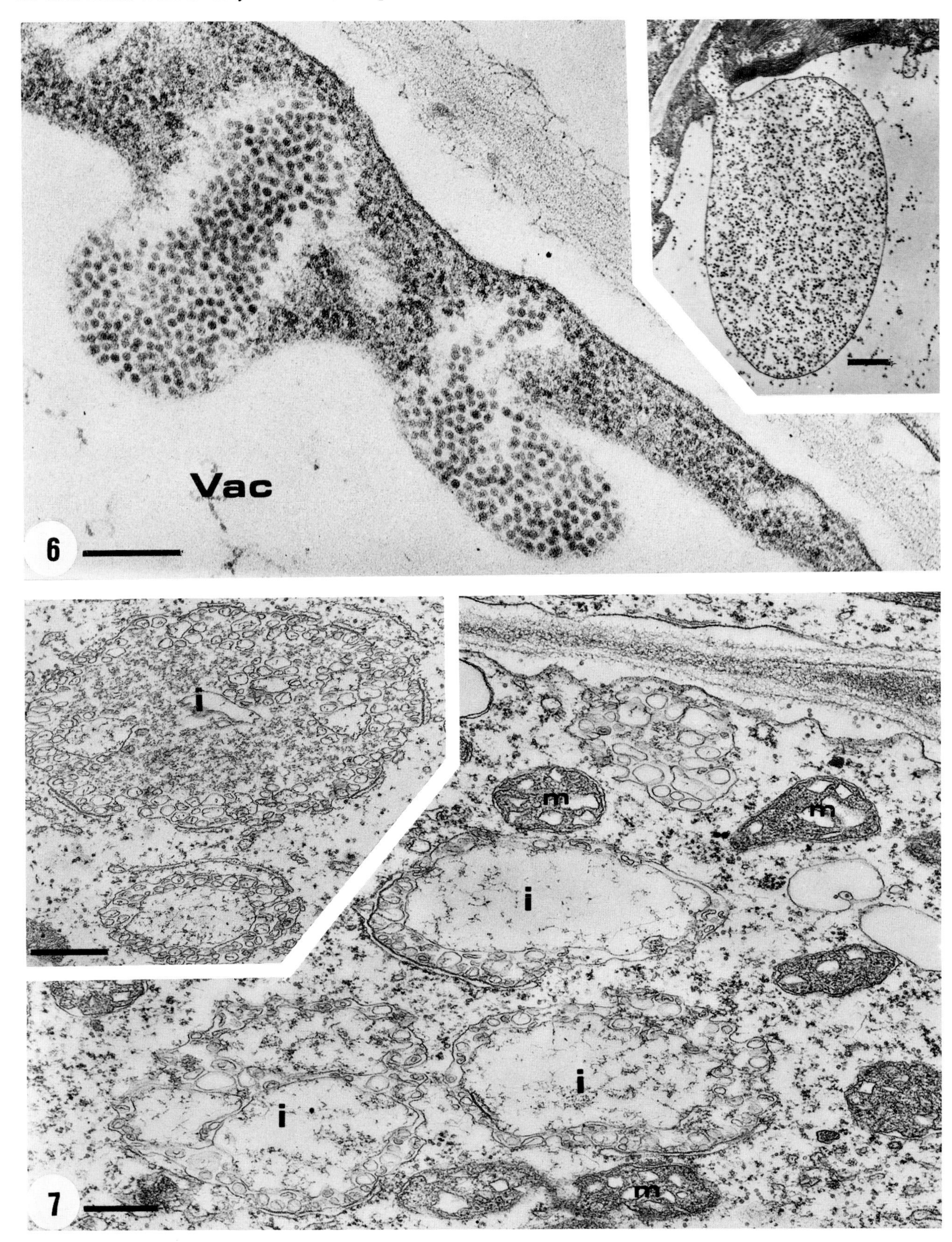
Vac
6
i
m
m
i
i
i
m
7

Fig. 8. Section through a parenchyma cell of pelargonium infected with leaf curl virus, showing two altered chloroplasts (Ch), which contain clusters of virus particles in the stroma (encircled areas) and several globose-elongated bodies surrounded by a double membrane, likely deriving from plastidial budding. Protuberances, possibly representing budding sites, are marked by arrows. Virus particles (V) are scattered throughout the cytoplasm. Bar = 500 nm. (Inset) A plastid in the process of budding with virus particles within the bud. Bar = 250 nm. (From Martelli and Russo, 1972, by kind permission of the Copyright holder Cambridge University Press.)

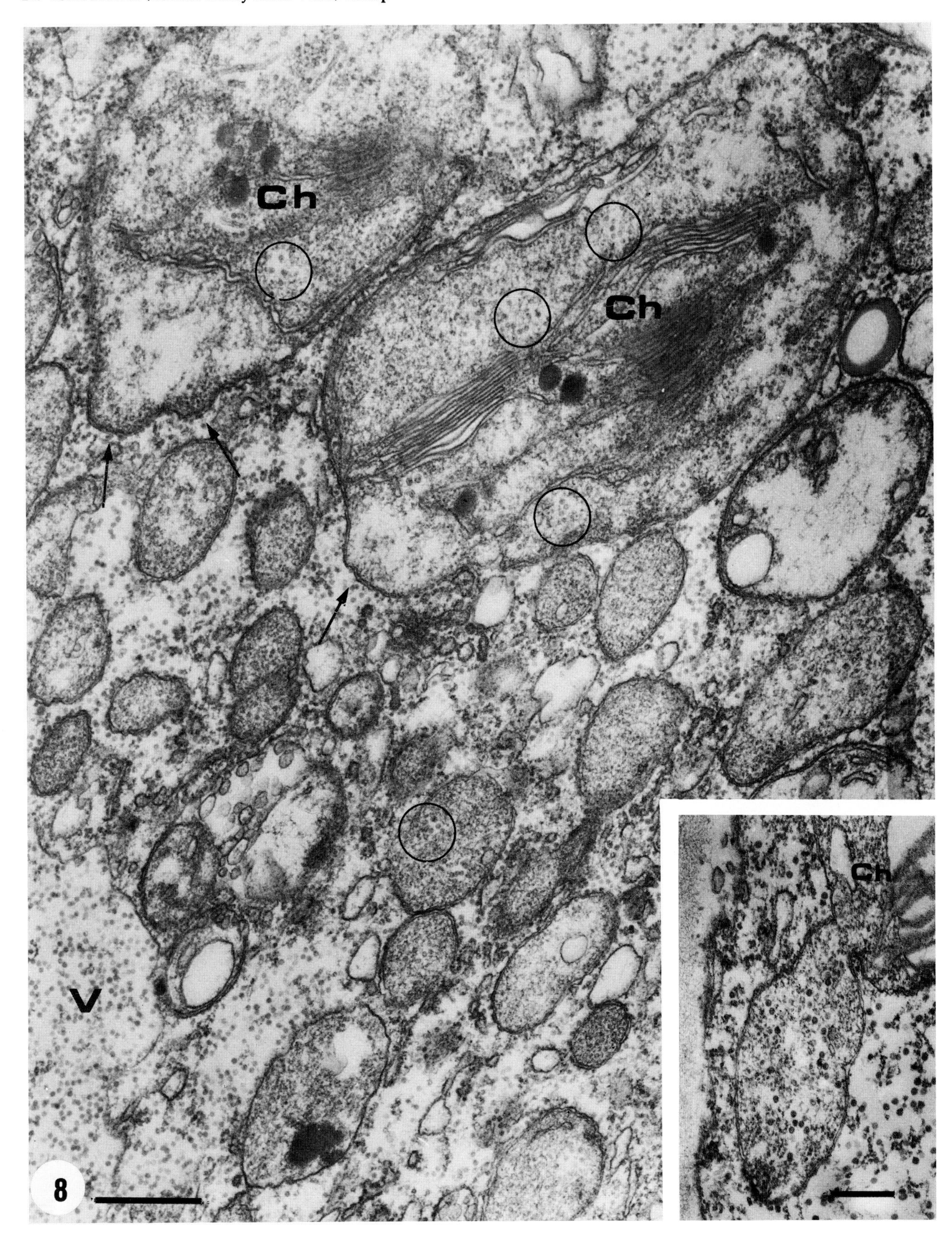
Ch
Ch
V
8
Ch

Fig. 9. Cross section through a nucleus (N) of a cell infected with mottled crinkle virus. The nucleus contains two membranous inclusions and a few virus particles scattered in the nucleoplasm. Clusters of virions (V) are visible in the surrounding cytoplasm. Bar = 500 nm.

Fig. 10. Close-up of a nucleus in a cell infected with tomato bushy stunt virus. Note the association of virus particles with the nucleolus (Nu). Bar = 250 nm.

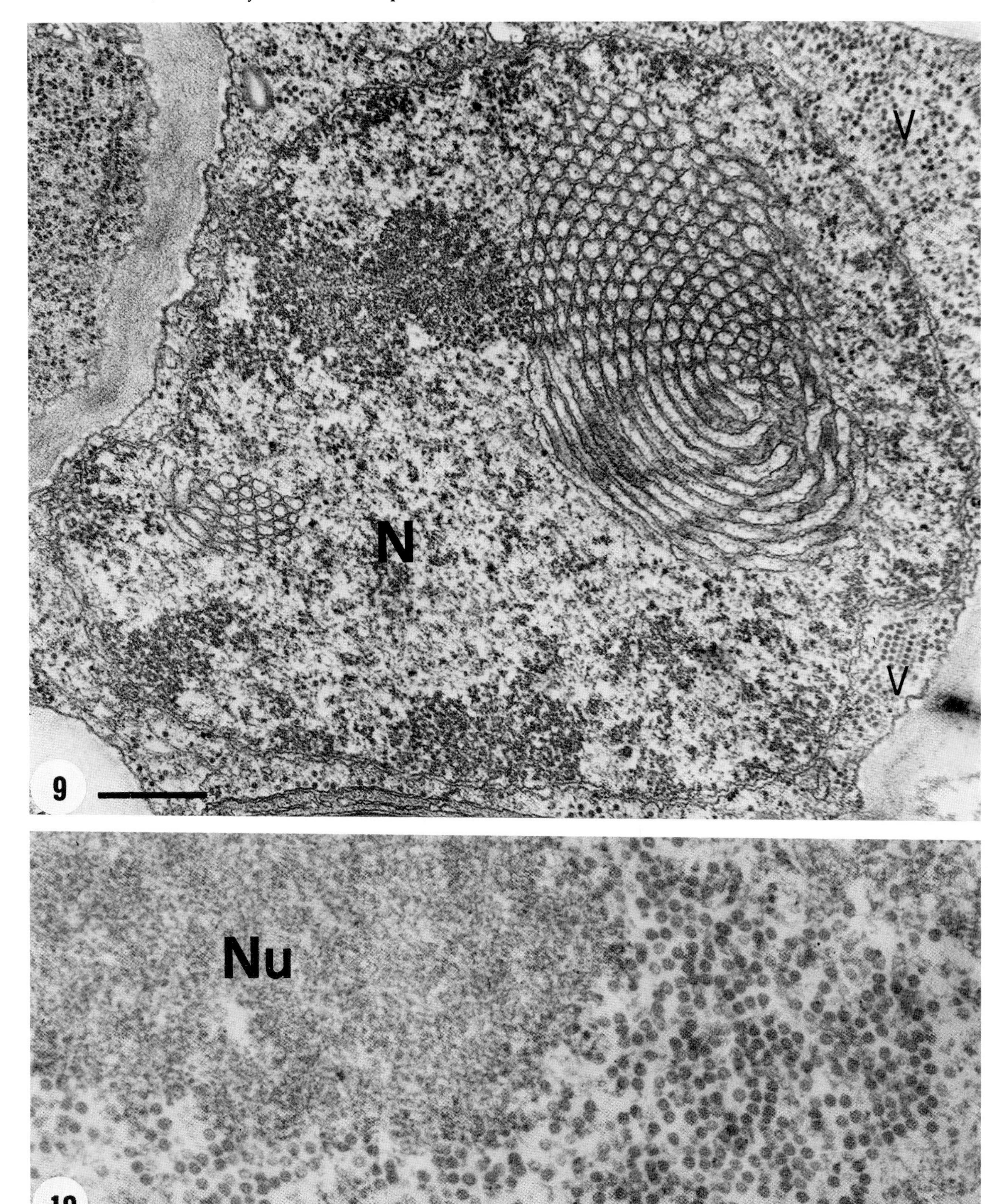
V
N
V
9
Nu
10

Chapter 15

Tobacco Necrosis Virus Group

B. KASSANIS

I. Introduction 281
II. Biological Properties 281
A. Strains and Serological Relationships 281
B. Activation of Satellite Virus 282
C. Transmission by *Olpidium brassicae* 282
D. Kinetics of Heat Inactivation 282
E. Inactivation by Ultraviolet Radiation 282
F. Unstable Variants 283
III. Physicochemical Properties 283
A. Virus Particles 283
B. Structural Virus Protein 283
C. Virus Nucleic Acid 284
References 284

I. INTRODUCTION

Tobacco necrosis virus (TNV) is commonly found in the roots of plants grown in irrigated fields or in glasshouses where unsterilized soil is used. It seldom causes disease in the leaves except for the Augusta disease of tulips (Kassanis, 1949) and stipple-streak of beans (Bawden and van der Want, 1949). Although the diseases it causes are not economically important, the virus is of considerable academic interest because it activates the satellite virus which is not able to multiply on its own (Kassanis, 1962). It can be transmitted mechanically to many plant species of several botanical families, but usually produces only necrotic local lesions and rarely invades the plant systemically. In nature it spreads in the soil from one root to another by the aid of the zoospores of the fungus *Olpidium brassicae* (Teakle, 1962).

II. BIOLOGICAL PROPERTIES

A. Strains and Serological Relationships

Several strains have been described which fall into two groups or serotypes. The strains of group A show wide serological differences from those of group D, but within each group the strains are serologically closely related, although they differ in other biological properties. The two groups of strains can be distinguished by the type of local lesions they form in the primary leaves of bean plants (*Phaseolus vulgaris*). Strains of group A form lesions that spread along the veins to form a necrotic net, while strains of group D form discrete round necrotic lesions. Strains of group A can be propagated in tobacco *Nicotiana tabacum*, those of group D in *Nicotiana clevelandii*. Infectivity tests can be made in the primary leaves of beans. In immunodiffusion tests all strains give a single line, but strain B of group A has to be treated with formaldehyde before it gives a visible line (Babos and Kassanis, 1963a; Kassanis and Phillips, 1970).

B. Activation of Satellite Virus

Undoubtedly the most interesting property of TNV is its ability to activate the satellite virus. Satellite virus is a small isometric virus, 17 nm in diameter, containing single stranded RNA of about 0.4×10^6 molecular weight. Probably because of its small amount of RNA the virus is unable to multiply unaided but does so in the presence of TNV. Satellite virus is serologically unrelated to TNV. Interesting biological relationships occur between the two viruses regarding the size and number of lesions formed by TNV and the amount of virus produced in mixed infections (Kassanis, 1962). Different strains of satellite virus are activated only by certain strains of TNV and not by others (Kassanis and Phillips, 1970).

C. Transmission by *Olpidium brassicae*

In nature, TNV is transmitted from one root to another by zoospores of the parasitic chytrid fungus *Olpidium brassicae* (Wor.) Dang. (Teakle, 1962). In experiments using seedlings grown in nutrient solutions, transmission by *Olpidium* zoospores was more efficient than mechanical inoculation of leaves. With 10^5 zoospores/ml of inoculum, transmission to lettuce roots was obtained with as little as 0.05 μg virus/liter. The virus particles are pushed into the root cells as the zoospores penetrate the cell wall and the method of transmission is therefore mechanical. However, specificities have been found among virus strains (serotypes A and D), fungus isolates, and host species (Kassanis and Macfarlane, 1964, 1965).

D. Kinetics of Heat Inactivation

The thermal inactivation point of different strains ranges between 85° and 95°. TNV is not inactivated at the same exponential rate at all survival levels. The virus is apparently inactivated at two different rates as though it consists of two components. The changing ratio of the two components depending on the temperature, and some other properties of the strains, show that the virus preparations are homogeneous and the two components are produced by heating. In this behavior, TNV resembles some animal viruses. Inactivation of the RNA of TNV is similar to that of the whole nucleoprotein, indicating that only changes in the RNA are responsible for thermal inactivation of the virus (Babos and Kassanis, 1963b).

E. Inactivation by Ultraviolet Radiation

Strain A of TNV and its infective nucleic acid are equally susceptible to inactivation by ultraviolet radiation at all wavelengths tested (230–290 nm) and are photoreactivated to the same extent when plants inoculated with UV-irradiated virus are exposed to daylight. The shape of the action spectrum for inactivation of TNV by ultraviolet and its RNA both follow the shape of the absorption spectrum of the RNA. Thus, unlike some other viruses, the RNA of TNV behaves, in all these respects, in the same way irrespective of whether it is inside or outside the virus particle. For 50% inactivation of TNV, each mg of RNA must absorb about 0.27 joules of radiation energy of any wavelength between 230 and 290 nm, which corresponds to a quantum yield of about 0.65×10^{-3} at 260 nm (Kassanis and Kleczkowski, 1965).

F. Unstable Variants

Different strains of TNV all produce about 5% of one particular mutant that is not able to synthesize the structural protein. In a mixed population the RNA of these defective mutants is enveloped with protein coded by the RNA of the good particles. When, however, a variant is isolated from a single lesion infection and propagated, it is found to contain naked RNA and is therefore unstable. Highly infective extracts of unstable variants are obtained by grinding 1 gm of leaf in 6 ml of 0.06 *M*, pH 8, phosphate buffer containing 25 mg/ml of bentonite (Kassanis and Welkie, 1963).

III. PHYSICOCHEMICAL PROPERTIES

A. Virus Particles

The virus particles are isometric and about 26 nm in diameter. They are negatively stained in phosphotungstate except for strain B of group A, which is degraded in the stain unless previously treated with 2% formaldehyde for 30 minutes. Structural details cannot be seen (Fig. 1). Purified preparations contain one type of particle except when there is contamination with satellite virus. The sedimentation coefficient at infinite dilution $s_{20,w}$ is about 118 S, though smaller and larger values have been reported. The molecular weight of the virus particle is about 7.6×10^6 when calculated from the RNA and protein sizes, assuming 180 protein subunits (Lesnaw and Reichmann, 1969a). The isoelectric point is about pH 4.5 and the electrophoretic mobility is -7.4×10^{-5} and -3×10^{-5} $cm^2 sec^{-1} V^{-1}$, respectively for strain B and A in 0.066 *M* phosphate buffer, pH 7.0. Absorbance at 260 nm (1 mg/ml, 1 cm light path) is 5.0–5.5 and the A_{260}/A_{280} ratio is 1.7. The buoyant density in CsCl is 1.399. All strains of the virus, except strain B, readily crystallize when a purified preparation is treated with a little ammonium sulfate and left at 4° (Babos and Kassanis, 1963a; Cesati and van Regenmortel, 1969).

B. Structural Virus Protein

About 80% of the particle weight is protein. According to Uyemoto and Grogan (1969), the number of amino acid residues (not including tryptophan) per subunit, determined by amino acid analysis and tryptic peptide chromatography, is 197 and the subunit molecular weight 22,600. They found no free N-terminal residue and the C-terminal residue was isoleucine. However, Lesnaw and Reichmann (1969b) found that the N-terminal amino acid residue was unsubstituted alanine. They also found from kinetic studies with carboxypeptidase that the C-terminal sequence is Ser-Val-Val-Met. The reason for the different results might be the fact that they used different strains. Lesnaw and

Reichmann (1969a,b) calculated the molecular weight of the protein subunit from amino acid analysis, assuming one residue of histidine per mole of protein, to be 33,300; they obtained the same value by polyacrylamide gel electrophoresis and by quantitative end-group analysis. By polyacrylamide gel electrophoresis, Dr. J. M. Carpenter (personal communication) obtained values very close to 29,000 from mobility and gel concentration plots. From the known reliability of the methods with well characterized proteins, it is unlikely that the subunit size could be as low as 22,600 or as high as 33,300. Uyemoto and Grogan (1969) suggested that the number of subunits per particle was 180($T = 3$) with possibly an internal protein core. Lesnaw and Reichmann (1969a) claimed to have seen in the electron microscope TNV particles composed of 32 morphological units and suggested an icosahedral lattice with $T = 3$. At Rothamsted, investigators have been unable to see any surface structure although over many years they have examined many preparations of different strains.

Convincing evidence for the structure of TNV has not been produced but by analogy with other viruses it may have 180 protein subunits in a $T = 3$ icosahedral lattice. Taking the protein subunit molecular weight as 30,000, RNA molecular weight as 1.3×10^6, and content as 19%, a virus particle molecular weight of about 6.7×10^6 is obtained.

C. Virus Nucleic Acid

RNA content of TNV has been reported as 18% by Bawden and Pirie (1945) and 19.3% by Uyemoto and Grogan (1969). The latter workers found the $s_{20,w}$ value for the RNA to be 24.1 ± 2 from which they calculated the molecular weight to be 1.32×10^6. They found the base composition (mole %) to be G, 24.2–24.5; A, 25.7–26.2; C, 21.0–22.0; U, 27.6–28.7. Lesnaw and Reichmann (1969a), on the other hand, found the sedimentation coefficient of the RNA to be 27 S from which they calculated the molecular weight to be 1.57×10^6. They reported the base composition as G, 23.4; A, 28.8; C, 23.5; U, 24.3. A molecular weight of 1.3×10^6 was obtained by gel electrophoresis (Bishop *et al.*, 1967).

REFERENCES

Babos, P., and Kassanis, B. (1963a). *J. Gen. Microbiol.* **32,** 135–144.
Babos, P., and Kassanis, B. (1963b). *Virology* **20,** 490–497.
Bawden, F. C., and Pirie, N. W. (1945). *Br. J. Exp. Pathol.* **26,** 277–286.
Bawden, F. C., and van der Want, J. P. H. (1949). *Tijdschr. Plantenziekten* **55,** 142–150.
Bishop, D. H. L., Claybrook, J. R., and Spiegelman, S. (1967). *J. Mol. Biol.* **26,** 373–387.
Cesati, R. R., and van Regenmortel, M. H. V. (1969). *Phytopathol. Z.* **64,** 362–366.
Kassanis, B. (1949). *Ann. Appl. Biol.* **36,** 14–17.
Kassanis, B. (1962). *J. Gen. Microbiol.* **27,** 477–488.
Kassanis, B., and Kleczkowski, A. (1965). *Photochem. Photobiol.* **4,** 209–214.
Kassanis, B., and Macfarlane, I. (1964). *J. Gen. Microbiol.* **36,** 79–93.
Kassanis, B., and Macfarlane, I. (1965). *Virology* **26,** 603–612.
Kassanis, B., and Phillips, M. P. (1970). *J. Gen. Virol.* **9,** 119–126.
Kassanis, B., and Welkie, G. W. (1963). *Virology* **21,** 540–550.
Lesnaw, J. A., and Reichmann, M. E. (1969a). *Virology* **39,** 729–737.
Lesnaw, J. A., and Reichmann, M. E. (1969b). *Virology* **39,** 738–745.
Teakle, D. S. (1962). *Virology* **18,** 224–231.
Uyemoto, J. K., and Grogan, R. G. (1969). *Virology* **39,** 79–89.

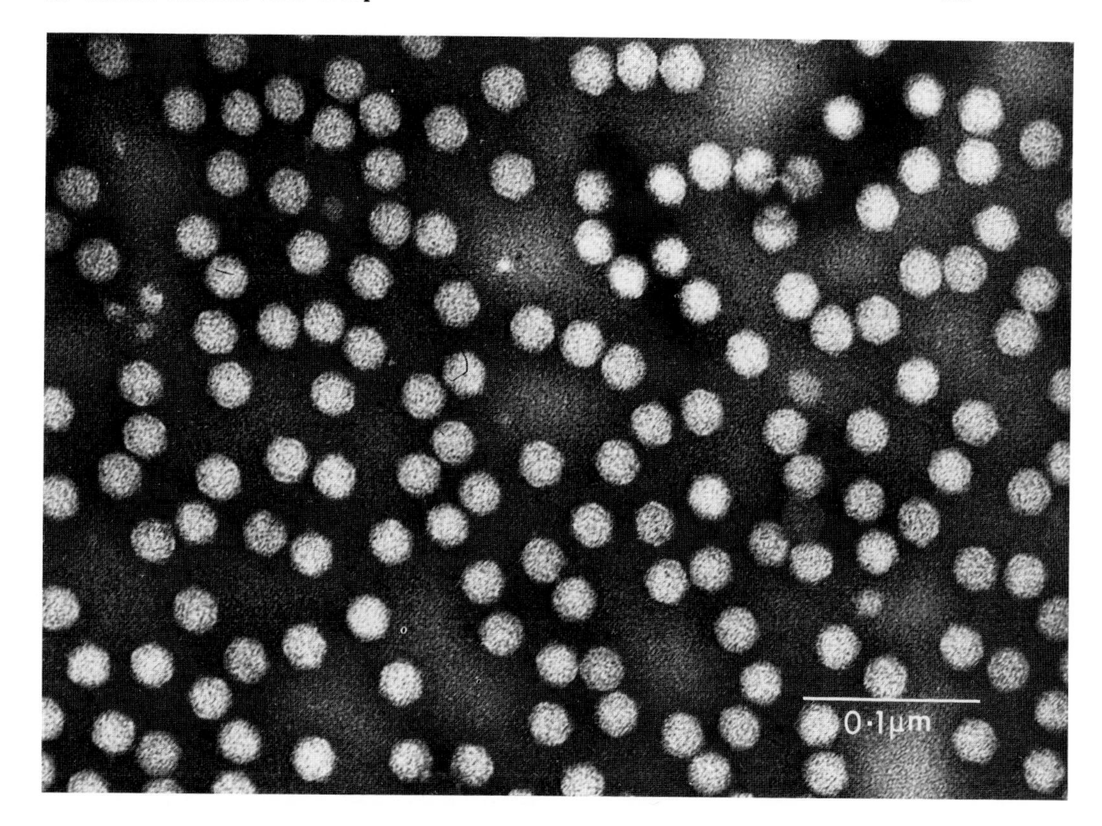

Fig. 1. Tobacco necrosis virus particles.

Chapter 16

Bromovirus (Brome Mosaic Virus) Group

J. B. BANCROFT AND R. W. HORNE

I. Introduction 287
II. Preparation and Characteristics of Bromoviruses 287
III. Translation *in Vitro* and Genetics 289
IV. Multiplication 290
V. Morphological Properties 290
References 291

I. INTRODUCTION

Cowpea chlorotic mottle virus (CCMV), brome mosaic virus (BMV), and broad bean mottle virus (BBMV) are small icosahedral RNA-containing plant viruses of about 25 to 28 nm, dry diameter, which collectively constitute the bromoviruses (Harrison *et al.*, 1971). The reader is referred to the extensive review of these viruses by Lane (1974).

II. PREPARATION AND CHARACTERISTICS OF BROMOVIRUSES

CCMV is normally grown in cowpea plants, BMV in barley and BBMV in broadbean. The three viruses are stable and soluble at 4 to 5, and can be purified by differential centrifugation at these pH levels (Bancroft, 1970a, 1971a; Gibbs, 1972). CCMV and BMV are not stable at pH 7.0 or higher in the absence of divalent cations. Yields from the three viruses are in the order of 0.5 to 2.0 mg/gm plant tissue.

The gross physical characteristics of the viruses are listed in Table I. It has been demonstrated that the bromoviruses "swell" reversibly at pH 7 (BBMV to a lesser extent), to give sedimentation coefficients about 10 S, slower than found at pH 5 ($s^{\circ}_{20,w} \sim 88$ S). The swollen form of the viruses at pH 7 can be dissociated in 1 *M* NaCl and reconstituted (Bancroft, 1970b); the capsid protein of

TABLE I

Some Physical Properties of the Bromoviruses

	Virus MW ($\times 10^6$)	% RNA	Protein subunit MW	Subunits
CCMV	4.6[a]	24	19,400	180
BMV	4.6[b]	21–22	20,300	180
BBMV	5.2[c]	21–23	20,900	180

[a] Bancroft *et al.* (1968).
[b] Bockstahler and Kaesberg (1962).
[c] Yamazaki *et al.* (1961). Calculated value from composition is 4.9×10^6.

CCMV can be induced to form various spherical and tubular forms (Bancroft, 1970b). The quarternary structural associations are controlled by carboxyl–carboxylate pairs (Bancroft *et al.*, 1967; Jacrot, 1975).

The bromoviruses are moderately good antigens. CCMV and BMV are slightly related serologically (Scott and Slack, 1971). Each virus contains a single species of coat protein and the compositions are listed in Table II. CCMV and BBMV have isoelectric points of pH 3.7 (Bancroft *et al.*, 1968) and 5.6 (Semancik, 1966), respectively, at $\mu = 0.1$, whereas that of BMV is 7.9 at the same ionic strength (Bockstahler and Kaesberg, 1962). The particles of CCMV and BMV have C-terminal tyrosinyl residues (Bancroft *et al.*, 1968; M. W. Rees, unpublished), whereas analine is found for BBMV (Yamazaki and Kaesberg, 1963). The N-terminal residues are acetylated, CCMV and BMV containing acetylserine (M. W. Rees, unpublished).

The molecular weights of RNA in the viruses are approximately 1 to 1.2 millions. The compositions are summarized in Table III. In the case of BMV, the 5′ terminal bases are ^{7}mGpppGUAC (Dasgupta *et al.*, 1975). The 3′ bases are

TABLE II

Amino Acid Compositions of the Bromovirus Coat Proteins

Amino acid	BMV	BBMV	CCMV
Lysine	13	15	12
Histidine	4	2	2
Arginine	13	12	9
Cysteine	1	2	2
Aspartic acid	10	13	11
Threonine	11	10	17
Serine	13	18	16
Glutamic acid	18	17	16
Proline	7	9	7
Glycine	10	10	10
Alanine	33	22	25
Valine	18	22	19
Methionine	3	2	1
Isoleucine	8	7	7
Leucine	15	18	16
Tyrosine	5	4	5
Phenylalanine	5	7	4
Tryptophan	2	0	3
	189[a]	190[b]	182[c]

[a] Stubbs and Kaesberg (1964).
[b] Yamazaki and Kaesberg (1963).
[c] Bancroft *et al.* (1971).

TABLE III

Base Ratios and Size Distribution of RNA from the Bromoviruses

	Base ratios	RNA molecular weights ($\times 10^6$)			
	A G C U	RNA 1	RNA 2	RNA 3	RNA 4
CCMV	0.25:0.27:0.20:0.28[a]	1.15	1.00	0.85	0.32[b]
BMV	0.27:0.28:0.21:0.24[c]	1.09	0.99	0.75	0.28[d]
BBMV	0.27:0.25:0.19:0.29[e]	1.10	1.03	0.90	0.36[f]

[a] Bancroft *et al.* (1968).
[b] Bancroft (1971b).
[c] Bockstahler and Kaesberg (1965).
[d] Lane and Kaesberg (1971).
[e] Yamazaki *et al.* (1961).
[f] Hull (1972).

ACCA-OH (Dasgupta *et al.*, 1975). The viral RNA's can be charged with tyrosine by bean leaf aminoacyl-tRNA synthetases (Hall *et al.*, 1972; Kohl and Hall, 1974).

Gel electrophoresis of the RNA's shows that each virus contains four size species, as shown in Table III. For CCMV and BMV, the RNA components 1 and 2 are separately encapsidated, whereas the components 3 and 4 are jointly encapsidated to give three kinds of nucleoproteins possessing slightly different densities in CsCl (Lane and Kaesberg, 1971; Bancroft and Flack, 1972). The densities for CCMV are 1.356, 1.360, and 1.364 gm/cm^3 for particles containing RNA 2, RNA's 3 + 4, and RNA 1, respectively. A mixture of RNA's 1 + 2 + 3 is required for infection with BMV and CCMV. Although the base compositions among the components are considered to be similar, sequences differ. The three larger RNA's of BMV and CCMV have distinct nucleotide sequences whereas that of RNA 4 is a subsequence of component 3 (Shih *et al.*, 1972). BBMV possesses two nucleoprotein populations in CsCl corresponding to the two larger classes of RNA, both of which are required for infection (Hull, 1972). Degradation is probably at least partly responsible for the two smaller classes of BBMV-RNA, the distribution of which is at present not clear.

III. TRANSLATION *IN VITRO* AND GENETICS

BMV and, to a lesser extent, CCMV and BBMV-RNA's have been examined *in vitro* in the wheat embryo protein-synthesizing system (Shih and Kaesberg, 1973; Davies and Kaesberg, 1974). In the case of BMV and CCMV, RNA 4 contains the capsid protein gene. RNA 3 contains two genes, one for the capsid protein and the other for a protein with a MW of 34,500 which is probably the virus-specified subunit of the replicase (Hariharasubramanian *et al.*, 1973). The RNA's 1 and 2 code for proteins with molecular weights of about 110,000 and 100,000, respectively (Davies and Kaesberg, 1975), but their function is at present not known.

Infectivity experiments involving wild-type and mutant RNA's have resulted in phenotypic assignments to the various components (Bancroft and Lane, 1973). The RNA 4, which is not required for infectivity, has no heritable effect. RNA 3 from CCMV and BMV determines the coat protein as would be expected from the *in vitro* translation results. The information in CCMV-RNA 3 can also rule on the type of systemic symptoms that develop and on the relative amounts of the RNA components. BMV-RNA 3 usually affects local lesion development as well, although RNA 2 from a single mutant of BMV will also do this. CCMV local lesions so far have only been affected by mutations on CCMV-RNA 2. One characteristic, that of lesion development at high temperatures, has been found for CCMV-RNA 1, but this has only been obtained once.

IV. MULTIPLICATION

CCMV and a strain of BMV (variant 5) multiply in tobacco protoplasts with a doubling time of 20 to 30 minutes at 24° to give yields of 10^6 to 10^7 particles per protoplast 24 hours after infection (Motoyoshi *et al.*, 1973, 1974). The virus appears in the cytoplasm and the cells do not lyse (Fig. 1).

Infection experiments in the presence of actinomycin D show that viral RNA appears within 7 hours after infection, and virus can be detected 2 hours later. RNA 3 is synthesized first and remains predominant throughout the infection period at least for CCMV (Bancroft *et al.*, 1975). Replicative forms (RF) have been identified for RNA's 1, 2, and 3 but not for 4, which is derived from part of RNA 3.

A BMV-RNA replicase with somewhat limited specifity has been partially purified. The 34,500 molecular weight protein is associated with the replicase activity (Hariharasubramanian *et al.*, 1973). This protein also occurs early in the infection of tobacco protoplasts, at least with CCMV but is soon superceded in quantity by coat protein (F. Sakai and J. W. Watts, unpublished). The control mechanism for this reaction is believed to center on the fact that RNA 4 inhibits *in vitro* translation of the other RNA's (Shih and Kaesberg, 1973) presumably by preferential binding to ribosomes. Early proteins are probably made in the absence of RNA 4 but, once made, stimulate its production which suppresses the translation of the early proteins while serving as coat messenger. Small amounts of the large proteins have been detected in extracts from infected protoplasts but their function and kinetics remain obscure (F. Sakai and J. W. Watts, unpublished).

V. MORPHOLOGICAL PROPERTIES

The capsid of the bromoviruses is composed of 180 structure units of about 19,400 MW assembled to form a 32 capsomere (T = 3) icosahedral shell. There are 12 groups of five peptides and 20 groups of six peptides (Figs. 2, 3, and 4). Electron micrographs of separated CCMV and BMV particles have shown that they are approximately 25 to 28 nm across depending on the staining and drying conditions used for electron microscopy (Anderegg *et al.*, 1963; Bancroft *et al.*, 1967). High resolution images of CCMV capsids have shown radially arranged components at their surfaces of about 4.0 nm long and 1.8 nm wide, (Figs. 4 and 5) and were interpreted as corresponding to the protein structure units (Horne *et al.*, 1975).

When highly purified and concentrated CCMV or BMV suspensions are spread onto the surfaces of freshly cleft mica in the presence of ammonium molybdate (Horne *et al.*, 1974, 1975), they rapidly form 2-dimensional or 3-dimensional crystalline assays (Figs. 6 and 7). Measurements from optical diffraction patterns obtained from the crystalline assays show a first order spacing of 26.0 nm (Figs. 8 and 9), corresponding to the particle center-to-center distance in crystal lattice. This value is in agreement with the X-ray scattering data and electron microscopic observations reported by Anderegg *et al.* (1963). The protein of CCMV can be reconstituted *in vitro* to form a variety of morphological products as shown in Figs. 10, 11, and in the diagram illustrated in Fig. 12 (Bancroft, 1970b).

REFERENCES

Anderegg, J. W., Wright, M., and Kaesberg, P. (1963). *Biophys. J.* **3,** 175–182.

Bancroft, J. B. (1970a). *CMI/AAB Descriptions Plant Viruses* No. 3.

Bancroft, J. B. (1970b). *Adv. Virus Res.* **16,** 99–135.

Bancroft, J. B. (1971a). *CMI/AAB Descriptions Plant Viruses* No. 49.

Bancroft, J. B. (1971b). *Virology* **45,** 830–834.

Bancroft, J. B., and Flack, J. H. (1972). *J. Gen. Virol.* **15,** 247–251.

Bancroft, J. B., and Lane, L. C. (1973). *J. Gen. Virol.* **19,** 381–389.

Bancroft, J. B., Hills, G. J., and Markham, R. (1967). *Virology* **31,** 354–379.

Bancroft, J. B., Hiebert, E., Rees, M. W., and Markham, R. (1968). *Virology* **34,** 224–239.

Bancroft, J. B., McLean, G. D., Rees, M. W., and Short, M. N. (1971). *Virology* **45,** 707–715.

Bancroft, J. B., Motoyoshi, F., Watts, J. W., and Dawson, J. R. O. 1975. *In* "Modification of the Information Content of Plant Cells" (R. Markham, D. R. Davies, D. A. Hopwood, and R. W. Horne, eds.), 2nd John Innes Symp., pp. 133–160. Elsevier, New York.

Bockstahler, L. E., and Kaesberg, P. (1962). *Biophys. J.* **2,** 1–9.

Bockstahler, L. E., and Kaesberg, P. (1965). *J. Mol. Biol.* **13,** 127–137.

Dasgupta, R., Shih, D. S., Saris, C., and Kaesberg, P. (1975). *Nature* (*London*) **256,** 624–628.

Davies, J. W., and Kaesberg, P. (1974). *J. Gen. Virol.* **25,** 11–20.

Davies, J. W., and Kaesberg, P. (1975). *Proc. Int. Congr. Virol., 3rd, 1975* p. 24.

Gibbs, A. J., (1972). *CMI/AAB Descriptions Plant Viruses* No. 101.

Hall, T. C., Shih, D. S., and Kaesberg, P. (1972). *Biochem. J.* **129,** 969–976.

Hariharasubramanian, V., Hadidi, A., Singer, B., and Fraenkel-Conrat, H. (1973). *Virology* **54,** 190–198.

Harrison, B. D., Finch, J. T., Gibbs, A. J., Hollings, M., Shepherd, R. J., Valento, V., and Wetter, C. (1971). *Virology* **45,** 356–363.

Horne, R. W., and Pasquali-Ronchetti, I. (1974). *J. Ultrastruct. Res.* **47,** 361–383.

Horne, R. W., Hobart, J. M., and Pasquali-Ronchetti, I. (1975). *J. Ultrastruct. Res.* **53,** 319–330.

Hull, R. (1972). *J. Gen. Virol.* **17,** 111–117.

Jacrot, B. (1975). *J. Mol. Biol.* **95,** 433–446.

Kohl, R. J., and Hall, T. C. (1974). *J. Gen. Virol.* **25,** 257–261.

Lane, L. C. (1974). *Adv. Virus Res.* **19,** 151–220.

Lane, L. C., and Kaesberg, P. (1971). *Nature* (*London*), *New Biol.* **232,** 40–43.

Motoyoshi, F., Bancroft, J. B., Watts, J. W., and Burgess, J. (1973). *J. Gen. Virol.* **20,** 177–193.

Motoyoshi, F., Bancroft, J. B., and Watts, J. W. (1974). *J. Gen. Virol.* **25,** 31–36.

Scott, H. A., and Slack, S. A. (1971). *Virology* **46,** 490–492.

Semancik, J. S. (1966). *Virology* **30,** 698–704.

Shih, D. S., and Kaesberg, P. (1973). *Proc. Natl. Acad. Sci. U.S.A.* **70,** 1799–1803.

Shih, D. S., Lane, L. C., and Kaesberg, P. (1972). *J. Mol. Biol.* **64,** 353–362.

Stubbs, J. D., and Kaesberg, P. (1964). *J. Mol. Biol.* **8,** 314–323.

Yamazaki, H., and Kaesberg, P. (1963). *J. Mol. Biol.* **6,** 465–473.

Yamazaki, H., Bancroft, J. B., and Kaesberg, P. (1961). *Proc. Natl. Acad. Sci. U.S.A.* **47,** 979–983.

Fig. 1. Electron micrograph from a thin section of a plant cell protoplast showing CCMV particles in the cytoplasm 48 hours after infection. ×40,000. (Motoyoshi *et al.*, 1973)

Fig. 2. CCMV particles negatively stained with uranyl acetate showing the morphological appearance of isolated particles. The hollow capsomeres are clearly visible at the surface of the virions when photographed in suitable orientations. ×950,000.

Fig. 3. High magnification of a single CCMV particle viewed along a twofold symmetry axis. The image was subjected to "averaging" by superimposing four views of the particle rotated 180° with the negative unreversed and reversed. ×3,900,000. (Bancroft, *et al.*, 1967)

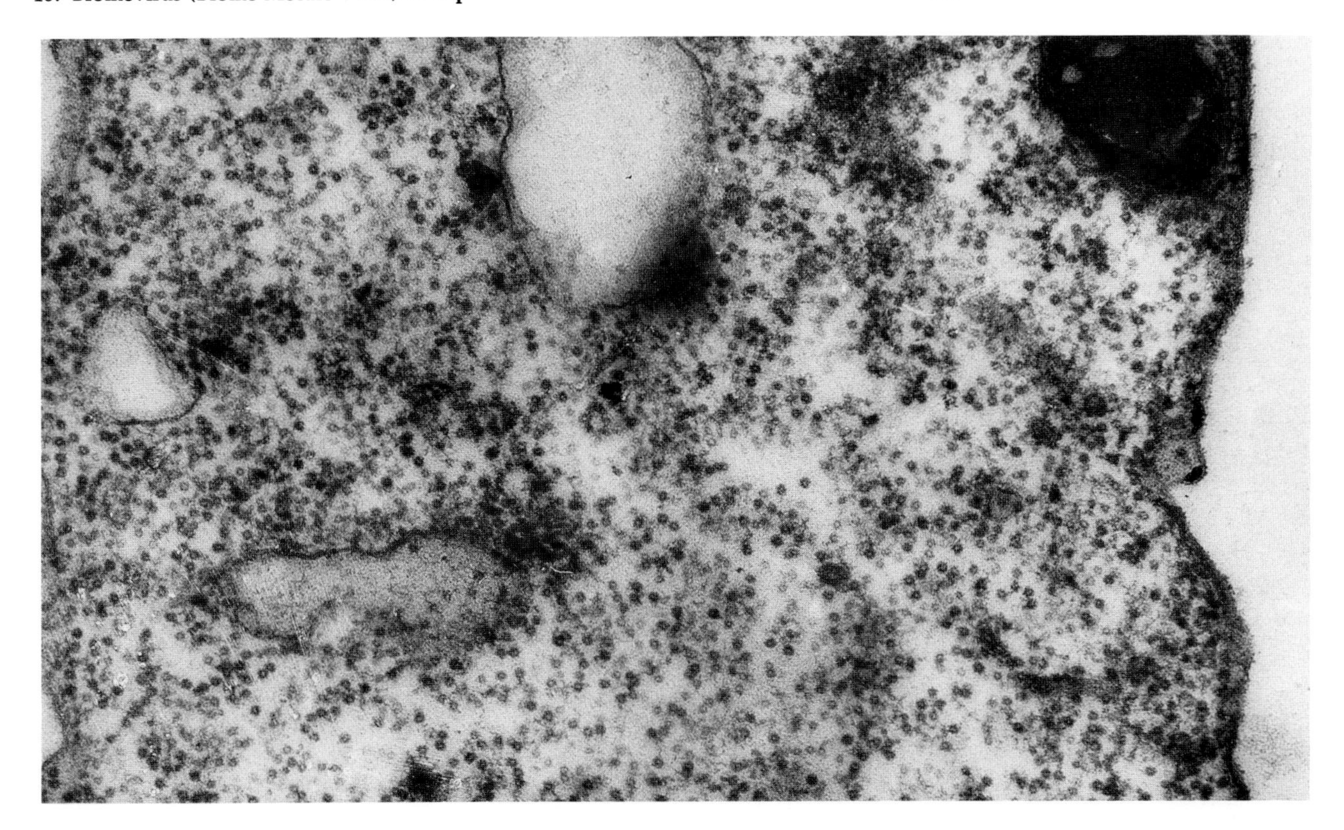

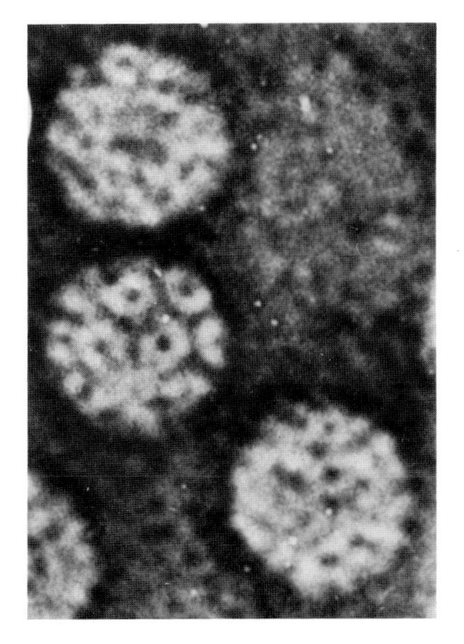

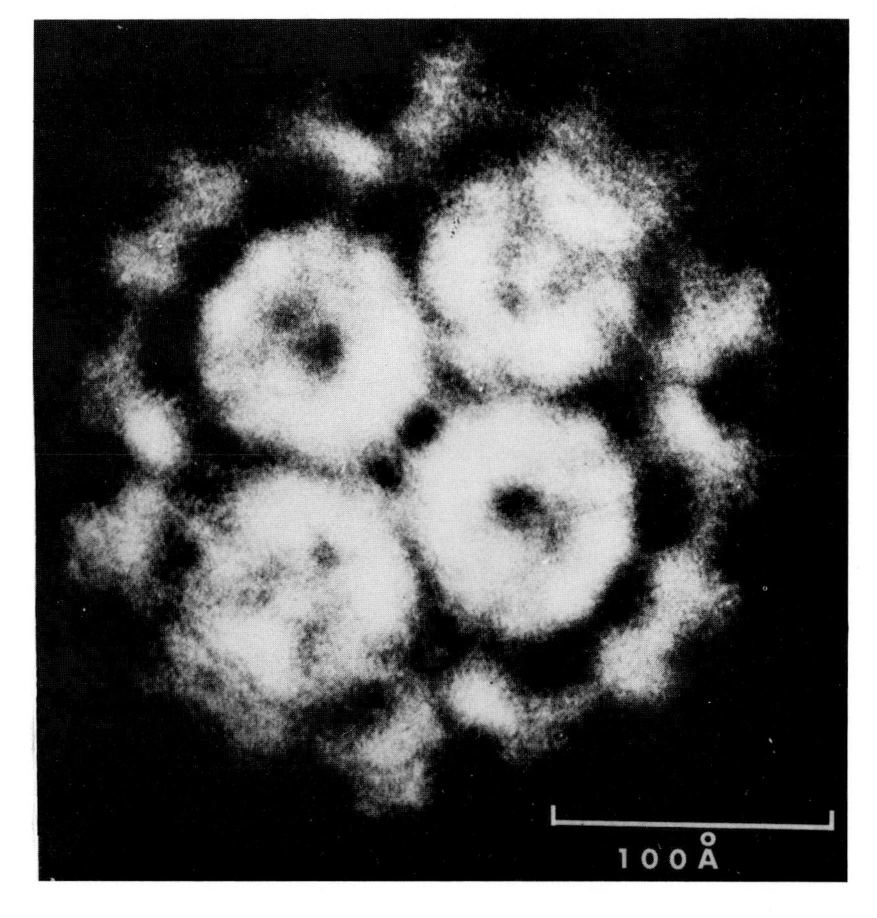
100 Å

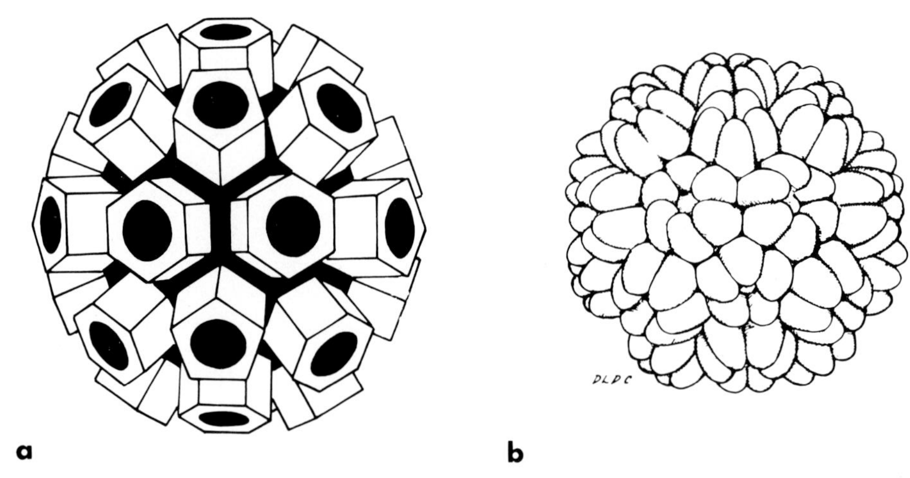

Fig. 4. (a) The drawing illustrates the 32 ($T = 3$) morphological arrangement of pentamer and hexamer capsomeres corresponding to the icosahedral capsid of CCMV viewed along a twofold axis. (Horne *et al.*, 1975). (b) The clustering of 180 asymmetrical protein structure units in groups of 5 and 6 assembled to form an icosahedral particle with 32 capsomeres. Each structure unit would correspond to a single protein of about 19,400 MW. (Reproduced by kind permission of D. L. D. Caspar, Brandeis University.)

Fig. 5. When CCMV particles are prepared according to the negative stain-carbon method (Horne and Pasquali-Ronchetti, 1974), fine surface detail can be seen extending radially (arrows) at the periphery of the particles. Analysis of the images has shown that these small features suggest a more complex arrangement similar to the type of structural pattern shown in Fig. 4b.

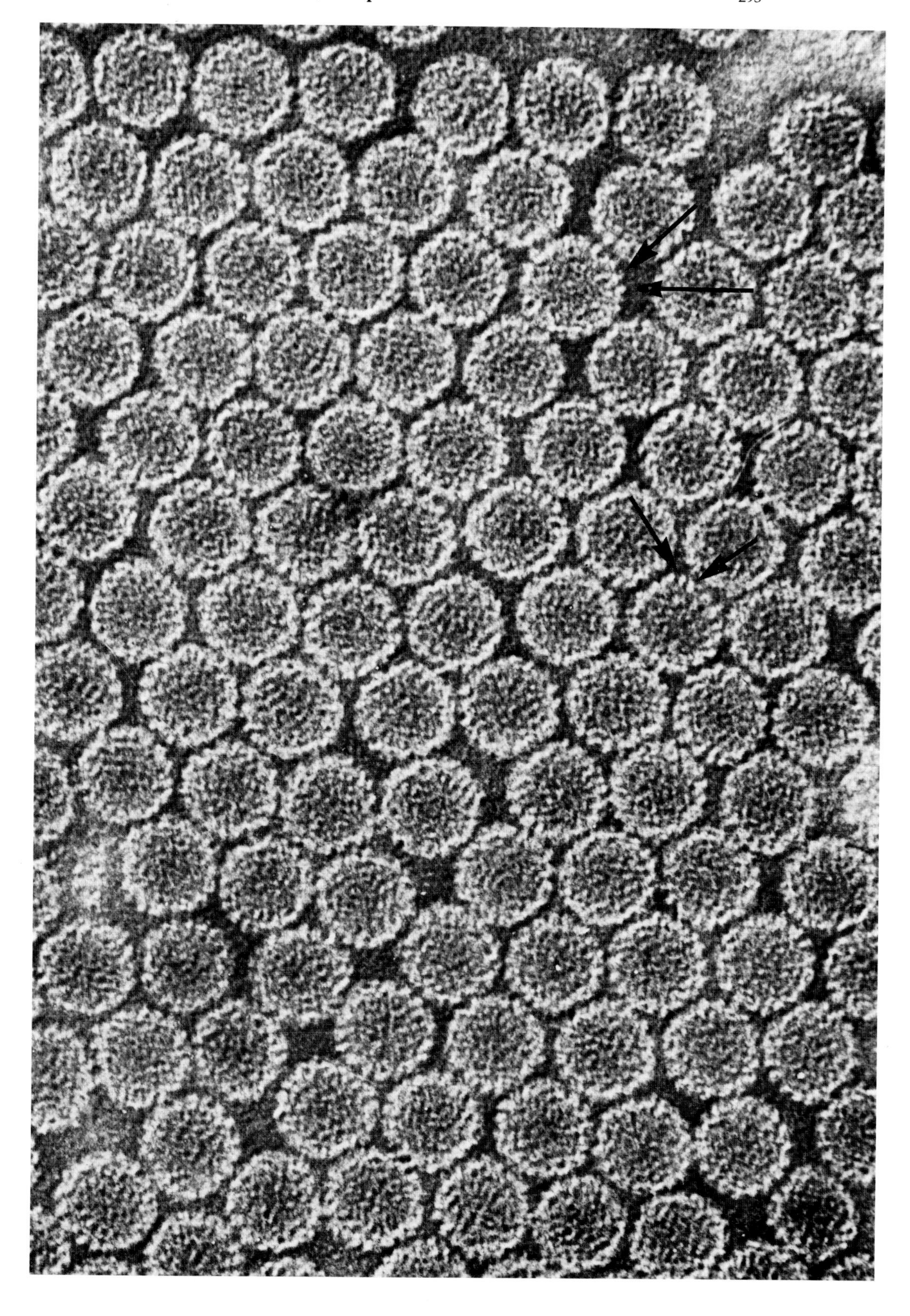

Fig. 6. Part of a two-dimensional CCMV crystalline array showing hexagonal packing of the particles. Small radial structural detail can be seen at the surfaces of the particles. ×150,000. (See also Fig. 8.)

Fig. 7. The upper part of the figure shows CCMV particles in square array. In this array the virus particles are viewed along a twofold axis (arrow) with each capsid rotated by 90° with respect to its neighbor. (See also Fig. 9.) × 150,000. View of the particles when prepared by the techniques used to form the CCMV crystalline arrays (Horne *et al.*, 1975), show evidence for the presence of capsomeres or morphological units normally seen in negatively stained specimens. The capsomere image can be obtained as illustrated in the lower figure by diagonally integrating the square array and superimposing particles within the lattice in the same twofold positions. ×2,200,000.

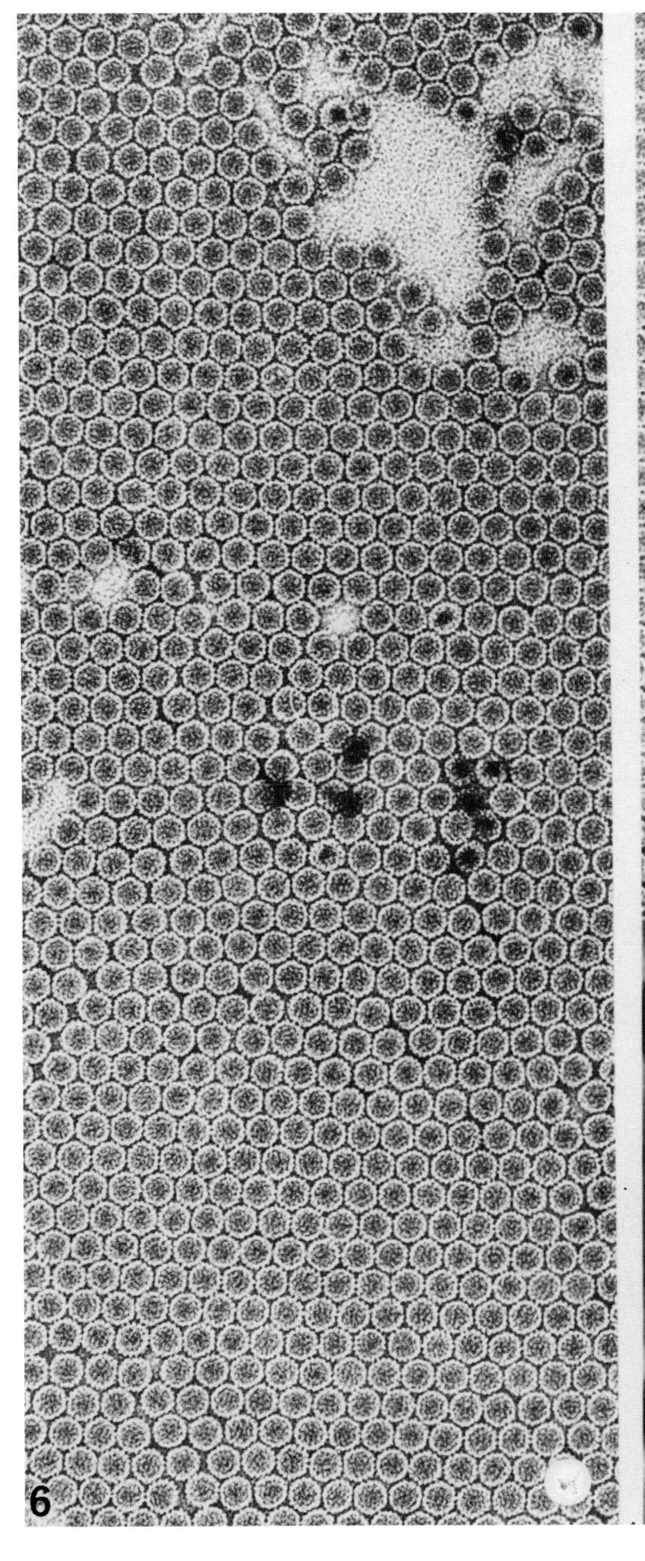
6

7

Fig. 8. (A) Optical diffraction pattern obtained from Fig. 6. (B) The first order spectra a–a correspond to the center-to-center particle spacing of about 26 nm. Diffraction spectra recorded on the original photographs extend out to positions b–b.

Fig. 9. (A) Optical diffraction from the square array shown in Fig. 7. (B) The center-to-center first order spacing a–a is about 26 nm. In the case of the square arrays the diffraction spectra is more extensive than in Fig. 8 and indicates a resolution extending out to about 2.0 nm as shown in the positions b–b on the tracing at the right.

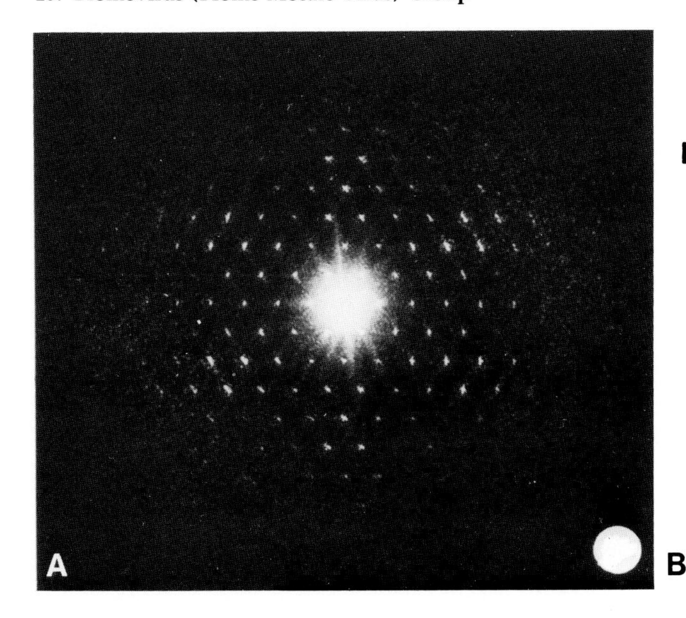

b a a b

B

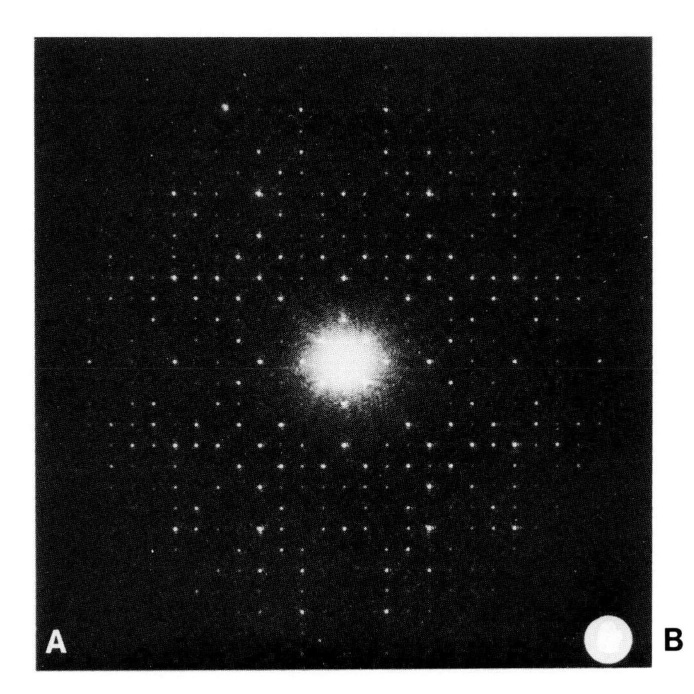

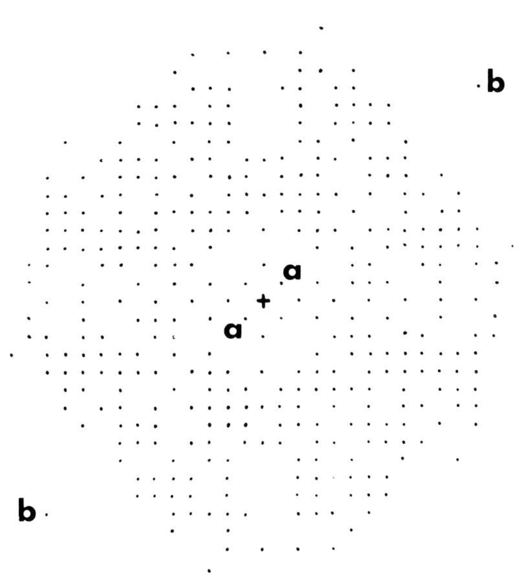

B

Fig. 10. The micrograph shows various forms and sizes of narrow tube-like structures assembled from CCMV protein. ×180,000. (Bancroft *et al.*, 1967)

Fig. 11. Wide tube-like structures formed from CCMV protein assembling on the surface of narrow tubes. ×180,000. (Bancroft *et al.*, 1967)

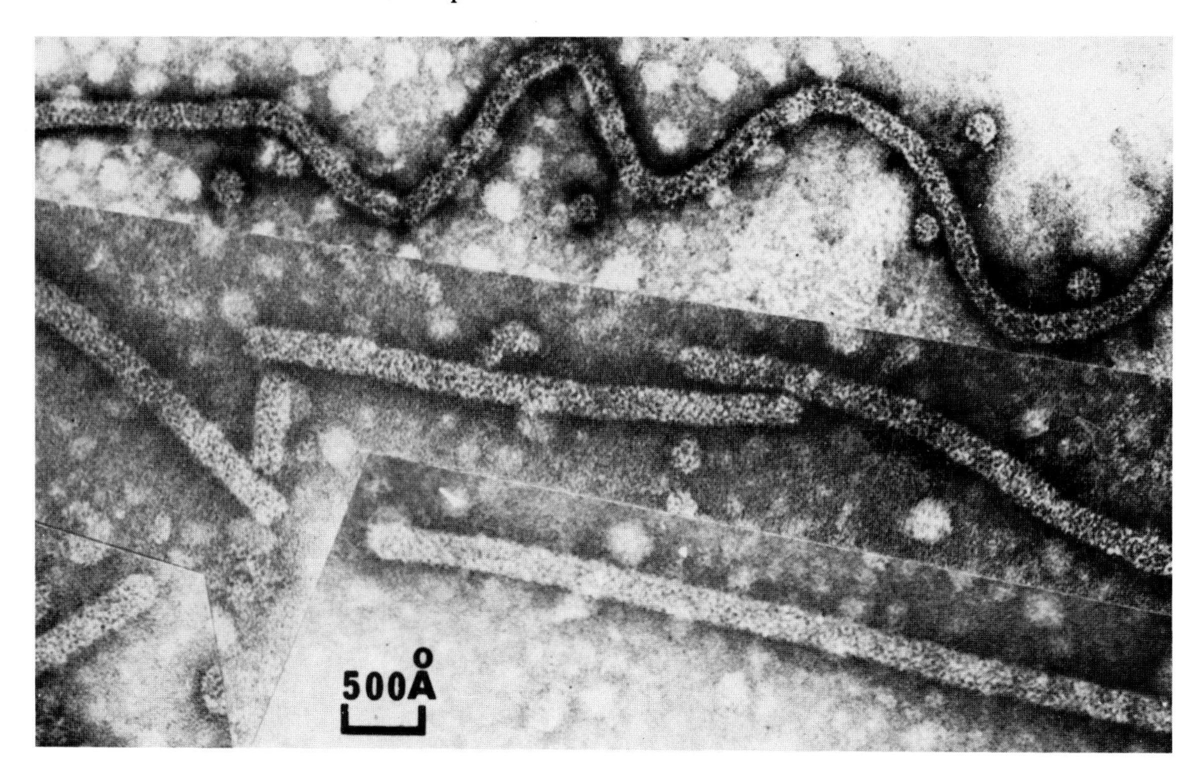
500Å

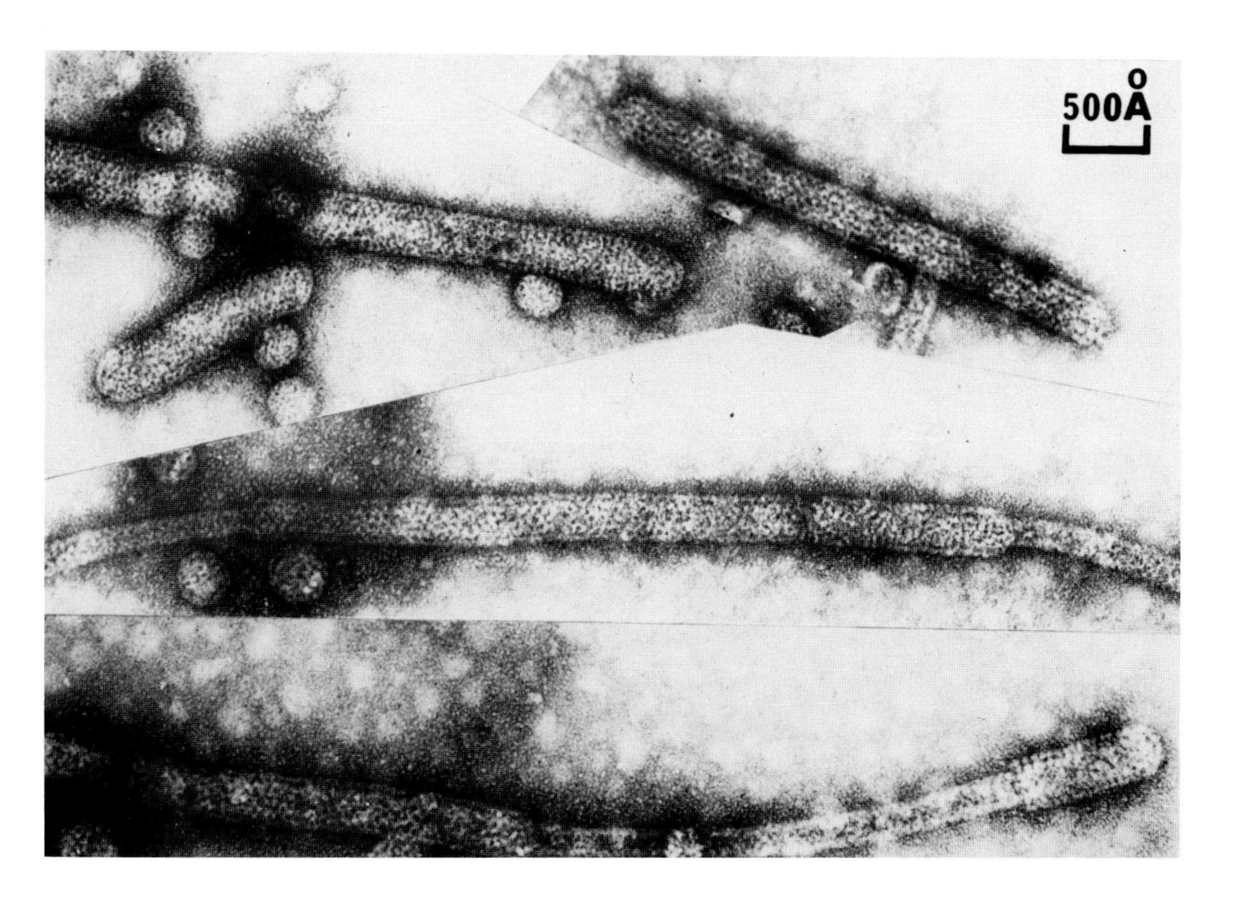
500Å

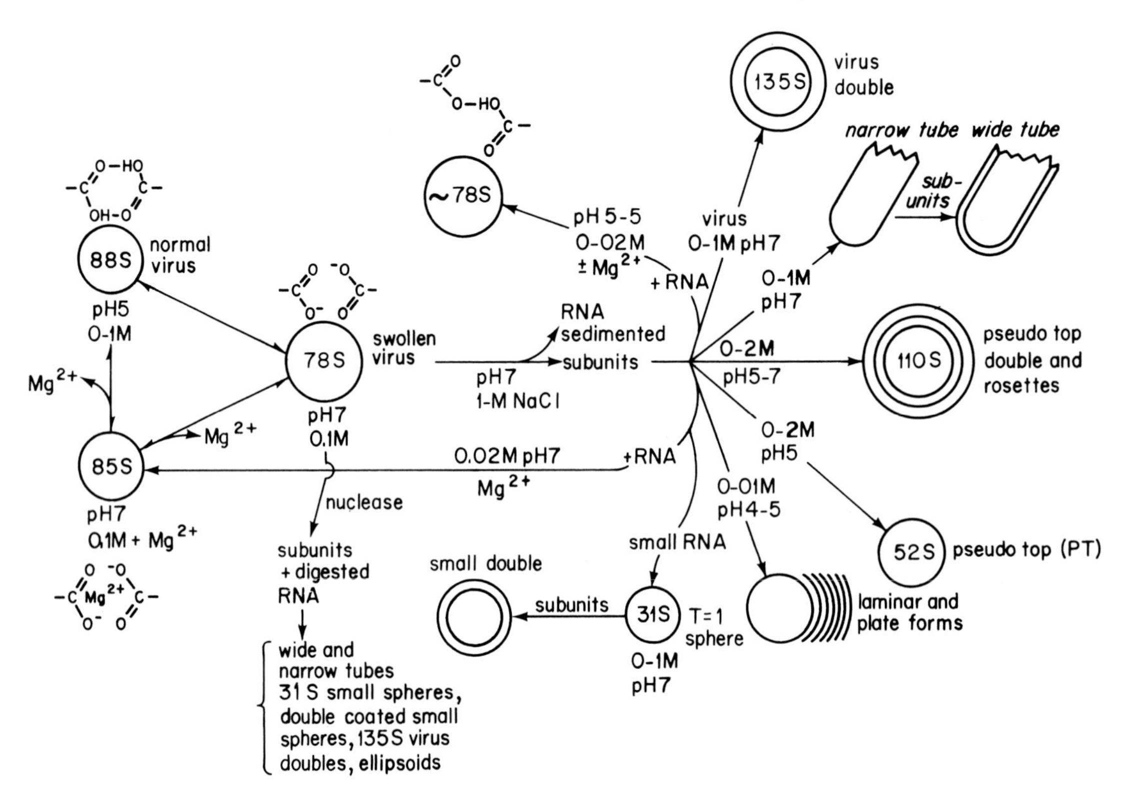

Fig. 12. The diagram illustrates a variety of morphological products assembled *in vitro* from CCMV when "swollen" and dissociated and the products reassembled under various conditions described by Bancroft, 1970b.

Chapter 17

Cucumovirus (Cucumber Mosaic Virus) Group

SUE A. TOLIN

I. Introduction 303
II. Members of the Cucumovirus Group 304
A. Cucumber Mosaic Virus 304
B. Tomato Aspermy Virus 304
C. Peanut Stunt Virus 305
III. Structure and Morphogenesis 305
A. Structure and Morphology 305
B. Morphogenesis 305
References 306

I. INTRODUCTION

Cucumoviruses constitute a group of viruses pathogenic to an extremely wide variety of herbaceous and woody plants. The name of the group is derived from the type member, cucumber mosaic virus (CMV). This name was proposed by the Plant Virus Subcommittee of the International Committee on the Nomenclature of Viruses (ICNV) and was accepted by the ICNV in 1970 (Harrison *et al.*, 1971).

Members of the group have a number of characteristics in common but differ primarily in their host specificity and serological relatedness. All cucumoviruses are transmitted by aphids in the stylet-borne manner and by mechanical inoculation of sap or purified virus. A serological relationship can be demonstrated among the strains within each member of the group and to some extent between representative strains of the different viruses. The degree of serological relatedness between the viruses has been interpreted differently by various authors (Grogan *et al.*, 1963; Lawson, 1967; Mink, 1969; Tolin and Boatman, 1972; Devergne and Cardin, 1973, 1976; Mink *et al.*, 1975; Richter *et al.*, 1975).

Cucumoviruses are naked, icosahedral particles that sediment uniformly at about 100 S and contain about 18% single-stranded RNA. The complete viral genome is divided into three molecules of RNA of different lengths which are encapsidated separately, similar to the Bromoviruses (Peden and Symons, 1973; Lot *et al.*, 1974). A fourth and, in some strains, a fifth segment of RNA (both smaller than the genome segments) are also encapsidated. Pseudorecombinant strains have been produced by combining genome RNA molecules isolated from different CMV strains (Marchoux *et al.*, 1974) and from different member viruses (Habili and Francki, 1974c; Marchoux *et al.*, 1975).

The viruses of the group and strains of each virus are similar in physicochemical properties but can be differentiated by various chemical probes (Mink *et al.*, 1969; Boatman *et al.*, 1973; Stace-Smith and Tremaine, 1973; Habili and Francki, 1974a,b; Boatman and Kaper, 1976; Kaper, 1976; Lot and Kaper, 1976). The amino acid composition, electrophoretic mobility, and the RNA component ratio also vary between viruses and strains (Marchoux *et al.*, 1972; Stace-Smith and Tremaine, 1973; Habili and Francki, 1974a; Monsion *et al.*, 1974; Wood and Coutts, 1975).

II. MEMBERS OF THE CUCUMOVIRUS GROUP

A. Cucumber Mosaic Virus

The first descriptions of cucumber mosaic virus were made before 1920 (Gibbs and Harrison, 1970; Smith, 1972). Since then, countless strains of CMV have been isolated throughout the world. Smith (1972) lists 27 families of dicots and four families of monocots that are infected by CMV strains. Nearly all strains infect tobacco and cucumber systemically. Only a few strains infect cowpea systemically; most induce necrotic local lesions on inoculated primary leaves and do not spread. CMV-induced diseases are probably most important economically in tomatoes, cucurbits, tobacco, flowers such as gladiolus and lilies, and in various ornamental plants. Strains are usually designated by a letter which may refer to a symptom (Y = yellow), to a geographical designation of the isolation site (Q = Queensland), or to a host from which the isolation was made (N = *Nicotiana;* R = *Ranunculus;* T = *Trandescantia*). Alternatively, strains have been numbered (Price's No. 6), or synonymous names such as yellow cucumber mosaic, spinach blight, or tomato fern leaf virus have been applied.

B. Tomato Aspermy Virus

The name tomato aspermy virus (TAV) was applied by Brierly *et al.* (1955) to the causal virus of a tomato disease described by Blencowe and Caldwell (1949) in England. The virus and other aspermy-type viruses have continued to be called TAV in spite of the fact that chrysanthemums were the original host. The virus is best known in connection with chrysanthemum flower distortion in England (Hollings and Stone, 1971) and has also been isolated from chrysanthemum in Victoria, Australia (TAV-V) (Habili and Francki, 1974a) as well as in the United States (Grogan *et al.*, 1963; Lawson, 1967). The strains described by Lawson (1967), particularly his CV-L, appear to differ from most TAV's and perhaps should be considered separately from TAV. Chrysanthemum mild mottle virus (Tochihara, 1970) is an additional aspermy-type virus described from Japan.

The host range of TAV is nearly as wide as that of CMV (Hollings and Stone, 1971; Smith, 1972). One general distinction is that TAV is less able to infect cucumber and other cucurbits than is CMV. TAV isolates do not infect cowpea systemically.

C. Peanut Stunt Virus

Peanut stunt virus (PSV) is the name given to a virus that caused a widespread epidemic in peanuts in Virginia and North Carolina in the early to mid 1960's (Miller and Troutman, 1966). It is also known as groundnut stunt because the common name of *Arachis hypogaea* in many parts of the world is groundnut. Like the other cucumoviruses, PSV has a wide host range, particularly among the Leguminosae. All PSV isolates infect cowpea and peanut systemically and may or may not infect cucumber. Natural infections resulting in disease losses have been reported in annual and perennial legumes (peanut, bean, cowpea, soybean, clover, crownvetch) and in tobacco (Mink, 1972). Reports of its occurrence have thus far been limited to North America and Japan, although some viruses isolated in Europe, Africa, and Central America, and described as CMV strains, have a number of properties in common with PSV. PSV isolates vary somewhat in their pathogenicity and stability and appear to be similar, but not identical, serologically (Tolin and Boatman, 1972).

III. STRUCTURE AND MORPHOGENESIS

A. Structure and Morphology

The identical morphology of three cucumoviruses can be seen in Figs. 1–3. The particles have a hollow center and frequently appear flattened, distorted, and disrupted. Finch *et al.* (1967) analyzed the fine structure of negatively stained preparations of CMV strain S and described a hexamer–pentamer grouping of 180 protein subunits in a T = 3 icosahedral surface lattice. The particles are indistinguishable from those of bromoviruses except for the slightly larger size of 30 nm. Like the bromoviruses, CMV has been dissociated and reassembled (Kaper and Geelen, 1971).

The particles of most cucumoviruses are disrupted in phosphotungstate (Murant, 1965; Francki *et al.*, 1966; Gibbs and Harrison, 1970; Hollings and Stone, 1971). Uranyl acetate (1–3% unbuffered) or uranyl formate stains have been used with more success with CMV (Francki *et al.*, 1966; Finch *et al.*, 1967), TAV (Hollings and Stone, 1971; Habili and Francki, 1974a), and PSV (Tolin, 1967; Mink, 1972; Waterworth *et al.*, 1973). Virions can also be detected in negatively stained leaf dip or epidermal strip preparations from a number of different species of infected plants (Lawson and Hearon, 1970) as shown in Fig. 3 for TAV (Canadian strain) in tobacco and Fig. 4 for PSV in cowpea (S. A. Tolin, unpublished).

B. Morphogenesis

There have been several reports of observations of CMV in infected cells, only one of TAV (Lawson and Hearon, 1970), and none of PSV. Relatively few of the cells examined have contained virus particles, which are difficult to distinguish from ribosomes. In cells, particles are nearly uniform in size with a diameter of about 25 nm, have a doughnut-like or dense uniform shape, are larger than ribosomes, and cannot be detected in cells of uninoculated plants. Honda and Matsui (1974) destroyed intracellular ribosomes by incubating CMV-infected leaf discs on phosphate buffer prior to fixation and confirmed that the doughnut-like particles were viruses. Virus particles have been observed in the cytoplasm and in the nucleus but not in mitochondria or chloroplasts (Gerola *et al.*, 1965; Lyons and Allen, 1969; Honda and Matsui, 1974). In some cells, aggregates of CMV particles have been observed (Gerola *et al.*, 1965; Honda and Matsui, 1968, 1974). In some of the aggregates the virus particles appeared to be randomly arranged whereas in other they were arranged in a face-centered cubic system or hexagonal array. These crystalline arrays have been observed in vacuoles, cytoplasm, and nuclei of CMV-infected tobacco leaves. In tobacco protoplasts inoculated with CMV, virions showed an increased tendency to aggregate and were often associated with the plasmalemma, tonoplast, and nucleolus, or were in the cytoplasm either free or attached to intracytoplasmic vacuoles (Honda *et al.*, 1974).

Regular arrays of particles have been observed *in vitro* in sections of purified CMV aggregates produced by ultracentrifugation or by polyethylene glycol precipitation but not in sections of virus aggregates produced by salting out (ammonium sulfate) or by acid pH precipitation (pH 3.5) (Ehara *et al.*, 1976). Virions and purified ribosomes were essentially the same size in sections of purified particles, but the virions were more electron dense. Regular arrangement of CMV was prevented by mixing virus with pure ribosomes prior to centrifugation, which may explain why CMV crystals are seldom observed in infected cells in the cytoplasm where ribosomes are abundant (Ehara *et al.*, 1976).

Cytological abnormalities in CMV-infected cells include increased cytoplasm density, myelin-like structures in the cytoplasm and chloroplasts, chloroplasts with degenerated stroma lamellae and grana, as well as crystalline inclusions and deep protrusions of cytoplasm into affected chloroplasts (Gerola *et al.*, 1965; Honda and Matsui, 1968, 1974; Ehara and Misawa, 1975). In protoplasts fixed within 48 hours of inoculation, chloroplasts were not affected even though considerable progeny virus could be seen (Honda *et al.*, 1974). Protoplast nuclei that contained virions were devoid of heterochromatin and had vacuolated nucleoli, suggesting that CMV synthesis may occur in the nucleus.

Some of the ultrastructural features of tobacco cells infected with the Car strain of CMV (Lovisolo *et al.*, 1968) are shown in Fig. 5. A small crystal of virus particles is seen within a vesicle. Protrusion of cytoplasm into the chloroplasts is evident. Virus particles cannot be readily distinguished from ribosomes in the dense cytoplasm. Figure 6 is an enlargement of a single CMV crystal showing the regular arrangement of virus particles.

ACKNOWLEDGMENT

The author wishes to thank Dr. R. G. Milne (Laboratory for Applied Plant Virology, Torino, Italy) for providing Figs. 5 and 6 of the Car strain of CMV, and Dr. S. Boatman (Chemistry Department, Hollins College, Va.) and Mrs. R. H. Ford (Department of Plant Pathology and Physiology, Blacksburg, Va.) for their help in preparation of the electron micrographs and virus preparations.

REFERENCES

Blencowe, J. W., and Caldwell, J. (1949). *Ann. Appl. Biol.* **36,** 320–326.
Boatman, S., and Kaper, J. M. (1976). *Virology* **70,** 1–16.
Boatman, S., Kaper, J. M., and Tolin, S. A. (1973). *Phytopathology* **63,** 801.
Brierly, P., Smith, F. F., and Doolittle, S. P. (1955). *Plant Dis. Rep.* **39,** 152–156.
Devergne, J. C., and Cardin, L. (1973). *Ann. Phytopathol.* **5,** 409–430.
Devergne, J. C., and Cardin, L. (1975). *Ann. Phytopathol.* **7,** 255–276.
Ehara, Y., and Misawa, T. (1975). *Phytopathol. Z.* **84,** 233–252.
Ehara, Y., Misawa, T., and Nagayama, H. (1976). *Phytopathol. Z.* **87,** 28–39.

Finch, J. T., Klug, A., and van Regenmortel, M. H. V. (1967). *J. Mol. Biol.* **24,** 303–305.
Francki, R. I. B., Randles, J. W., Chambers, T. C., and Wilson, S. B. (1966). *Virology* **28,** 729–741.
Gerola, F. M., Bassi, M., and Belli, G. (1965). *Caryologia* **18,** 567–597.
Gibbs, A. J., and Harrison, B. D. (1970). *CMI/AAB Descriptions Plant Viruses* No. 1.
Grogan, R. G., Uyemoto, J. K., and Kimble, K. (1963). *Virology* **21,** 36–42.
Habili, N., and Francki, R. I. B. (1974a). *Virology* **57,** 392–401.
Habili, N., and Francki, R. I. B. (1974b). *Virology* **60,** 29–36.
Habili, N., and Francki, R. I. B. (1974c). *Virology* **61,** 443–449.
Harrison, B. D., Finch, J. T., Gibbs, A. J., Hollings, M., Shepherd, R. J., Valenta, V., and Wetter, C. (1971). *Virology* **45,** 356–363.
Hollings, M., and Stone, O. M. (1971). *CMI/AAB Descriptions Plant Viruses* No. 79.
Honda, Y., and Matsui, C. (1968). *Phytopathology* **58,** 1230–1235.
Honda, Y., and Matsui, C. (1974). *Phytopathology* **64,** 534–539.
Honda, Y., Matsui, C., Otsuki, Y., and Takebe, I. (1974). *Phytopathology* **64,** 30–34.
Kaper, J. M. (1976). *Virology* **71,** 185–198.
Kaper, J. M., and Geelen, J. L. M. C. (1971). *J. Mol. Biol.* **56,** 277–294.
Lawson, R. H. (1967). *Virology* **32,** 357–362.
Lawson, R. H., and Hearon, S. (1970). *Virology* **41,** 30–37.
Lot, H., and Kaper, J. M. (1976). *Virology* **74,** 209–222.
Lot, H., Marchoux, G., Marrou, J., Kaper, J. M., West, C. K., van Vloten-Doting, L., and Hull, R. (1974). *J. Gen. Virol.* **22,** 81–93.
Lovisolo, O., Conti, M., and Luisoni, E. (1968). *Phytopathol. Mediterr.* **7,** 71–76.
Lyons, A. R., and Allen, T. C., Jr. (1969). *J. Ultrastruct. Res.* **27,** 198–204.
Marchoux, G., Douine, L., Marrou, J., and Devergne, J. C. (1972). *Ann. Phytopathol.* **4,** 363–365.
Marchoux, G., Marrou, J., Douine, L., Lot, H., and Quiot, J. B. (1974). *Ann. Phytopathol.* **6,** 225.
Marchoux, G., Devergne, J. C., and Douine, L. (1975). *Int. Virol.* **3,** 90.
Miller, L. I., and Troutman, J. L. (1966). *Plant Dis. Rep.* **50,** 139–143.
Mink, G. I. (1969). *Phytopathology* **59,** 1889–1893.
Mink, G. I. (1972). *CMT/AAB Descriptions Plant Viruses* No. 92.
Mink, G. I., Silbernagel, M. J., and Saksena, K. N. (1969). *Phytopathology* **59,** 1625–1631.
Mink, G. I., Iizuka, N., and Kiriyama, K. (1975). *Phytopathology* **65,** 65–68.
Monsion, M., Bachelier, J. C., and Dunez, J. (1974). *Ann. Phytopathol.* **6,** 226.
Murant, A. F. (1965). *Virology* **26,** 538–544.
Peden, K. W. C., and Symons, R. H. (1973). *Virology* **53,** 487–492.
Richter, J., Oertel, C., and Proll, E. (1975). *Arch. Phytopathol. Pflanzenchutz* **11,** 189–196.
Smith, K. M. (1972). "A Textbook of Plant Virus Diseases," 3rd ed. Academic Press, New York.
Stace-Smith, R., and Tremaine, J. H. (1973). *Virology* **51,** 401–408.
Tochihara, H. (1970). *Ann. Phytopathol. Soc. Jap.* **36,** 1–10.
Tolin, S. A. (1967). *Phytopathology* **57,** 834.
Tolin, S. A., and Boatman, S. (1972). *Phytopathology* **62,** 793.
Waterworth, H. E., Monroe, R. L., and Kahn, R. P. (1973). *Phytopathology* **63,** 93–98.
Wood, K. R., and Coutts, R. H. A. (1975). *Physiol. Plant Pathol.* **7,** 139–146.

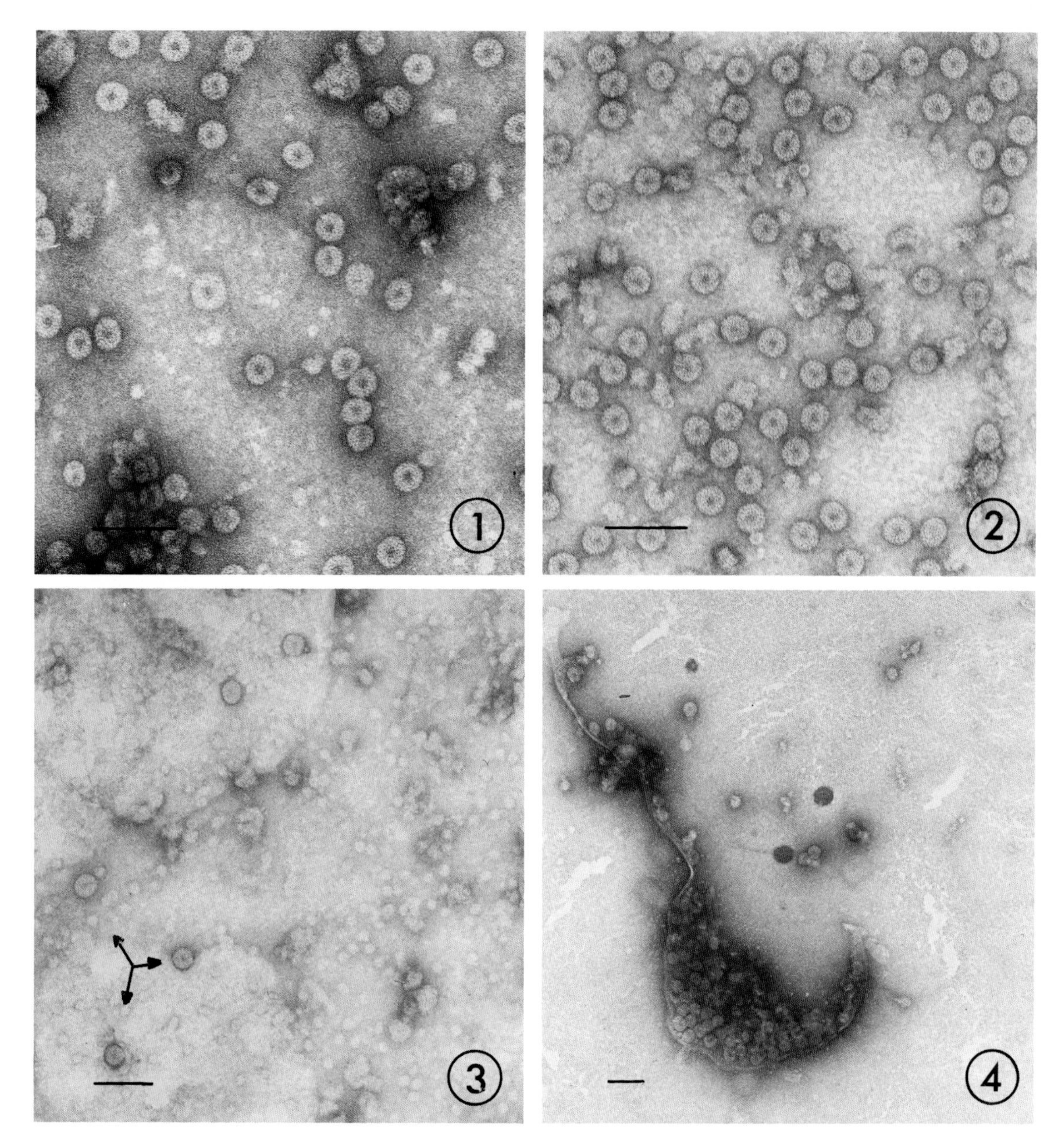

Figs. 1–4. Electron micrographs of negatively stained (1% uranyl acetate) members of the cucumoviruses. Horizontal bars equal 100 nm.

Fig. 1. Purified cucumber mosaic virus, strain S.

Fig. 2. Purified peanut stunt virus.

Fig. 3. Tomato aspermy virus, Canadian strain, in an epidermal strip preparation from infected tobacco. Virus particles (arrows) have a rigid shape and are easily distinguished from smaller ribosomes.

Fig. 4. Peanut stunt virus in an epidermal strip preparation from infected cowpea. Virus particles appear to have been associated with a membrane.

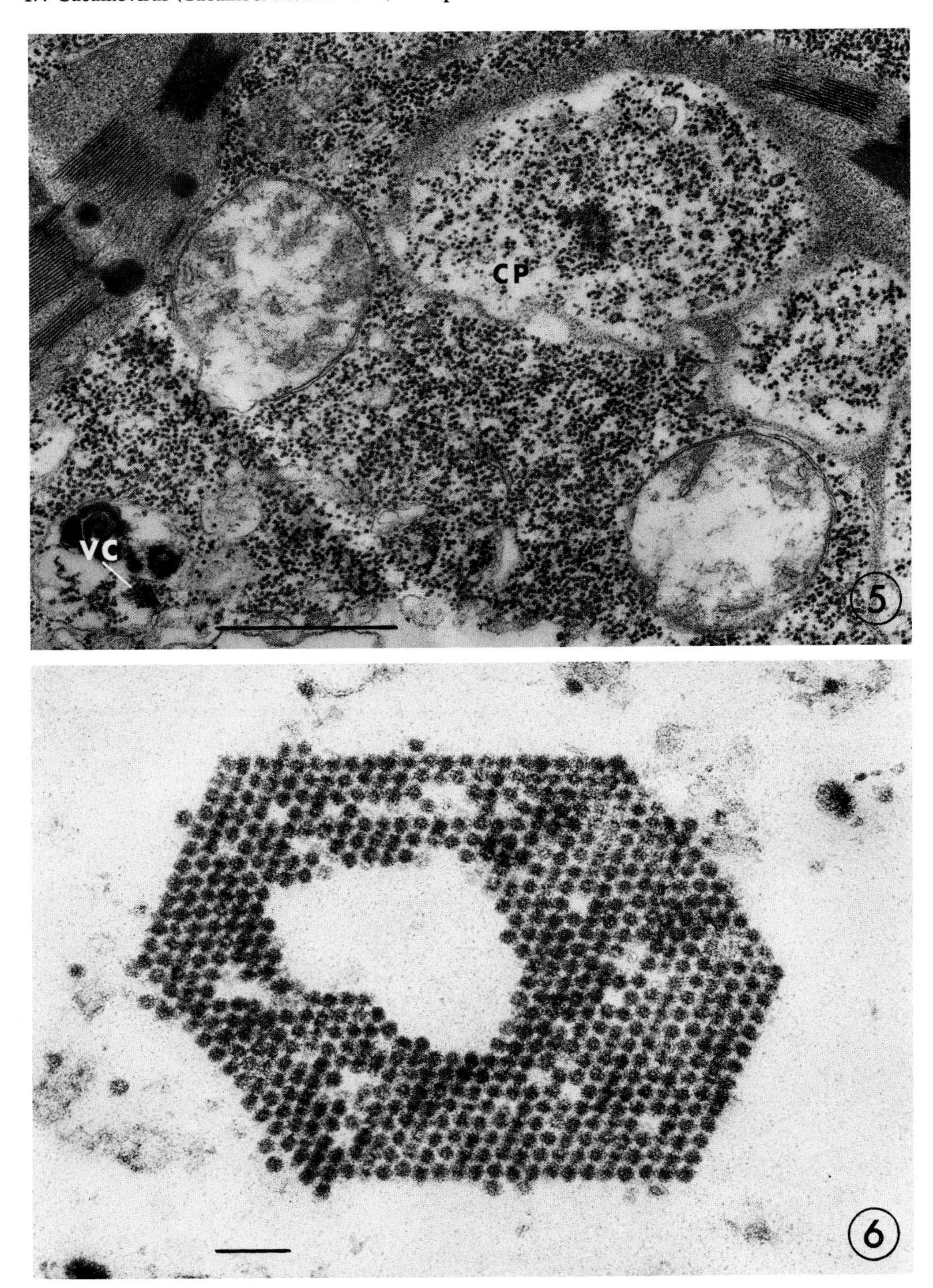

Fig. 5. Section of a leaf of *Nicotiana clevelandii* inoculated with cucumber mosaic virus, strain Car. VC, virus crystal; CP, cytoplasmic protrusion into nucleus. Bar equals 1 μm.

Fig. 6. Crystal of cucumber mosaic virus, strain Car, in inoculated leaf of *Nicotiana glutinosa*. Bar equals 100 nm.

Chapter 18

Potyvirus (Potato Virus Y) Group

D. S. TEAKLE AND R. D. PARES

I. Introduction . . . 311
II. Members of the Potyvirus Group . . . 312
III. General Properties . . . 312
A. Chemical Composition . . . 312
B. Transmission . . . 312
C. Serology . . . 312
D. Morphology . . . 313
E. Ultrastructure . . . 313
F. Morphogenesis . . . 313
G. Inclusions . . . 313
IV. Conclusions . . . 314
References . . . 315

I. INTRODUCTION

The word potyvirus is an acronym derived from the type member, potato virus Y, which commonly infects potatoes and some other solanaceous crops. Potyviruses have been found in a wide range of seed plants and probably even ferns (Nienhaus *et al.*, 1974).

The potyvirus group, as originally described by Brandes (1964) and redescribed by Harrison *et al.* (1971), is characterized by flexuous, rod-shaped virions which measure 720–800 nm × 12 nm and have helical symmetry with a pitch of 3.4 nm. Other properties include a thermal inactivation point of 50–60° C, a concentration in sap often reaching 5–25 mg/liter, a longevity *in vitro* of a few days, and a distant serological relationship between members. The symptoms produced are usually mosaics and the host range is rather narrow, while transmission is by mechanical means or by aphids.

Another characteristic shared by potyviruses but not listed by Brandes (1964) or Harrison *et al.* (1971) is the ability to induce the production of cylindrical protein inclusions in the cytoplasm of infected cells (Edwardson, 1966).

II. MEMBERS OF THE POTYVIRUS GROUP

About 100 viruses or virus strains with all or most properties typical of the potyvirus group have been described (Edwardson, 1974), making the potyvirus group the largest among plant viruses. Separation of members has usually been done on the basis of (1) a substantial difference in host range and/or symptoms, and (2) serological unrelatedness or only a distant serological relationship.

In the past, these criteria have allowed the delineation of many important members of the group, for instance, beet mosaic virus in beets, bean common mosaic virus in French beans, potato virus Y in solanaceous plants, passionfruit woodiness virus (Fig. 1) in passionfruit, sugarcane mosaic virus (Fig. 2) in sugarcane and certain other grasses, and watermelon mosaic virus in cucurbits. Taxonomic problems arise because host range and serological relationships may vary independently. For instance, although bean common mosaic and bean yellow mosaic viruses are serologically related, bean yellow mosaic virus has a wider natural host range, including some nonlegumes such as gladiolus. On the other hand, some strains of sugarcane mosaic virus have similar host ranges among the grasses but are not serologically related (Snazelle *et al.*, 1971). Presumably potyviruses are evolving independently for host range, coat protein, and other properties.

III. GENERAL PROPERTIES

A. Chemical Composition

There are few reports of chemical analyses of potyviruses, probably because these viruses usually occur in relatively low concentrations in extracted sap and may be difficult to purify because of aggregation and insolubility problems (Fig. 3). If the few viruses which have been studied are representative of the group, then potyviruses are nucleoproteins comprising about 5% RNA and 95% protein. The RNA is single-stranded and has a weight of about 3×10^6 daltons, while the protein consists of several thousand identical subunits of about 22,000 daltons and 194 amino acid residues each (e.g., Damirdagh and Shepherd, 1970a,b).

B. Transmission

Potyviruses are transmitted naturally by aphids in a "nonpersistent" or "stylet-borne" manner and artificially by mechanical inoculation of leaves with infective sap extracts. Virus acquisition and inoculation each occur within a few minutes during aphid probing of plants, and the persistence (carry-over) of virus in the insect is limited. There is no evidence of virus multiplication in the vector.

Eriophyid mites transmit several viruses, such as wheat streak mosaic virus, which resemble potyviruses in properties other than the typical vector. With mites, acquisition times (15 minutes to 2 hours) and persistence (hours or days) are longer than are usual for aphid transmitted potyviruses. In addition, certain soil-transmitted viruses, such as oat mosaic virus, resemble potyviruses in most properties except transmission by aphids.

C. Serology

Edwardson (1974) lists 38 potyviruses which are serologically related to one or more other members of the group. Since sera often have only a low or moderate homologous titer, the determination of distant serological relationships may be difficult.

D. Morphology

Although potyviruses typically have flexuous virions of 720–800 nm × 12 nm (Figs. 1–4), flexibility and size are somewhat variable characters. Govier and Woods (1971) found that particles of four potyviruses were 50–100 nm shorter and more flexuous in the absence of magnesium ions than in their presence. Also, some potyviruses may break during purification (Fig. 5) and, thus, have apparently shorter lengths than if taken directly from the host plant (Taylor and Pares, 1968). These occurrences together with errors introduced during measurement by electron microscopy lead to rather different lengths being obtained for the same virus. With passionfruit woodiness virus, normal or modal lengths of between 670 and 765 nm have been reported for Australian isolates (Taylor and Kimble, 1964; Leggat and Teakle, 1975), while a serologically related strain from the Ivory Coast had a modal length of about 820 nm (de Wijs, 1974). Atropa mild mosaic virus may have a modal length of 925 nm (Harrison and Roberts, 1971).

E. Ultrastructure

Potyvirus virions have indistinct, verrucous surface structure when stained with uranyl formate and examined in the electron microscope, whereas under similar conditions potexviruses and carlaviruses show distinct and indistinct crossbanding, respectively (Varma *et al.*, 1968). Using optical diffraction patterns, these workers found that potyviruses have helically arranged subunits with a pitch of 3.3–3.5 nm.

Leggat (1972) stored purified passionfruit woodiness virus in 0.01 *M* sodium citrate, pH 6.7, at 4° for 6–11 weeks, and when the particles were phosphotungstate-stained she found that one end sometimes showed helical structure. The helical portions were of varying lengths (Figs. 6–8), and had width of 13–17 nm, pitch of 6.2 nm and darkly staining cores. Possibly these structures are portions of disintegrating virions showing their substructure clearly, or alternatively, are aggregated viral protein taking on a helical substructure.

Presumably the RNA in the virion is embedded in the protein and is helically wound with the same pitch as that of the protein subunits.

F. Morphogenesis

As exemplified by passionfruit woodiness virus, virions of potyviruses are found in the cytoplasm of infected cells, either aggregated laterally (Figs. 9–12) or occurring as individual particles. Although the virions are neither membrane bound nor associated with any cytoplasmic organelle, aggregates of virions are frequently attached at one end to part of the cell endomenbrane system, such as the tonoplast (Figs. 9–11), the endoplasmic reticulum (Fig. 12), or the plasma membrane (Pares and McGechan, 1975).

Although virions accumulate in the cytoplasm, the site of RNA replication may be in or near chloroplasts (Mayhew and Ford, 1974).

G. Inclusions

Cylindrical inclusions of unknown function are produced in the cytoplasm of plant cells infected by potyviruses (Edwardson, 1974), and may be detected in extracts (Fig. 13) or sections (Figs. 11 and 14–21) of diseased tissues. Inclusions are striated with a periodicity of about 5 nm (Fig. 13), are dissolved by pepsin and trypsin, and when purified have ultraviolet light absorption spectra typical of proteins. In addition, they are immunogenic and different viruses produce serologically different inclusions, which in turn are serologically different from their respective virion and host proteins. It is possible, therefore, that inclusion formation is virus coded.

The inclusions induced by passionfruit woodiness virus appear in transverse section as pinwheels (Figs. 11, 14, and 15) and in longitudinal section as series of parallel, electron-opaque lines known as bundle inclusions (Figs. 16 and 17). Tubular and laminated aggregate inclusions (Figs. 15, 18) are associated with the pinwheels, and the pinwheel and tubular inclusions are frequently associated at one end with part of the cell endomembrane system (Figs. 16, 17, and 19).

As is also evident in published micrographs involving other potyviruses, pinwheel inclusions induced by passionfruit woodiness virus generally occur in groups. All the pinwheels in a particular group appear oriented parallel to one another, while adjacent groups are often perpendicular to each other (Fig. 20).

Not all the potyviruses have both tubular and laminated aggregate inclusions associated with the pinwheels, and potyviruses can be divided into three subgroups based on the types of inclusion produced (Edwardson, 1974). The subgroups have either (1) pinwheel and tubular inclusions, (2) pinwheel and laminated aggregate inclusions, or (3) pinwheel, tubular, and laminated aggregate inclusions.

Inclusions of passionfruit woodiness virus form in areas of a complex of membranes and ribosomes, somewhat like those described for wheat spindle streak virus (Hooper and Wiese, 1972). There is little published work on the early stages of inclusion development of any potyvirus, but Edwardson (1974) states that the inclusions first form as small cylinders at the periphery of epidermal cells and gradually increase in size and number. Massive inclusions may form later away from the periphery by a congregation of cylindrical inclusions and cytoplasmic organelles.

A few potyviruses, such as tobacco etch virus, also produce nuclear inclusions in infected cells.

IV. CONCLUSIONS

The potyviruses must be considered a successful group of plant viruses because they are numerous, widespread, and common. Their flexuous, rod-shaped virions, measuring approximately 750 × 12 nm, are readily detected in negatively stained preparations, but usually no substructure can be resolved. In sections of infected host tissues, aggregates of virions and pinwheel inclusions are identified more readily than individual virions.

ACKNOWLEDGMENTS

The authors thank Mrs. F. W. Leggat, Department of Microbiology, University of Queensland, for permission to use Figs. 6–8. Also, we thank Mr. A. McGregor, Department of Microbiology and the Staff of the Electron Microscope Centre, University of Queensland, and Miss W. Beard, Electron Microscope Unit, Biological and Chemical Research Institute, Rydalmere, New South Wales, for assistance in the preparation of the micrographs.

REFERENCES

Brandes, J. (1964). *Mitt. Biol. Bundesanst. Land-Forstwirtsch., Berlin-Dahlem* **110,** 1–130.
Damirdagh, I. S., and Shepherd, R. J. (1970a). *Virology* **40,** 84–89.
Damirdagh, I. S., and Shepherd, R. J. (1970b). *Phytopathology* **60,** 132–142.
de Wijs, J. J. (1974). *Ann. Appl. Biol.* **77,** 33–40.
Edwardson, J. R. (1966). *Am. J. Bot.* **53,** 359–364.
Edwardson, J. R. (1974). *Fl. Agric. Exp. Stn., Monogr. Ser.* No. 4.
Govier, D. A., and Woods, R. D. (1971). *J. Gen. Virol.* **13,** 127–132.
Harrison, B. D., and Roberts, I. M. (1971). *J. Gen. Virol.* **10,** 71–78.
Harrison, B. D., Finch, J. T., Gibbs, A. J., Hollings, M., Shepherd, R. J., Valenta, V., and Wetter, C. (1971). *Virology* **45,** 356–363.
Hooper, G. R., and Wiese, M. V. (1972). *Virology* **47,** 644–672.
Leggat, F. W. (1972). B. Sc. Hons. Report, Dept. of Microbiology, University of Queensland.
Leggat, F. W., and Teakle, D. S. (1975). *Aust. Plant Pathol. Soc. Newsl.* **4,** 22–23.
Mayhew, D. E., and Ford, R. E. (1974). *Virology* **57,** 503–509.
Nienhaus, F., Mack, C., and Schinzer, U. (1974). *Z. Pflanzenkr. Pflanzenschutz* **81,** 533–537.
Pares, R. D., and McGechan, J. K. (1975). *Aust. J. Bot.* **23,** 905–914.
Snazelle, T. E., Bancroft, J. B., and Ullstrup, A. J. (1971). *Phytopathology* **61,** 1059–1063.
Taylor, R. H., and Kimble, K. A. (1964). *Aust. J. Agric. Res.* **15,** 560–570.
Taylor, R. H., and Pares, R. D. (1968). *Aust. J. Agric. Res.* **19,** 767–773.
Varma, A., Gibbs, A. J., Woods, R. D., and Finch, J. T. (1968). *J. Gen. Virol.* **2,** 107–114.

Figs. 1–8. Negatively stained (phosphotungstic acid) virions of various members of the potyvirus group. All the potyviruses have flexuous, rod-shaped virions, usually with little or no sub-structure being visible. Bar = 250 nm.

Fig. 1. Passionfruit woodiness virus, which has been partially purified, together with several virions of tobacco mosaic virus (arrow), which can be distinguished by the darkly staining central canal. (×30,000.)

Fig. 2. Sugarcane mosaic virus, Queensland blue couch grass strain, which has been partially purified. (×30,000.)

Fig. 3. Sugarcane mosaic virus, Australian Johnson grass strain, showing aggregation of virions following partial purification. (×30,000.)

Fig. 4. An unnamed potyvirus from Siratro (*Macroptilium atropurpureum*), obtained from leaves without purification. Note the uniformity in the length of the virions as compared with virions of potyviruses which have been partially purified. (×30,000.)

Fig. 5. Sugarcane mosaic virus, Queensland blue couch grass strain, showing many fragmented virions following partial purification. (×30,000.)

Figs. 6–8. Virions of passionfruit woodiness virus showing helical substructure at one end (arrows) following partial purification and storage at 4°. The helical portion may reflect either partial disintegration of the virion or aggregation of viral protein. (Figs. 6 and 8, ×160,000; Fig. 7, ×100,000.)

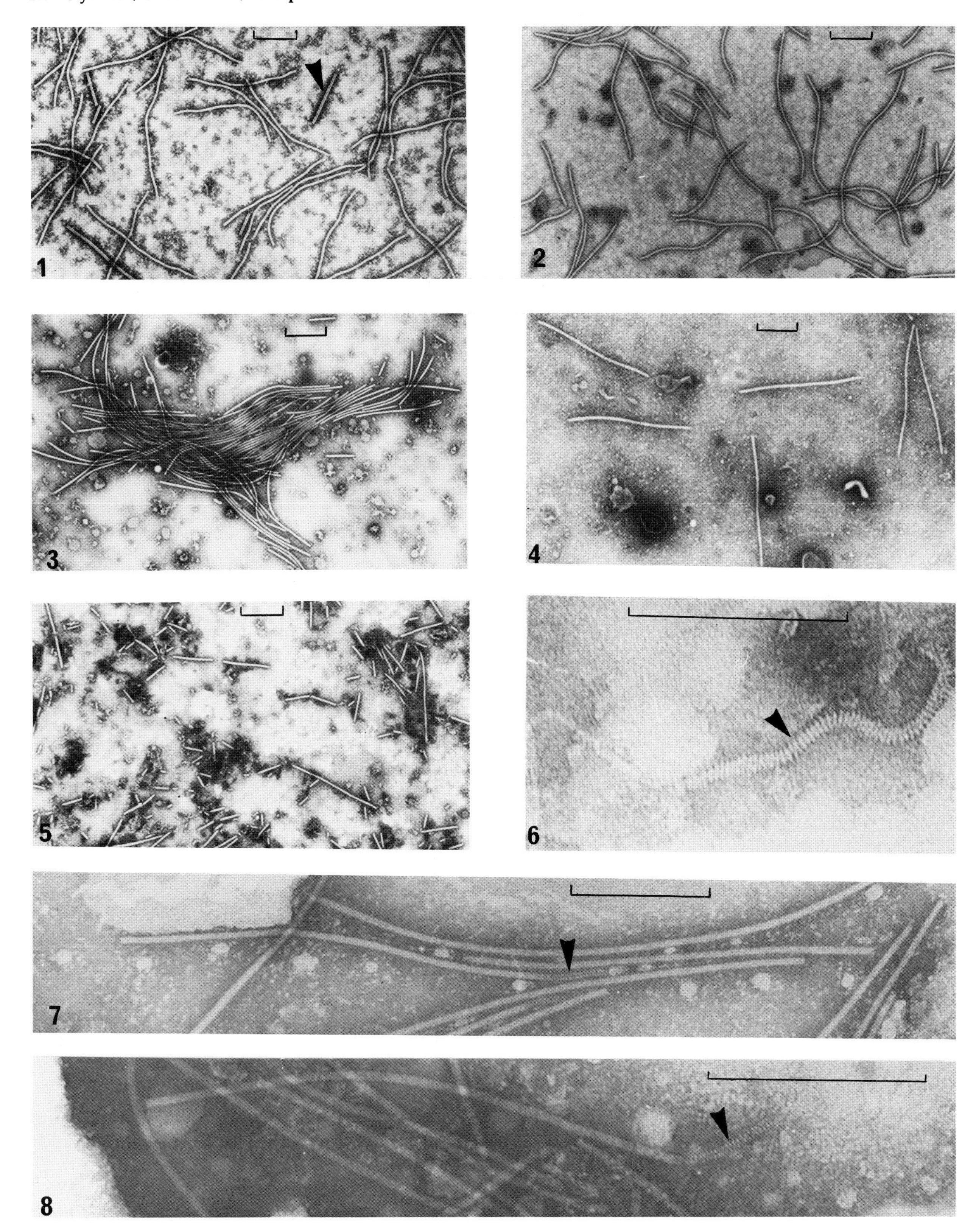
1
2
3
4
5
6
7
8

Figs. 9–12; 14–17. Sections of French bean leaves infected with passionfruit woodiness virus (PWV). VA, virus aggregate; Tp, tonoplast; LA, laminated aggregate; PW, pinwheel inclusion; ER, endoplasmic reticulum; PM, plasma membrane; GA, Golgi apparatus. Bar = 250 nm.

Fig. 9. Aggregate of PWV attached at one end to the tonoplast. (×40,000.)

Fig. 10. Detail of the association of a PWV aggregate and the tonoplast. (×110,000.)

Fig. 11. Area of cytoplasm infected by PWV showing association of tonoplast, virus aggregate, laminated aggregate and pinwheel inclusions. (×52,000.)

Fig. 12. Aggregate of PWV associated at one end with the endoplasmic reticulum. (×98,000.)

Fig. 13. Negatively stained cylindrical inclusion in sap from a French bean leaf infected with PWV. Bar = 250 nm. (×125,000.)

Fig. 14. Area of cytoplasm showing an association of PWV particles, pinwheels, and a Golgi apparatus. (×41,000.)

Fig. 15. Cylindrical inclusions showing the frequently observed intimate association with laminated aggregates. The arrows indicate areas where the layers of the aggregate separate to become arms of pinwheels. (×38,000.)

Fig. 16. Longitudinal section of a pinwheel inclusion showing association with the plasma membrane. (×68,000.)

Fig. 17. Longitudinal section of a pinwheel inclusion showing association with the endoplasmic reticulum. (×41,000.)

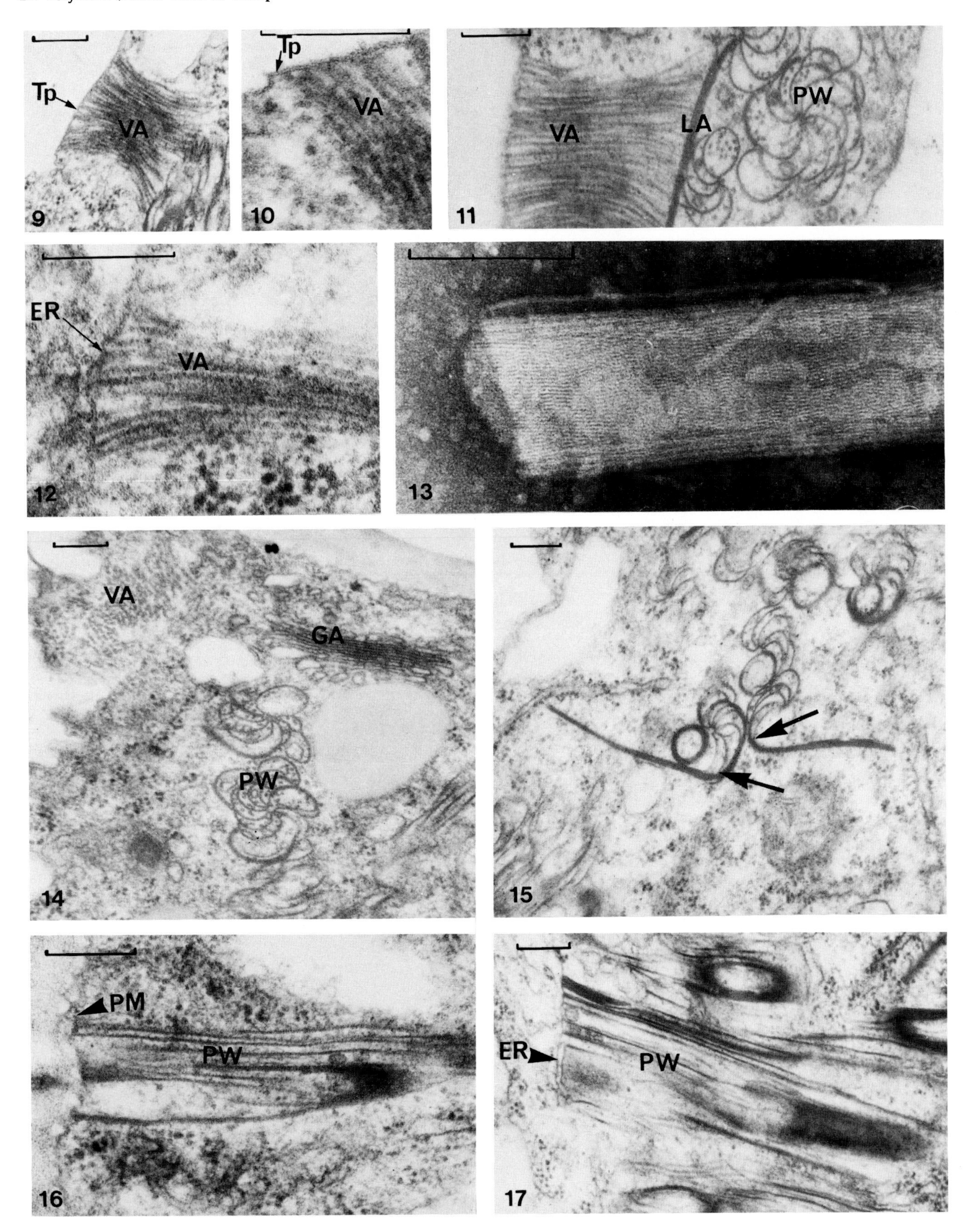
Tp
VA
9
Tp
VA
10
VA
LA
PW
11
ER
VA
12
13
VA
GA
PW
14
15
PM
PW
16
ER
PW
17

Figs. 18–21. Sections of French bean leaves infected with passionfruit woodiness virus (PWV). Bar = 250 nm. LA, laminated aggregate inclusion; T, tubular inclusion; PW, pinwheel inclusion.

Fig. 18. A group of laminated aggregate and tubular inclusions showing association with a partially formed pinwheel inclusion (arrow). (× 101,000.)

Fig. 19. Area of infected cytoplasm showing two tubular inclusions sectioned tangentially. One of these tubular inclusions can be seen to be associated at the end with endoplasmic reticulum, as is the pinwheel inclusion adjacent to it (arrows). (× 47,000.)

Fig. 20. Area of cytoplasm infected with PWV showing three groups of pinwheels, each group being oriented at right angles to the adjacent group. (× 38,000.)

Fig. 21. Laminated aggregate inclusion of PWV showing the layers making up the aggregate. (× 70,000.)

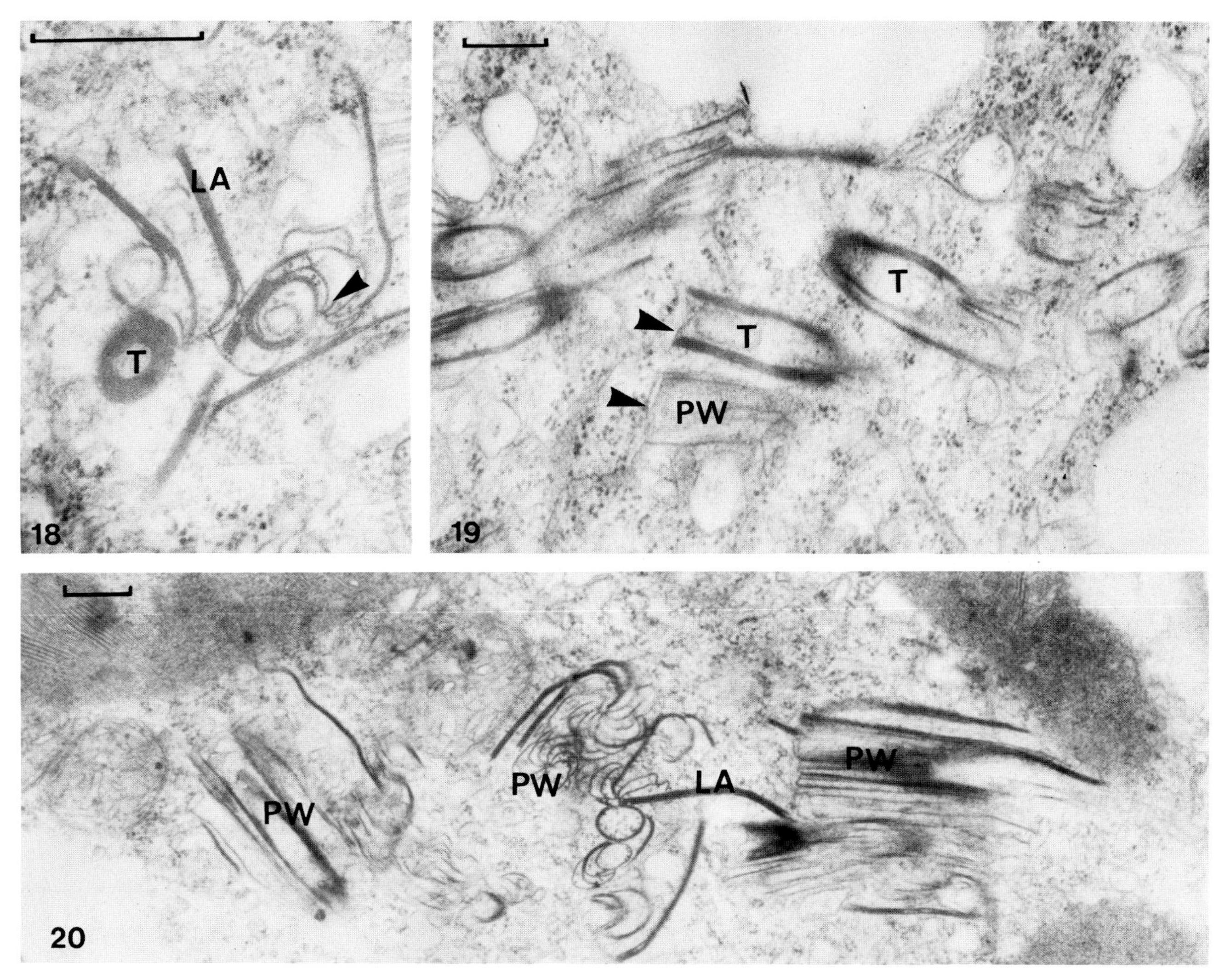
LA
T
18
T
T
PW
19
PW
PW
LA
PW
20

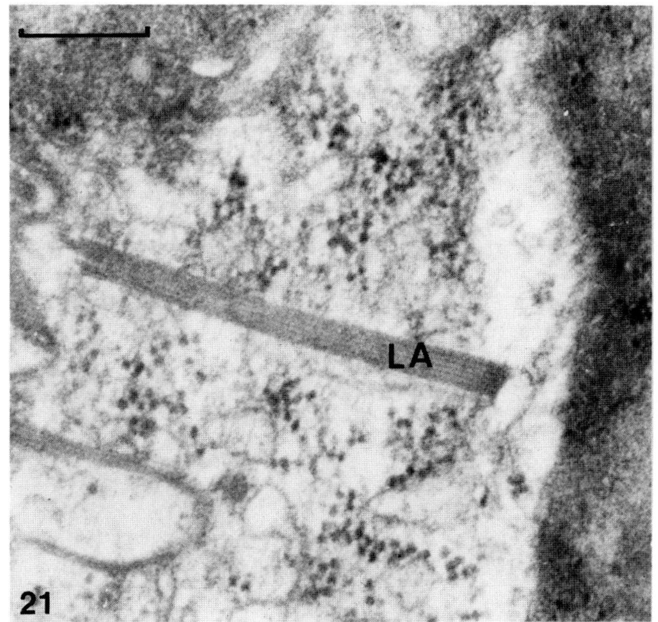
LA
21

Chapter 19

Watermelon Mosaic Virus Group*

M. H. V. VAN REGENMORTEL

I.	Introduction	323
II.	Biological Properties	324
III.	Morphology	324
	References	324

I. INTRODUCTION

Watermelon mosaic virus (WMV) was first described by Anderson (1954) and was found by van Regenmortel (1960) to consist of flexuous rods 700–800 nm long. It was subsequently identified as a member of the largest group of plant viruses, the potato virus Y or potyvirus group (van Regenmortel *et al.*, 1962). This group of viruses is characterized by the following properties: flexible particles with a normal length, in the range of 700–950 nm, presence of cytoplasmic cylindrical inclusions in infected tissue, widespread serological relationships between members of the group, mechanical and aphid transmission, and thermal inactivation point in the range of 50°–65° C (Brandes and Bercks, 1965; Harrison *et al.*, 1971; Edwardson, 1974a). When antisera to the dissociated subunits of the virions were used for comparative studies, a closer serological relationship between different members of the group was found than when antisera to the complete virus particles were used (Shepard *et al.*, 1974). When examined by polyacrylamide gel electrophoresis, the coat proteins of potyviruses appear to be very similar; however, in view of their electrophoretic heterogeneity it is still uncertain whether their molecular weight is 28,000 or 34,000 daltons (Hiebert and McDonald, 1973; Huttinga and Mosch, 1974). In view of the similar physicochemical properties of different potyviruses, the following cryptogram of potato virus Y may tentatively be assigned to the whole group = R/I : 3/6 : E/E : S/Ap (Makkouk and Gumpf, 1974).

* This group is usually incorporated into to potyvirus group (Chapter 18).

II. BIOLOGICAL PROPERTIES

WMV causes chlorosis, mottle, leaf distortion, and stunt in various cucurbits and is symptomless in most other hosts outside the Cucurbitaceae family. Some strains, often referred to as WMV-I, infect only cucurbits while others (WMV-2) have a wider host range. The WMV-2 isolates differ in the range of noncucurbitaceous plants they infect (Grogan *et al.*, 1959; van Regenmortel *et al.*, 1962; Molnar and Schmelzer, 1964). Edwardson (1974b) has listed hosts of WMV in 21 additional botanical families, a large number of hosts being of the Leguminosae, Chenopodiaceae, and Malvaceae families. WMV-1 and WMV-2 isolates were originally thought to belong to separate viruses (Webb and Scott, 1965), but there is little justification to consider them as two different viruses on the basis of host range variations since they cannot be differentiated by cross-absorption serological tests (Milne and Grogan, 1969). WMV is serologically related to potato virus Y and bean yellow mosaic virus (van Regenmortel *et al.*, 1962) to papaya ringspot virus (Milne and Grogan, 1969) and blackeye cowpea mosaic virus (Uyemoto *et al.*, 1973).

III. MORPHOLOGY

WMV particles are flexuous filaments (Fig. 1) with a normal length reported to range from 725 nm by van Regenmortel *et al.* (1962) to 700–765 nm by Edwardson (1974a). Large masses of virus particles aggregated around membranes have been observed in extracts obtained by the dip method (Purcifull *et al.*, 1968). Virus particles aggregate readily during purification (van Regenmortel, 1971). A hundredfold increase in infectivity endpoint has been obtained by dispersing aggregated particles by ultrasonic treatment (van Regenmortel, 1964).

Cytoplasmic inclusions typical of those formed by potyviruses are found in thin sections of WMV-infected tissue. These are cylindrical inclusions (Fig. 2) composed of curved plates radiating from a central axis, which appear as pinwheels in cross section and as bundles in longitudinal section (Edwardson, 1966; Purcifull and Edwardson, 1967; Edwardson *et al.*, 1968), and tubular inclusions which are scroll-like structures composed of a curved thin plate. Direct reconstructions of such inclusions from their appearance in serial sections have been described by Edwardson (1966) and Purcifull and Edwardson (1967). The cytoplasmic inclusions are composed of protein and are striated with a periodicity of 5 nm (Fig. 2A). This protein has a molecular weight of approximately 68,000 daltons and is serologically unrelated to both normal host protein and virus particles. According to Edwardson (1974a) WMV-1 and WMV-2 isolates can be distinguished cytologically by the presence, only in WMV-2-infected tissue, of nuclear inclusions, sandwich inclusions (Fig. 2D), and thin flat structures known as laminated aggregates.

ACKNOWLEDGMENTS

Electron micrographs were kindly supplied by Dr. D. Lesemann (Fig. 1) and Dr. J. R. Edwardson (Fig. 2).

REFERENCES

Anderson, C. W. (1954). *Phytopathology* **44,** 198–202.
Brandes, J., and Bercks, R. (1965). *Adv. Virus Res.* **11,** 1–24.

Edwardson, J. R. (1966). *Am. J. Bot.* **53,** 359–364.
Edwardson, J. R. (1974a). *Fla., Agric. Exp. Stn., Monogr.* **4,** 1–398.
Edwardson, J. R. (1974b). *Fla., Agric. Exp. Stn., Monogr.* **5,** 1–225.
Edwardson, J. R., Purcifull, D. E., and Christie, R. G. (1968). *Virology* **34,** 250–263.
Grogan, R. G., Hall, D. H., and Kimble, K. A. (1959). *Phytopathology* **49,** 366–376.
Harrison, B. D., Finch, J. T., Hollings, M., Shepherd, R. J., Valenta, V., and Wetter, C. (1971). *Virology* **45,** 356–363.
Hiebert, E., and McDonald, J. G. (1973). *Virology* **56,** 349–361.
Huttinga, H., and Mosch, W. H. M. (1974). *Neth. J. Plant Pathol.* **80,** 19–27.
Makkouk, K. M., and Gumpf, D. J. (1974). *Phytopathology* **64,** 1115–1118.
Milne, K. S., and Grogan, R. G. (1969). *Phytopathology* **59,** 809–818.
Molnar, A., and Schmelzer, K. (1964). *Phytopathol. Z.* **51,** 361–84.
Purcifull, D. E., and Edwardson, J. R. (1967). *Virology* **32,** 393–401.
Purcifull, D. E., Edwardson, J. R., and Christie, S. R. (1968). *Virology* **35,** 478–482.
Shepard, J. F., Secor, G., and Purcifull, D. E. (1974). *Virology* **58,** 464–475.
Uyemoto, J. K., Provvidenti, R., and Purcifull, D. E. (1973). *Phytopathology* **63,** 209.
van Regenmortel, M. H. V. (1960). *Virology* **12,** 127–130.
van Regenmortel, M. H. V. (1964). *Virology* **23,** 495–502.
van Regenmortel, M. H. V. (1971). *CMI/AAB Descriptions Plant Viruses* No. 63.
van Regenmortel, M. H. V., Brandes, J., and Bercks, R. (1962). *Phytopathol. Z.* **45,** 205–216.
Webb, R. E., and Scott, H. A. (1965). *Phytopathology* **55,** 895–900.

Fig. 1. Particles of watermelon mosaic virus prepared by the dip method; Pd-shadowed. (A) Single particles; (B) particles mostly aggregated. Bar = 1000 nm.

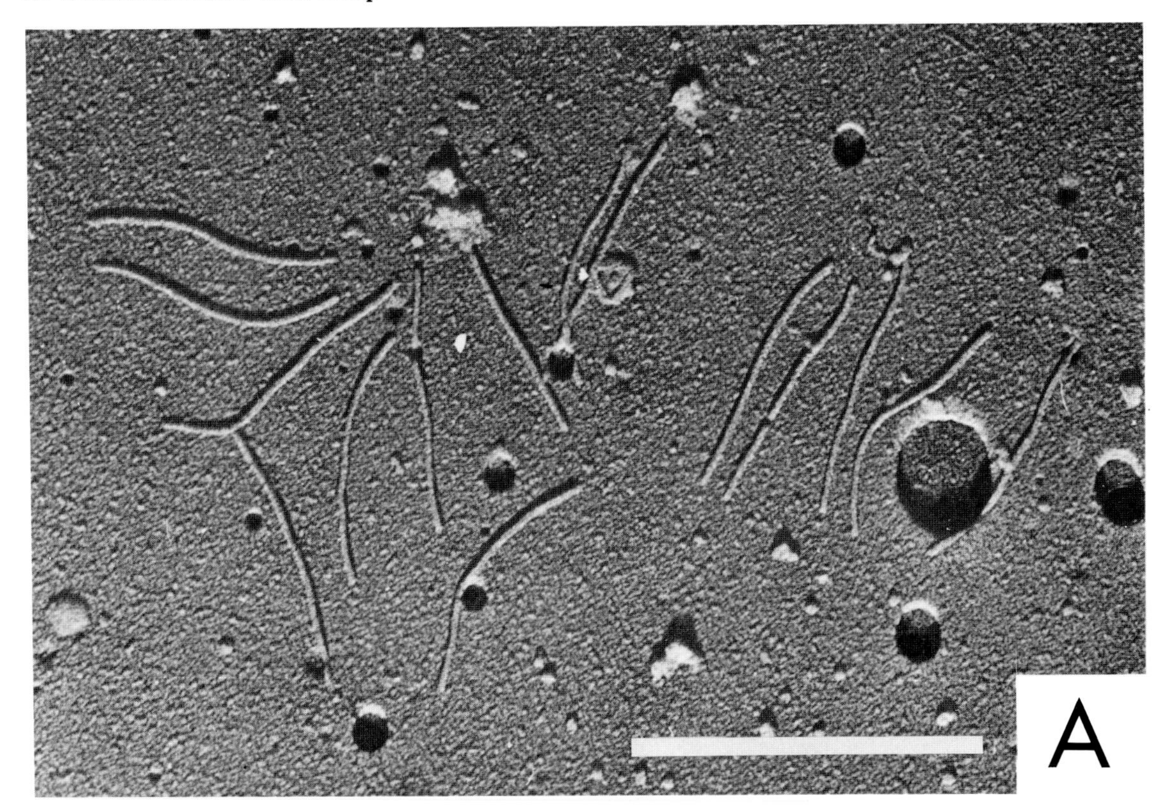
A

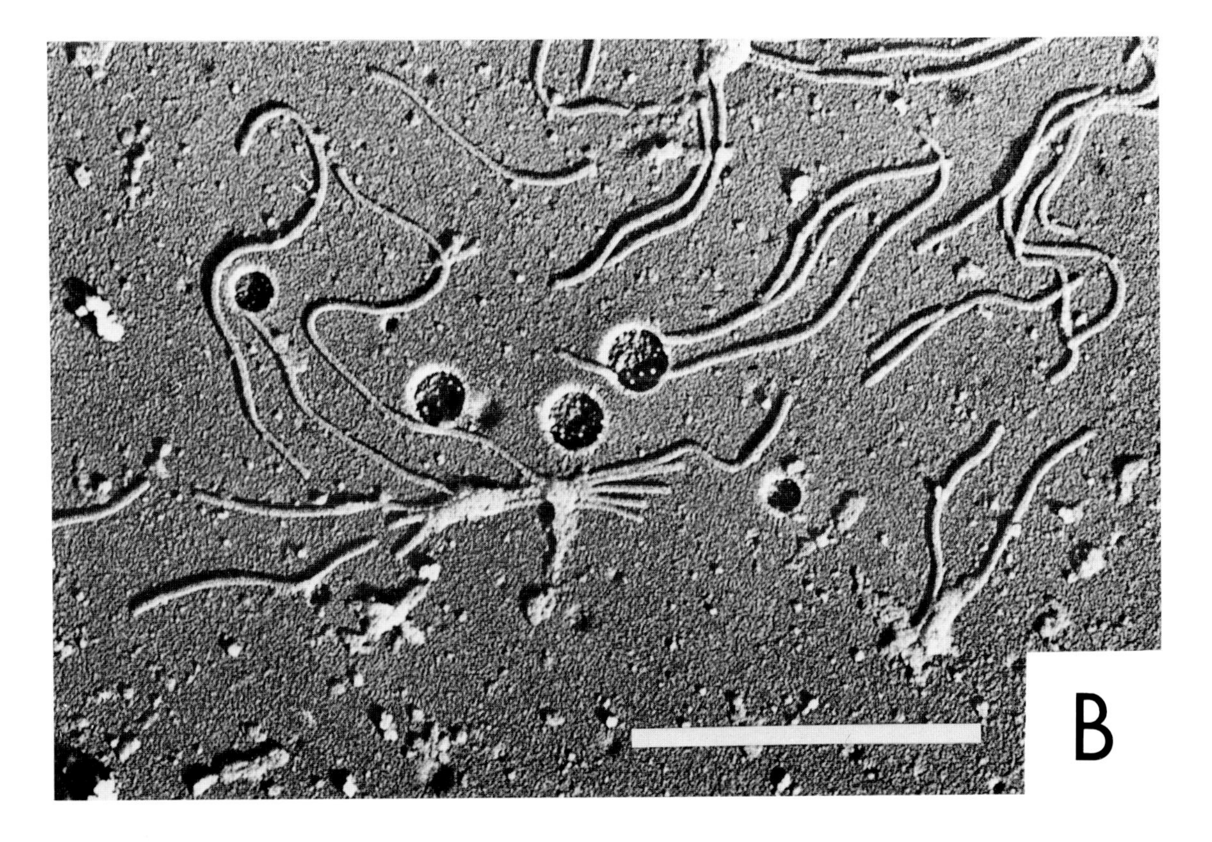
B

Fig. 2. (A) Diagrammatic representation of cylindrical inclusions induced by watermelon mosaic virus, reconstructed from their appearance in serial sections. In cross section these inclusions appear as pinwheels and circles and in longitudinal section as bundles. The inset represents the striation pattern with an average spacing of 5 nm. The rod-shaped particles associated with the inclusions are probably virus particles. (Purcifull and Edwardson, 1967; Edwardson *et al.*, 1968.) (B) Cylindrical inclusions of WMV-infected lupin cell showing cross sections and flat plates. Bar = 200 nm. (C) Cross section of cylindrical inclusions in cytoplasm of WMV-infected pumpkin cell. Bar = 200 nm. (D) Sandwich inclusion found in cytoplasm of WMV-infected pumpkin cell. According to Edwardson (1974a) this type of structure is only found in tissue infected with WMV-2 type isolates. Bar = 200 nm.

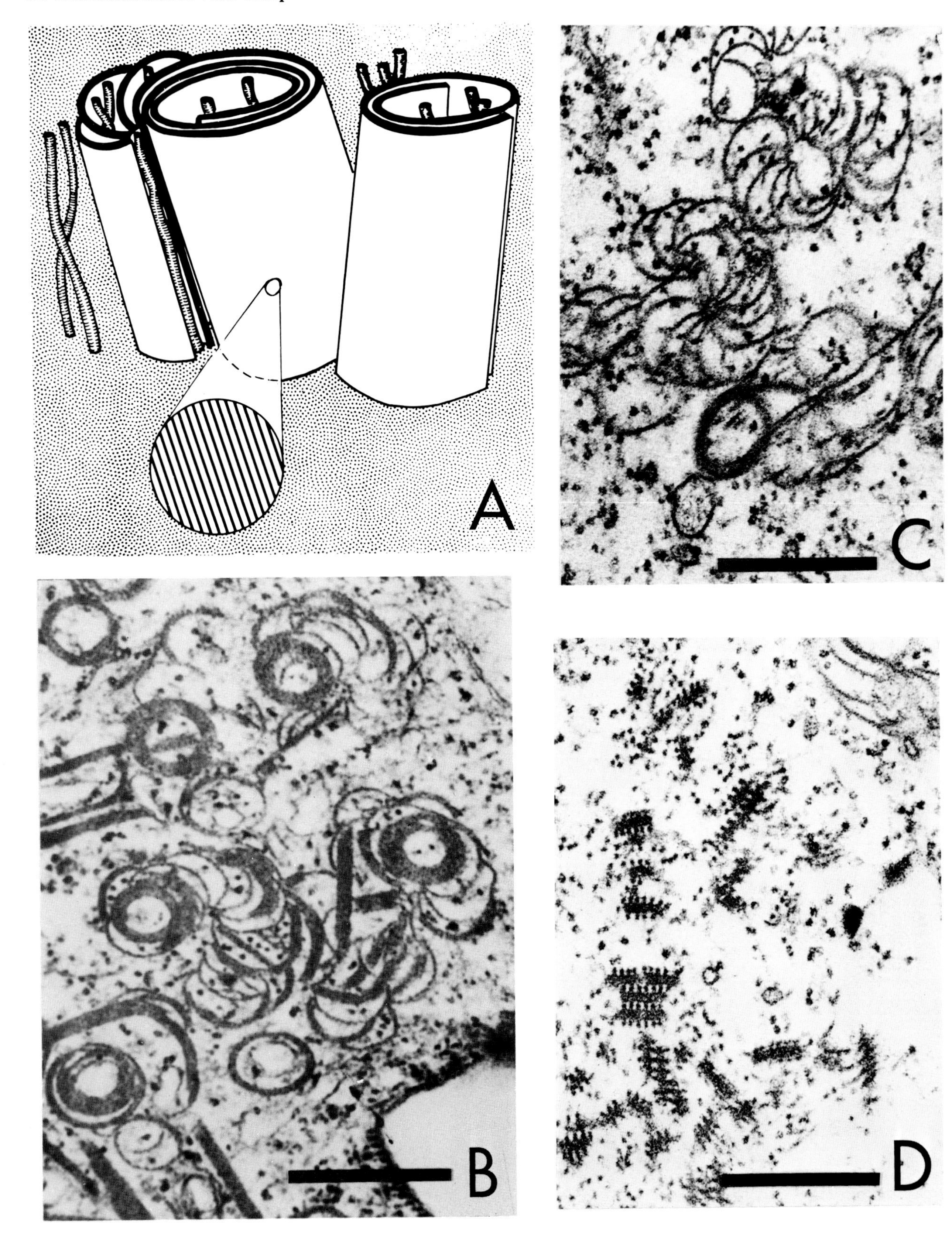
A
B
C
D

Chapter 20

Potexvirus (Potato Virus X) Group

D.-E. LESEMANN AND RENATE KOENIG

I. Structure of Particles 331
II. Physical and Chemical Properties of Particles 335
III. Structural Alterations Induced by
Potexviruses in Host Cells 336
References 337

The potato virus X group (potexvirus group, cryptogram [R/I:2.1/6:E/E:S/(Fu, Ap)]) has been defined by Brandes (1964) as a group of elongated flexuous viruses which are about 480–580 nm long and about 13 nm wide. The particles have no envelopes. Most of the viruses in this group are rather distantly interrelated serologically, the available information on the results of serological cross tests is summarized in Fig. 1.

Potexviruses usually occur in high concentrations in systemically infected hosts. They are causing diseases in a wide range of monocotyledoneous and dicotyledoneous plants, but the natural host range of individual viruses is rather limited. The viruses are easily transmitted mechanically, the thermal inactivation points are between 60° and 80° C. The typical members of the group are normally not transmitted by vectors. A few possible exceptions have been reported, however, for PVX and WClMV (for abbreviation of virus names see Table I). Potato virus X may be transmitted by fungi (Nienhaus and Stille, 1965) and by grasshoppers—probably mechanically on the insect's mouthparts (e.g., Schmutterer, 1961). A low percentage of aphid transmission has been reported in one instance for WClMV (Goth, 1962).

Several viruses with flexuous particles that share certain properties with the typical members of the group are known, but they may differ from the typical members in other properties. Some of these viruses have not been studied in detail. Typical and possible members of the potexvirus group are listed in Table I.

I. STRUCTURE OF PARTICLES

Exact determinations of particle lengths may be influenced by the virus source, preparation procedure, calibration of instrument magnification, mea-

TABLE I

Typical and Possible Members of the Potexvirus Group[a]

NL[b] (nm)	Virus name	Abbreviation	Typical (T) or possible (P) member	Properties of possible members as compared with properties of typical members: Similar properties	Different properties
395–445	*Dioscorea* latent virus[1]	DLV	P	No aphid transmission, high concentration in host plants	TIP[c] > 55° < 60°
475	*Cymbidium* mosaic virus	CybMV	T		
478	Rhubarb virus I and II[2]	RhuV I and II	P	TIP > 60° < 65° and > 70° < 75°	DEP[d] 10^{-3} -10^{-4}
479	Parsley virus 5[3]	PalV 5	P	Serological relationship with PVX	Transmission by aphids in persistent manner in the presence of helper virus
480	White clover mosaic virus	WClMV	T		
490	*Hydrangea* ringspot virus	HyRSV	T		
495	Cassava common mosaic virus	CCMV	P	TIP 65°–70° DEP 10^{-5}–10^{-6}	No serological relationship with PVX found
500	Filamentous particles from *Boletus edulis*[4]		P		Central canal clearly visible
similar to PVX	Parsnip virus 3[5]	PanV 3	P	Serological relationship to PVX	
515	Potato virus X	PVX	T		
520	Cactus virus X	CaVX	T		
525	*Malva* veinal necrosis virus[6]	MVNV	P	No aphid transmission, TIP 70°	
530	Papaya mosaic virus	PaMV	T		
530–545	*Nerine* virus X[7]	NeVX	P	Distant serological relationship with NaMV and ClYMV	
540	Clover yellow mosaic virus	ClYMV	T		
550	*Narcissus* mosaic virus	NaMV	T		
550–580	Negro coffee mosaic virus[8]	NCoMV	P		Transmitted by aphids, particle width 21 nm

580	Potato aucuba mosaic virus	PoAMV	P	Base composition of RNA similar to PVX, clear crossbanding in uranyl salts	No serological relationship with other potexviruses found, transmitted by aphids in presence of helper virus
580	*Zygocactus* virus[9]	ZyCV	P	TIP 72°–74°	Low concentration in host plants, no serological relationship found with PVX, CaVX and PoAMV
580	Artichoke curley dwarf virus[10]	ACDV	P		DEP 10^{-3}–10^{-4}, TIP 55°–60°
580	*Centrosema* mosaic virus[11]	CenMV	P	Particles show clear crossbanding with uranyl salts	Transmitted by aphids, DEP 10^{-3}, TIP 55°–58°
596	*Hippeastrum* latent virus[12]	HiLV	P	TIP 70°–80°	DEP 10^{-3}–10^{-4}

[a] Superscript numbers indicate references for those viruses for which C.M.I./A.A.B. descriptions have not yet been published: (1) Waterworth *et al.* (1974); (2) MacLachlan (1960); (3) Frowd and Tomlinson (1974); (4) Huttinga *et al.* (1975); (5) Garrett and Tomlinson (1966); (6) Costa and Kitajima (1970); (7) Maat (1976); (8) Verma and Niazi (1974); (9) Casper and Brandes (1969); (10) Morton (1961); (11) Varma *et al.* (1968); (12) Brölman-Hupkes (1975).

[b] NL, normal length.

[c] TIP, thermal inactivation point.

[d] DEP infectivity dilution endpoint.

surement procedure, analysis of the results, and individual errors which are introduced by the person doing the measurements. These difficulties have been summarized by Brandes (1964) and are probably the reason that for some potexviruses different normal length values have been reported from different laboratories. The values in Table I are approximate ones.

The reported data on particle diameters are also rather variable. This is not unexpected in view of different preparation methods, e.g., metal-shadowing or negative-staining, and in view of possible flattening or shrinking effects. At present the most reliable values seem to be derived from negatively stained preparations. Varma *et al.* (1968) reported diameters between 11.5 and 13.3 nm for four different potexviruses. The interparticle separation as determined by X-ray diffraction for PVX and NaMV is smaller than the particle diameter measured in electron micrographs due to interpenetration when particles pack together (Tollin *et al.*, 1967).

After negative staining with uranyl salts potexviruses often show a clear crossbanding which is illustrated in Fig. 2 for particles of PVX. In addition, some longitudinal lines are seen. Similar crossbanding has been reported for CybMV (Francki, 1966), NaMV (Tollin *et al.*, 1967), CenMV, HyRSV, PoAMV, PVX, and WClMV (Varma *et al.*, 1968). The appearance of negatively stained particles varies considerably due to variations in the thickness and penetration of the stain, to superposition of the images of both sides of the particles, and also to the flexibility of the particles (Varma *et al.*, 1968). Thus, on the same grid, particles of CybMV showed either crossbanding (Fig. 3), or oblique lines (Fig. 4). The fine structure of particles after staining with uranyl salts can be an aid in the differentiation of potexviruses and carlaviruses. The latter exhibit only indistinct cross-banding but clear lines parallel to the long axis (Varma *et al.*, 1968).

A central canal is usually not seen in particles of potexviruses, however, in some instances it has been demonstrated (Francki, 1966; Varma *et al.*, 1968). With PVX its diameter was calculated to be about 3.1 nm (Wilson and Tollin, 1969). A canal is clearly visible in particles of the filamentous *Boletus* virus (Table I) after staining with phosphotungstate. This may indicate a more fundamental difference between the fine structure of this virus and that of the viruses of similar length in higher plants.

Optical and X-ray diffraction measurements indicate a helical arrangement of protein subunits in several potexviruses (Tollin *et al.*, 1967, 1975; Varma *et al.*, 1968; Wilson and Tollin, 1969; Kaftanova *et al.*, 1975; Goodman *et al.*, 1975). The reported values for the pitch vary between 3.3 nm and 3.7 nm (Tollin *et al.*, 1967, 1975; Varma *et al.*, 1968; Goodman *et al.*, 1975). In NaMV and PVX the layer-line spacings as determined by X-ray diffraction depend on the water content. This indicates that the interactions between successive turns of the helix are less strong than, for example, in tobacco mosaic virus (TMV). This may be the reason that the particles of potexviruses are more flexible than those of TMV (Tollin *et al.*, 1967). Goodman *et al.* (1975) suggested that the assembly and structural integrity of PVX may depend upon ionic bonds between protein subunits and RNA, whereas in TMV they depend upon protein to protein interactions. In NaMV the repeat distance of the helix is five times the pitch and there are probably 6.8 subunits per turn of the helix (Tollin *et al.*, 1968, 1975). For PVX, Wilson and Tollin (1969) suggested an integral number of subunits (probably 10) per turn of the helix, but Goodman *et al.* (1975) concluded that the helix in PVX is repeated every eight turns. Eleven subunits per turn were determined for WClMV (Varma *et al.*, 1968).

In preparations of PVX and NaMV proteins a reassembly to rodlike particles was recently achieved. Protein reassembly products of PVX revealed a stacked disk structure different from the structure of native particles. Nucleoprotein reassembly products, however, were similar or identical to native virus when analyzed by optical diffraction methods (Kaftanova *et al.*, 1975; Goodman *et al.*, 1975). In NaMV-protein reassembly products, the subunits are in a double-helical arrangement (Robinson *et al.*, 1975).

II. PHYSICAL AND CHEMICAL PROPERTIES OF PARTICLES

In the analytical ultracentrifuge, potexviruses sediment as single components, provided that they are present as monomers. The reported sedimentation coefficients, at infinite dilution, range from 114 S to 130 S. Higher values have been determined for tobamoviruses (190 S) and for carlaviruses and potyviruses (140 S to 170 S).

About 6% single stranded RNA has been found in all potexviruses that have been checked for this property, i.e., CybMV, WClMV, CCMV, PVX, PaMV, and PoAMV. The RNA of PVX has a base composition G22; A32; C24; U22 (molar percentages). Rather similar values have been reported for the RNA's of CybMV, WClMV, PaMV, and PoAMV. Adenine at about 32% is always the most common base.

Determinations of the molecular weight of the capsid proteins of potexviruses have been done in recent years with SDS-polyacrylamide electrophoresis, gel chromatography in the presence of SDS or guanidine hydrochloride, and with amino acid analysis combined with peptide mapping. For the capsid protein of PVX the reported values range from 22,300 to 31,500 daltons. Since the undegraded protein of PVX behaves anomalously in SDS-polyacrylamide electrophoresis (Koenig, 1972), molecular weight estimates based on this method tend to be too high. Further complications arise from the fact that the capsid protein of PVX may become partially degraded in the assembled virus by plant proteases during purification or storage (Koenig *et al.*, 1970; Shepard and Secor, 1972; Tung and Knight, 1972). This degradation may proceed from the N terminal as well as from the C terminal of the peptide chain (J. H. Tremaine and R. Koenig, unpublished, 1976). The correct value for the molecular weight of the undegraded PVX protein is probably around 26,000 daltons (Tung and Knight, 1972; J. H. Tremaine and R. Koenig, unpublished, 1976).

For several other potexviruses, e.g., WClMV, ClYMV, PaMV, and CaVX, protein molecular weights around 20,000 daltons have been reported (e.g., Lesnaw and Reichmann, 1970; Koenig *et al.*, 1970; Koenig, 1972; Hill and Shepherd, 1972; Tung and Knight, 1972). The proteins of some of these viruses may form double or multiple bands in SDS-polyacrylamide electrophoresis, which suggests that they, too, may be susceptible to partial degradation in the assembled viruses. Possibly, all these forms are already degradation products of larger proteins with sizes more similar to that of undegraded PVX protein. Alternatively, there may be a true difference in the size of the coat protein gene of PVX and other potexviruses. A somewhat exceptional position of PVX in the potexvirus group, which may be more heterogenous than originally thought, is also suggested by its unique ability to induce the formation of specific laminated inclusion components (see Section III).

Amino acid compositions have been reported for the capsid proteins of PVX and WClMV. By cluster analysis, Tremaine and Argyle (1970) found that these two proteins are more similar to each other than to the capsid proteins of any other plant viruses.

III. STRUCTURAL ALTERATIONS INDUCED BY POTEXVIRUSES IN HOST CELLS

Information on the relationships of potexviruses to host cells—although sometimes sparse—exists for about half of the viruses listed in Table I. These data are in part derived from light microscopical studies but mostly from electron microscopy of ultrathin sections.

Like most virus-infected cells, those infected with potexviruses often also exhibit nonspecific signs of cell damage, e.g., accumulation of starch or lipid in the plastids, or disorganization of plastidal and cytoplasmic membranes. In the surroundings of local lesions induced by PVX in *Gomphrena* a heavy deposition of callose along the cell walls was detected (Allison and Shalla, 1974).

The distribution of infected cells within different host tissues has not reportedly been examined in most of the publications on potexviruses. Potato virus X (PVX) (Kikumoto and Matsui, 1961; Pennazio and Appiano, 1975; S. Doraiswamy and D. Lesemann, unpublished), CCMV (Kitajima and Costa, 1966), CybMV (Lawson and Hearon, 1974), and CaVX (S. Doraiswamy and D. Lesemann, unpublished) are known to induce inclusion bodies in various tissues, possibly only with the exception of tracheids and sieve tubes. With PVX, meristematic dome cells already contain virus aggregates (Pennazio and Appiano, 1975), whereas with CybMV meristematic dome cells of *Cattleya* contained no virus (Lawson and Hearon, 1974). Often, virus-induced inclusions are most clearly seen within epidermal cells and have therefore frequently been studied at this site by means of light microscopy.

The most prominent structural alterations seen in infected cells are inclusion bodies that are in fact aggregates of virus particles. These aggregates are very variable in size, sometimes they are almost occluding the cell lumen. The shape of the aggregates is mostly irregular as reported with DLV (Lawson *et al.*, 1973), CybMV (Lawson and Hearon, 1974; Doraiswamy and Lesemann, 1974; Hanchey *et al.*, 1975), WClMV (Iizuka and Iida, 1965; Tapio, 1970; Lesemann, 1975, unpublished), HyRSV (Zeyen, 1973), CCMV (Kitajima and Costa, 1966), PVX (Kikumoto and Matsui, 1961; Doraiswamy and Lesemann, 1974; Pennazio and Appiano, 1975), PaMV (Zettler *et al.*, 1968), ClYMV (Purcifull *et al.*, 1966; Moline and Ford, 1974), and PoAMV (Turner, cited in Kassanis and Govier, 1972). In some cases, however, inclusions have a characteristic shape. Thus, the aggregates of CaVX and NaMV are often spindle shaped (Fig. 5), circular, or sledge-like (Fig. 6) (Amelunxen, 1958, and references therein; Štefanac and Lubešić, 1974). The plant species may (Amelunxen, 1958), or may not (e.g., Štefanac and Lubešić, 1974; Doraiswamy and Lesemann, 1974) have an influence on the shape of the inclusions. The virus aggregates are not membrane bounded and are usually located within the cytoplasm, often in an evagination of the cytoplasm into the vacuole (Amelunxen and Thaler, 1967). Aggregates of CaVX and NaMV, however, occasionally occur within the nuclei (Amelunxen, 1958, and references therein; Turner, 1971; Štefanac and Lubešić, 1974).

In spindle-shaped inclusions, virus particles are aggregated longitudinally and laterally and are oriented with their longitudinal axis roughly parallel with that of the spindle (Fig. 7). The individual particles appear mostly not straight but slightly flexuous (Fig. 7). With a few exceptions the filamentous particles are also arranged roughly parallel in aggregates with shapes other than spindles. Several patterns of arrangement can then be found. For example, the orientation of the parallel particles can vary gradually within one plane of an aggregate, so that in cross sections "spiral"- or "whirl"-like patterns are seen (Purcifull *et al.*, 1966) (Fig. 8). Such patterns have been found with WClMV, ClYMV, PaMV, CaVX, and CybMV (for references, see preceding paragraph).

Virus particles can also be arranged side by side with all ends more or less in one plane, thus building up plate-like aggregates which appear as tiers or bands in section (Fig. 9). The plates can have the height of one (Figs. 9 and 10), or of two particle lengths. They can be constructed very accurately (WClMV, PaMV) (Fig. 10) but also quite disorderly (CybMV, PVX, ClYMV) (Fig. 9). The packing density of particles in the aggregates is mostly similar to Fig. 7, but is sometimes so dense that paracrystalline arrays are formed (Fig. 11). On the other side, with CybMV and CaVX (D. Lesemann, unpublished), very loose aggregates have been found in which the particles are not arranged in parallel but at random (Fig. 12) and appear strongly curved.

Light microscopic studies on the development of the inclusions of CaVX have been summarized by Thaler (1966). So called X-bodies preceding the virus aggregates have not yet been analyzed by ultrathin sections. With the electron microscope, the formation of aggregates of PVX was followed up by Honda *et al.* (1975). In protoplasts the first progeny virus was found scattered in the cytoplasm, larger aggregates formed after higher concentrations of particles had been reached in later stages of infection.

Three potexviruses, i.e., PVX, HyRSV, and ClYMV induce the formation of further distinct inclusions in addition to virus aggregates. PVX-infected cells in all host plants studied contain 5- to 10-μm thick inclusions containing "laminated inclusion components" (LIC), which have been most intensively analyzed by Shalla and Shepard (1972) (Fig. 13). These thin sheets of proteinaceous material may (Fig. 13), or may not (Fig. 14), bear bead-like structures on both surfaces. The beads are different from ribosomes in size and are, as the sheets, not antigenically related to PVX and its subunits. The sheets often form thick files or are concentrically arranged to form scrolls (Figs. 13 and 14). LIC seem not to be correlated with the virus synthesis since they appear after the virus in PVX-infected protoplasts (Honda *et al.*, 1975). Also, in *Gomphrena*, virus particles are found far more distant from the necrotic region of local lesions than LIC (Allison and Shalla, 1974). In addition to LIC, Honda *et al.* (1975) detected in early infection stages suspicious proliferated endoplasmic reticulum appearing at the same time as the nascent LIC. HyRSV has been reported to induce "cylindrical" inclusions without a central core (Zeyen, 1973). It is not clear whether these inclusions are comparable to LIC of PVX. ClYMV induces amorphous inclusions in *Vicia faba* (Fig. 15) which contain viral antigen (Schlegel and Delisle, 1971). These inclusions are located in the cytoplasm and in the vacuoles; their function is unknown.

ACKNOWLEDGMENT

Part of the work reported here was supported by the Deutsche Forschungsgemeinschaft.

REFERENCES*

Allison, A. V., and Shalla, T. A. (1974). *Phytopathology* **64,** 784–793.
Amelunxen, F. (1958). *Protoplasma* **49,** 140–178.
Amelunxen, F., and Thaler, I. (1967). *Z. Pflanzenphysiol.* **57,** 269–279.
Bercks, R., and Brandes, J. (1961). *Phytopathol. Z.* **42,** 45–56.
Bercks, R., and Brandes, J. (1963). *Phytopathol. Z.* **47,** 381–390.

* Data for which no references are given have been reviewed in the C.M.I./A.A.B. descriptions of the particular viruses, i.e., No. 4 (PVX, 1970), No. 27 (CybMV, 1970), No. 41 (WClMV, 1971), No. 45 (NaMV, 1971), No. 56 (PaMV, 1971), No. 58 (CaVX, 1971), No. 90 (CCMV, 1972), No. 98 (PoAMV, 1972), No. 111 (ClYMV, 1973), No. 114 (HyRSV, 1973).

Brandes, J. (1964). *Mitt. Biol. Bundesanst. Land- Forstwirtsch., Berlin-Dahlem* **110,** 1–130.
Brandes, J., and Bercks, R. (1962-1963). *Phytopathol. Z.* **46,** 291–300.
Brölman-Hupkes, J. E. (1975). *Neth. J. Plant Pathol.* **81,** 226–236.
Casper, R., and Brandes, J. (1969). *J. Gen. Virol.* **5,** 155–156.
de Bokx, J. A. (1965). *Plant Dis. Rep.* **49,** 742–746.
Doraiswamy, S., and Lesemann, D. (1974). *Phytopathol. Z.* **81,** 314–319.
Francki, R. I. B. (1966). *Aust. J. Biol. Sci.* **19,** 555–564.
Frowd, J. A., and Tomlinson, J. A. (1972). *Ann. Appl. Biol.* **72,** 177–188.
Frowd, J. A., and Tremaine, J. H. (1974). *Proc. Can. Phytopathol. Soc.* **41,** 25–26.
Garrett, R. G., and Tomlinson, J. A. (1966). *Rep. Nat. Veg. Res. Stn., 1965* p. 74.
Goodman, R. M., Horne, R. W., and Hobart, J. M. (1975). *Virology* **68,** 299–308.
Goth, R. W. (1962). *Phytopathology* **52,** 1228.
Hanchey, P., Livingstone, C. H., and Reeves, F. B. (1975). *Physiol. Plant Pathol.* **6,** 227–231.
Hill, J. H., and Shepherd, R. J. (1972). *Virology* **47,** 817–822.
Honda, Y., Kajita, S., Matsui, C., Otsuki, Y., and Takebe, I. (1975). *Phytopathol. Z.* **84,** 66–74.
Huttinga, H., Wichers, H. J., and Dieleman-van-Zaayen, A. (1975). *Neth. J. Plant Pathol.* **81,** 102–106.
Iizuka, N., and Iida, W. (1965). *Ann. Phytopathol. Soc. Jpn.* **30,** 46–53.
Kaftanova, A. S., Kiselev, N. A., Novikov, V. K., and Atabekov, J. G. (1975). *Virology* **65,** 283–287.
Kassanis, B., and Govier, D. A. (1972) *CMI/AAB Descriptions Plant Viruses* No. 98.
Kikumoto, T., and Matsui, C. (1961). *Virology* **13,** 294–299.
Kitajima, E. W., and Costa, A. S. (1966). *Bragantia* **25,** XXIII-XXVIII.
Koenig, R. (1972). *Virology* **50,** 263–266.
Koenig, R. (1975). *Phytopathol. Z.* **84,** 193–200.
Koenig, R., and Bercks, R. (1968). *Phytopathol. Z.* **61,** 382–397.
Koenig, R., Stegemann, H., Francksen, H., and Paul, H. L. (1970). *Biochim. Biophys. Acta* **207,** 184–189.
Lawson, R. H., and Hearon, S. S. (1974). *Int. Symp. Virus Dis. Ornam. Plants, 3rd, 1972* pp. 195–206.
Lawson, R. H., Hearon, S. S., Smith, F. F., and Kahn, R. P. (1973). *Phytopathology* **63,** 1435.
Lesnaw, J. A., and Reichmann, M. E. (1970). *Virology* **42,** 724–731.
Maat, D. Z. (1976). *Neth. J. Plant Pathol.* **82,** 95–102.
MacLachlan, D. S. (1960). *Can. J. Plant Sci.* **40,** 104–109.
Moline, H. E., and Ford, R. E. (1974). *Physiol. Plant Pathol.* **4,** 219–228.
Morton, D. J. (1961). *Phytopathology* **51,** 731–734.
Nienhaus, F., and Stille, B. (1965). *Phytopathol. Z.* **54,** 335–337.
Pennazio, S., and Appiano, A. (1975). *Phytopathol. Mediterr.* **14,** 12–15.
Purcifull, D. E., Edwardson, J. R., and Christie, R. G. (1966). *Virology* **29,** 276–284.
Robinson, D. J., Hutcheson, A., Tollin, P., and Wilson, H. R. (1975). *J. Gen. Virol.* **29,** 325–330.
Schlegel, D. E., and Deslisle, D. E. (1971). *Virology* **45,** 747–754.
Schmutterer, H. (1961). *Z. Angew. Entomol.* **47,** 277–301.
Shalla, T. A., and Shepard, J. F. (1972). *Virology* **49,** 654–667.
Shepard, J. F., and Secor, G. A. (1972). *Phytopathology* **62,** 1154–1160.
Štefanac, Z., and Lubešić, N. (1974). *Phytopathol. Z.* **80,** 148–152.
Tapio, E. (1970). *Ann. Agric. Fenn.* **9,** 1–97.
Thaler, J. (1966). *Protoplasmatologia* **IIB2bγ,** 1–85.
Tollin, P., Wilson, H. R., Young, D. W., Clatro, J., and Mowat, W. P. (1967). *J. Mol. Biol.* **26,** 352–355.
Tollin, P., Wilson, H. R., and Young, D. W. (1968). *J. Mol. Biol.* **34,** 189–192.
Tollin, P., Wilson, H. R., and Mowat, W. P. (1975). *J. Gen. Virol.* **29,** 331–333.
Tremaine, J. H., and Argyle, E. (1970). *Phytopathology* **60,** 654–659.
Tung, J. S., and Knight, C. A. (1972). *Virology* **49,** 214–223.
Turner, R. H. (1971). *J. Gen. Virol.* **13,** 177–179.
Varma, A., Gibbs, A. J., Woods, R. D., and Finch, J. T. (1968). *J. Gen. Virol.* **2,** 107–114.
Varma, A., Gibbs, A. J., and Woods, R. D. (1970). *J. Gen. Virol.* **8,** 21–32.
Verma, V. S., and Niazi, F. R. (1974). *Z. Pflanzenkr. Pflanzenschutz* **81,** 608–610.
Waterworth, H. E., Lawson, R. H., and Kahn, R. P. (1974). *J. Agric. Univ. P. R.* **58,** 351–357.
Wilson, H. R., and Tollin, P. (1969). *J. Gen. Virol.* **5,** 151–154.
Zettler, F. W., Edwardson, J. R., and Purcifull, D. E. (1968). *Phytopathology* **58,** 332–335.
Zeyen, R. J. (1973). *Proc. Int. Congr. Plant Pathol., 2nd, 1973* Abstract No. 0275.

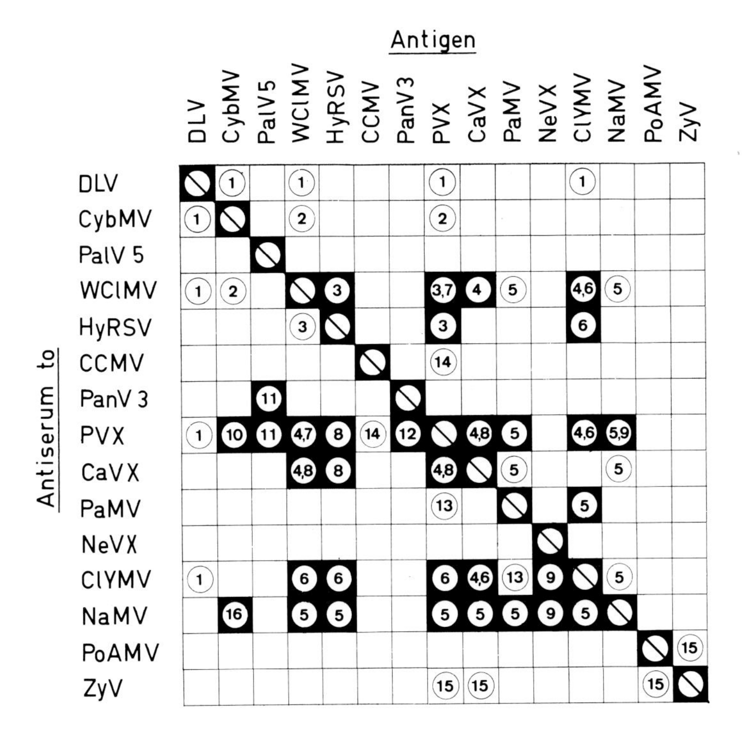

Fig. 1. Serological relationships in the potexvirus group. For abbreviations of virus names see Table I. Only those viruses are listed for which serological tests have been reported. Encircled numbers indicate the reference: (1) Waterworth *et al.* (1974); (2) Frowd and Tremaine (1974); (3) Bercks and Brandes (1961); (4) Koenig and Bercks (1968); (5) Koenig (1975); (6) Bercks and Brandes (1963); (7) Varma *et al.* (1970); (8) Brandes and Bercks (1962-1963); (9) Maat (1976); (10) Brandes (1964); (11) Frowd and Tomlinson (1972); (12) Garrett and Tomlinson (1966); (13) de Bokx (1965); (14) Kitajima and Costa (1966); (15) Casper and Brandes (1969); (16) R. Koenig (unpublished). Black squares containing encircled numbers indicate that serological tests have given positive results at least with some antisera. White squares with encircled numbers indicate that serological tests have given negative results. Since most relationships among potexviruses are rather distant, it seems possible that in the future with larger numbers of antisera positive results may nevertheless be obtained (Koenig and Bercks, 1968; Koenig, 1975). White empty squares indicate that no results of serological tests have been published.

Fig. 2. Particles of PVX in crude extract of infected tobacco leaf negatively stained in uranyl formate.

Figs. 3 and 4. Particles of CybMV in crude extract of infected leaf of *Epidendrum* sp. negatively stained with uranyl formate showing crossbanding (Fig. 3) and oblique lines (Fig. 4); both particles are from the same grid.

Figs. 5 and 6. Spindle-shaped (Fig. 5) and sledge-like (Fig. 6) inclusion bodies in unstained epidermal cuttings of CaVX-infected *Zygocactus* sp.

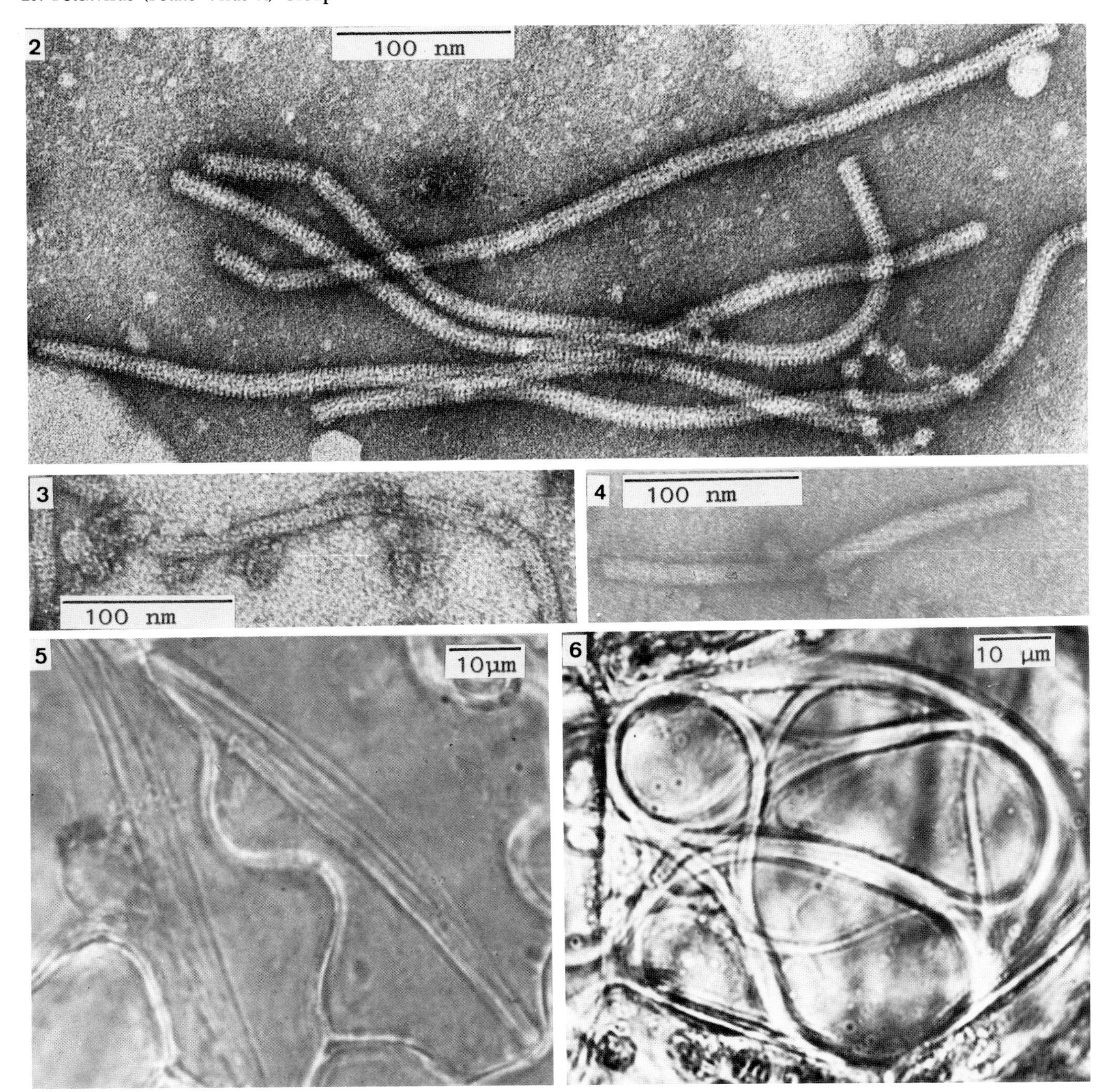
2
100 nm
3
100 nm
4
100 nm
5
10μm
6
10 μm

Fig. 7. Ultrathin section of leaf parenchyma cell of CaVX-infected *Chenopodium quinoa*. Spindle-shaped virus aggregates in longitudinal and cross section.

Figs. 8, 10, and 11. Ultrathin sections of WClMV-infected *Phaseolus vulgaris*.

Fig. 8. Particle aggregate with several whirl-like regions.

Fig. 10. Side-by-side aggregate with particle ends in one plane.

Fig. 11. Dense-particle aggregates with paracrystalline regions in longitudinal and cross section.

Fig. 9. Ultrathin section of PVX-infected meristematic dome cells of *Datura stramonium*, with stacked, plate-like virus aggregates. Reproduced from Pennazio and Appiano (1975), courtesy of *Phytopathologia Mediterranea*.

Fig. 12. Ultrathin section of CaVX-infected *Cereus* sp. showing in a parenchyma cell viral aggregate of randomly oriented, strongly curved particles.

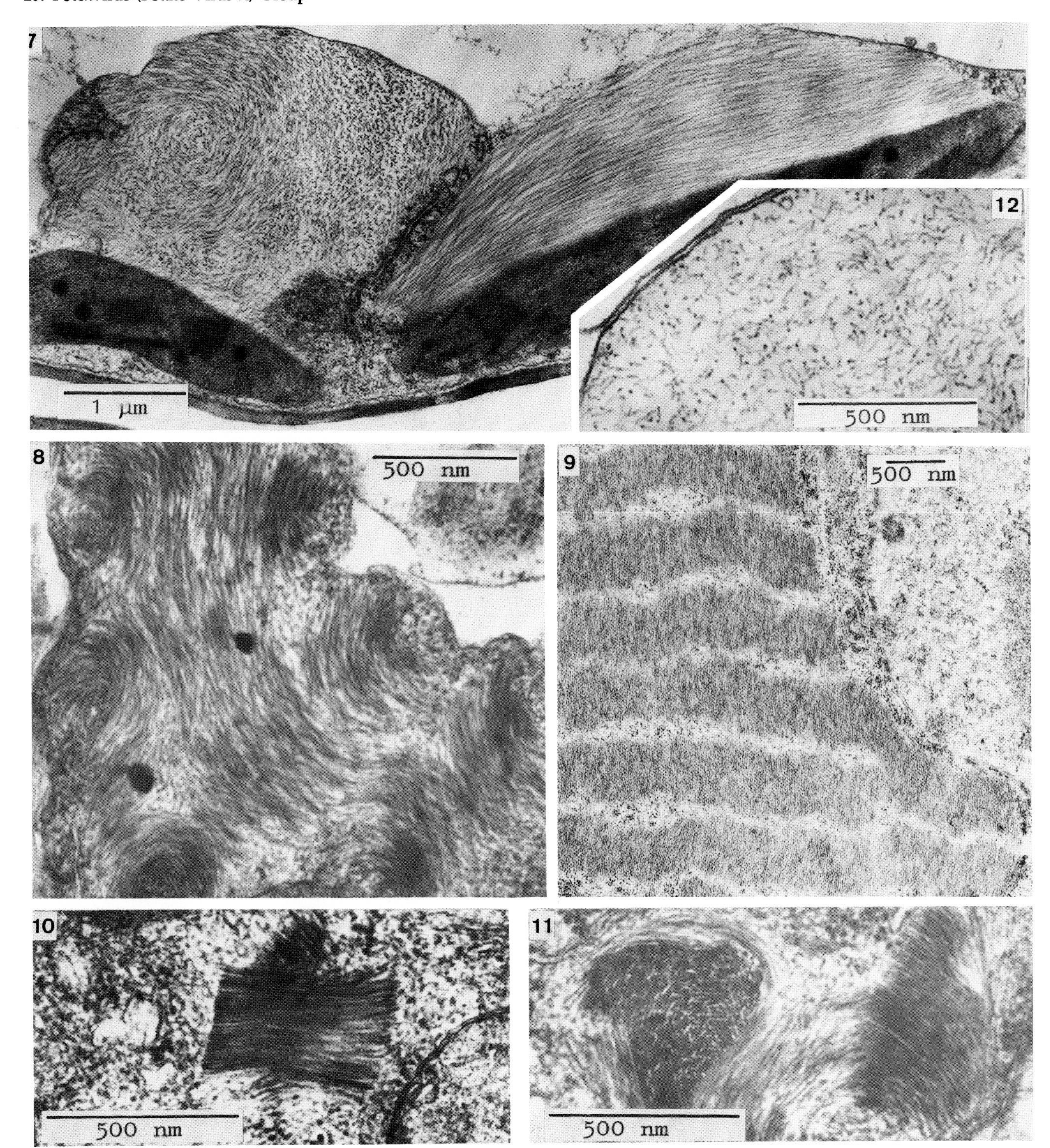
7
1 μm
12
500 nm
8
500 nm
9
500 nm
10
500 nm
11
500 nm

Figs. 13 and 14. Ultrathin sections of parenchyma cells of PVX-infected *Datura stramonium* leaf.

Fig. 13. Inclusion body area with several systems of "laminated inclusion components" (LIC) covered with beads (inset); viral aggregates mostly in contact with LIC.

Fig. 14. Whirl-like aggregate of smooth (compare with inset) LIC.

Fig. 15. Ultrathin section of ClYMV-infected *Vicia faba* showing an amorphous inclusion body in the vacuole. Micrograph courtesy of Dr. D. E. Schlegel.

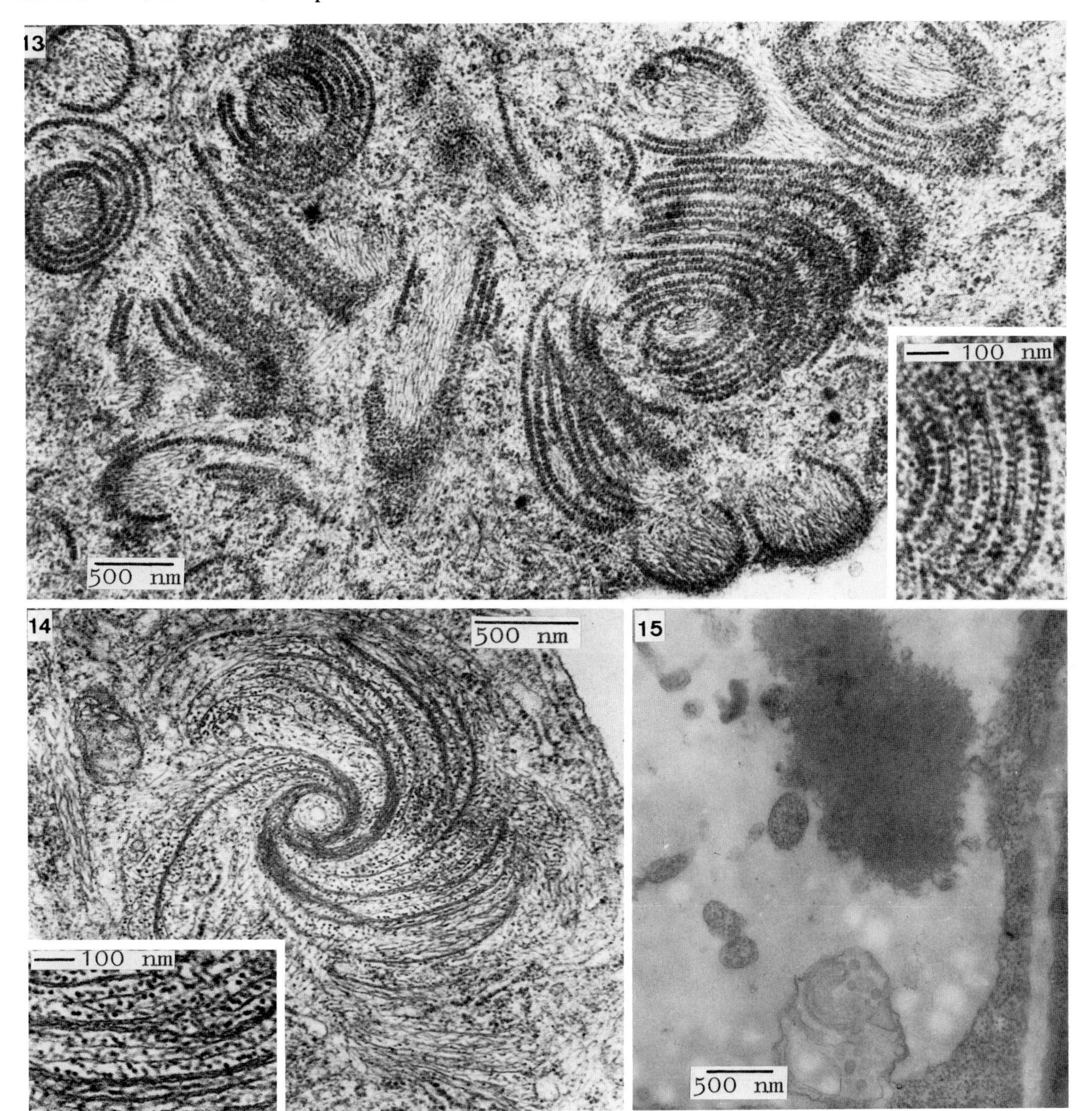
13
500 nm
100 nm
14
500 nm
100 nm
15
500 nm

Chapter 21

Tymovirus (Turnip Yellow Mosaic Virus) Group

R. E. F. MATTHEWS

I. Introduction 347
II. Members of the Group 348
III. Components of the Virus 348
A. The Coat Protein 348
B. The RNA 348
C. Polyamines 348
IV. Structure of the Particle 350
V. Empty Protein Shells and Minor Nucleoproteins 351
VI. Cytological Evidence concerning Virus Replication 351
References 352

I. INTRODUCTION

The name tymovirus was introduced by Harrison *et al.* (1971) for the group of small icosahedral plant viruses that have turnip yellow mosaic virus (TYMV) as the type member. The name tymovirus was approved by the International Committee on Taxonomy of viruses in 1975.

Members of the group have particles about 30 nm in diameter made up of 180 identical protein subunits of $\approx$20,000 daltons MW that are arranged with icosahedral symmetry. Clustering of subunits into pentamers and hexamers gives 32 morphological subunits. In purified preparations of tymoviruses two major classes of particles are found: noninfectious empty protein shells and infectious nucleoprotein particles containing a single-stranded RNA of $\approx 1.9 \times 10^6$ daltons MW, making up about 34% by weight of the particle, and having a high content of cytidylic acid.

Biologically, viruses in this group are characterized by the following properties: (1) Transmission is by flea beetles (Coleoptera, Insecta) and by mechanical means. (2) Mosaic disease is normally produced. (3) In infected cells small vesicles are formed near the periphery of the chloroplasts. These appear as invaginations of the inner and outer chloroplast membranes. The vesicles open to the cytoplasm through a narrow neck. Vesicles having this structure appear to be diagnostic for the tymovirus group and are concerned with viral RNA replication.

II. MEMBERS OF THE GROUP

At a stage when only about seven tymoviruses had been described, two subgroups were suggested on the basis of serological relationships and base composition of their RNA (Gibbs, 1969; Harrison *et al.*, 1971). A more detailed study of eleven viruses has since revealed a continuous range of serological relationships within the group (Koenig and Givord, 1974). At present there seems to be no sound basis for the formation of subgroups. Table I lists some properties of 11 tymoviruses while Fig. 1 summarises their serological relationships.

III. COMPONENTS OF THE VIRUS

A. The Coat Protein

The TYMV coat protein has proved difficult to isolate in native monomer form. It is usually studied as the oxidized, carboxymethylated or aminoethylated protein. The 189 residues in the protein give a MW = 20,133 and are made up as follows: Asp 7 residues, Asn 4, Thr 26, Ser 17, Glu 6, Gln 8, Pro 20, Gly 8, Ala 15, Val 14, Cys 4, Met 4, Ile 16, Leu 17, Tyr 3, Phe 5, Lys 7, His 3, Trp 2, Arg 3. The primary structure has been established by Peter *et al.* (1972) (Table II).

B. The RNA

The RNA of TYMV consists of one single-stranded piece of MW = 1.9×10^6 (Markham, 1959; Mitra and Kaesberg, 1965). This makes up about 34% by weight of the virus. The $s_{20,w}$ in 0.1 *M* KCl—phosphate buffer pH 7.0—is 28.4 ± 0.5 S (Mitra and Kaesberg, 1965). The RNA contains about 6400 nucleotides. Base compositions are given in Table I.

TYMV RNA appears to have a tRNA-like structure at or near the 3′-end since it is recognized by at least three different enzymes catalyzing reactions involving tRNA's. The amino acid is added to the 3′-CCA terminus of the RNA. TYMV-valyl RNA appears to donate its valine for the synthesis of polypeptide chains in *in vitro* studies (Haenni *et al.*, 1973).

A base-paired double-stranded form of the viral RNA can be isolated from infected plants (Ralph *et al.*, 1965).

C. Polyamines

Two polyamines have been identified in purified preparations of TYMV (Beer and Kosuge, 1970). These are spermidine (about 1% by weight of the virus) and spermine (less than 0.04% by weight):

spermidine: $H_2N—CH_2—CH_2—CH_2—NH—CH_2—CH_2—CH_2—CH_2—NH_2$
spermine: $H_2N—CH_2—CH_2—CH_2—NH—CH_2—CH_2—CH_2—CH_2—NH—CH_2—CH_2—CH_2—NH_2$

The content of the amines was somewhat variable in different preparations of the virus. The work of Johnson and Markham (1962) suggests that the amines

TABLE I

Some Properties of Tymoviruses[a]

Name, abbreviation, and reference	Known geographical distribution	Susceptible family	Base composition of RNA			
			G	A	C	U
Eggplant mosaic virus (EMV) Gibbs and Harrison (1969); Koenig and Givord (1974)	Trinidad	Chenopodiaceae, Solanaceae	16	22	38	25
Andean potato latent virus (APLV) Gibbs *et al.* (1966)	Colombia, Bolivia, Peru	Chenopodiaceae, Solanaceae	15	22	34	29
Belladonna mottle virus (BMV) Paul (1971)	Europe, U.S.A.	Solanaceae	17	23	33	27
Dulcamara mottle virus (DMV) Gibbs *et al.* (1966)	England	Solanaceae	17	23	32	28
Ononis yellow mosaic virus (OYMV) Gibbs *et al.* (1966)	England	Papilionaceae, Solanaceae	16	21	34	29
Scrophularia mottle virus (ScrMV) Berck *et al.* (1971)	Germany	Caryophyllaceae, Labiatae, Papilionaceae, Scrophulariaceae, Solonaceae, Umbelliferae, Valerionaceae	17	22	32	29
Wild cucumber mosaic virus (WCMV) van Regenmortel (1972)	U.S.A.	Cucurbitaceae	16	17	41	26
Turnip yellow mosaic virus (TYMV) Matthews (1970)	Europe	Cruciferae, Resedaceae	17	22	38	22
Okra mosaic virus (OkMV) Givord and Hirth (1973)	Africa (Ivory Coast)	23 families including all those listed for other viruses except Valerianaceae	17	17	40	26
Desmodium yellow mosaic virus (DeYMV) Walters and Scott (1972); Scott and Moore (1972)	U.S.A.	Papilionaceae	16	22	37	24
Cacao yellow mosaic virus (CaYMV) Brunt (1970); Brunt *et al.* (1965)	Africa (Sierra Leone)	Apocynaceae, Bombacaceae, Chenopodiaceae, Solanaceae, Sterculiaceae	16	22	33	29

[a] Viruses listed in order of serological relationship as determined by Koenig and Givord (1974).

TABLE II

Complete Sequence of Turnip Yellow Mosaic Virus Coat Protein[a]

Residues	Sequence
1 5 10 15	Acetyl-Met-Glu-Ile-Asp-Lys-Glu-Leu-Ala-Pro-Gln-Asp-Arg-Thr-Val-Thr-
16 20 25 30	Val-Ala-Thr-Val-Leu-Pro-Ala-Val-Pro-Gly-Pro-Ser-Pro-Leu-Thr-
31 35 40 45	Ile-Lys-Gln-Pro-Phe-Gln-Ser-Glu-Val-Leu-Phe-Ala-Gly-Thr-Lys-
46 50 55 60	Asp-Ala-Glu-Ala-Ser-Leu-Thr-Ile-Ala-Asn-Ile-Asp-Ser-Val-Ser-
61 65 70 75	Thr-Leu-Thr-Thr-Phe-Tyr-Arg-His-Ala-Ser-Leu-Glu-Ser-Leu-Trp-
76 80 85 90	Val-Thr-Ile-His-Pro-Thr-Leu-Gln-Ala-Pro-Thr-Phe-Pro-Thr-Thr-
91 95 100 105	Val-Gly-Val-Cys-Trp-Val-Pro-Ala-Asn-Ser-Pro-Val-Thr-Pro-Ala-
106 110 115 120	Gln-Ile-Thr-Lys-Thr-Tyr-Gly-Gly-Gln-Ile-Phe-Cys-Ile-Gly-Gly-
121 125 130 135	Ala-Ile-Asn-Thr-Leu-Ser-Pro-Leu-Ile-Val-Lys-Cys-Pro-Leu-Glu
136 140 145 150	Met-Met-Asn-Pro-Arg-Val-Lys-Asp-Ser-Ile-Gln-Tyr-Leu-Asp-Ser-
151 155 160 165	Pro-Lys-Leu-Leu-Ile-Ser-Ile-Thr-Ala-Gln-Pro-Thr-Ala-Pro-Pro-
166 170 175 180	Ala-Ser-Thr-Cys-Ile-Ile-Thr-Val-Ser-Gly-Thr-Leu-Ser-Met-His-
181 185 189	Ser-Pro-Leu-Ile-Thr-Asp-Thr-Ser-Thr-COOH

[a] From Peter *et al.*, 1972.

may be absent from the empty protein shells. Beer and Kosuge (1970) calculated that there were sufficient polyamines present in the virus to neutralize about 20% of the phosphate groups of the RNA.

IV. STRUCTURE OF THE PARTICLE

TYMV is a remarkably stable icosahedral particle. Protein–protein interactions must be strong as evidenced by the existence of the stable empty protein shells discussed below, but the detailed nature of the bonding holding the shell together has not been established.

Protein–RNA interactions in the particle probably involve hydrogen bonds between protein carboxylate and nucleotide amino groups (Kaper, 1969, 1971; Jonard *et al.*, 1972).

TYMV has the following physical parameters: $s_{20,w}$, 116; MW, 5.4×10^6; diameter close to 28 nm (see below); isoelectric point pH 3.75; partial specific volume, 0.661; effective buoyant density in CsCl, 1.42 gm/cm^3.

TYMV was the first virus for which an icosahedral structure with clustering of subunits, was established. Almost all structural studies within the tymovirus group have been carried out on TYMV. Figure 2 summarizes knowledge of the structure of TYMV (see also Figs. 3–6).

There are 180 identical coat protein subunits. The triangulation number T = 3 (P = 3; f = 1). The subunits are clustered into 12 groups of five and 20 groups of six. These 32 morphological subunits are readily identified in negatively stained preparations of TYMV (Fig. 3). The RNA strand is entirely enclosed within the protein shell. Some of the RNA is intimately associated with the protein, but the exact arrangement of all the RNA in the particle has not been established.

No central hole in the particle was detected by X-ray analysis (Klug *et al.*, 1966) so that most of the space within the shell is probably occupied by RNA. From hypochromicity studies on the RNA, Haselkorn (1962) concluded that about two-thirds of the bases within the virus are ordered sufficiently to produce the degree of hypochromicity observed in a polynucleotide helix. The location of the polyamine within the particle is not established, but it is presumably associated with the RNA.

The diameter of the virus determined by X-ray diffraction is 28 nm (Klug *et al.*, 1966). By electron microscopy the diameter ranged from 30 nm at the tips of the subunit clusters to a 26 nm minimum in the valleys between clusters (Finch and Klug, 1966). The diffusion coefficient measured by laser-beam spectroscopy gave the hydrodynamic diameter as 29.5 ± 2% (Harvey, 1973). The close correspondence in these values suggests that there is little shrinkage of TYMV particles on drying and this is supported by the relatively low water content (0.7 gm/gm particle).

V. EMPTY PROTEIN SHELLS AND MINOR NUCLEOPROTEINS

The presence of empty viral protein shells is a characteristic of the tymovirus group. The ratio of empty protein shells to complete TYMV found in leaf extracts falls from about 0.5 to 0.2 over the first 4 weeks following inoculation (Francki and Matthews, 1962). The proportion of empty protein shells for various tymoviruses varies widely (e.g., Fig. 11).

The structure of the TYMV empty shell is identical to that in the virus except for the absence of the RNA (and presumably the polyamines). The empty protein shells have the following physical parameters: $s_{20,w}$ 53; MW 3.6×10^6 daltons; isoelectric point pH 3.75; partial specific volume 0.733; effective buoyant density in CsCl 1.29 gm/cm^3.

Artificial empty shells can be prepared from TYMV by treatment of the virus at pH 11.55 with 1 *M* KCl at 30° for 1 hour (Kaper, 1969). The RNA which escapes from the particle is usually in a degraded state.

When TYMV preparations are fractionated in CsCl density gradients a series of minor nucleoprotein fractions can be isolated as well as empty shells and virus (Matthews, 1960). None of these minor components appear to have any biological activity and they represent only a few percent of the total viral nucleoprotein. If the CsCl density gradients are relatively shallow and formed in long tubes at pH 5–7, double banding of all nucleoprotein components is apparent (Matthews, 1974). The basis for this phenomenon is not established.

VI. CYTOLOGICAL EVIDENCE CONCERNING VIRUS REPLICATION

When thin sections are prepared from infected cells and stained by standard procedures the tymovirus particles have an apparent diameter of about 21 nm, corresponding to the RNA within the particle. This is very similar to the stained RNA diameter in cytoplasmic ribosomes. Thus, there are considerable difficulties in identifying any particular particle in a cell as a tymovirus particle. This problem has been studied recently in detail using TYMV as a model (Hatta, 1975).

In thin sections of purified crystals of TYMV the measured interparticle distance is always less than the diameter of the particle. Particles form various regular arrays depending on the plane of the section in relation to the crystal lattice (Figs. 7–9). Empty protein shells form arrays in which the staining density is reversed compared with nucleoprotein (Figs. 7 and 10). TYMV can be induced to form microcrystalline arrays within infected cells by plasmolysis of the tissue in a sucrose solution. In such crystalline arrays the interparticle distance is usually larger than expected, presumably because cellular components become trapped within the arrays. Figure 14 shows arrays of okra mosaic particles in a infected cell.

Cytological changes induced by infection can be found before virus production begins. The earliest cytological change detected in Chinese cabbage cells infected with TYMV is the formation of peripheral vesicles about 50–60 nm diameter in the chloroplasts. These are formed by invagination of both inner and outer chloroplast membranes. They have a narrow neck (10–13 nm) opening to the cytoplasm (Figs. 12, 13, 15, and 16). These vesicles appear to be characteristic for all tymoviruses in all the hosts that have been investigated. Several lines of evidence strongly suggest that these vesicles are the major site of viral RNA production (Laflèche and Bové, 1969; Bové *et al.*, 1972; Ralph *et al.*, 1971; Ushiyama and Matthews, 1970). In Chinese cabbage leaves infected with TYMV we have recently established a sequence of cytological changes that lead to the production of complete virus particles (Hatta and Matthews, 1974, 1976). These are described in Figs. 17–24.

REFERENCES

Beer, S. V., and Kosuge, T. (1970). *Virology* **40,** 930–938.

Bercks, R., Huth, W., Koenig, R., Lesemann, D., Paul, H. L., and Querfurth, G. (1971). *Phytopathol. Z.* **71,** 341–356.

Bové, C., Mocquot, B., and Bové, J. M. (1972). *Symp. Biol. Hung.* **13,** 43–59.

Brunt, A. A. (1970). *CMI/AAB Descriptions Plant Viruses* No. 11.

Brunt, A. A., Kenten, R. H., Gibbs, A. J., and Nixon, H. L. (1965). *J. Gen. Microbiol.* **38,** 81–90.

Finch, J. T., and Klug, A. (1966). *J. Mol. Biol.* **15,** 344–364.

Francki, R. I. B., and Matthews, R. E. F. (1962). *Virology* **17,** 367–380.

Gibbs, A. J. (1969). *Ad. Virus Res.* **14,** 263–328.

Gibbs, A. J., and Harrison, B. D. (1969). *Ann. Appl. Biol.* **64,** 225–231.

Gibbs, A. J., Hecht-Poinar, E., Woods, R. D., and McKee, R. K. (1966). *J. Gen. Microbiol.* **44,** 177–193.

Givord, L., and Hirth, L. (1973). *Ann. Appl. Biol.* **74,** 359–370.

Haenni, A. L., Prochiantz, A., Bernard, O., and Chapeville, F. (1973). *Nature (London), New Biol.* **241,** 166–168.

Harrison, B. D., Finch, J. T., Gibbs, A. J., Hollings, M., Shepherd, R. J., Valenta, V., and Wetter, C. (1971). *Virology* **45,** 356–363.

Harvey, J. D. (1973). *Virology* **56,** 365–368.

Haselkorn, R. (1962). *J. Mol. Biol.* **4,** 357–367.

Hatta, T. (1975). *Virology* **69,** 237–245.

Hatta, T., and Matthews, R. E. F. (1974). *Virology* **59,** 383–396.

Hatta, T., and Matthews, R. E. F. (1976). *Virology* **73,** 1–16.

Hatta, T., Bullivant, S., and Matthews, R. E. F. (1973). *J. Gen. Virol.* **20,** 37–50.

Johnson, M. W., and Markham, R. (1962). *Virology* **17,** 276–281.

Jonard, G., Witz, J., and Hirth, L. (1972). *J. Mol. Biol.* **66,** 165–169.

Kaper, J. M. (1969). *Science* **166,** 248–250.

Kaper, J. M. (1971). *J. Mol. Biol.* **56,** 259–276.

Klug, A., Longley, W., and Leberman, R. (1966). *J. Mol. Biol.* **15,** 315–343.

Koenig, R., and Givord, L. (1974). *Virology* **58,** 119–125.

Laflèche, D., Bové, J. M. (1969). *Prog. Photosynth. Res.* **1,** 74–83.

Markham, R. (1959). *In* "The Viruses" (F. M. Burnet and W. M. Stanley, eds.), Vol. 2, pp. 33–125. Academic Press, New York.

Matthews, R. E. F. (1960). *Virology* **12,** 521–539.

Matthews, R. E. F. (1970). *CMI/AAB Descriptions Plant Viruses* No. 2.
Matthews, R. E. F. (1974). *Virology* **60,** 54–64.
Mitra, S., and Kaesberg, P. (1965). *J. Mol. Biol.* **14,** 558–571.
Paul, H. L. (1971). *CMI/AAB Descriptions Plant Viruses* No. 52.
Peter, R., Stehelin, D., Reinbolt, J., Collot, D., and Duranton, H. (1972). *Virology* **49,** 615–617.
Ralph, R. K., Matthews, R. E. F., Matus, A., and Mandel, H. (1965). *J. Mol. Biol.* **11,** 202–212.
Ralph, R. K., Bullivant, S., and Wojcik, S. J., (1971). *Virology* **44,** 473–479.
Scott, H. A., and Moore, B. J. (1972). *Virology* **50,** 613–614.
Ushiyama, R., and Matthews, R. E. F. (1970). *Virology* **42,** 293–303.
van Regenmortel, M. H. V. (1972). *CMI/AAB Descriptions Plant Viruses* No. 105.
Walters, H. J., and Scott, H. A. (1972). *Phytopathology* **62,** 125–128.

Fig. 1. Serological interrelationships in the tymovirus group. The first number following each antiserum indicates the number of rabbits immunized; the second indicates the number of bleedings tested. The viruses are arranged according to the approximate closeness of their serological relationships. The white squares indicate the average number of two-fold dilution steps by which the heterologous titers of the antisera tested were below the homologous ones (all squares black), e.g., ▬▬▬▬▬ signifies that the heterologous titer was on the average five two-fold dilution steps below the homologous titer; i.e., if the homologous titer was 1:4096 the average heterologous titer was 1:128; if the homologous titer was 1:512 the average heterologous titer was 1:16. ☐☐☐☐◪ signifies that the heterologous titer was on the average more than nine two-fold dilution steps below the homologous titer. Frequently, only a few antisera showed this reactivity. ☐☐☐☐☐ signifies that no heterologous reaction was found (from Koenig and Givord, 1974).

Fig. 2. Summary of knowledge of the structure of TYMV. (Left) Drawing of outside of particle showing clustering of subunits into groups of 5 and 6. (From Finch and Klug, 1966.) (Center) Diagram incorporating the significant radii deduced from X-ray analysis. (From Klug *et al.*, 1966.) Bottom right quarter: Spherically averaged dimensions for the protein shell. a = outer radius (140 Å) and b = inner radius (105 Å). Bottom left quarter: Spherically averaged radius for the RNA (e = 117 Å) deduced from salt change studies. Top right quarter: Schematic fluctuation in density of the protein shell. c = 150 Å, the interparticle distance in the crystal; d = 145 Å, the effective scattering radius of the protuberances corresponding to the protein subunits. This distance is not significantly different from the outer radius. Top left quarter: Schematic fluctuation in density of RNA, f = 125 Å, the effective radius of the 32 bumps of RNA. (Right) Schematic drawing, based on the data in the center drawing indicating the relation of the gross RNA distribution in the virus to the arrangement of the protein subunits. A section through the diametral plane of the virus is shown. The precise shape of the subunits and the detailed path of the RNA are not known. The distribution of protein and RNA toward the center of the particle are likewise unknown. (From Klug *et al.*, 1966.)

Antigen

EMV APLV BMV DMV OYMV ScrMV WCuMV TYMV OkMV DeYMV CoYMV

Antiserum

EMV 3 / 28

APLV 3 / 37

BMV 6 / 33

DMV 3 / 27

OYMV 3 / 27

ScrMV 6 / 35

WCuMV 6 / 13

TYMV 7 / 27

OkMV 3 / 25

DeYMV 4 / 21

CoYMV 4 / 13

1

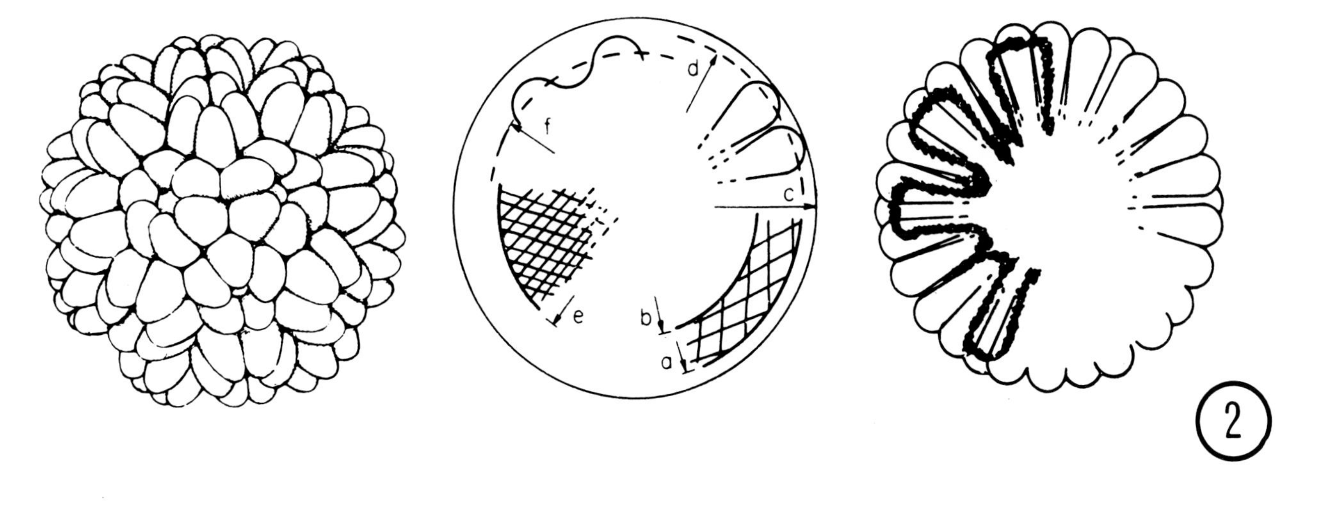

2

Fig. 3. TYMV Purified nucleoprotein particles negatively stained with uranyl acetate. Particles that show clear detail of five-fold and six-fold clusters lie with a twofold axis perpendicular to the plane of the support film. (×261,000.) Photo: S. Bullivant.

Fig. 4. TYMV Purified empty protein shells, negatively stained with uranyl acetate. Particles show penetration of the stain into the interior to a variable degree. (×261,000.) Photo: S. Bullivant.

Figs. 5 and 6. Octahedral crystals of purified TYMV grown in 0.75 *M* ammonium sulfate. (×675.) Photo: T. Hatta.

Figs. 7–10. Thin sections of crystals of TYMV nucleoprotein and empty protein shells grown in 1.5 *M* ammonium sulphate and stained with uranyl acetate and lead citrate. (×210,000.) Photos: T. Hatta.

Fig. 7. Nucleoprotein particles in rod-shaped array (view is approximately down a twofold axis).

Fig. 8. Nucleoprotein particles in hexagonal array (view is approximately down a threefold axis).

Fig. 9. Nucleoprotein particles in tetragonal array. (View is approximately down a fourfold axis.)

Fig. 10. Empty protein shells. (View approximately down a twofold axis) (cf., Fig. 7).

Fig. 11. The proportion of empty protein shells and nucleoprotein particles for two tymoviruses. Undiluted expressed sap from infected leaves was heated to 55° for 5 minutes and then subjected to centrifugation in the analytical centrifuge for 12 minutes at 40,000 rpm and 6°. Upper schlieren pattern = TYMV in Chinese cabbage leaves. Lower pattern = okra mosaic virus in cucumber. The right hand peak in each pattern is the virus nucleoprotein. The next most rapidly sedimenting components are the empty protein shells. Note the very much greater concentration of these for okra mosaic virus.

Fig. 12. A small peripheral vesicle induced in a Chinese cabbage chloroplast by TYMV infection showing the continuity of inner chloroplast and outer vesicle membranes and stranded material inside the vesicle with the staining properties of double-stranded nucleic acid. (×235,000.) Photo: S. Bullivant.

Fig. 13. Small peripheral vesicles in a chloroplast isolated from TYMV-infected Chinese cabbage. Fine structure of membranes revealed by freeze-fracturing. (×92,000.) (From Hatta *et al.*, 1973.)

Fig. 14. Arrays of okra mosaic virus particles in an infected cucumber cell. The tissue was subjected to plasmolysis with 40% sucrose before fixation. The nuclear membranes run down across the field from left to right. Nucleoprotein particles occur in the cytoplasm. Particles in the nucleus appear to consist entirely of empty shells. (×101,000.) (From Hatta and Matthews, 1976.)

Fig. 15. Scattered and clustered vesicles in a chloroplast from a Chinese cabbage leaf infected with TYMV. The fracture plane shows the A face of the outer chloroplast membrane. The distribution of fractured vesicle necks is clearly shown. (×44,800.) Photo: T. Hatta.

Fig. 16. Diagramatic representation of fine structure of vesicle membranes in chloroplasts of TYMV infected cells. Distribution of the normal (larger) and virus-induced (smaller) particles within the membranes is indicated. The size of a TYMV particle is indicated for comparison. (Drawing by T. Hatta.)

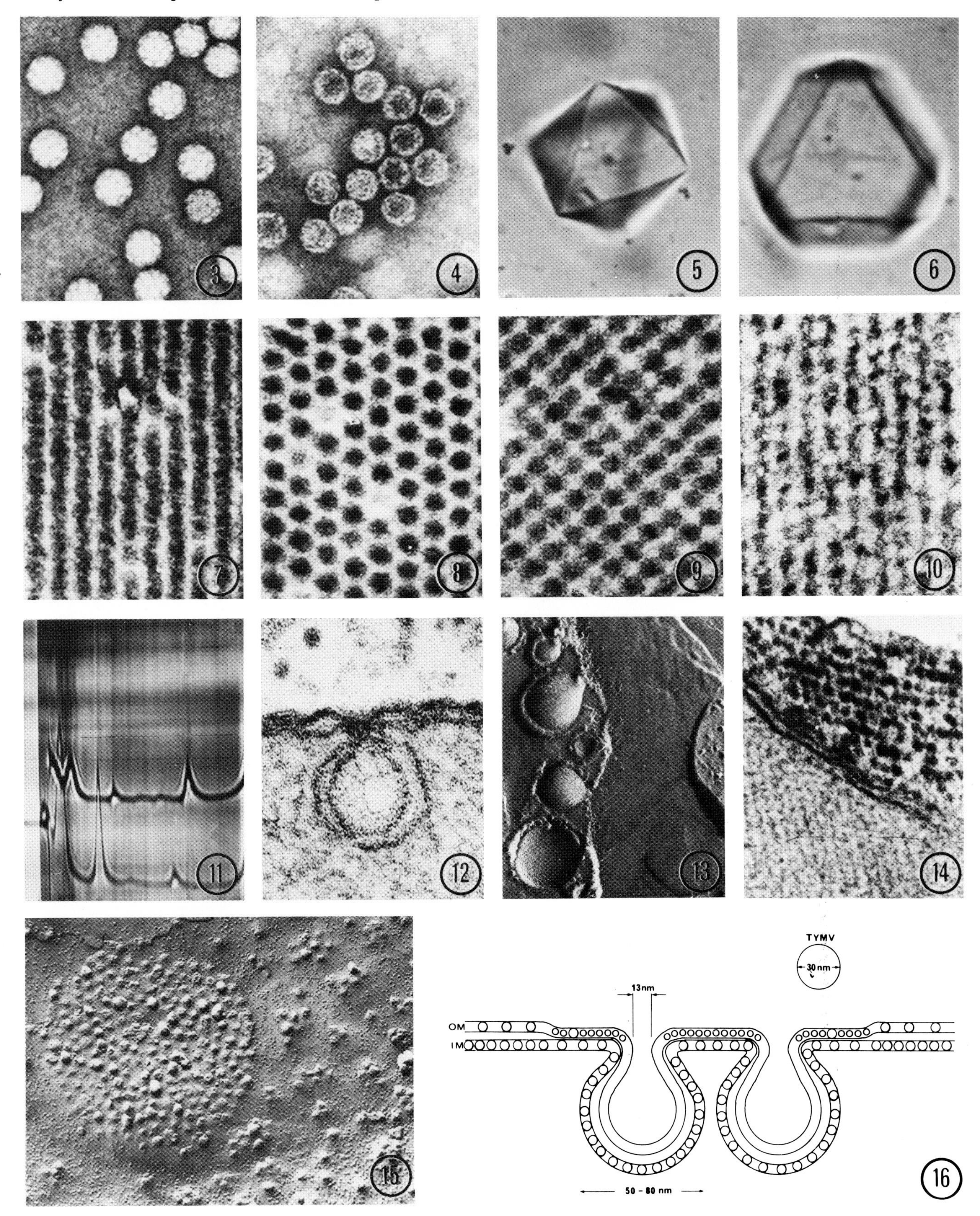
3
4
5
6
7
8
9
10
11
12
13
14
15
TYMV
30 nm
13nm
OM
IM
50 - 80 nm
16

Figs. 17–24. Cytological stages in the infection of Chinese cabbage leaf cells by TYMV. (From Hatta and Matthews, 1974.)

Fig. 17. Stage A. The first detectable change is the appearance of scattered peripheral vesicles in the chloroplasts, which otherwise appear normal. (×24,000.) Inset shows a small group of vesicles around a sunken region of the surface. (×72,000.)

Fig. 18. Stage B. The same as stage A except that the chloroplasts are swollen to a near spherical shape. (×24,000.)

Figs. 19–20. Stage C. Large clusters of small peripheral vesicles have appeared in addition to the scattered vesicles. Endoplasmic reticulum in the cytoplasm overlies these clustered vesicles. (19, ×24,000; 20, ×77,000.)

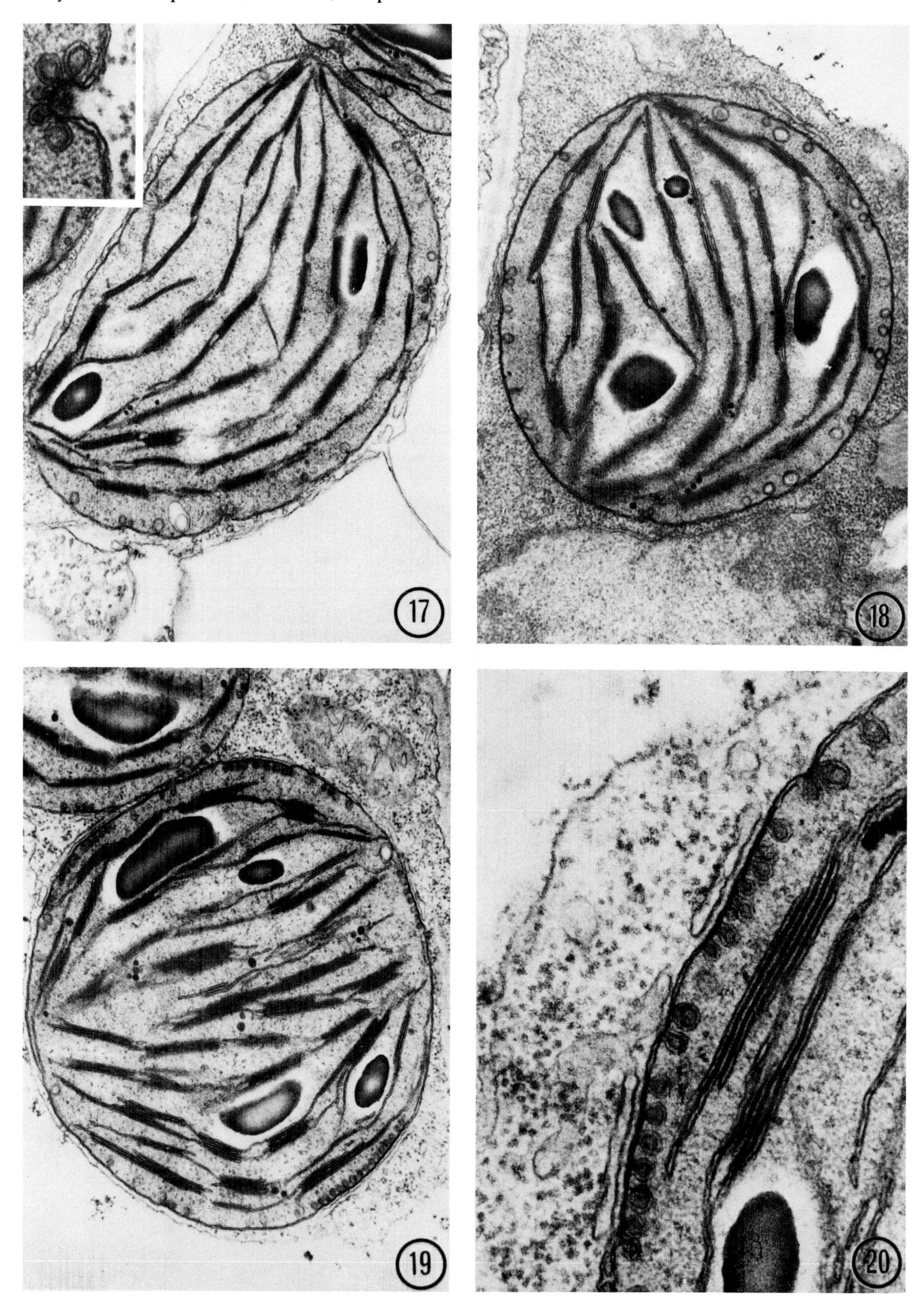
17
18
19
20

Fig. 21. Stage D. The endoplasmic reticulum overlying the clustered vesicles has disappeared to be replaced by zones of relatively electron lucent material in the cytoplasm. (×24,000.)

Fig. 22. Stage E. The chloroplasts become clumped together, joined in the region of the electron lucent zones. (×24,000.)

Fig. 23. Stage F. The electron lucent material between the clumped chloroplasts disappears, to be replaced by numerous virus-like particles (×24,000.)

Fig. 24. Intact virus particles can be detected first at stage D in the cytoplasm just outside the electron lucent zone. The tissue has been plasmolyzed to induce virus particles to form ordered arrays. Other evidence indicates that the electron lucent material consists of pentamers and hexamers of viral coat protein (Hatta and Matthews, 1976). We assume that the virus particles are assembled from RNA strands produced in the vesicles and the pentamers and hexamers of the electron lucent zones. (×84,000.)

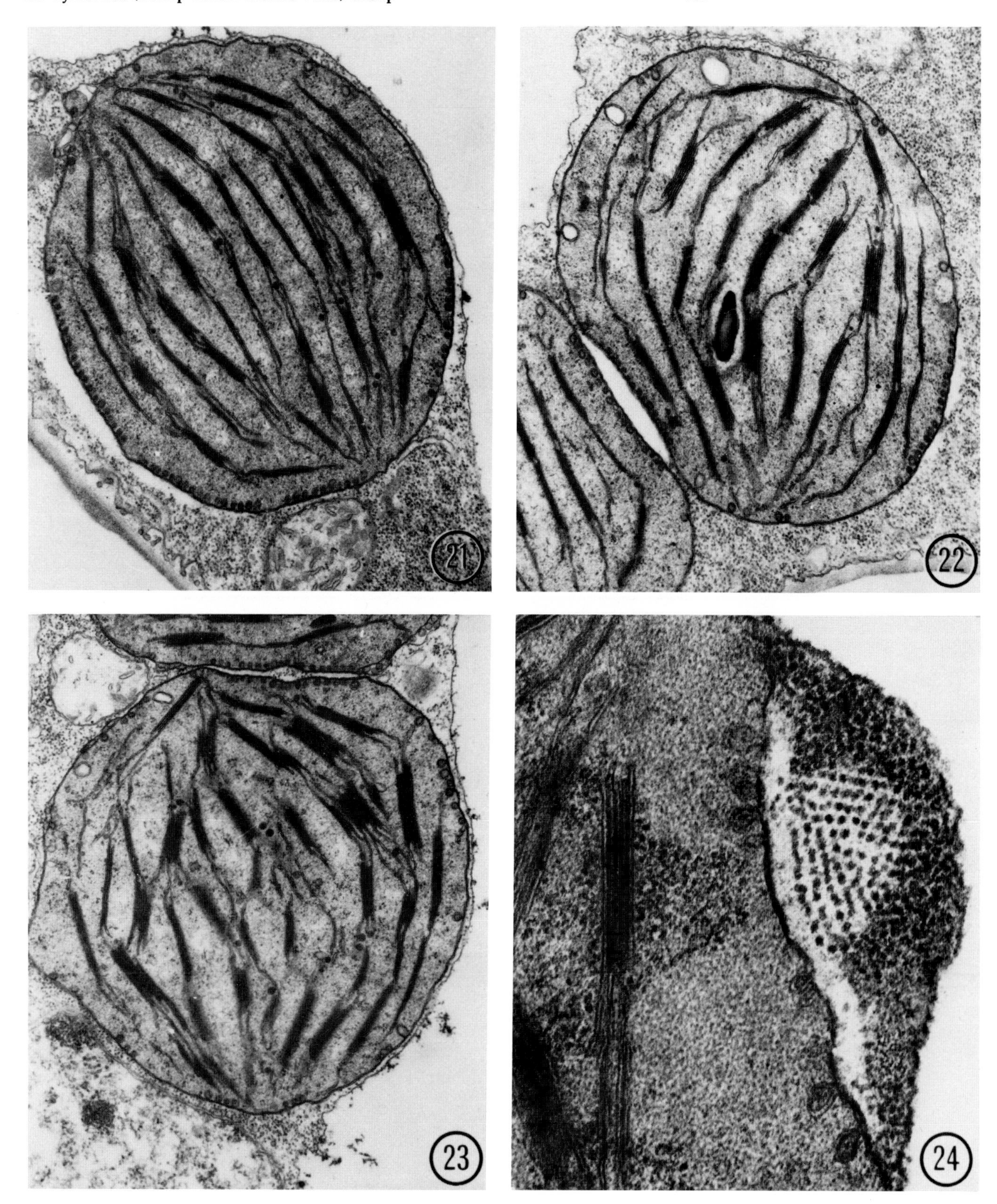
21
22
23
24

Chapter 22

Luteovirus (Barley Yellow Dwarf Virus) Group

W. F. ROCHOW AND H. W. ISRAEL

I. Introduction 363
II. Members of the Group 364
III. General Properties 364
IV. Virion Structure 365
V. Cell–Virus Interactions 365
References 366

I. INTRODUCTION

The luteovirus group (type member, barley yellow dwarf virus) was one of four new groups of plant viruses approved by the International Committee on Taxonomy of Viruses in September, 1975 (Shepherd *et al.*, 1976). Because of their economic importance, these viruses have been studied for a long time. The disparity between the many years of study and the very recent addition to the list of characterized viruses results from restrictions the biological properties of luteoviruses place on research. These limiting properties include the lack of mechanical virus transmission to plants, the need to use aphid vectors to transmit viruses to plants, and the confinement of these viruses to the phloem tissue of their host plants. This tissue specificity is one reason for the extremely low titers in virus preparations made from infected plants. Aphids transmit these viruses in the persistent, circulative manner associated with their feeding in phloem tissue of plants (Rochow, 1969a). This persistent aphid–virus relationship features the ability of aphids to transmit for many days after acquiring virus. There is no evidence for virus replication within the vector (Paliwal and Sinha, 1970).

Studies of luteoviruses are possible because two bioassay techniques can be applied to many members of the group (Rochow, 1969b). One technique is based on injection of virus into hemolymph of aphid vectors, which, in turn, are able to transmit virus to the plants on which they feed. The other bioassay involves acquisition of viruses by aphids fed through membranes on preparations mixed with sucrose, and the subsequent transmission of virus to plants on which such aphids feed. Combination of the membrane feeding method with use of virus-specific antiserum provides a useful kind of infectivity neutralization test (Duffus and Gold, 1969; Rochow *et al.*, 1971).

II. MEMBERS OF THE GROUP

The luteovirus group includes five characterized isolates of barley yellow dwarf virus (BYDV) that have been studied in New York (Rochow, 1969b; Johnson and Rochow, 1972). Similar isolates have been studied in Manitoba (Gill, 1967, 1969). These variants of BYDV were first distinguished by the pronounced specificity between isolates of BYDV and the aphid species that transmit them (Rochow, 1969a). Other luteoviruses include beet western yellows virus and soybean dwarf virus.

Because so many viruses similar to BYDV are poorly characterized, the list of possible luteoviruses is rather long (Shepherd *et al.*, 1976). These possible members of the group include beet mild yellowing virus, banana bunchy top virus, turnip yellows virus, bean leaf roll virus, carrot red leaf virus, cotton anthocyanosis virus, Filaree red leaf virus, Physalis mild chlorosis virus, strawberry mild yellow edge virus, subterranean clover stunt virus, tomato yellow top virus, tobacco vein distorting virus, and turnip latent virus.

On the basis of biological properties and particle morphology, potato leafroll virus would also be a member of this group. But potato leafroll virus appears to be distinct in two ways: virus replication in the aphid vector (*Myzus persicae*) has been reported (Stegwee and Ponsen, 1958), and potato leafroll virus contains DNA, not RNA (Sarkar, 1976). Potato leafroll virus and beet western yellows virus were not closely related in a study using serological and cross-protection tests (Duffus and Gold, 1969).

III. GENERAL PROPERTIES

BYDV has been identified as a small isometric particle because of a positive correlation between the particle and virus infectivity in rate-zonal centrifugations, in equilibrium-zonal centrifugations, and in gel-filtration tests. A similar particle was also obtained from viruliferous aphids (Rochow and Brakke, 1964). Most studies, as well as this discussion, focus on the comparison between two vector specific isolates of BYDV: MAV, transmitted specifically by *Macrosiphum avenae* (Fabricius), and RPV, transmitted specifically by *Rhopalosiphum padi* (Linnaeus) (Rochow, 1969b; Rochow *et al.*, 1971). The isometric virus particles of both MAV and RPV have a sedimentation coefficient of 115–118 S (Rochow and Brakke, 1964; Rochow *et al.*, 1971). Both RPV and MAV contain a single component of single-stranded RNA of molecular weight 2.0×10^6 daltons (Brakke and Rochow, 1974). Untreated RNA of RPV may have a slightly more compact configuration than that of MAV. Both virions seem to have a high content of RNA (more than 20%), but the actual percentage composition is not known. Current studies of the virus protein, using polyacrylamide-SDS gel electrophoresis, by Dr. R. Scalla (Station de Physiopathologie Végétale, Dijon, France) indicate that both RPV and MAV have one major polypeptide subunit. All comparisons consistently showed that RPV protein subunit is slightly larger than that of MAV. The tentative molecular weight for the subunit of MAV is 23.5×10^3 daltons; that of RPV is 24.5×10^3.

Many luteoviruses are effective antigens in rabbits. Several kinds of serological tests have shown a close relationship between MAV and PAV, a third BYDV isolate transmitted nonspecifically by both *R. padi* and *M. avenae*. Similar serological tests revealed that RPV was distinct from MAV and PAV (Aapola and Rochow, 1971; Rochow *et al.*, 1971). Antisera for all three of these BYDV isolates neutralized several isolates of beet western yellows virus (Duffus and Rochow, 1973), although RPV appears to be more closely related to beet western yellows virus than are MAV and PAV (Rochow and Duffus, 1973). Other serological studies have indicated a close relationship between beet western yellows virus and turnip yellows virus (Duffus and Russell, 1972). The long-suspected relationship between beet western yellows virus and beet mild yellowing virus has recently been shown by serology (Duffus and Russell, 1975).

IV. VIRION STRUCTURE

In some of our early micrographs of BYDV, the small, isometric virion seemed to have fewer faces than a typical icosahedral particle. Also, BYDV does not aggregate in regular arrays on air-dried grids (Figs. 1–6), as do many icosahedral viruses. These observations suggested that BYDV might have octahedral symmetry. But most observations favor the common icosahedral virus form (Figs. 1–9). High-voltage electron microscope studies, carried out currently by Dr. G. A. deZoeten (University of Wisconsin at Madison) of preparations of BYDV sent from Ithaca, support an icosahedral symmetry. Diameter of the virus particles has been reported to be about 30 nm in shadowed preparations (Rochow and Brakke, 1964); about 24 nm (Jensen, 1969), 22–25 nm (Paliwal and Sinha, 1970) and 23–26 nm (Gill and Chong, 1975) in thin sections of infected plant tissue. In some negatively stained preparations (Figs. 1–6 and 8) the diameter varies between 20 and 24 nm. Particles of all four isolates of BYDV that we have examined are similar (Rochow, 1970), but RPV is less stable than MAV in phosphotungstate at pH 6.85 (Figs. 1–6).

V. CELL–VIRUS INTERACTIONS

Gill and Chong (1975) recently investigated the kinds of plant cells involved in BYDV replication and studied ultrastructure of infected cells. They confirmed the restriction of BYDV to phloem parenchyma, companion cells, and sieve elements of oat leaves. Cellular inclusions were detected as early as 2 days after inoculation of a leaf with an isolate of BYDV similar to MAV. Inclusions observed were virus-like particles, slender filaments, small vesicles containing fibrils, and amorphous material. A marked cytological feature of the infection was the presence of densely staining filaments in sieve elements and in the cytoplasm and nucleoplasm of other phloem cells (Figs. 10–13). The nuclear alterations in BYDV-infected cells were considered unusual for plant virus infections. Alterations included massive clumping of the heterochromatin and appearance of filaments in the nucleoplasm (Fig. 13). But BYDV particles did not appear to be assembled within the nucleus; Gill and Chong (1975) suggested that replication of BYDV occurs in cytoplasm.

ACKNOWLEDGMENTS

Our work, a cooperative investigation by the Agricultural Research Service, U. S. Department of Agriculture, and Cornell University Agricultural Experiment Station, is supported in part by NSF grant BMS74-19814, and by CSRS grant #316-15-53 awarded to J. R. Aist and H. W. Israel. We thank Dr. C. C. Gill for supplying the material for Fig. 10–13. We also appreciate the permission of Dr. R. Scalla and Dr. G. A. deZoeten to discuss unpublished data from current joint research.

REFERENCES

Aapola, A. I. E., and Rochow, W. F. (1971). *Virology* **46,** 127–141.
Brakke, M. K., and Rochow, W. F. (1974). *Virology* **61,** 240–248.
Duffus, J. E., and Gold, A. H. (1969). *Virology* **37,** 150–153.
Duffus, J. E., and Rochow, W. F. (1973). *Abstr. Pap. Int. Cong. Plant Pathol., 2nd, 1973* Abstract No. 0895.
Duffus, J. E., and Russell, G. E. (1972). *Phytopathology* **62,** 1274–1277.
Duffus, J. E., and Russell, G. E. (1975). *Phytopathology* **65,** 811–815.
Gill, C. C. (1967). *Phytopathology* **57,** 713–718.
Gill, C. C. (1969). *Can. J. Bot.* **47,** 1277–1283.
Gill, C. C., and Chong, J. (1975). *Virology* **66,** 440–453.
Jensen, S. G. (1969). *Virology* **38,** 83–91.
Johnson, R. A., and Rochow, W. F. (1972). *Phytopathology* **62,** 921–925.
Paliwal, Y. C., and Sinha, R. C. (1970). *Virology* **42,** 668–680.
Rochow, W. F. (1969a). *In* "Viruses, Vectors, and Vegetation" (K. Maramorosch, ed.), pp. 175–198. Wiley (Interscience), New York.
Rochow, W. F. (1969b). *Phytopathology* **59,** 1580–1589.
Rochow, W. F. (1970). *CMI/AAB Descriptions Plant Viruses* No. 32.
Rochow, W. F., and Brakke, M. K. (1964). *Virology* **24,** 310–322.
Rochow, W. F., and Duffus, J. E. (1973). *Abstr. Pap. Int. Cong. Plant Pathol., 2nd, 1973* Abstract No. 0898.
Rochow, W. F., Aapola, A. I. E., Brakke, M. K., and Carmichael, L. E. (1971). *Virology* **46,** 117–126.
Sarkar, S. (1976). *Virology* **70,** 265–273.
Shepherd, R. J., Francki, R. I. B., Hirth, L., Hollings, M., Inouye, T., MacLeod, R., Purcifull, D. E., Sinha, R. C., Tremaine, J. H., Valenta, V., and Wetter, C. (1976). *Intervirology* **6,** 181–184.
Stegwee, D., and Ponsen, M. B. (1958). *Entomol. Exp. Appl.* **1,** 291–300.

Figs. 1–9. Barley yellow dwarf virus (BYDV) in preparations made from infected plants.

Figs. 1–6. Two vector-specific isolates are compared and contrasted in purified (MAV, Figs. 1 and 4; RPV, Figs. 2 and 5) and in mixed (Figs. 3 and 6) preparations. (×250,000; bar = 0.1 μm.)

Figs. 1–3. Virions negatively stained with vanadatomolybdate, pH 3.2, which penetrates to the center of the particles (Fig. 1) and in which capsids of both isolates are stable.

Figs. 4–6. Virions stained with potassium phosphotungstate, which does not enter the intact MAV capsids but does penetrate those of RPV, which at this pH (6.85) are typically disrupted (Fig. 5).

Figs. 7–9. Other preparations of the MAV isolate. (×125,000; bar = 0.1 μm.)

Fig. 7. Section through pelleted virions showing their unstained centers and the loose packing characteristic of all BYDV isolates under these and other conditions.

Fig. 8. The MAV isolate is here mixed with type strain of tobacco mosaic virus and stained with vanadatomolybdate to illustrate relative sizes.

Fig. 9. A positive print of a molybdenum replicate showing typical icosahedral shadow patterns.

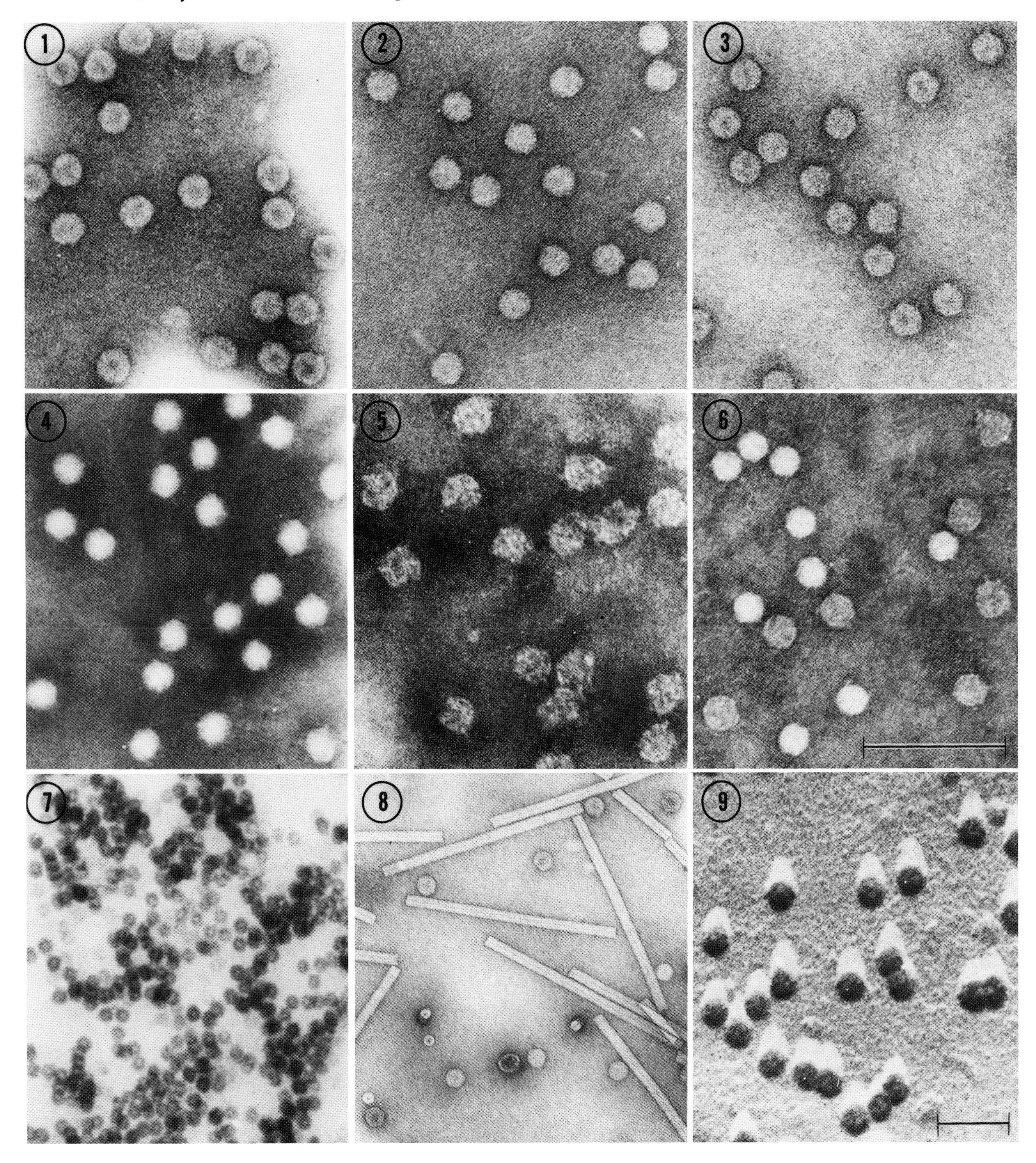
1
2
3
4
5
6
7
8
9

Figs. 10–13. Barley yellow dwarf virus (an isolate similar to MAV) in immature sieve elements of a host plant (*Avena sativa* L.). Figures 10–13 courtesy of C. C. Gill (Gill and Chong, 1975) and reproduced by permission of Academic Press, Inc.

Fig. 10. A mass of filaments and virus particles in the cell cytoplasm, adjacent to a pitfield (top), at an early stage of infection. (×42,000; bar = 0.5 μm.)

Figs. 11–13. Portions of cells in advanced stages of infection.

Fig. 11. Enlarged view of virus particles and associated filaments. (×87,000; bar = 0.25 μm.)

Fig. 12. Virus particles arranged in circles or whorls within membranes. (×42,000; bar = 0.5 μm.)

Fig. 13. Filaments and virus particles in the nucleoplasm; heterochromatin is darkly stained. (×65,000; bar = 0.25 μm.)

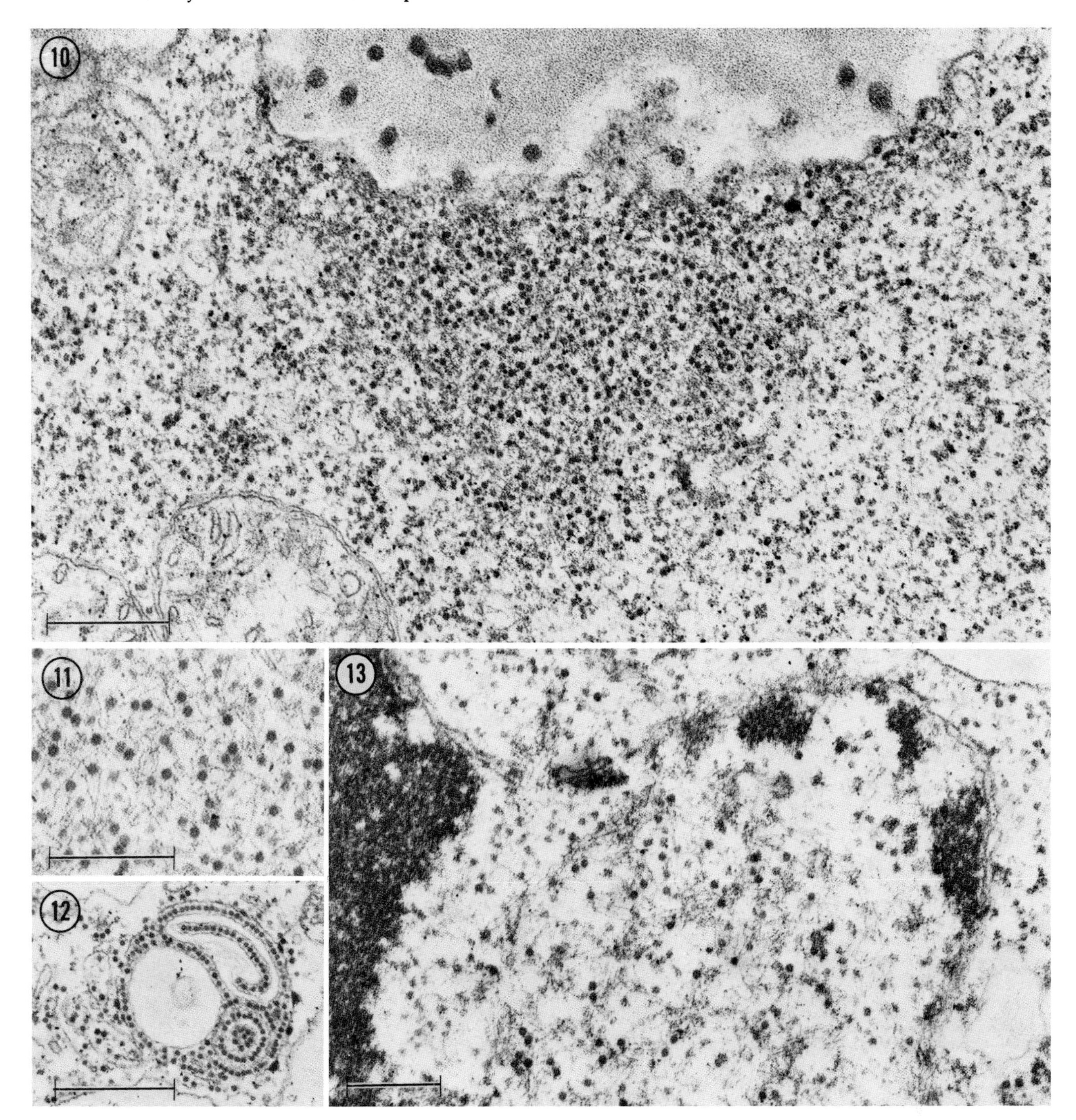
10
11
13
12

Chapter 23

Ilarvirus (Necrotic Ringspot) Group, Subgroup B

L. S. LOESCH-FRIES, E. L. HALK, AND R. W. FULTON

The ringspot viruses of *Prunus* were originally grouped together (Fulton, 1968a) because of similarities in particle morphology, stability characteristics, and symptoms produced in diseased plants, hence, the name ILAR (Isometric LAbile Ringspot). This group included *Prunus* necrotic ringspot (NRSV), prune dwarf (PDV), apple mosaic (ApMV), North American plum line pattern, and Danish plum line pattern viruses. These, with tobacco streak, elm mottle, and other viruses have been proposed as a taxonomic group (Shepherd *et al.*, 1976). These latter viruses are serologically distinct and are considered as subgroup A. The *Prunus* viruses and apple mosaic virus are considered subgroup B. Only subgroup B viruses will be discussed here, particularly NRSV, PDV, and ApMV which have been studied in most detail.

The woody and herbaceous host ranges of these viruses vary, but they typically produce ringspot symptoms often followed by latency. These viruses are multiparticulate with particles from 14 to 30 nm in diameter. They have a maximum longevity in diluted plant sap of 9 to 18 hours (Boyle *et al.*, 1954; Fulton, 1957; Paulsen and Fulton, 1969) and have thermal inactivation points between 44° and 62° (Waterworth and Fulton, 1964). The instability in undiluted plant sap is primarily caused by sensitivity to oxidized polyphenols (Hampton and Fulton, 1961). Infectivity can be extended by including in extracts antioxidants such as 2-mercaptoethanol or copper chelating compounds such as diethyl dithiocarbamate. The A_{260}/A_{280} ratios (1.45 to 1.6) and the buoyant densities (about 1.35 in CsCl, except for PDV) of these viruses suggest a relatively low content of RNA. The RNA is surrounded by a protein coat; no envelope is present. ILAR viruses can be separated into several particle types by analytical or sucrose rate zonal density gradient centrifugation (Fulton, 1967; Tolin, 1963; Paulsen and Fulton, 1969; Loesch and Fulton, 1975). Particle types differ in size, which presumably reflects differences in the RNA content and in infectivity characteristics.

It is difficult to preserve the ultrastructure of these viruses for electron microscopy, therefore, some of the particles often appear flattened and distorted. Thus far, it has not been possible to distinguish these viruses in infected cells, probably because they are small and in relatively low concentration.

Recent evidence indicates particle types of subgroup B viruses differ in size (Lister and Saksena, 1976). The particle types have different sedimentation coefficients (de Sequeira, 1967; Loesch and Fulton, 1975; E. L. Halk, unpublished). Purified virus preparations show only one peak upon free boundary electrophoresis (van Regenmortel, 1964; Fulton, 1967; E. L. Halk, unpublished), but when subjected to electrophoresis in polyacrylamide gels, they separate into several bands corresponding to the centrifugal components (E. L. Halk and L. S. Loesch-Fries, unpublished). Separation of particle types, therefore, must be primarily on the basis of size. The relative sizes of the main particle types estimated from electron micrograph measurements are: for NRSV, 23 nm, 25 nm, and 27 nm; for PDV, 14 nm, 17 nm, and 23 nm; for ApMV, 21 nm, 24 nm, and 26 nm (Fig. 1) (E. L. Halk and L. S. Loesch-Fries, unpublished). Measurements of both of the middle two of four particle types of ApMV averaged 24 nm. Because polyacrylamide gel electrophoresis and sucrose density gradient centrifugation indicated four particle types, it seems probably that these measurements were made on incompletely separated particle types.

Presumably, the genome of these viruses is divided, as with other multiparticle plant viruses, among the particles. The viruses contain 14–16% RNA (Barnett and Fulton, 1969; E. L. Halk, unpublished). Four or five pieces of RNA are present in preparations of nucleic acid from these viruses. Their estimated molecular weights from polyacrylamide gel electrophoresis are 1.30×10^6, 0.89×10^6, 0.67×10^6, and 0.27×10^6 for NRSV (Loesch and Fulton, 1975); 1.26×10^6, 0.95×10^6, 0.76×10^6, 0.68×10^6, and 0.30×10^6 for PDV (E. L. Halk, unpublished); 1.23×10^6, 1.00×10^6, 0.69×10^6, and 0.31×10^6 for ApMV (L. S. Loesch-Fries, unpublished). There is no information as to how the pieces are distributed among particle types. Few, if any, particle types have an entire complement of RNA segments since infectivity data indicate two or more particles are necessary for infectivity (Loesch and Fulton, 1975; E. L. Halk, unpublished; Fulton, 1962).

These viruses can be grouped into several antigenically distinct types. One group includes NRSV, Danish plum line pattern virus, and a virus isolated from hop. These viruses cross react with antiserum to other members of the group, but are not serologically identical. Rose mosaic virus, ApMV, and another virus isolated from hop compose a second group. Members of this group are serologically identical and have a few epitopes in common with those of the NRSV serological group (Fulton, 1968b; Casper, 1973). PDV and North American plum line-pattern viruses are serologically distinct from each other and the other two groups.

Little is presently known about the coat protein. NRSV protein subunits have 196 amino acid residues and a molecular weight of 2.5×10^4 (Barnett and Fulton, 1969). The molecular weight of PDV coat protein is about 2.4×10^4 (E. L. Halk, unpublished). Evidence indicates that subgroup B viruses require all RNA's for infectivity; however, coat protein can substitute for the smallest RNA (Gonsalves and Fulton, 1977; E. L. Halk, unpublished).

REFERENCES

Barnett, O. W., and Fulton, R. W. (1969). *Virology* **39,** 556–561.
Boyle, J. S., Moore, J. D., and Keitt, G. W. (1954). *Phytopathology* **44,** 303–312.

Casper, R. (1973). *Phytopathology* **63,** 238–240.
de Sequeira, O. (1967). *Virology* **31,** 314–322.
Fulton, R. W. (1957). *Phytopathology* **47,** 683–687.
Fulton, R. W. (1962). *Virology* **18,** 477–485.
Fulton, R. W. (1967). *Phytopathology* **57,** 1197–1201.
Fulton, R. W. (1968a). *Tagunsber., Dtsch. Akad. Landwirtschaftswiss.* **97,** 123–138.
Fulton, R. W. (1968b). *Phytopathology* **58,** 635–638.
Gonsalves, D., and Fulton, R. W. (1977). *Virology* (in press).
Hampton, R. E., and Fulton, R. W. (1961). *Virology* **13,** 44–52.
Lister, R. M., and Saksena. K. N. (1976). *Virology* **70,** 440–450.
Loesch, L. S., and Fulton, R. W. (1975). *Virology* **68,** 71–78.
Paulsen, A. Q., and Fulton, R. W. (1969). *Ann. Appl. Biol.* **63,** 233–240.
Shepherd, R. J., Francki, R. I. B., Hirth, L., Hollings, M., Inouye, T., MacLeod, R., Purcifull, D. E., Sinha, R. C., Tremaine, J. H., Valenta, V., and Wetter, C. (1976). *Intervirology* **6,** 181–184.
Tolin, S. A. (1963). *Phytopathology* **53,** 891.
van Regenmortel, M. H. V. (1964). *Virology* **23,** 495–503.
Waterworth, H. E., and Fulton, R. W. (1964). *Phytopathology* **54,** 1155–1160.

Fig. 1. Top, middle, and bottom particles of PDV, NRSV, and ApMV fixed with 0.5% glutaraldehyde, 0.5% acrolein and negatively stained with 1% sodium phosphotungstate. (A) PDV top particles; (B) PDV middle particles; (C) PDV bottom particles; (D) NRSV top particles; (E) NRSV middle particles; (F) NRSV bottom particles; (G) ApMV top particles; (H) both types of ApMV middle particles; (I) ApMV bottom particles. Bar = 100 nm.

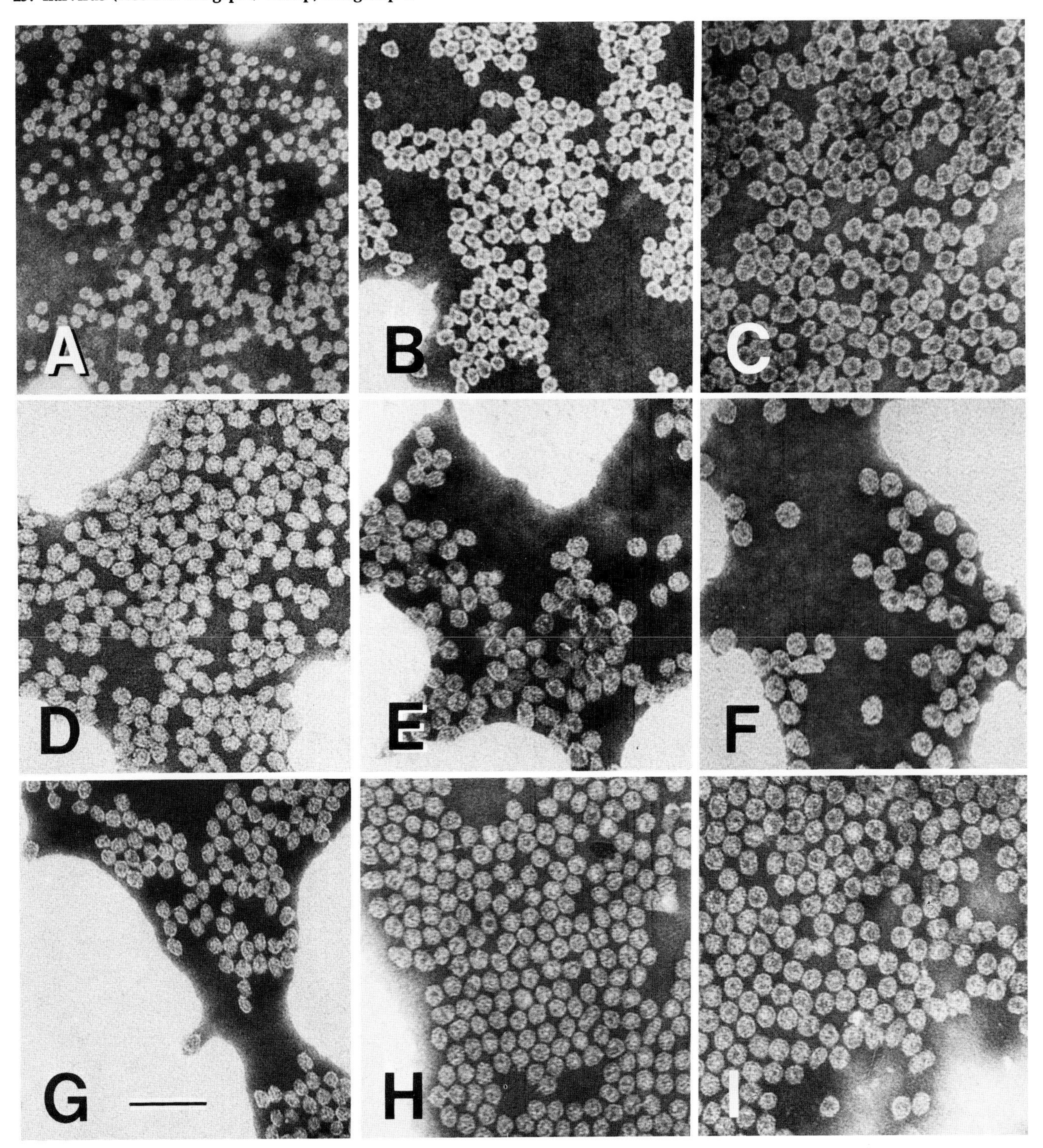
A
B
C
D
E
F
G
H
I

Chapter 24

Plant Reovirus Group

EISHIRO SHIKATA

I. General Description 377
II. Structure of the Virions 378
III. Localization of the Virus Particles *in Situ* 379
References 381

I. GENERAL DESCRIPTION

Common features of the viruses belonging to this group are (1) large isometric particles of 70–80 nm in diameter; (2) double-stranded RNA of viral genome; (3) not sap transmissible; (4) transmitted by leafhoppers or planthoppers in persistent (propagative) manner; (5) multiplication of the virus either in plant or insect hosts; (6) characteristic cytopathic changes, namely, "viroplasm" formed in the infected cells of both hosts.

This group has not yet been clearly established, because of lack of data on the composition and structure of some viruses. Wound tumor virus (WTV) (Black, 1970), rice dwarf virus (RDV) (Iida *et al.*, 1972), maize rough dwarf virus (MRDV) (Lovisolo, 1971), rice black-streaked dwarf virus (RBSDV) (Shikata, 1974), sugarcane Fiji disease virus (FDV) (Hutchinson and Francki, 1973), cereal tillering disease virus (CTDV) (Lindstein *et al.*, 1973), oat sterile dwarf virus (OSDV) (Lindstein *et al.*, 1973), and pangola stunt virus (PSV) (Kitajima and Costa, 1971) are tentatively listed as possible members of this group.

WTV and RDV are transmitted by leafhoppers, and MRDV, RBSDV, FDV, CTDV, and PSV are transmitted by planthoppers. All viruses in this group are known to multiply either in plant and insect hosts as demonstrated by insect transmission experiments, artificial injection techniques, and electron microscopy. Transovarial passage of the viruses to the progeny insects has been found in RDV (Fukushi, 1933), WTV (Black, 1953), and MRDV (Harpaz, 1972). Except for WTV and PSV, most viruses listed above infect only plants belonging to the family Gramineae.

II. STRUCTURE OF THE VIRIONS

Morphology of WTV was first reported as icosahedral particles of 60 nm in diameter with 92 capsomeres (Bills and Hall, 1962). In 1968, Kimura and Shikata carried out electron microscopic studies on the fine structure of RDV and demonstrated an icosahedron model, 75 nm in diameter, consisting of 32 capsomeres. Each capsomere is composed of five or six structural units (Figs. 1 and 2). A total of 180 structural units cover the surface of a particle. Recently, I. Kimura (personal communication; Black, 1970) noted that electron micrographs of purified WTV showed fine structure details identical with those of RDV. Thus, these two viruses supposedly have similar morphological characteristics as indicated in (Figs. 1 and 3).

Initial studies on the purification and morphology of MRDV and RBSDV indicated that the purified particles of those viruses obtained either from diseased plant or viruliferous insect hosts were similar in shape, about 60 nm in diameter without any outer projections (Kitagawa and Shikata, 1969; Wetter *et al.*, 1969) (Fig. 4). These two viruses cause similar diseases, have similar host range and insect vectors (Lovisolo, 1971; Harpaz, 1972; Shikata, 1974), and appear to be closely related according to serological reactions (Luisoni *et al.*, 1973). Most likely MRDV and RBSDV represent different strains of the same virus.

Lesemann (1972) and Milne *et al.* (1973) pointed out that some outer projections or spikes were present on MRDV particles 75 nm in diameter when the particles were carefully purified or fixed prior to negative staining (Figs. 9a–e). The outer projections or spikes were not visible with some purification procedures and PTA staining. The possible structure of complete MRDV has been indicated by Milne *et al.* (Figs. 9a–d). Dip preparations in PTA of diseased leaves of RBSDV, fixed in osmic acid or paraformaldehyde prior to dipping, clearly revealed the larger particles, 80 to 85 nm in diameter, with outer projections (Figs. 5 and 6). There are no such particles in dip preparations in PTA without prefixation, only smaller particles 60 nm in diameter.

The size of Fiji disease virus in dip preparations, and in plant and insect hosts is 70 nm in diameter (Figs. 21, 22, and 23), but the purified preparations revealed a diameter of 55–60 nm (Teakle and Steindle, 1969; Francki and Grivell, 1972). Pangola stunt virus about 60–70 nm in diameter with a dense core of 40 nm has been found infected cells of both hosts. Purified virus from diseased plant and insect hosts revealed a diameter of 70 nm (Kitajima and Costa, 1971).

No serological relationship was observed between reovirus and WTV (Gamez *et al.*, 1967) nor among viral protein of FDV, RDV, or MRDV (Ikegami and Francki, 1973), whereas MRDV and RBSDV (Luisoni *et al.*, 1973), MRDV, and pangola stunt virus (R. G. Milne, personal communication) showed a close serological relationship.

Sedimentation coefficients of the virus particles are 514 ± 10 S for WTV (Kalmakoff *et al.*, 1969) and 510 S for RDV (Iida *et al.*, 1972).

Double-stranded RNA of the viral genome was first demonstrated in WTV (Black and Markham, 1963; Gomatos and Tamm, 1963). RDV (Miura *et al.*, 1966), MRDV (Ikegami and Francki, 1973), RBSDV (Y. Kitagawa and T. Kodama, personal communication), and FDV (Ikegami and Francki, 1973, 1975) all proved to have double-stranded RNA genomes. dsRNA content is 22% for WTV (Gomatos and Tamm, 1963) and 11% for RDV (Miura *et al.*, 1966). The molecular weight of each viral RNA is $15–17 \times 10^6$ for WTV (Reddy and Black, 1973; Reddy *et al.*, 1974), $16–17 \times 10^6$ for RDV (Reddy *et al.*, 1974), $19–20 \times 10^6$ for MRDV (Reddy *et al.*, 1975b), $19–20 \times 10^6$ for RBSDV (Reddy *et al.*, 1975b), and $19–20 \times 10^6$ for FDV (Reddy *et al.*, 1975a).

Polyacrylamide gel coelectrophoretic separation of dsRNA genome segments was carried out by Reddy and Black (1973) and Reddy *et al.* (1974; 1975a,b) on WTV, RDV, FDV, MRDV, and RBSDV. The results indicated that WTV and RDV genomes were composed of 12 separate segments, whereas 10 segments were obtained from FDV, MRDV, RBSDV, and reovirus type 2. It is very interesting to note that the number of segments separated from WTV and RDV genomes that belong to leafhopper-borne subgroup differs from that of RDV, MRDV, and RBSDV genomes that belong to a planthopper-borne subgroup.

Base composition of dsRNA of WTV is G:A:C:U = 18.6:31.1:19.1:31.3 (Gomatos and Tamm, 1966), and that of RDV is G:A:C:U = 22.8:27.8:21.4:20.8 (Miura *et al.*, 1966).

III. LOCALIZATION OF THE VIRUS PARTICLES *IN SITU*

In addition to the difference in the number of dsRNA segments, the leafhopper-borne subgroup showed characteristic cytopathic changes in the infected cells of both hosts when examined by electron microscopy. The planthopper-borne subgroup produces similar cytopathic changes but these are slightly different from the leafhopper-borne subgroup.

After the first demonstration of multiplication of RDV in its insect vectors (Fukushi, 1939, 1940) and subsequent confirmation of this finding with WTV by artificial injection (Black and Brakke, 1952), electron microscope observations provided direct evidence for virus multiplication in both hosts (Fukushi *et al.*, 1962; Shikata, 1966; Shikata and Maramorosch, 1967b).

In general, four types of virus accumulation are seen in cells of both hosts infected with the viruses of this group. They are virus accumulation (1) in phagocytic structures which are defined structures surrounded by myelin-like membranes (Fig. 14), (2) in tubular structures which surround a row of virus particles (Figs. 16 and 17), (3) in the viroplasm (Figs. 13 and 15), and (4) in crystalline arrays or accumulations in the cytoplasm (Figs. 10, 12, and 14).

Differences in cytopathic changes caused by the leafhopper-borne subgroup and the planthopper-borne subgroup are in the size and shape of the viroplasm appearing in the infected cells of both hosts. As shown in Figs. 11, 13, and 15, typical spherical viroplasms were usually formed by the leafhopper-borne subgroup surrounded by the virus particles at the periphery, and their size did not exceed the size of the nuclei. Spreading of immature particles within the viroplasm was scarcely seen. On the other hand, large and wide areas of amorphous shaped viroplasms were observed in the infected cells of the planthopper-borne subgroup (Figs. 18–21 and 23). The immature smaller particles were intermingled in the viroplasm, but the mature, larger particles were seen at the crevices or at the periphery of the viroplasm. According to the interpretation of experiments carried out by Shikata and Maramorosch (1967a), who used abdominally injected WTV, the virions could have been captured by pinocytosis in the cells. The first progeny virus was seen around the viroplasm at the initial stage of infection, then the viroplasm and the mature virus particles around the viroplasm increased in size and number within the infected cells (Shikata and Maramorosch, 1967a,b).

The viroplasm consists of electron dense granular or fine threadlike structures and no surrounding membranes, so that the viroplasm is in direct contact with the cytoplasm organelles (Figs. 11 and 15). The phagocytic structures in which virus particles accumulate at the later stage of infection are enclosed by layers of myelin-like membranes (Fig. 14). The tubular structures around a row of mature virus particles are seen only at the later stage of infection (Figs. 16 and 17). Crystalline arrays of virus particles are formed from the accumulation of the virus in the phagocytic structures (Fig. 14) or in the vacuolated parts of the cytoplasm as a result of virus assembly. Accordingly, it was concluded that the viroplasm supposedly represented the "factory" for WTV assembly. This view was supported by the study of RDV infected cells of monolayer cultures of *Nephotettix cincticeps,* RDV insect vectors, using autoradiography and immunoelectron microscopy with ferritin antibody (Nasu and Mitsuhashi, 1968).

There are two different sized particles in the infected cells of MRDV and RBSDV. The large mature particles, 75–85 nm in diameter with a dense core of 50 nm in diameter, were arranged at the crevice or at the periphery of the viroplasm, and scattered in the cytoplasm outside the viroplasm (Fig. 18–20). Their size corresponds to the complete virions of MRDV and to that of the larger particles obtained in the fixed preparations of RBSDV (Figs., 5, 6, and 8). Crystalline arrays of MRDV and RBSDV are always composed of the larger mature particles (Figs. 18 and 20).

The smaller particles scattered in the viroplasm have a diameter of 50 nm, which approximately corresponds to the size of purified MRDV and RBSDV. This seems to be the inner core of the virus particles (Figs. 4, 7, and 8).

Localization of WTV is restricted to the tumorous cells of diseased plants which arise from the phloem (Figs. 12 and 13) (Shikata and Maramorosch, 1966), whereas RDV was found in the cells of chlorotic portions of the diseased leaves of rice plants and in the mesophyl cells adjacent to the vascular bundle (Fig. 10) (Shikata, 1966). MRDV, RBSDV, and FDV were localized in the proliferating cells originated from the phloem of infected plants (Figs. 16, 17, and 21) (Shikata and Kitagawa, 1977; Gerola and Bassi, 1966; Teakle and Steindle, 1969). Some of the organs of the infected insects, such as fat body, intestine, Malpighian tubules, trachea, muscle, epidermis, mycetome, salivary gland, blood, nervous system, and ovarian tubules, were attacked by the viruses of this group (Shikata, 1966; Shikata and Maramorosch, 1965; Hirumi *et al.*, 1967; Vidano, 1967; Granados *et al.*, 1968; Francki and Grivell, 1972).

The finding of RDV and WTV multiplication in insects by means of electron microscopy confirmed that the virus can be carried from the intestines to the salivary glands by the blood cells, while the virus multiplied in different organs. Thus, the transmission continued efficiently after the incubation period. Fat body and other organs continuously supply the virus to the blood cells that act as a carrier of the virus in the insect. The mycetome was found, by means of electron microscopy, to carry RDV to the eggs at the early stage of ovary maturation (Nasu, 1965).

ACKNOWLEDGMENTS

The author wishes to express sincere appreciation to Dr. I. Kimura, Institute for Plant Virus Research, Chiba, Japan; Dr. O. Lovisolo, Dr. A. Appiano, and Dr. R. G. Milne, Laboratorio di Fito-virologia Applicata, Torino, Italy; Dr. R. Francki and Dr. T. Hatta, Waite Agricultural Institute, Glenn Osmond, South Australia; Dr. D. S. Teakle, Department of Microbiology, University of Queensland, Brisbane, Australia for their kind help in supplying the enlarged prints of electron micrographs and for the irvaluable suggestions in preparing this manuscript. He also is indebted to Dr. K. Maramorosch for reading the manuscript.

REFERENCES

Bills, R. F., and Hall, C. E. (1962). *Virology* **17,** 123–130.
Black, L. M. (1953). *Phytopathology* **43,** 9–10.
Black, L. M. (1970). *CMI/AAB Descriptions Plant Viruses* No. 34.
Black, L. M., and Brakke, M. K. (1952). *Phytopathology* **42,** 269–273.
Black, L. M., and Markham, R. (1963). *Neth. J. Plant Pathol.* **69,** 215.
Francki, R. I. B., and Grivell, C. J. (1972). *Virology* **48,** 305–307.
Fukushi, T. (1933). *Proc. Imp. Acad. (Tokyo)* **9,** 457–460.
Fukushi, T. (1939). *Proc. Imp. Acad. (Tokyo)* **15,** 142–145.
Fukushi, T. (1940). *J. Fac. Agric., Hokkaido Univ.* **45,** 83–154.
Fukushi, T., Shikata, E., and Kimura, I. (1962). *Virology* **18,** 192–205.
Gamez, R., Black, L. M., and MacLeod, R. (1967). *Virology* **32,** 163–165.
Gerola, F. M., and Bassi, M. (1966). *Caryologia* **19,** 13–40.
Gomatos, P. J., and Tamm, I. (1963). *Proc. Natl. Acad. Sci. U.S.A.* **50,** 878–885.
Granados, R. R., Ward, L., and Maramorosch, K. (1968). *Virology* **34,** 790–796.
Harpaz, I. (1972). "Maize Rough Dwarf Virus." Isr. Univ. Press, Jerusalem.
Hirumi, H., Granados, R. R., and Maramorosch, K. (1967). *J. Virol.* **1,** 430–444.
Hutchinson, P. B., and Francki, R. I. B. (1973). *CMI/AAB Descriptions Plant Viruses* No. 119.
Iida, T., Shinkai, A., and Kimura, I. (1972). *CMI/AAB Descriptions Plant Viruses* No. 102.
Ikegami, M., and Francki, R. I. B. (1973). *Virology* **56,** 404–406.
Ikegami, M., and Francki, R. I. B. (1975). *Virology* **64,** 464–470.
Kalmakoff, J., Lewandowski, L. J., and Black, D. R. (1969). *J. Virol.* **4,** 851–856.
Kimura, I., and Shikata, E. (1968). *Proc. Jpn. Acad.* **44,** 538–543.
Kitagawa, Y., and Shikata, E. (1969). *Mem. Fac. Agric., Hokkaido Univ.* **6,** 446–451.
Kitajima, E. W., and Costa, A. S. (1971). *Electron Microsc., Proc. Congr. Int. Congr., 7th 1970* Vol. **3,** pp. 323–324.
Lesemann, D. (1972). *J. Gen. Virol.* **16,** 273–284.
Lindstein, K., Gerhardson, B., and Pettersson, J. (1973). *Natl. Swed. Inst. Plant Prot., Contrib.* **15,** 375–397.
Lovisolo, O. (1971). *CMI/AAB Descriptions Plant Viruses* No. 72.
Luisoni, E., Lovisolo, O., Kitagawa, Y., and Shikata, E. (1973). *Virology* **52,** 281–283.
Milne, R. G., Conti, M., and Lisa, V. (1973). *Virology* **53,** 130–141.
Miura, K., Kimura, I., and Suzuki, N. (1966). *Virology* **28,** 571–579.
Nasu, S. (1965). *Jpn. J. Appl. Entomol. Zool.* **9,** 225–237.
Nasu, S., and Mitsuhashi, J. (1968). *Virus (Kyoto)* **18,** 40–42.
Reddy, D. V. R., and Black, L. M. (1973). *Virology* **54,** 557–562.
Reddy, D. V. R., Kimura, I., and Black, L. M. (1974). *Virology* **60,** 293–296.
Reddy, D. V. R., Boccardo, G., Outridge, R., Teakle, D. S., and Black, L. M. (1975a). *Virology* **63,** 287–291.
Reddy, D. V. R., Shikata, E., Boccardo, G., and Black, L. M. (1975b). *Virology* **67,** 279–282.
Shikata, E. (1966). *J. Fac. Agric., Hokkaido Univ.* **55,** 1–110.
Shikata, E. (1974). *CMI/AAB Descriptions Plant Viruses* No. 135.
Shikata, E., and Kitagawa, Y. (1977). *Virology* **77,** 826–842.
Shikata, E., and Maramorosch, K. (1965). *Virology* **27,** 461–475.
Shikata, E., and Maramorosch, K. (1966). *J. Natl. Cancer Inst.* **36,** 97–116.
Shikata, E., and Maramorosch, K. (1967a). *J. Virol.* **1,** 1052–1073.
Shikata, E., and Maramorosch, K. (1967b). *Virology* **32,** 363–377.
Teakle, D. S., and Steindle, D. R. L. (1969). *Virology* **37,** 139–145.
Vidano, C. (1967). *Atti Accad. Sci. Torino, Cl. Sci. Fis., Mat. Nat.* **101,** 717–733.
Wetter, C., Luisoni, E., Conti, M., and Lovisolo, O. (1969). *Phytopathol. Z.* **66,** 197–212.

Fig. 1. Purified particles of RDV, negatively stained in 2% PTA. (Courtesy of Dr. I. Kimura.)

Fig. 2. A model of RDV virion showing fivefold (top), threefold (middle), and twofold (bottom) symmetry and their respective super position. (Courtesy of Dr. I. Kimura.)

Fig. 3. Purified particles of WTV, negatively stained in 2% PTA. (Courtesy of Dr. I. Kimura.)

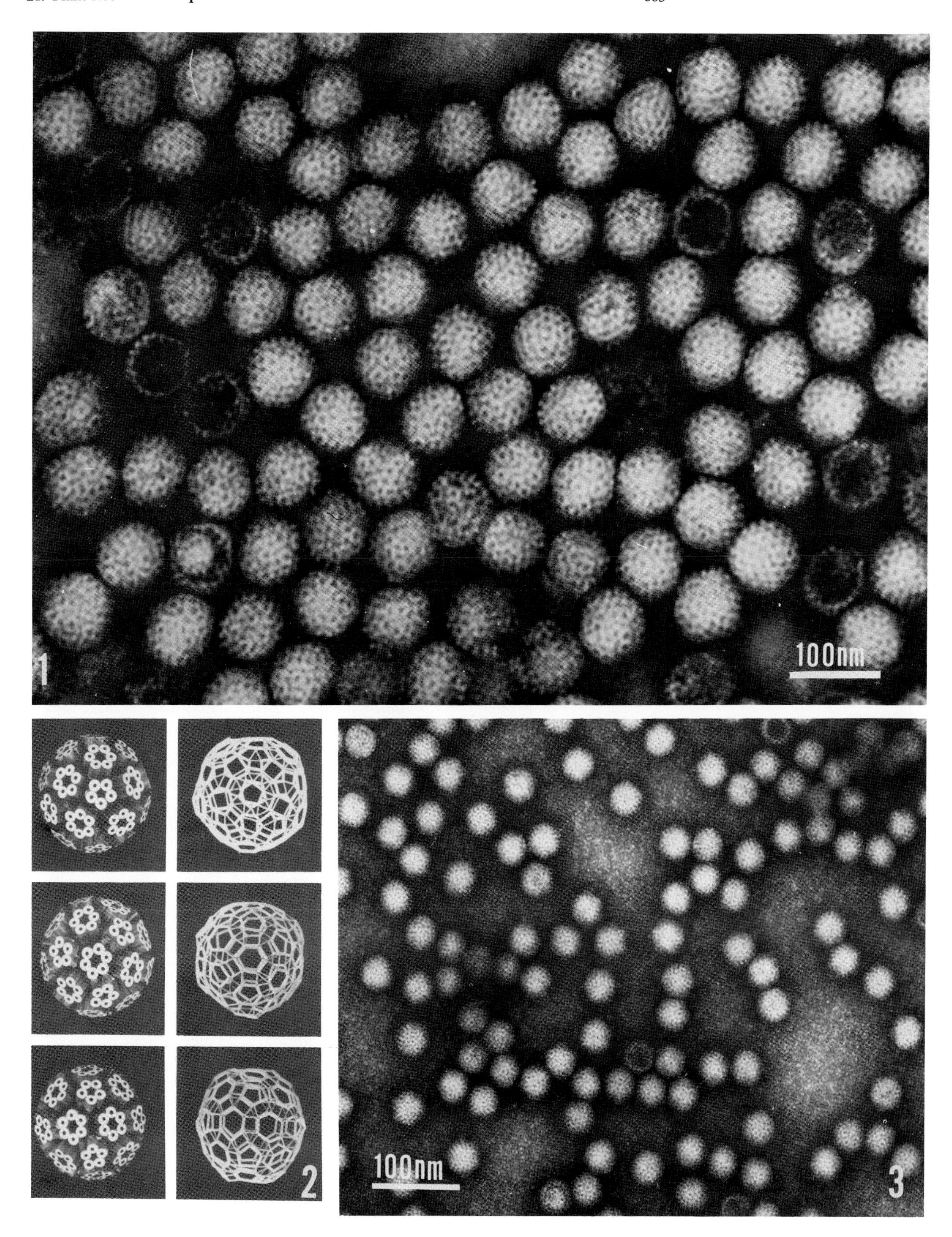
100nm
1
2
100nm
3

Fig. 4. Purified subviral particles of RBSDV, negatively stained in 2% PTA.

Fig. 5. A dip preparation in 1% PTA of RBSDV infected rice leaf prefixed in 2% osmic acid for 2 hours.

Fig. 6. A dip preparation in 1% PTA of RBSDV infected rice leaf prefixed in 2% paraformaldehyde for 2 hours.

Fig. 7. High magnification of an ultrathin section showing viroplasm in an RBSDV-infected cell of the insect vector, *Unkanodes albifascia*. Note that smaller particles which correspond to subviral particles are scattered within the viroplasm.

Fig. 8. High magnification of an ultrathin section showing accumulation of larger particles which correspond to complete virions in the cell cytoplasm from RBSDV-infected *U. albifascia*.

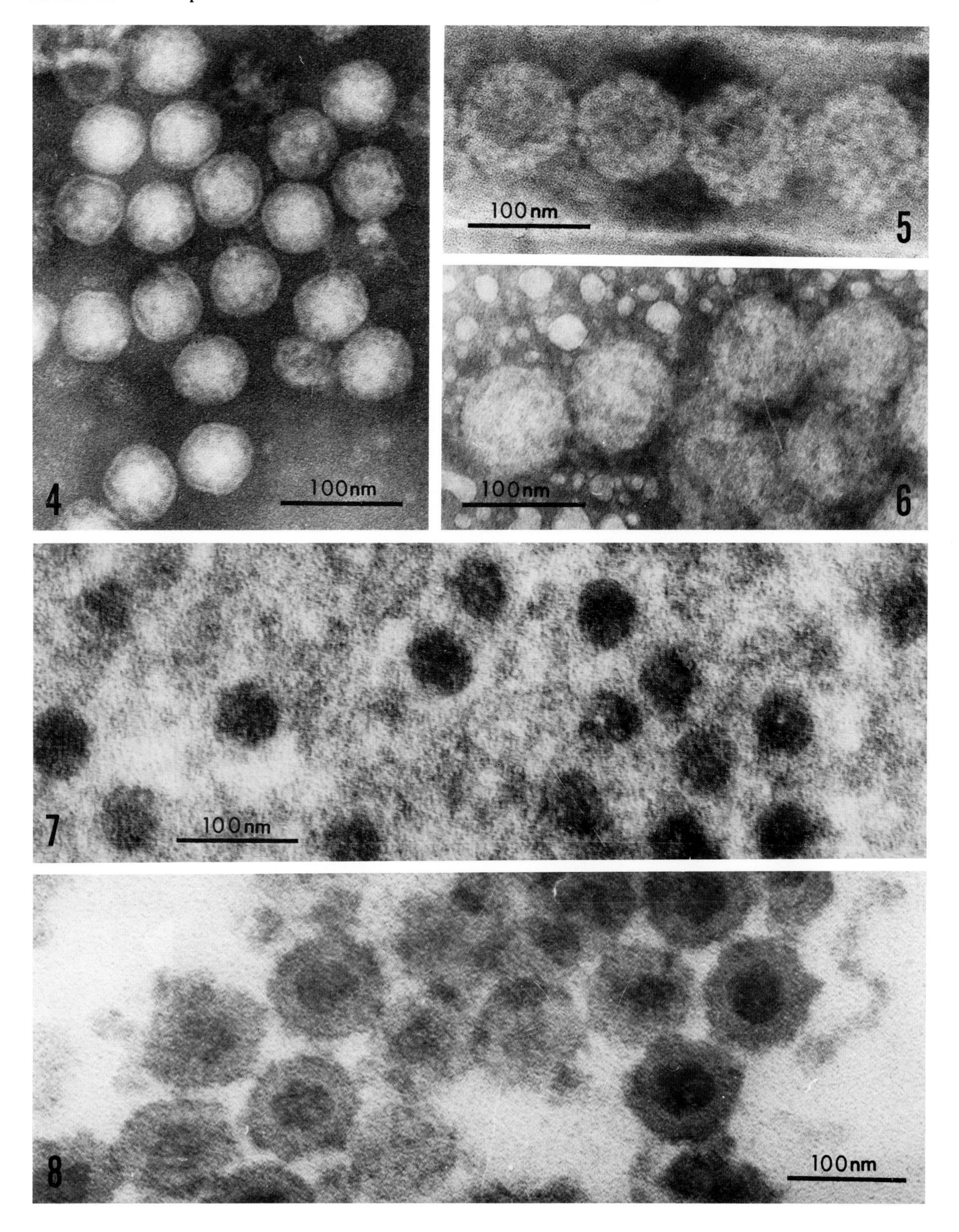
100 nm
4
100 nm
5
100 nm
6
100 nm
7
100 nm
8

Figs. 9a–e and A–C. MRDV particles negatively stained (a–e), and their models (A–C). (Courtesy of Dr. R. G. Milne.) Figs. 9a–c are the same magnification as indicated in e.

Fig. 9a. Complete virion fixed in glutaraldehyde and negatively stained in potassium phosphtungstate (KPT) showing the external A spike.

Fig. 9b. Partially stripped virions negatively stained with uranyl acetate (UA), compared with model B.

Fig. 9c. Inner capsid plus B spikes, stained in UA, compared with model C.

Fig. 9d. Inner capsid alone, stained in KPT.

Fig. 9e. Complete virions stained in UA, showing A spikes, compared with model A. Note that particles appear larger in KPT than in UA.

Fig. 9A. Model of virion, assuming icosahedral arrangement of 92 capsomeres in the outer capsid.

Fig. 9B. Partly stripped model.

Fig. 9C. Inner capsid with B spikes retained.

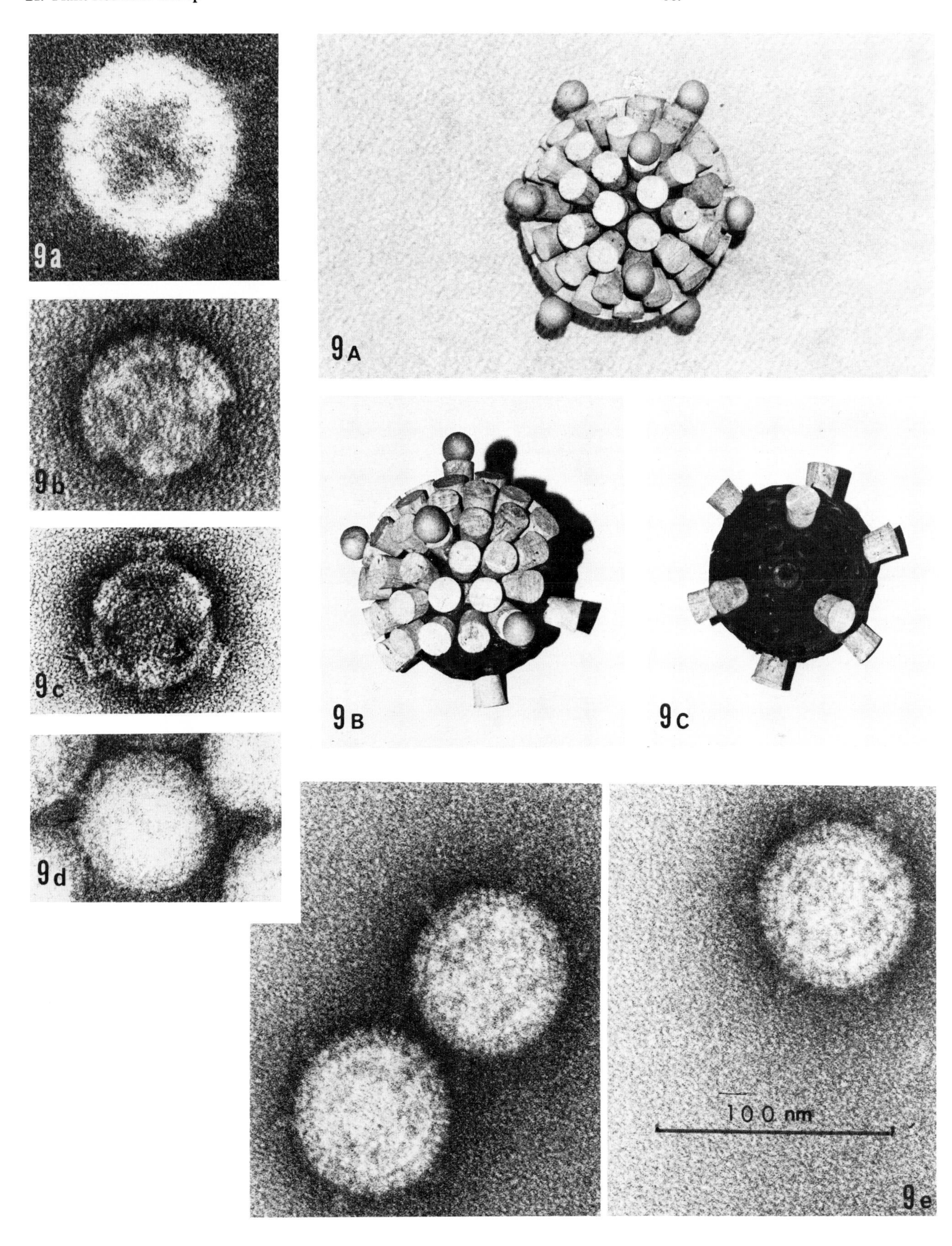
9a
9b
9c
9d
9A
9B
9C
100 nm
9e

Fig. 10. Accumulation of RDV particles in the chlorotic portion of the rice leaf.

Fig. 11. Virus particles arranged at the perphery of viroplasm (P) in an RDV infected cell of insect vector *Nephotettix cincticeps*.

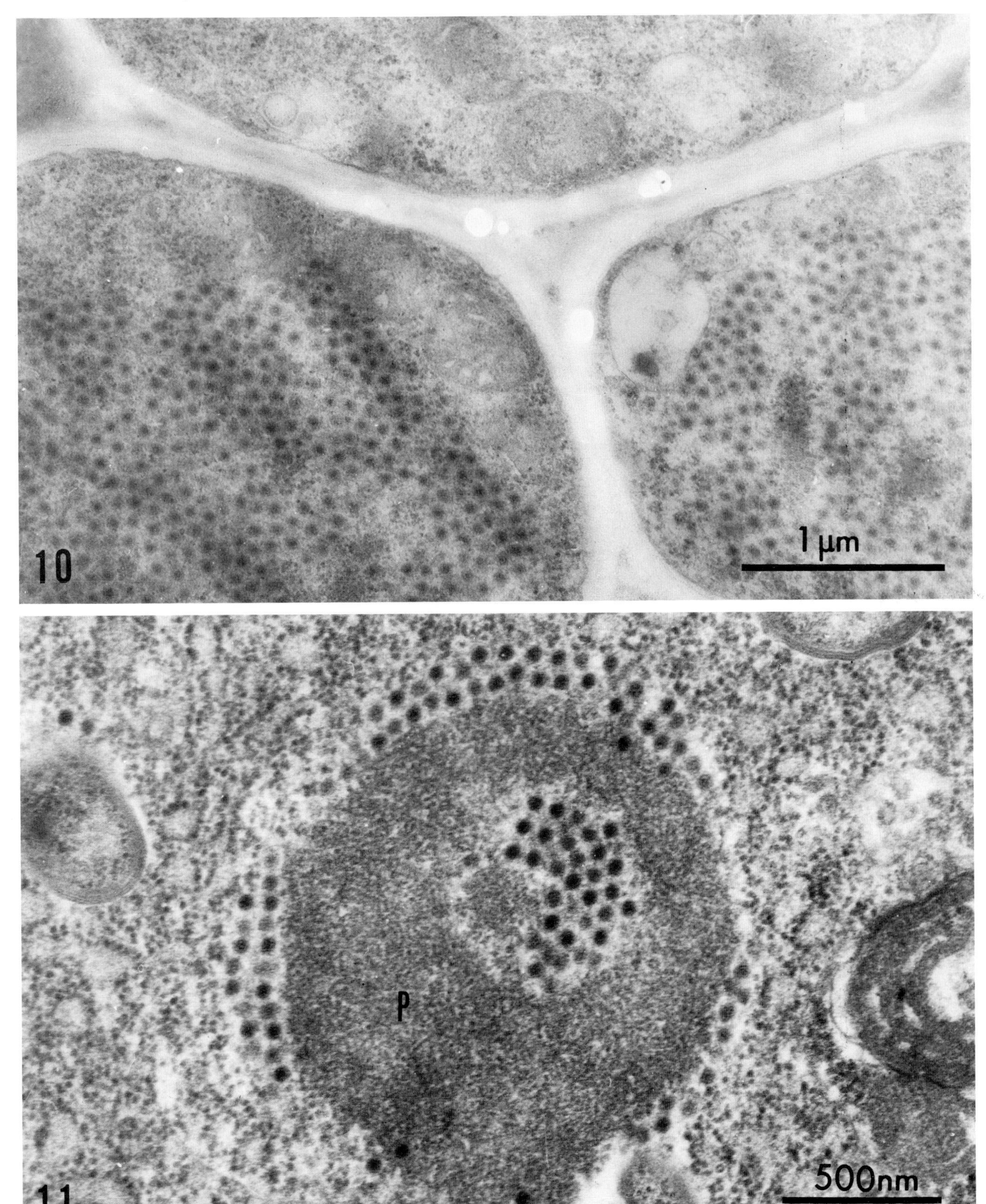
1 µm
10
P
500nm
11

Fig. 12. An ultrathin section of vein enation on infected crimson clover, *Trifolium incarnatum*. Note thick cell wall and WTV accumulation (arrows) in the tumorous cells.

Fig. 13. A round viroplasm (P) near the nucleus (N) in leaf veinlet enation of WTV infected crimson clover. (From Shikata and Maramorosch, 1967b, Fig. 14; Academic Press.)

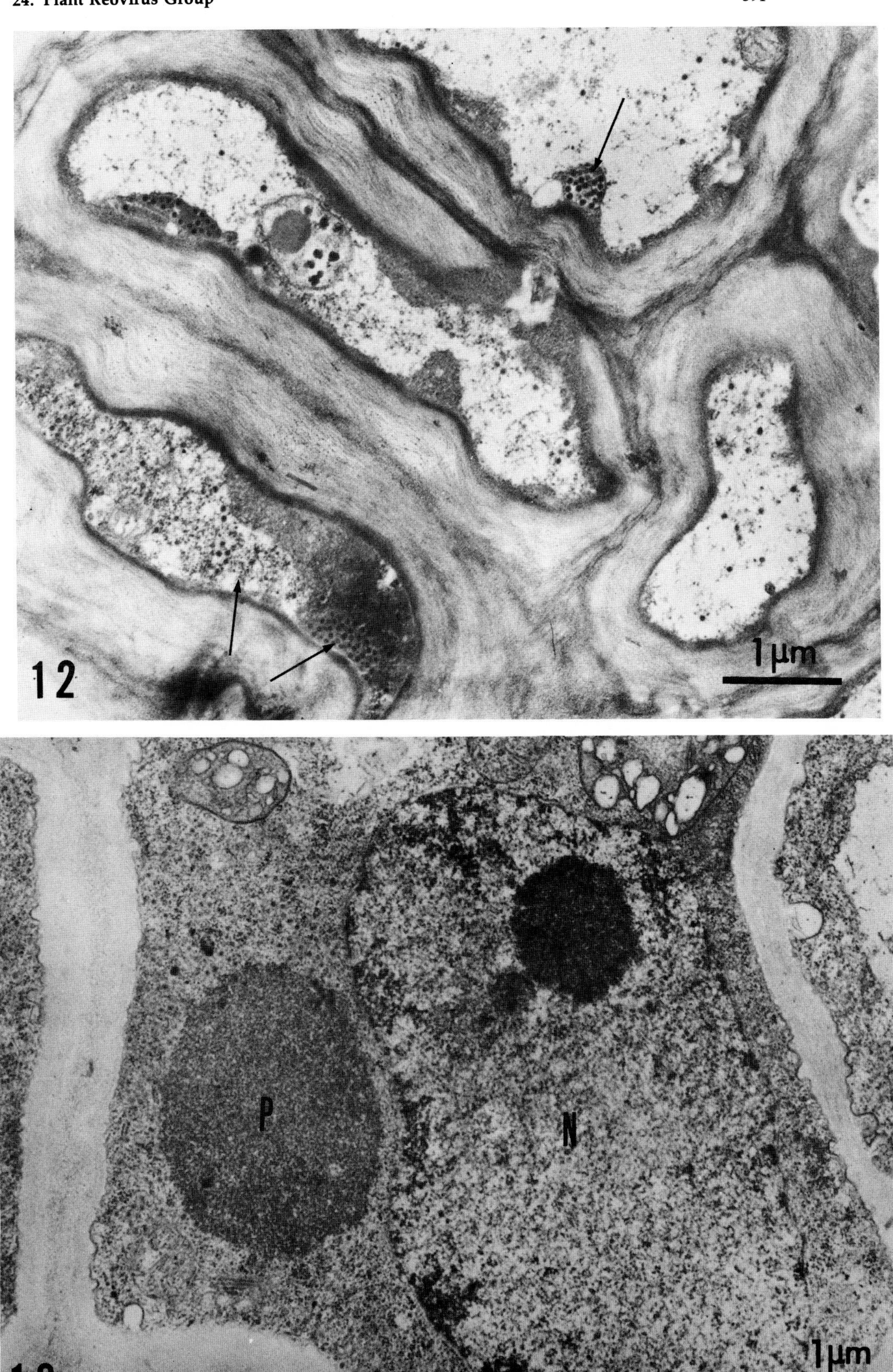
12
1 μm
13
P
N
1 μm

Fig. 14. Accumulation and crystalline formation of WTV in phagocytic structure (arrows) within cell cytoplasm of the insect vector, *Agallia constricta*. (From Shikata and Maramorosch, 1967a, Fig. 20; Americal Society for Microbiology.)

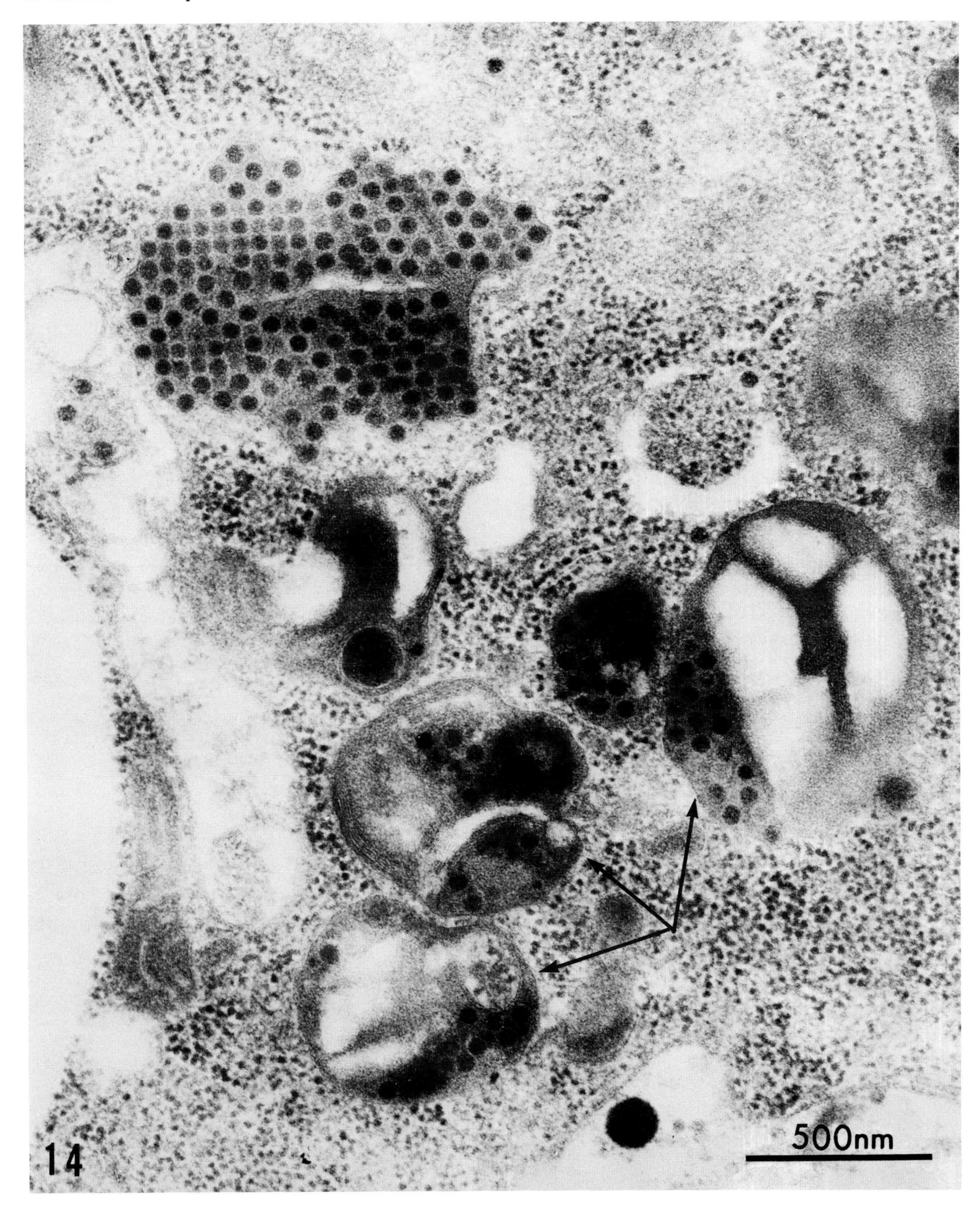

500nm
14

Fig. 15. A large viroplasm (P) observed in the cytoplasm of a WTV infected fat body cell from *A. constricta*. Note that the WTV particles are arranged at the periphery of the viroplasm. (From Shikata and Maramorosch, 1967a, Fig. 11; Americal Society for Microbiology.)

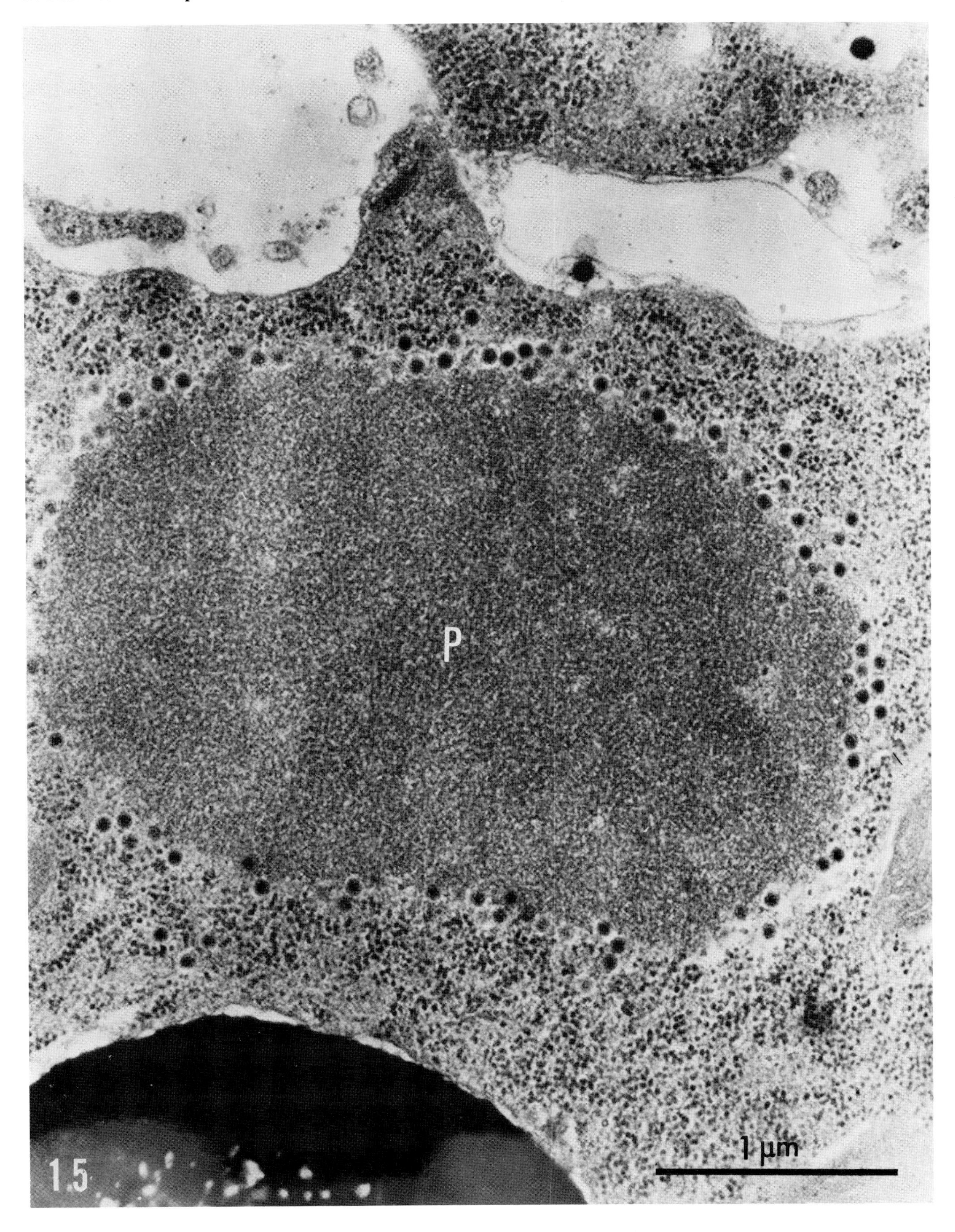
P
1 μm
15

Fig. 16. An ultrathin section of a corn plant infected with RBSDV. Virus particles appear in fibrillar viroplasm and in tubular structure in the cytoplasm.

Fig. 17. An ultrathin section of root neoplasia of a corn plant infected with MRDV reveals the virus particles within viroplasm (P) or in tubular structure in the cytoplasm. N, nucleus. (Courtesy of Dr. A. Appiano.)

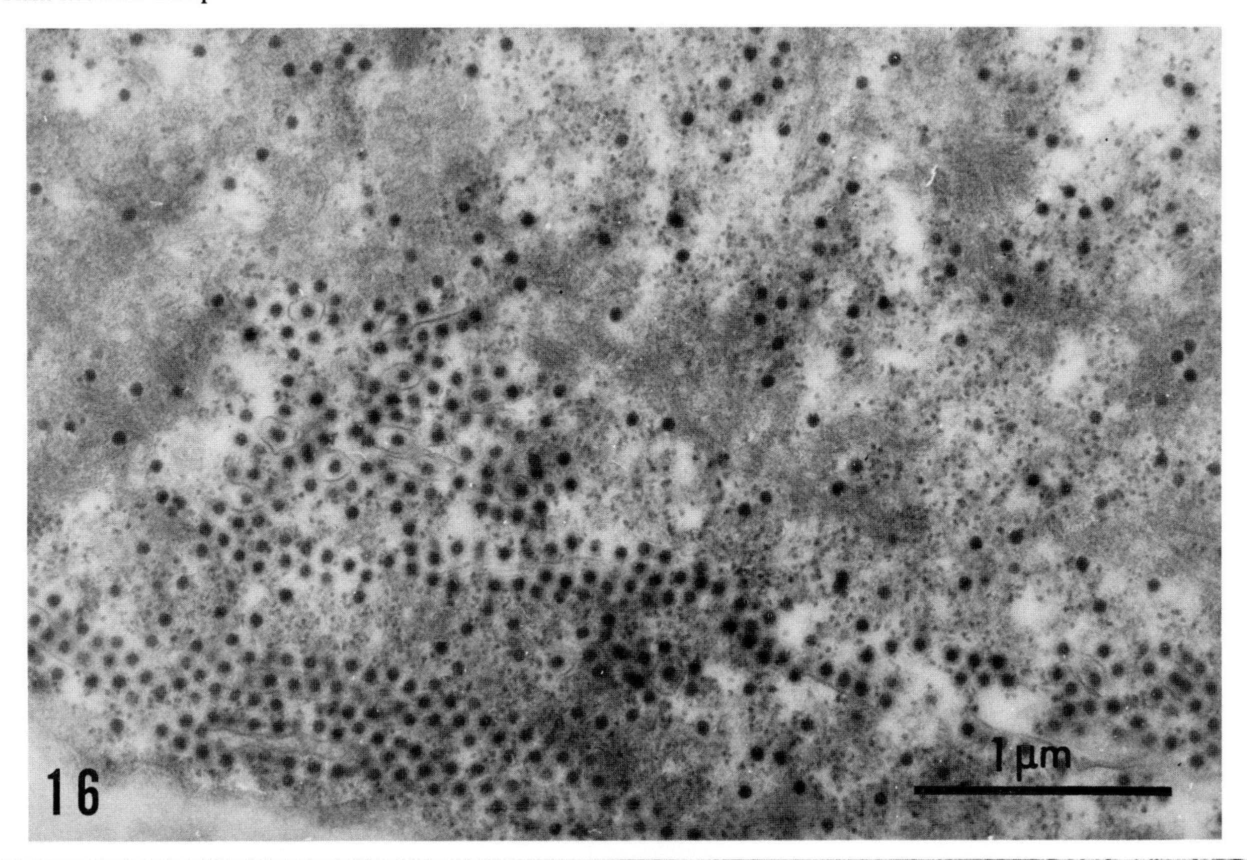
1 μm
16

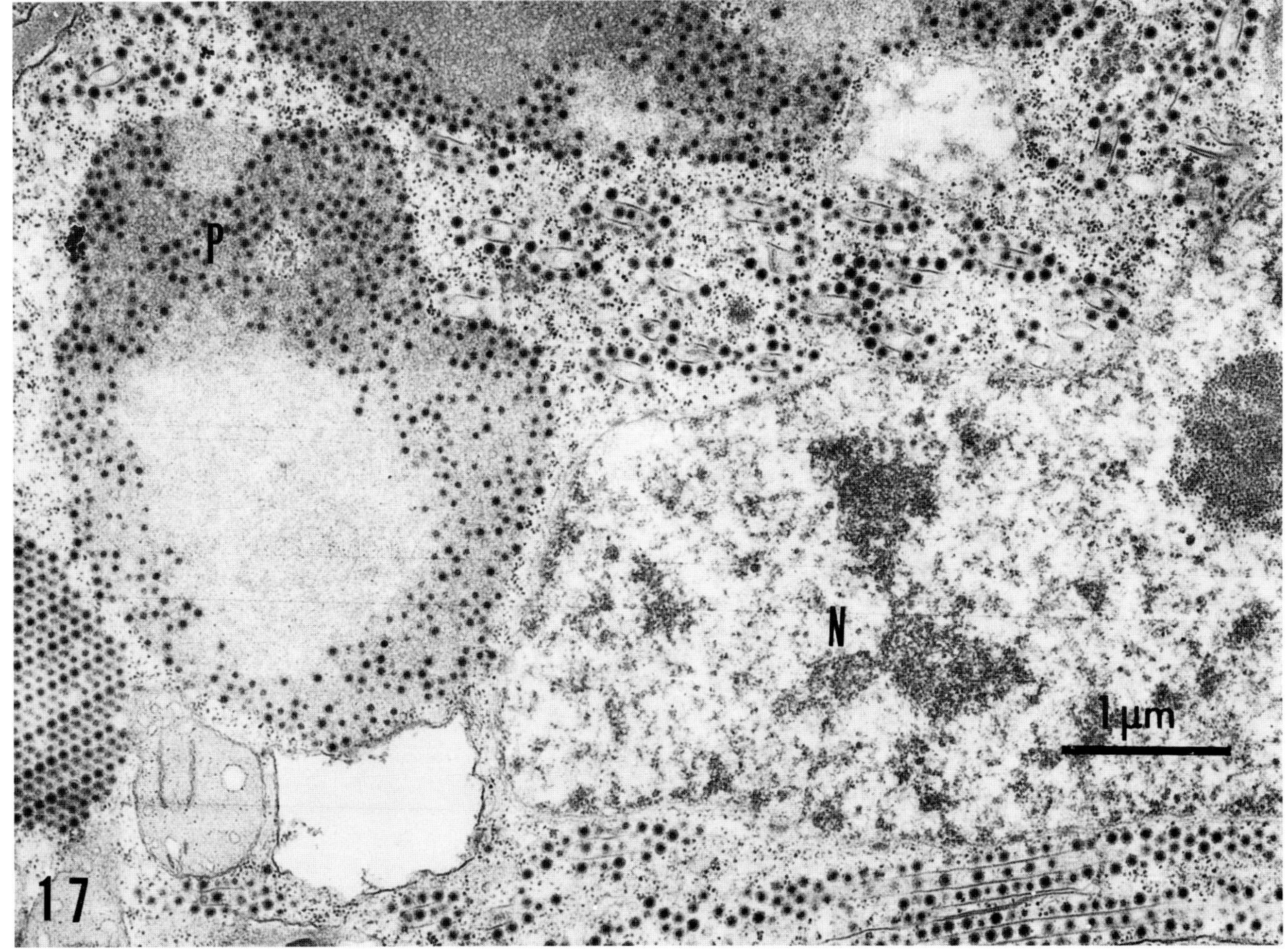
P
N
1 μm
17

Fig. 18. A crystalline arrangement of RBSDV which is composed of larger particles (complete virions) alone within large viroplasmic area. Some scattered particles are also observed. Note the fine fibrillar structure forming the viroplasm.

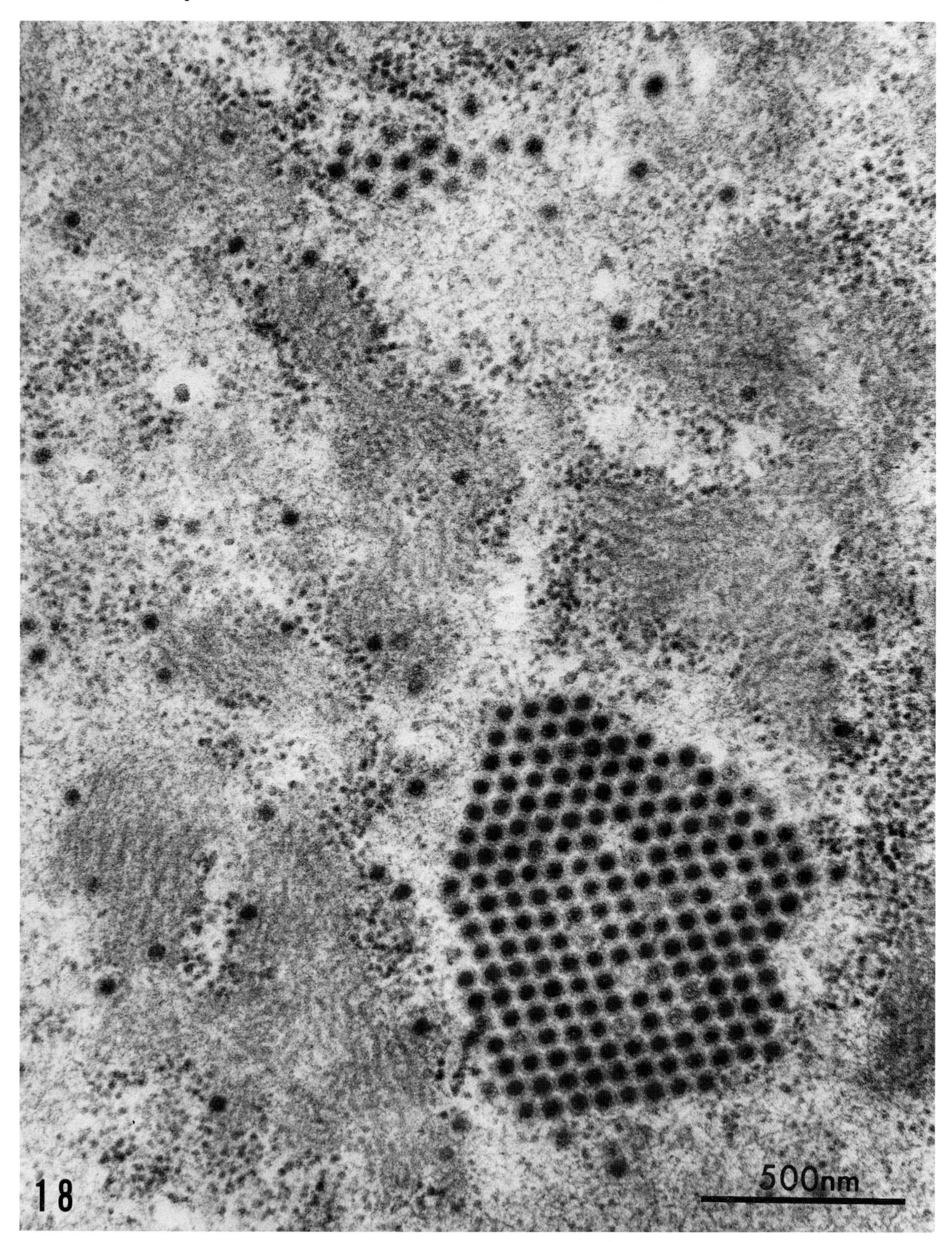
18
500nm

Fig. 19. A part of viroplasm induced by RBSDV infection of *U. albifascia*. Smaller particles (subviral particles) of RBSDV appeared within electron dense granular viroplasm.

Fig. 20. Granular and fibrillar structure of viroplasm induced by RBSDV infection of *U. albifascia*. Subviral particles are observed in the viroplasm and a crystalline form is composed of only complete particles.

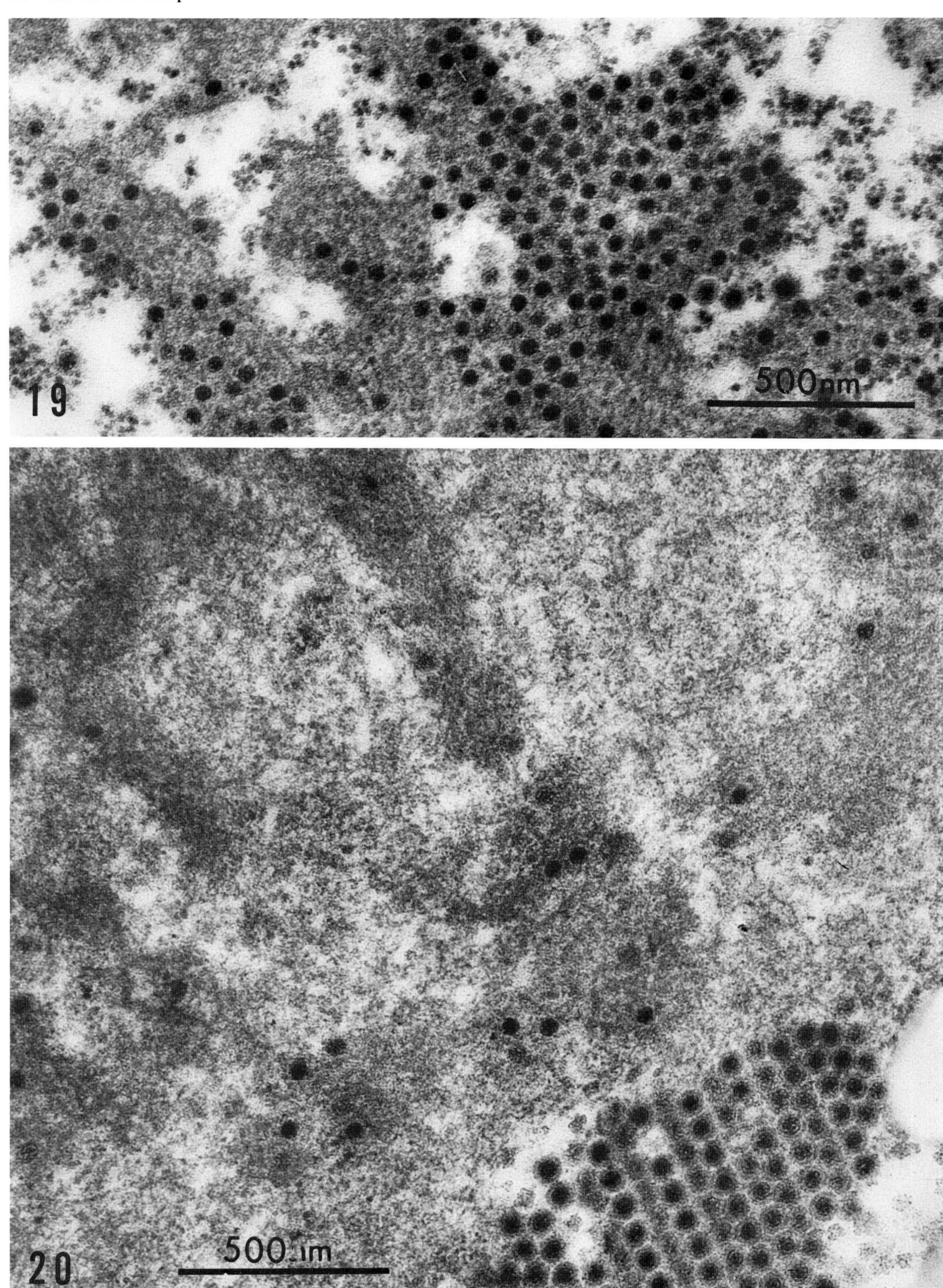
19
500nm
20
500 nm

Fig. 21. An ultrathin section of sugarcane gall cell reveals viroplasm (P) and FDV particles in the cytoplasm. (Courtesy of Dr. R. I. B. Francki and Dr. T. Hatta.)

Fig. 22. Leaf dip preparation of FDV particles stained with uranyl acetate. (Courtesy of Dr. R. I. B. Francki and Dr. T. Hatta.)

Fig. 23. An ultrathin section of an insect vector cell, *Perkinsiella saccharicida,* infected with FDV, showing viroplasm (P) and virus particles forming paracrystals in the cytoplasm. (Courtesy of Dr. R. I. B. Francki.)

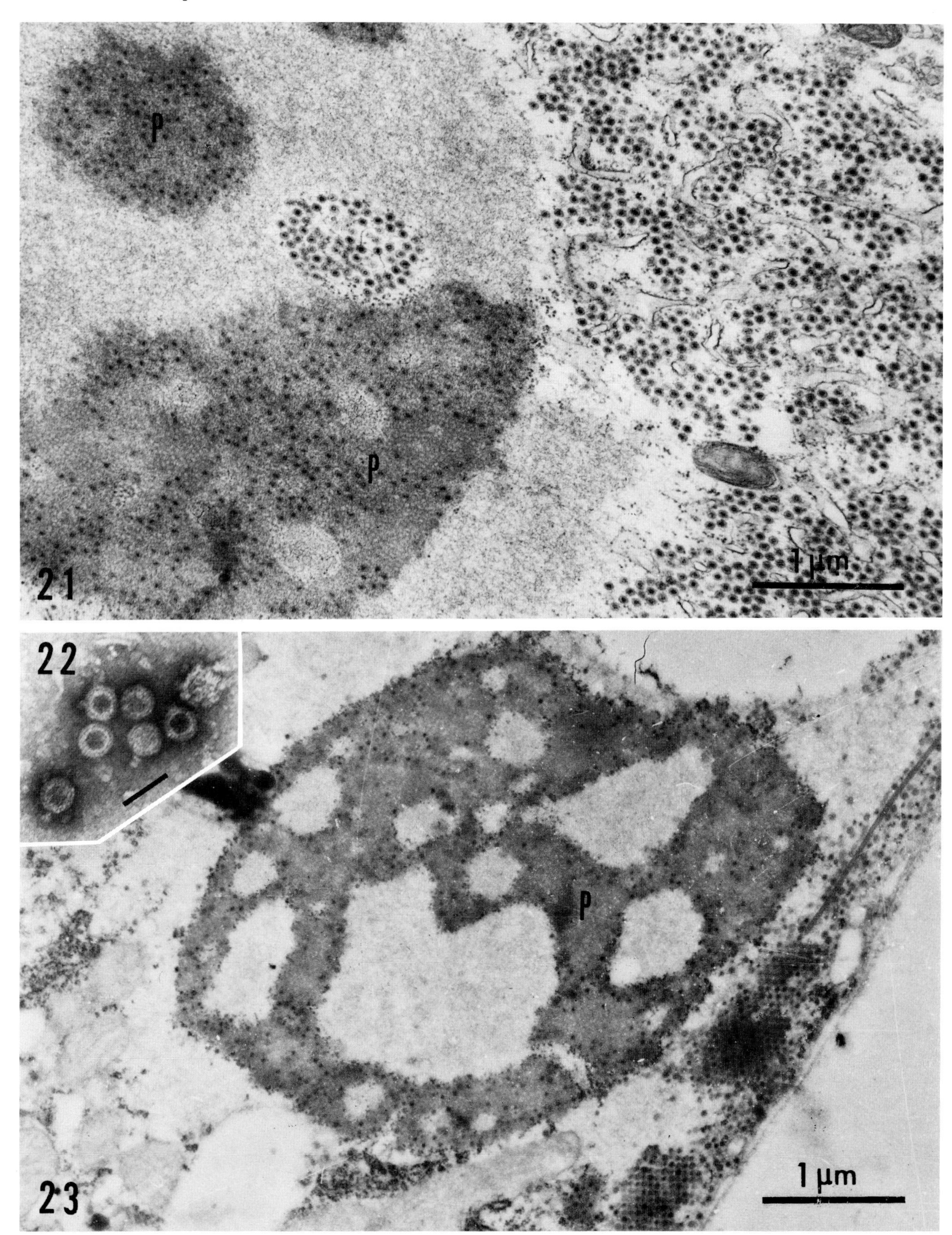

P
P
21
1 μm
22
P
23
1 μm

Chapter 25

Pinwheel Inclusions and Plant Viruses

H. W. ISRAEL AND H. J. WILSON

I. Introduction . . . 405
II. Morphology . . . 406
III. Morphogenesis . . . 407
A. Formation . . . 407
B. Transformation . . . 407
IV. Structure and Composition . . . 407
V. Serology . . . 408
VI. Conclusions . . . 408
References . . . 409

I. INTRODUCTION

Assorted nuclear and cytoplasmic inclusions often accompany plant virus infections. Sometimes both host and virus contribute to the composition of inclusions; other times, only one or the other participates (Rubio-Huertos, 1972). Moreover, the presence of either inclusion or virus in a cell may not require the coincidence of the other, for not all plant viruses induce special inclusions, just as some inclusions may occur independently of viruses (Matthews, 1970).

Pinwheel inclusions, so-named (Edwardson, 1966) because in certain profiles they have obviously reminiscent features (Figs. 11–13), were likely constituents of the amorphous aggregates shown some time ago by Bawden and Kassanis (1941) and Sheffield (1931, 1934) to be circumstantially associated with specific viruses. However, subsequent virological, pathological, and morphogenetic studies have revealed that pinwheels are present in plant cells under circumstances where viruses may or may not be implicated.

Pinwheels occur as strictly cytoplasmic inclusions in various tissues and organs of many different plants. While early work with light microscopy did little to uncover the form, function, or fate of pinwheels, initial investigations with electron microscopy, begun about 2 decades ago (Borges, 1958), showed that these inclusions were not amorphous or crystalline like some others (Sheffield, 1941), but they were rather complex bodies, exhibiting an array of sizes, forms, and shapes. Pinwheels are often found in different spatial attitudes clustered together in large, localized aggregates (Figs. 4–6, 11, 12, and 20) identifiable also by light microscopy. These amorphous aggregates are variously composed also of virus particles, host cell organelles and elements of unknown source and function.

A variety of descriptive terms have resulted from the many electron microscope studies done on pinwheel morphology. Some descriptions, e.g., "dense bands" and "looped profiles" (Matsui and Yamaguchi, 1964); "curled inclusions" (Lee, 1965), "pinwheels" and "bundles" (Edwardson, 1966), "laminated" or "circular inclusions" (Edwardson *et al.*, 1968), and "rotate inclusions" (Wilson *et al.*, 1974a) account for two-dimensional views of the inclusions. Other expressions, like "plate-like inclusions" (Lee, 1965), "tubes," "scrolls," and "cylindrical inclusions" (Edwardson *et al.*, 1968), and "lamellar inclusions" (Kitajima *et al.*, 1968) accommodate somewhat for the third dimension also. Considering both precedence and convenience, we use *pinwheel inclusion* or, simply, *pinwheel* to denote these structures.

Edwardson (1974) has reported that pinwheels are present solely in association with the many viruses of the potato virus Y group (PVY group or potyviruses) (see also, Teakle and Pares, this volume), which are known to infect well over 1000 different species in over 50 families of angiosperms. He has correlated pinwheel occurrence with over 75 different potyviruses and, thus, maintains that pinwheels are diagnostic for the group. Other authors (Edwardson *et al.*, 1972; Chamberlain, 1974; Tu, 1974) have correlated different pinwheel features, such as size, number, form, and associated particles, with specific potyviruses. However, similar, if not identical, inclusions also occur in association with viruses other than potyviruses (Albouy and Lapierre, 1972; Shalla and Shepard, 1972) and in apparently uninfected tissues where known viruses have not been demonstrated (Wilson *et al.*, 1974a,b).

II. MORPHOLOGY

Fully formed pinwheel inclusions are composed of 10–20 thin, finely striated, triangular (Andrews and Shalla, 1974) or rectangular (Edwardson *et al.*, 1968) curved plates. These plates are uniquely and somewhat equivalently bound to a central core in such a way that they diverge at their outer edges (Figs. 8, 11 and 13). The composite structure, thus, appears rotate in transverse section (Figs. 8, 11, 12 and 13) and lineate in longitudinal section (Figs. 1–3, 7, 9, and 19). Individual inclusions may also have one or more of their elementary curved plates extended into closely appressed, flattened or coiled, configurations (Edwardson *et al.*, 1968). The flattened extensions appear laminate and straight in most views (Figs. 5, 11, 12, and 19); coiled extensions are also laminate, but in transverse sections the laminae appear as tightly wound spirals (Fig. 4). According to Edwardson (1974), the proportion of flattened to coiled plate-extensions in any given inclusion is likely to be governed by the genome of the potyvirus involved. Extracted, yet somewhat collapsed, inclusions composed chiefly of flattened plates are basically triangular in negatively stained preparations; by contrast, coiled inclusions, similarly prepared, are rectangular (Edwardson *et al.*, 1968; Teakle and Pares, this volume). Both forms of pinwheels may measure several micrometers across and are fully visible as plates (flattened extensions) or tubes (coiled extensions) by light microscopy (Edwardson *et al.*, 1972). Occasionally, individual pinwheels are bound together by shared plates (Fig. 13).

III. MORPHOGENESIS

A. Formation

With time the pinwheel inclusions in any particular cell change in size, shape, number, and position. They are usually formed in perpendicular contact with cytoplasmic membranes, particularly the endoplasmic reticulum (Langenberg and Schroeder, 1973) (Figs. 1–3, 7, and 9), and the plasmalemma (Lawson *et al.*, 1971; Andrews and Shalla, 1974) (Figs. 1 and 7). Formation on plasmalemmas often occurs near connections with plasmodesmata (Figs. 2 and 3). In some instances the bases of inclusions even extend into and through the plasmodesmata (Langenberg and Schroeder, 1973; Weintraub *et al.*, 1974). After some time inclusion size and number increase in the cells and pinwheels are found dissociated from membranes and randomly oriented in the cytoplasm (Lawson *et al.*, 1971; Andrews and Shalla, 1974) (Figs. 4–6, 10–12, 19, and 20). While it is not known how many pinwheels a given cell can produce, the estimates are very large (Edwardson *et al.*, 1968; Wilson *et al.*, 1974a). When and how pinwheel formation is initiated or terminated is not known, but exogenous factors such as light (Lawson *et al.*, 1971) (Fig. 6) and growth-promoting substances (Wilson *et al.*, 1974a; Steward *et al.*, 1975) (Figs. 13–17) markedly affect the inclusions. Elaborate and complex formations like pinwheels are not common to living cells and no one has proposed mechanisms for their development. However, recent studies on rotating chemical reactions, in which the two separate processes of molecular diffusion and chemical reaction combine to give rise to scrolls and pinwheels in almost lifelike fashion (Winfree, 1974), may provide bases for future experimental work.

B. Transformation

Under conditions which control somatic embryogenesis in aseptically cultured carrot cells (Steward *et al.*, 1975), pinwheel inclusions can be made to form, transform, and ultimately dissipate (Wilson *et al.*, 1974a,b, 1976). These pinwheels represent one of several types of cytoplasmic inclusions which, among other things, distinguish individual stages in the development of embryos and plants from initially somatic tissues. The pinwheels are present only in cells of proliferating callus (Figs. 7–13); they are absent from quiescent cells of the preceding explant stage and are quickly replaced in successive stages by cratile inclusions (in differentiating cells in loose aggregates; Fig. 14) or by multifibrillar bundles (in dense cells of nodules or in cells of proembryonic globules; Fig. 15). As these latter stages continue development, first into proembryos, then into embryos and plants, *none* of the special inclusions are retained by the tissues. Moreover, both the morphogenetic propensities of the cells and the form of the special inclusions in their cytoplasts can be reversibly manipulated (Wilson *et al.*, 1976) (Figs. 16 and 17). Because the conditions which govern the occurrence of one or the other of the special inclusions do so to the exclusion of the others, it has been suggested that they are all alternative expressions of the same basic subunit (Wilson *et al.*, 1974a).

IV. STRUCTURE AND COMPOSITION

The plates that comprise pinwheel inclusions have prominent unidirectional striations about 5 nm apart in negatively stained preparations (Purcifull and Edwardson, 1967). By the vacuum dehydration technique the plates appear *in situ* as linear arrays of subunits with a spacing of 5 nm (Edwardson *et al.*, 1968). Each plate, in freeze-etched preparations of pinwheels, *in situ* or in isolation, is composed of subunits packed in rectangular array having a mean spacing of 6 nm (McDonald and Hiebert, 1974). These subunits are thought to correspond to a protein with a molecular weight of around 67,000–70,000 daltons which has been isolated by polyacrylamide gel electrophoresis from purified inclusions dissociated by treatment with sodium dodecyl sulfate (Hiebert and McDonald, 1973).

That pinwheels are essentially protein is not without additional support. Shepard (1968) found that while different components of the inclusions were differentially sensitive to the proteolytic enzyme subtilisin, even the least sensitive elements, the curved plates, could with time be digested from thin sections (Fig. 18). Additionally, Weintraub and Ragetli (1968) reported that nucleases had no effect on pinwheels, however, the inclusions were completely digested in thin sections by pepsin and trypsin. Furthermore, ultraviolet absorption spectra of purified inclusions indicate that the major constituent is protein (Hiebert *et al.*, 1971).

Whether or not other kinds of molecules contribute to pinwheel composition is not known. Our recent unpublished observation that radioactivity from [^{3}H]*myo*-inositol or its metabolic products coincides with pinwheels *in situ* (Fig. 20) may indicate that the inclusions are not exclusively protein. *Myo*-inositol, as it was applied in these experiments, promotes intense cell proliferation (Degani and Steward, 1969).

V. SEROLOGY

Purified pinwheel inclusions are immunogenic and this feature has been used in attempts to identify their constituent protein(s). Shepard and Shalla (1969), using direct immunoferritin procedures, showed that pinwheels formed in association with tobacco etch virus (TEV) were antigenically unrelated to TEV; by indirect immunoferritin techniques, Shepard *et al.* (1974) have confirmed the findings (Fig. 19). Hiebert *et al.* (1971) reported that purified inclusions associated with TEV and PVY were serologically distinct from each other, from their respective purified viruses and from host protein. Because Purcifull *et al.* (1973) found the pinwheels from five different potyviruses to be morphologically distinct and immunogenically unique—despite the host plants used to propagate the respective viruses—they opined that the inclusions were coded by the viral genomes in question; hence, inclusions could serve in virus classification and diagnosis. However, recent work by McDonald and Hiebert (1975) showed that although morphologically different pinwheel inclusions characterized infections by three strains of turnip mosaic virus, the inclusions were antigenically alike.

VI. CONCLUSIONS

Pinwheel inclusions have been shown to be real structures in the cytoplasts of plant cells. Although unique in form, they exhibit a range in size, shape, and complexity. Their formation in cells seems, thus far, to result from infection by any of a number of plant viruses and/or application of different combinations of plant growth-promoting substances. What relationships obtain between the pinwheels brought about by these two different kinds of factors are not known.

Equally obscure are any clues to the function of these unusual cellular formations. Whether given pinwheels are sufficiently unique to serve in virus classification and diagnosis is an open matter.

ACKNOWLEDGMENTS

This work was made possible in part by grants to Professor F. C. Steward, Director of the former Laboratory of Cell Physiology, Growth and Development, Cornell University, from the National Institutes of Health (GM 09609) and from the National Aeronautical and Space Administration (NASA Contract 2-6065), to Professor (*emeritus*) A. F. Ross, Department of Plant Pathology, Cornell University, from the United States Public Health Service (AI 02540), and to J. R. Aist and H. W. Israel, Department of Plant Pathology, Cornell University, from the United States Department of Agriculture (CSRS 316-15-53) and the National Science Foundation (PCM76–17209).

REFERENCES

Albouy, J., and Lapierre, H. (1972). *Ann. Phytopathol.* **4,** 353–358.
Andrews, J. H., and Shalla, T. A. (1974). *Phytopathology* **64,** 1234–1243.
Bawden, F. C., and Kassanis, B. (1941). *Ann. Appl. Biol.* **28,** 107–118.
Borges, M. de L. V. (1958). *Agron. Lusit.* **20,** 283–294.
Chamberlain, J. A. (1974). *J. Gen. Virol.* **23,** 201–204.
Degani, N., and Steward, F. C. (1969). *Ann. Bot. (London)* [N. S.] **33,** 483–504.
Edwardson, J. R. (1966). *Am. J. Bot.* **53,** 359–364.
Edwardson, J. R. (1974). *Fla., Agric. Exp. Stn., Monogr. Ser.* No. 4.
Edwardson, J. R., Purcifull, D. E., and Christie, R. G. (1968). *Virology* **34,** 250–263.
Edwardson, J. R., Zettler, F. W., Christie, R. G., and Evans, I. R. (1972). *J. Gen. Virol.* **15,** 113–118.
Hiebert, E., and McDonald, J. G. (1973). *Virology* **56,** 349–361.
Hiebert, E., Purcifull, D. E., Christie, R. G., and Christie, S. R. (1971). *Virology* **43,** 638–646.
Kitajima, E. W., Camargo, I. J. B., and Costa, A. S. (1968). *J. Electron Microsc.* **17,** 144–153.
Langenberg, W. G., and Schroeder, H. F. (1973). *Virology* **55,** 218–223.
Lawson, R. H., Hearon, S. S., and Smith, F. F. (1971). *Virology* **46,** 453–463.
Lee, P. E. (1965). *J. Ultrastruct. Res.* **13,** 359–366.
McDonald, J. G., and Hiebert, E. (1974). *Virology* **58,** 200–208.
McDonald, J. G., and Hiebert, E. (1975). *Virology* **63,** 295–303.
Matsui, C., and Yamaguchi, A. (1964). *Virology* **22,** 40–47.
Matthews, R. E. F. (1970). "Principles of Plant Virology," pp. 273–297. Academic Press, New York.
Purcifull, D. E., and Edwardson, J. R. (1967). *Virology* **32,** 393–401.
Purcifull, D. E., Hiebert, E., and McDonald, J. G. (1973). *Virology* **55,** 275–279.
Rubio-Huertos, M. (1972). *In* "Principles and Techniques in Plant Virology" (C. I. Kado and H. O. Agrawal, eds.), pp. 62–75. Van Nostrand-Reinhold, Princeton, New Jersey.
Shalla, T. A., and Shepard, J. F. (1972). *Virology* **49,** 654–667.
Sheffield, F. M. L. (1931). *Ann. Appl. Biol.* **18,** 471–493.
Sheffield, F. M. L. (1934). *Ann. Appl. Biol.* **21,** 430–453.
Sheffield, F. M. L. (1941). *J. R. Microsc. Soc.* [3] **61,** 30–45.
Shepard, J. F. (1968). *Virology* **36,** 20–29.
Shepard, J. F., and Shalla, T. A. (1969). *Virology* **38,** 185–188.
Shepard, J. F., Gaard, G., and Purcifull, D. E. (1974). *Phytopathology* **64,** 418–425.
Steward, F. C., Israel, H. W., Mott, R. L., Wilson, H. J., and Krikorian, A. D. (1975). *Philos. Trans. R. Soc. London, Ser. B* **273,** 33–53.
Tu, J. C. (1974). *Cytobios* **10,** 37–43.
Weintraub, M., and Ragetli, H. W. J. (1968). *J. Cell Biol.* **38,** 316–328.
Weintraub, M., Ragetli, H. W. J., and Lo, E. (1974). *J. Ultrastruct. Res.* **46,** 131–148.
Wilson, H. J., Israel, H. W., and Steward, F. C. (1974a). *J. Cell Sci.* **15,** 57–73.
Wilson, H. J., Goodman, R. M., and Israel, H. W. (1974b). *Proc. Am. Phytopathol. Soc.* **1,** 147.
Wilson, H. J., Goodman, R. M., and Israel, H. W. (1976). *Arch. Virol.* **51,** 347–354.
Winfree, A. T. (1974). *Sci. Am.* **230,** 82–95.

Figs. 1–6. Formation of pinwheel inclusions associated with infection of morning glory (*Ipomoea setosa* Ker.) leaves by sweet potato russet crack virus. Bars = 0.5 μm. [Courtesy of R. H. Lawson (Lawson *et al.*, 1971) and reproduced by permission of Academic Press, Inc.]

Figs. 1–3. Longitudinal sections through pinwheels in perpendicular contact with plasmalemmas near plasmodesmata (arrows) and endoplasmic reticulum during early stages of infection (Fig. 1, ×38,000; Figs. 2 and 3, ×42,000).

Fig. 4. Pinwheels with coiled plate-extensions dissociated from membranes during late stages of infection and randomly oriented in the cytoplasm. (×34,000.)

Fig. 5. Pinwheels composed chiefly of flattened plate-extensions in cells infected for several months. (×32,000.)

Fig. 6. Pinwheels in cells infected for several months, but from leaves given supplemental shade. (×24,000.)

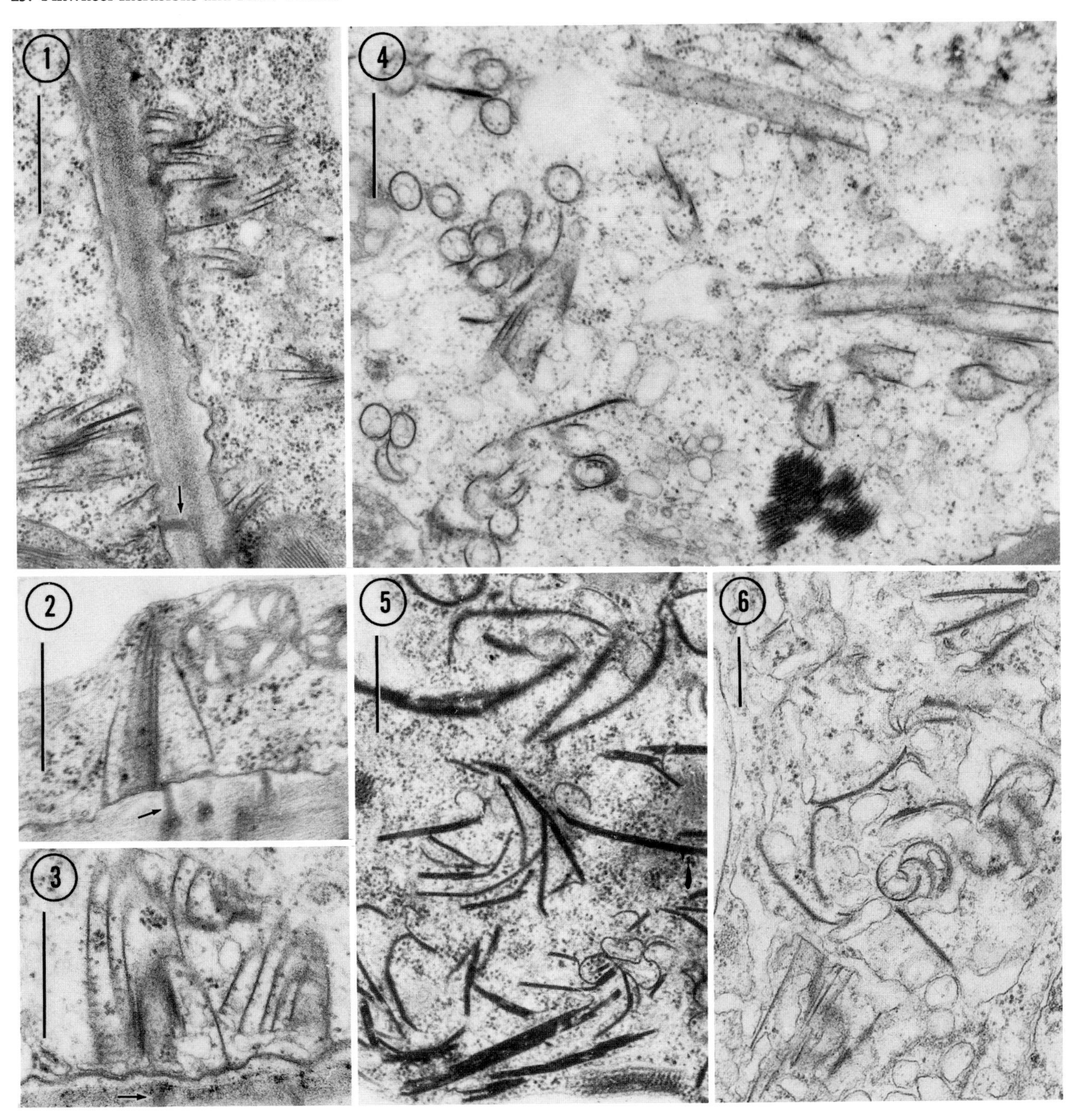
1
2
3
4
5
6

Figs. 7–12. Formation of pinwheel inclusions associated with proliferation of cultured carrot (*Daucus carota* L.) cells (Figs. 7–9, bars = 0.25 μm; Figs. 10–12, bars = 0.5 μm).

Figs. 7–9. Pinwheels in cells sampled shortly after application of appropriate exogenous growth factors.

Fig. 7. Longitudinal section of an inclusion in contact with a plasmalemma. (×60,000.)

Fig. 8. Transverse sections through very small, forming pinwheels. (×60,000.)

Fig. 9. Longitudinal sections of pinwheels in contact with endoplasmic reticulum. (×55,000.)

Fig. 10. Forming pinwheels clustered randomly in the parietal cytoplasm of a young callus cell. (×30,000.)

Figs. 11 and 12. Clustered, fully formed pinwheels in old callus cells (×30,000.) For further details see Wilson *et al.* (1974a) and Steward *et al.* (1975).

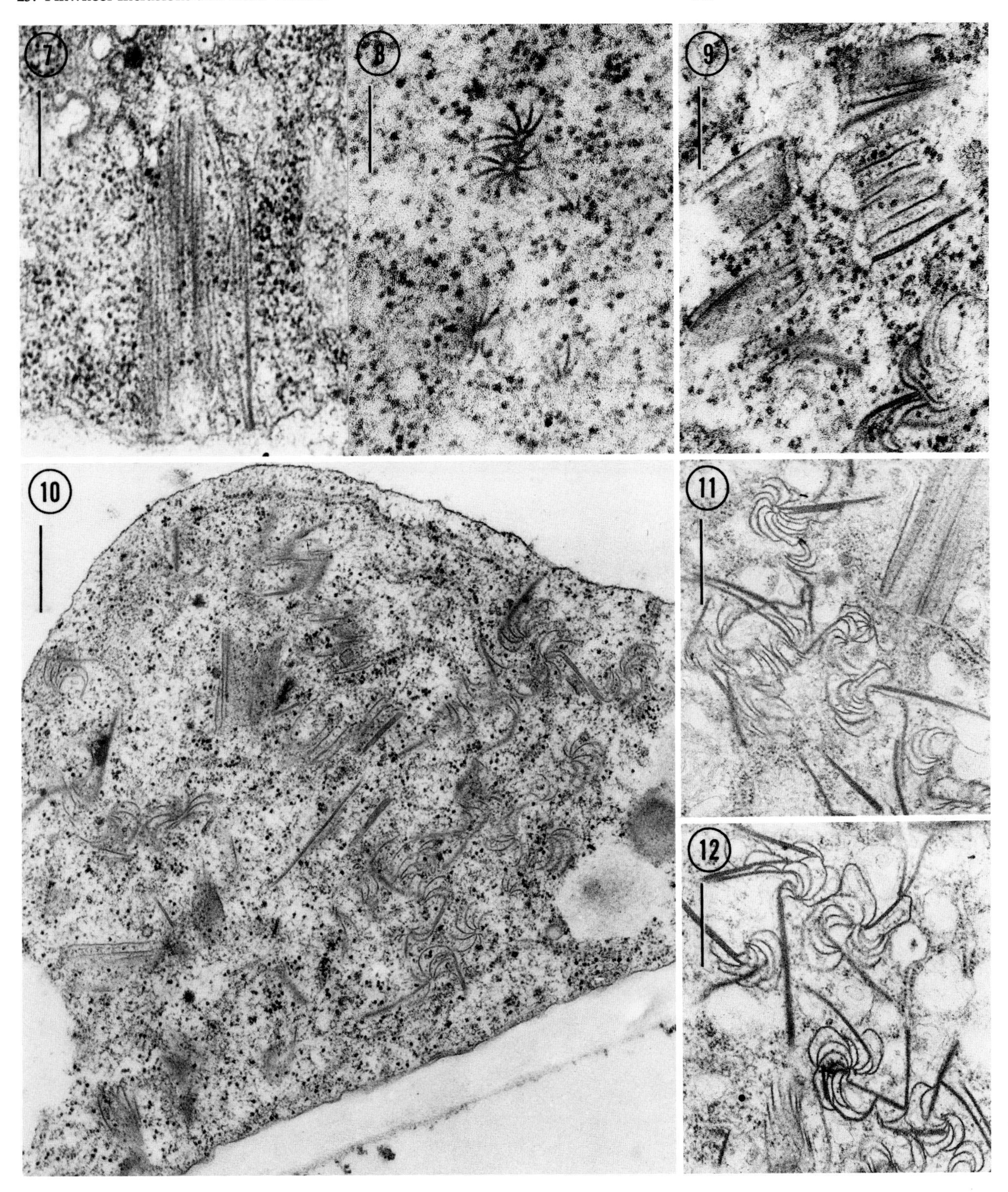
7
8
9
10
11
12

Figs. 13–17. Special inclusions associated with morphogenesis in cultured carrot (*Daucus carota* L.) tissues (bars = 0.25 μm).

Fig. 13. Pinwheel inclusions in a proliferating callus cell. (×75,000.)

Fig. 14. Cratile inclusion in a differentiating cell of a loose aggregate. (×40,000.)

Fig. 15. Multifibrillar bundle in a small, dense cell of a proembryonic globule. (×85,000.)

Figs. 16 and 17. Formation of pinwheel inclusions from multifibrillar bundles in initially proembryonic cells which were sampled during morphogenetic transformation to undifferentiated callus (Fig. 16, ×70,000; Fig. 17, ×100,000). See also Wilson *et al.* (1974a,b), Steward *et al.* (1975), and Wilson *et al.* (1976).

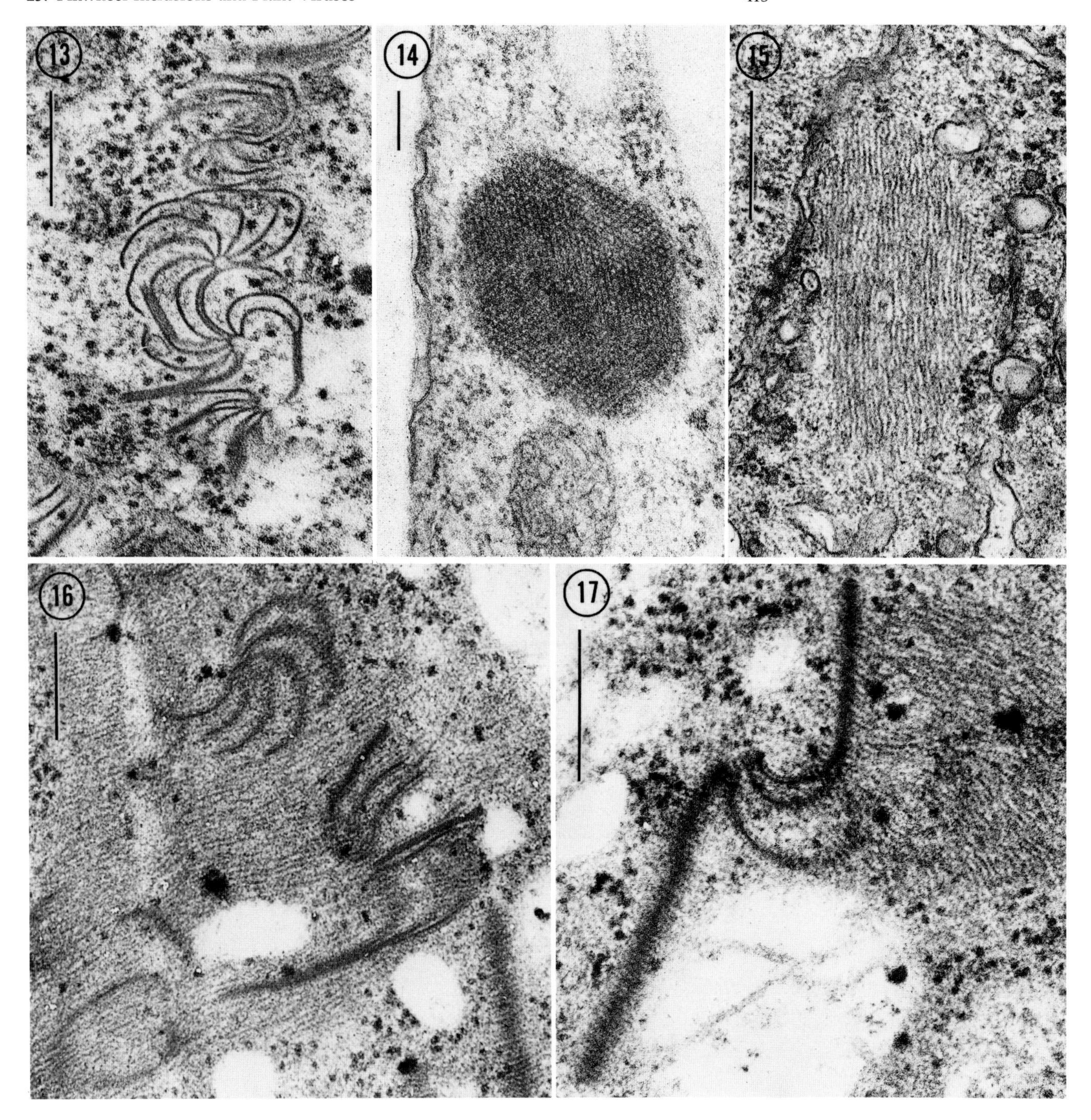
13
14
15
16
17

Figs. 18–20. Composition of pinwheel inclusions. Bars = 0.5 μm. Figure 18 courtesy of J. F. Shepard (Shepard, 1968) and reproduced by permission of Academic Press, Inc.; Fig. 19 courtesy of G. Gaard (Shepard *et al.*, 1974); Fig. 20 from work of H. W. Israel *et al.* (unpublished).

Fig. 18. Electron transparent voids in a section which resulted from complete digestion of a pinwheel inclusion by treatment with subtilisin. (×40,000.)

Fig. 19. Section through a tobacco etch virus (TEV)-infected cell where ferritin-tagged anti-TEV antibody is combined with virus particles (arrows) but not with the darkly stained pinwheel inclusions. (×60,000.)

Fig. 20. High-resolution autoradiograph of a section through a carrot callus cell exposed to [^{3}H]*myo*-inositol, in combination with other growth factors, showing developed grains concentrated over the pinwheel inclusions. (×24,000.)

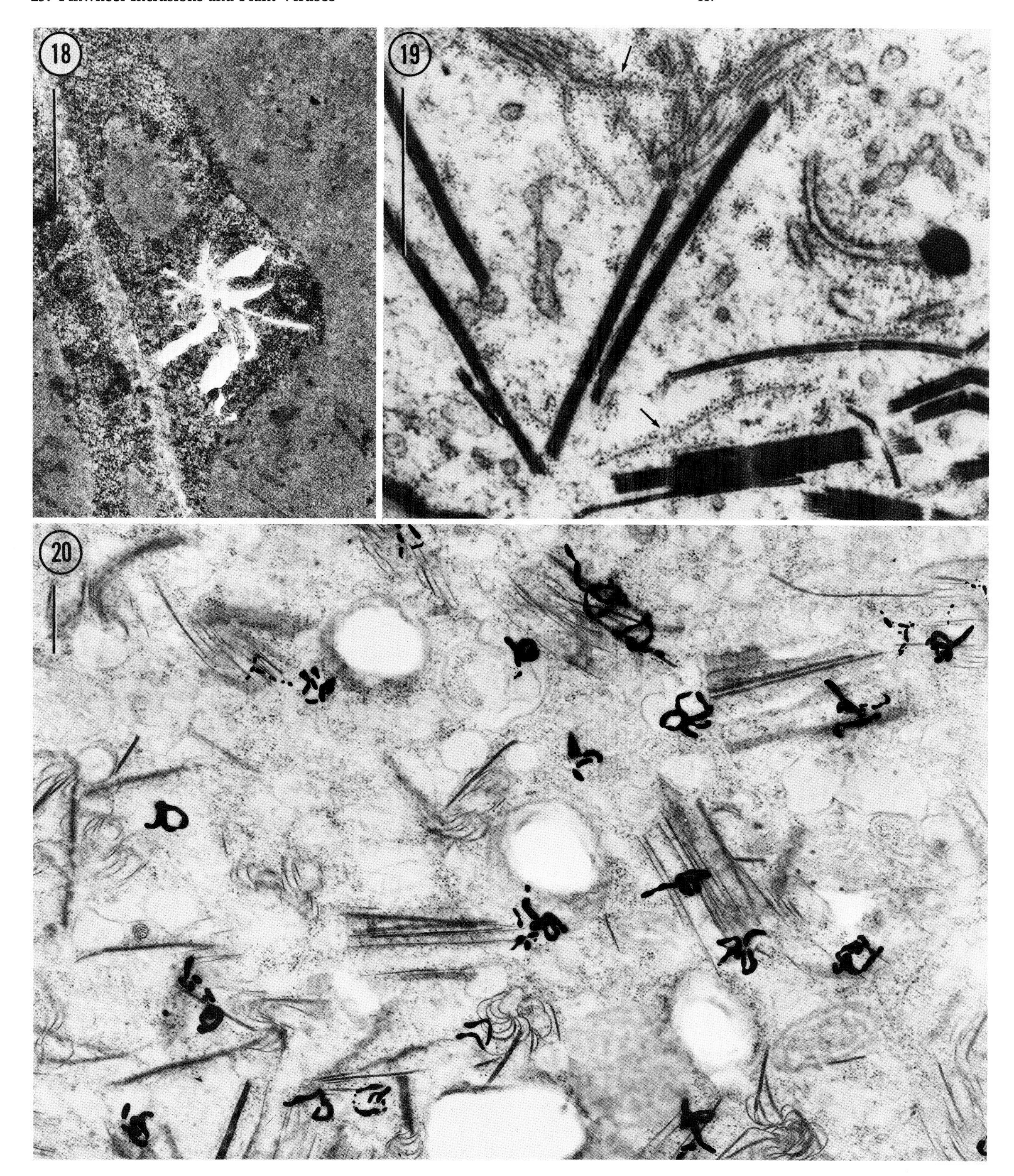
18
19
20

PART III

Mycoviruses and Viroids

Chapter 26

Mycoviruses

K. N. SAKSENA

I. Introduction 421
II. Biological Implications 421
III. Physicochemical Properties 422
IV. Ultrastructural Aspects 423
V. Concluding Remarks 424
References 424

I. INTRODUCTION

Viruses infecting fungi are called mycoviruses. Following early evidence for the presence of viruses in the cultivated mushroom (Hollings, 1962) numerous reports of viruses and virus-like particles in fungi have appeared. However, only a few of these particles have been well characterized and still fewer have been transmitted or shown to be infectious. An attempt is made here to survey briefly the salient features of mycoviruses and to present an overview of their ultrastructural aspects. For detailed information, interested readers are referred to more extensive reviews (Bozarth, 1972; Hollings and Stone, 1971; Lemke and Nash, 1974).

II. BIOLOGICAL IMPLICATIONS

The majority of known mycoviruses are latent, a feature that probably eluded their early recognition. An exception, however, is the infectious, degenerative disease of the cultivated mushroom, *Agaricus bisporus* (Sinden and Hauser, 1950), caused by a complex of viruses (Hollings, 1962; Hollings and Stone, 1971; Dieleman-van Zaayen and Temmink, 1968). At least three distinct viruses, e.g., spherical particles 25–27 nm (AbV-25) (Fig. 1), and 34–36 nm (AbV-34) (Fig. 2), and bacilliform particles about 19 × 50 nm (AbV-19/50) (Fig. 3) have consistently been isolated (Dieleman-van Zaayen, 1972; Saksena, 1975), but their individual role in pathogenecity is unclear. Both AbV-25 and AbV-34 contain double-stranded ribonucleic acid (dsRNA) (Barton, 1975) and AbV-19/50 contains single-stranded RNA (ssRNA) (Molin and Lapierre, 1973).

Although mushroom viruses are transmitted by hyphal anastomosis and carried by spores, quantification of transmission and infectivity of mycoviruses generally remains difficult due to the lack of appropriate bioassay methods. Artificial infection methods, including infection of fungal protoplasts, have mostly been ineffective.

An apparent correlation exists between mycoviral dsRNA and the cytoplasmically inherited killer phenomena in *Saccharomyces cerevisiae* (Bevan *et al.*, 1973) and *Ustilago maydis* (Day and Anagnostakis, 1973). Killer strains of *S. cerevisiae* possess two species of dsRNA referred to as P1 and P2 (2.5 and 1.4 $\times$ 10^6 daltons, respectively). Sensitive strains either have P1 dsRNA only or lack dsRNA completely. Nuclear genes are known to influence the replication of P2 dsRNA in killer cells. Isometric virus particles (Fig. 4) have been isolated from *S. cerevisiae* (Adler, 1975; Buck *et al.*, 1973a; Herring and Bevan, 1974). Similarly, in *U. maydis* a virus (Fig. 5) containing dsRNA genome segments (0.06 to 2.87 $\times$ 10^6 daltons) has been correlated with a killer system (Wood and Bozarth, 1973).

Other biological implications of mycoviruses and viral dsRNA have been discussed elsewhere (Hollings and Stone, 1971; Lemke and Nash, 1974). These considerations include the role of mycoviral dsRNA in interferon induction and the possible role of mycoviruses in synthesis of secondary metabolites or in the control of pathogenic fungi.

III. PHYSICOCHEMICAL PROPERTIES

Most mycoviruses characterized thus far are isometric particles containing a segmented *ds*RNA genome. They are multicomponent systems, exhibiting centrifugal and density heterogeneity.

The *Penicillium chrysogenum* virus (PcV) (Fig. 6) has been well characterized (Buck *et al.*, 1971); Nash *et al.*, 1973; Wood and Bozarth, 1972). PcV has a sedimentation coefficient of 145–150 S, a buoyant density of 1.27 gm/cm^3 in potassium tartarate and 1.35 gm/cm^3 in CsCl, a diffusion coefficient of 1.03 $\times$ 10^{-7} cm^2/sec, a partial specific volume of 0.728 gm/ml, an estimated molecular weight of 13.0 $\times$ 10^6, and a single capsid protein of 125,000 molecular weight. PcV contains dsRNA separable by electrophoresis into three species, 1.89, 1.99, and 2.18 $\times$ 10^6 daltons. This PcV dsRNA has a sedimentation coefficient of 13 S, a density of 1.60 gm/cm^3 in $CsSO_4$, a thermal melting point (T_m) of 100° in standard saline citrate buffer (SSC) and 88°–92° in 0.1 SSC, and contains approximately equimolar ratios of G:C and A:U. Molecules of the dsRNA (Fig. 7) have an average contour length of 0.86 μm, which corresponds to a molecular weight of 2.0 $\times$ 10^6. Each molecule is separately encapsidated in this multicomponent virus system. PcV is serologically related to a virus from *P. brevicompactum* (Wood *et al.*, 1971).

Virus particles present in *P. stoloniferum* represent a complex of two distinct isometric viruses each of which is multicomponent (Bozarth *et al.*, 1971; Buck and Kempson-Jones, 1970, 1973, 1974; Buck, 1975). They are separable by ion exchange chromatography or by electrophoresis and are designated as fast (PsV-F) and slow (PsV-S) viruses. PSV-F and PsV-S have two (one major and one minor) capsid polypeptides each, but none in common. Both viruses are heterogeneous for centrifugal (66 S to 113 S) and density components (1.29 to 1.376 gm/cm^3 in CsCl). In both cases, particles contain dsRNA or ssRNA, but for PsV-F, individual particle fractions have not been analyzed. The physical properties of PsV-S and their constituent RNA's clearly show that heterogeneity in this virus is due to different amounts of RNA within particles. PsV-S contains four major particle classes, e.g., E particles (contain no nucleic acid), M particles (contain only ssRNA), L particles (contain only dsRNA), and H particles (contain both the dsRNA and ssRNA). The M, L, and H particles are further resolved into two subclasses. The M1 and M2 particles contain one molecule of ssRNA, 0.47 and 0.56 × 10^6 daltons, respectively; the L1 and L2 particles have one molecule of dsRNA, 0.94 and 1.11 × 10^6 daltons, respectively; and the H1 and H2 particles contain one molecule of dsRNA, 0.94 and 1.11 × 10^6 daltons, respectively, plus some ssRNA. Amino acid composition, polypeptide ratios, and electrophoretic mobilities of M, L, and H particles are similar.

Viruses in *Aspergillus foetidus* (Ratti and Buck, 1972; Buck and Ratti, 1975a) also represent a complex of two serologically distinct isometric particles distinguishable as fast (AfV-F) and slow (AfV-S) by their relative electrophoretic mobility. Both AfV-F and AfV-S are multicomponent systems. AfV-F is comprised of five particle classes, FO, F1, F2, and F3, and F4, each containing a single dsRNA component, 1.24, 1.44, 1.70, 1.87 and 2.31 × 10^6 daltons, respectively. AfV-S has four major particle classes, S1, S2, S3, and S4, but S1 and S2 resolve further into the sublcasses S1a, S1b, S2a, and S2b. Physicochemical analysis of AfV-S particles show differences in sedimentation values corresponding with differences in RNA content, and three species of dsRNA, 0.1, 2.24, and 2.76 × 10^6 daltons are involved. The subclasses S1a and S2a contain a molecule of 2.24 and 2.76 × 10^6 daltons, respectively, whereas S1b and S2b contain a molecule of 2.24 and 2.76 × 10^6 daltons, respectively, together with an additional molecule of 0.1 × 10^6 daltons. S4 particles contain two molecules of dsRNA, both of 2.24 × 10^6 daltons, and S3 particles have an equivalent of 1.5 molecules of this dsRNA. AfV-F and AfV-S differ in their amino acid composition and have no polypeptides in common. Viruses identical to AfV-F and AfV-S have been found in another species, *A. niger* (Buck *et al.*, 1973b).

Association of RNA-dependent RNA polymerase activity with viruses from *P. chrysogenum* (Nash *et al.*, 1973), *P. stoloniferum* (Buck, 1975), and *A. foetidus* (Ratti and Buck, 1975) has been reported. In PsV-S, the polymerase activity is associated exclusively with H particles and yields one new molecule of dsRNA (or two complementary single strands of RNA) per particle. Product particles, thus, contain two molecules of dsRNA or its equivalent. Therefore, the particle-associated polymerase is a replicase. However, in the AfV-S system, the major product of polymerase activity is ssRNA. It has been proposed that dsRNA mycoviruses replicate simply by doubling (Buck and Ratti, 1975b), a concept that differs significantly from the replication of dsRNA viruses of other organisms (Wood, 1973).

IV. ULTRASTRUCTURAL ASPECTS

The majority of mycoviruses described are isometric but particles of other morphological types occur as well. These uncommon types reflect considerable diversity and include bacilliform particles (Hollings, 1962), rigid rods (Figs. 8 and 9) (Dieleman-van Zaayen *et al.*, 1970; Ushiyama and Hashioka, 1973), flexuous rods (Fig. 10) (Huttinga *et al.*, 1975), herpes-type virus particles (Fig. 11) (Kazama and Schornstein, 1973), and bacteriophage-like particles (Tikchonenko *et al.*, 1974).

Detailed ultrastructural studies of isometric viruses in fungal cells reveal a basic pattern of their intracellular localization. In the case of *A. bisporus* (Dieleman-van Zaayen, 1972; Dieleman-van Zaayen and Igesz, 1969), dense aggregates of virus particles occur in the cytoplasm, sometimes near septa of the vegetative cells (Fig. 12). However, in the mushroom tissue, AbV-34 particles are found scattered in cytoplasm or are present in vacuoles (Fig. 13). The presence of AbV-34 particles in and around septal pores suggests intercellular migration (Fig. 14). In spores, groups of AbV-34 particles occur in vacuoles or in the cytoplasm (Fig. 15). Often crystalline arrays of AbV-34 particles are observed in tubular or linear arrangements (Fig. 16). The AbV-25 particles are difficult to discern because of their resemblance to ribosomes, but they are found dispersed or aggregated in vacuoles. The AbV-19/50 particles are seldom seen in thin sections. In *P. chrysogenum* (Yamashita *et al.*, 1973) virus particles occur singly or in aggregates, again in cytoplasm or in vacuoles (Figs. 17 and 18). Occasionally crystalline arrays of particles are observed (Figs. 19 and 20). In older cells containing an excessive amount of virus, some degeneration of cytoplasm and of cellular organelles has been observed. In *P. stoloniferum* (Hooper *et al.*, 1972) virus particles occur in abundance in mycelium or in spores, occasionally in paracrystalline arrays. Intracellular appearance of these particles is quite comparable to that observed in *P. chrysogenum*.

The ultrastructure of a herpes-type virus in a species of *Thraustochytrium* has been studied in detail (Kazama and Schornstein, 1973). The replicative cycle of this virus is similar to that of the known herpesviruses.

V. CONCLUDING REMARKS

Despite many inherent problems, substantial progress has been made in the understanding of mycoviruses. They are generally latent, and are essentially isometric viruses with segmentally encapsidated dsRNA genomes. Apparently, multicomponent viruses are favored in fungi since their replication parallels that of fungal growth and, as no cellular lysis occurs, virus persists indefinitely in the infected strain. Clearly, the mycoviruses reveal certain unique biological features and their study should result in a better understanding of the complexity of eukaryotic cell.

ACKNOWLEDGMENTS

Sincere appreciation and thanks are due to J. P. Adler, R. F. Bozarth, A. Dieleman-van Zaayen, H. Huttinga, F. Y. Kazama, C. H. Nash, R. Ushiyama, H. A. Wood, and S. Yamashita for contributions of electron micrographs; to M. N. Haller for assistance with photography; and to P. A. Lemke for review of the manuscript. This work was supported by a fellowship from the Butler County Mushroom Farm, Inc., Worthington, Pennsylvania to the Carnegie-Mellon Institute of Research.

REFERENCES

Adler, J. P. (1975). *Dev. Ind. Microbiol.* **16,** 152–157.
Barton, R. J. (1975). *Proc. Int. Congr. Virol., 3rd, 1975 Abstract,* p. 147.
Bevan, E. A., Herring, A. J., and Mitchell, D. J. (1973). *Nature (London)* **245,** 81–86.
Bozarth, R. F. (1972). *Environ. Health Perspect.* **2,** 23–29.
Bozarth, R. F., Wood, H. A., and Mandelbrot, A. (1971). *Virology* **45,** 516–523.
Buck, K. W. (1975). *Nucleic Acids Res.* **2,** 1889–1902.
Buck, K. W., and Kempson-Jones, G. F. (1970). *Nature (London)* **225,** 945–946.
Buck, K. W., and Kempson-Jones, G. F. (1973). *J. Gen. Virol.* **18,** 223–235.
Buck, K. W., and Kempson-Jones, G. F. (1974). *J. Gen. Virol.* **22,** 441–445.
Buck, K. W., and Ratti, G. (1975a) *J. Gen. Virol.* **27,** 211–224.
Buck, K. W., and Ratti, G. (1975b). *Biochem. Soc. Trans.* **3,** 542–544.
Buck, K. W., Chain, E. B., and Himmelweit, F. (1971). *J. Gen. Virol.* **12,** 131–139.
Buck, K. W., Girvan, R. F., and Ratti, G. (1973a). *Biochem. Soc. Trans.* **1,** 1138–1140.

Buck, K. W., Lhoas, P., and Border, D. J. (1973b). *Biochem. Soc. Trans.* **1,** 1141–1142.
Day, P. R., and Anagnostakis, S. L. (1973). *Phytopathology* **63,** 1017–1018.
Dieleman-van Zaayen, A. (1972). *Virology* **47,** 94–104.
Dieleman-van Zaayen, A., and Igesz, O. (1969). *Virology* **39,** 147–152.
Dieleman-van Zaayen, A., and Temmink, J. H. M. (1968). *Neth. J. Plant Pathol.* **74,** 48–51.
Dieleman-van Zaayen, A., Igesz, O., and Finch, J. T. (1970). *Virology* **42,** 534–537.
Herring, A. J., and Bevan, E. A. (1974). *J. Gen. Virol.* **22,** 387–394.
Hollings, M. (1962). *Nature (London)* **196,** 962–965.
Hollings, M., and Stone, O. M. (1971). *Annu. Rev. Phytopathol.* **9,** 93–118.
Hooper, G. R., Wood, H. A., Myers, R., and Bozarth, R. F. (1972). *Phytopathology* **62,** 823–825.
Huttinga, H., Wichers, H. J., and Dieleman-van Zaayen, A. (1975). *Neth. J. Plant Pathol.* **81,** 102–106.
Kazama, F. Y., and Schornstein, K. L. (1973). *Virology* **52,** 478–487.
Lemke, P. A., and Nash, C. H. (1974). *Bacteriol. Rev.* **38,** 29–56.
Molin, G., and Lapierre, H. (1973). *Ann. Phytopathol.* **5,** 233–240.
Nash, C. H., Douthart, R. J., Ellis, L. F., Van Frank, R. M., Burnett, J. P., and Lemke, P. A. (1973). *Can. J. Microbiol.* **19,** 97–103.
Ratti, G., and Buck, K. W. (1972). *J. Gen. Virol.* **14,** 165–175.
Ratti, G., and Buck, K. W. (1975). *Biochem. Biophys. Res. Commun.* **66,** 706–711.
Saksena, K. N. (1975). *Dev. Ind. Microbiol.* **16,** 134–144.
Sinden, J. W., and Hauser, E. (1950). *Mushroom Sci.* **1,** 96–100.
Tikchonenko, T. I., Velikodvorskaya, G. A., Bobkova, A. F., Vartoshevich, Y. E., Lebed, E. P., Chaplygina, N. M., and Maksimova, T. S. (1974). *Nature (London)* **249,** 454–456.
Ushiyama, R., and Hashioka, Y. (1973). *Rep. Tottori Mycol. Inst.* (Jpn.) **10,** 797–805.
Wood, H. A. (1973). *J. Gen. Virol.* **20,** 61–85.
Wood, H. A., and Bozarth, R. F. (1972). *Virology* **47,** 604–609.
Wood, H. A., and Bozarth, R. F. (1973). *Phytopathology* **63,** 1019–1021.
Wood, H. A., Bozarth, R. F., and Mislivec, P. B. (1971). *Virology* **44,** 592–598.
Yamashita, S., Doi, Y., and Yora, K. (1973). *Virology* **55,** 445–452.

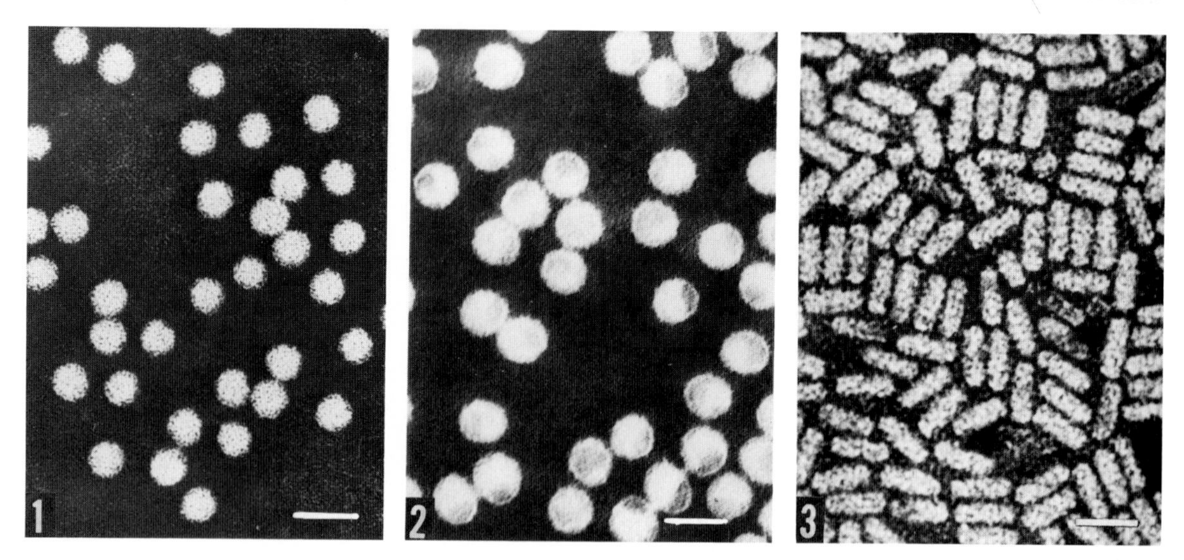

Figs. 1–3. Electron micrographs of viruses purified from the cultivated mushroom, *Agaricus bisporus*, stained with 1% PTA. (K. N. Saksena, unpublished.)

Fig. 1. Spherical particles (AbV-25), approximately 26 nm in diameter. Bar = 50 nm.

Fig. 2. Spherical particles (AbV-34) approximately 35 nm in diameter. Bar = 50 nm.

Fig. 3. Bacilliform particles (AbV-19/50), approximately 19 × 50 nm. Bar = 50 nm.

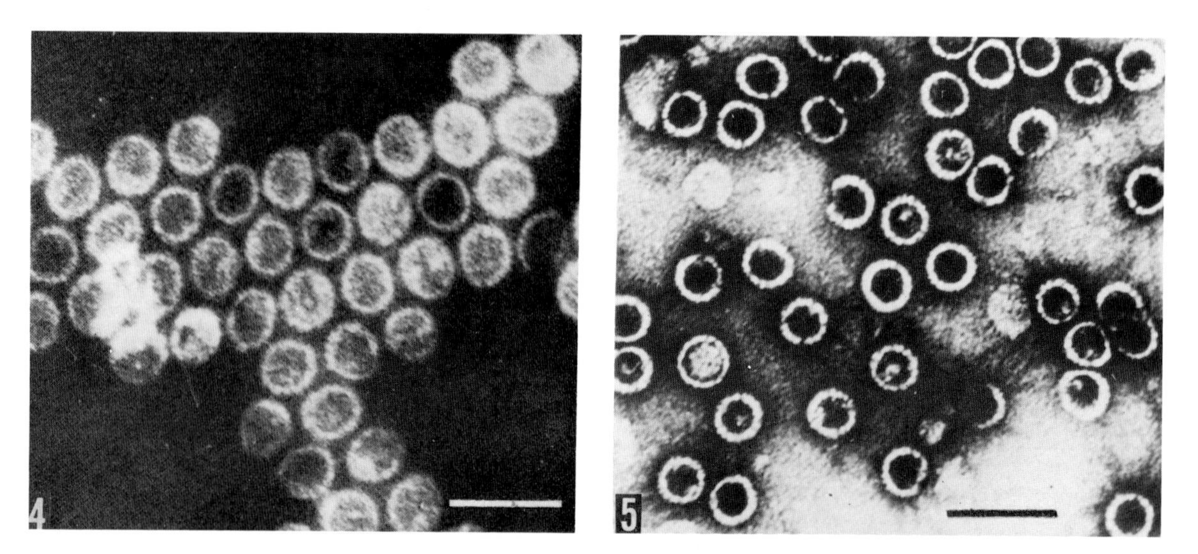

Fig. 4. Electron micrograph of isometric virus particles, approximately 39 nm in diameter, isolated from *Saccharomyces cerevisiae* stained with 2% PTA. Bar = 100 nm. [Courtesy of Adler (1975), with permission of the American Institute of Biological Sciences.]

Fig. 5. Electron micrograph of virus-like particles, approximately 41 nm in diameter, from *Ustilago maydis* (P1), stained with 1% uranyl acetate. Bar = 100 nm. [Courtesy of Wood and Bozarth (1973), with permission of the American Institute of Biological Sciences.]

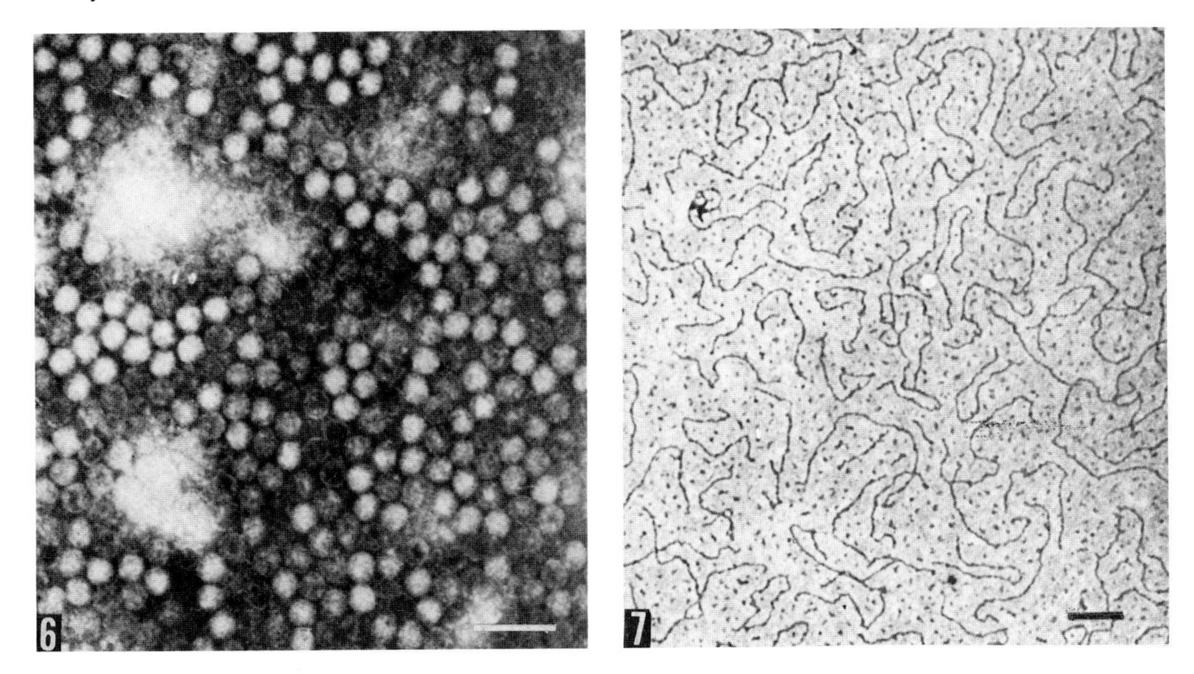

Figs. 6 and 7. Electron micrograph of purified virus from *Penicillium chrysogenum* (Fig. 6, bar = 100 nm) and of dsRNA derived from this virus (Fig. 7, bar = 200 nm). [Reproduced by permission of the National Research Council of Canada from C. H. Nash *et al.*, *Can J. Microbiol.* **19,** 97–103 (1973).]

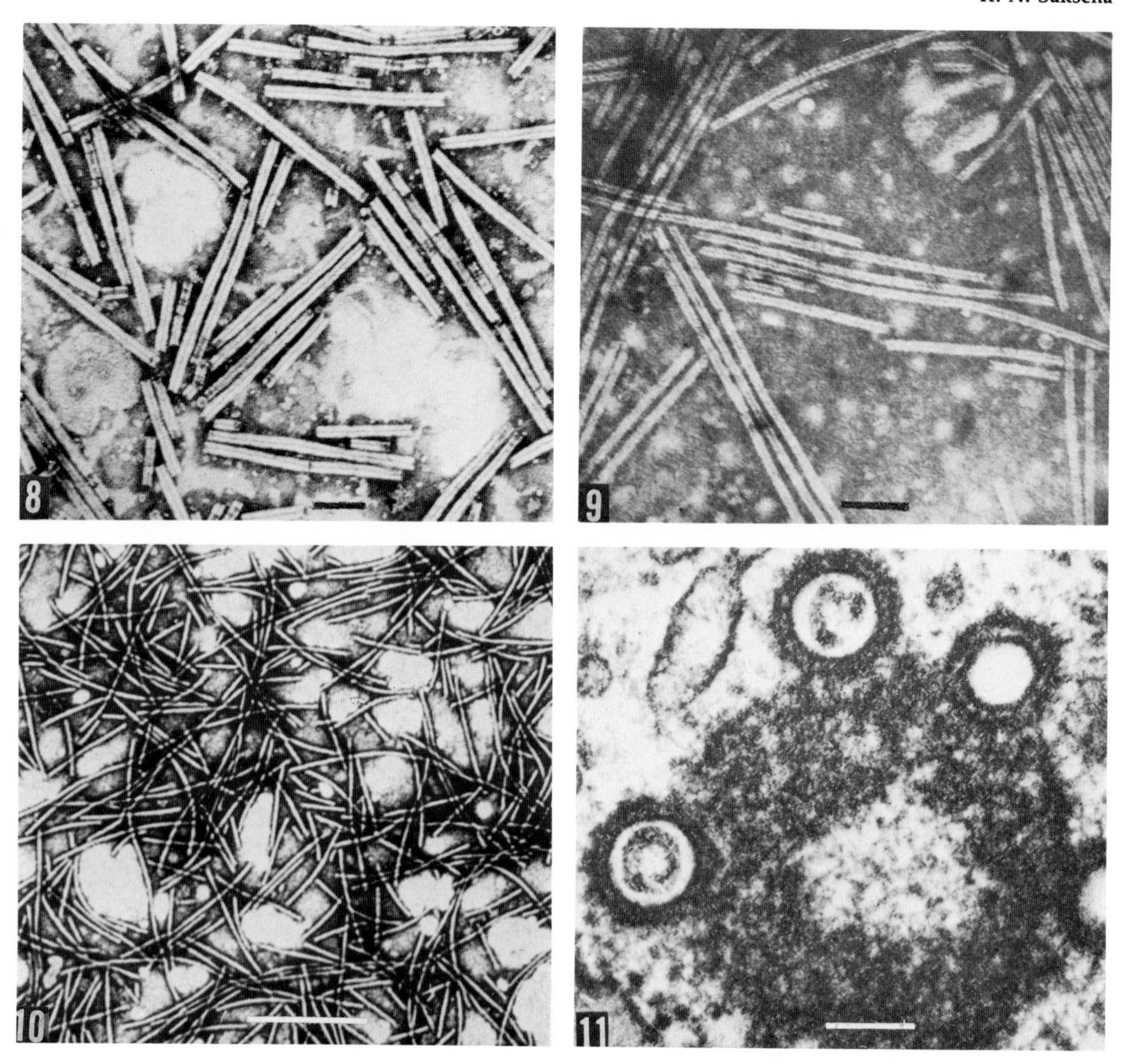

Fig. 8. Electron micrograph of rod-shaped virus-like particles, approximately 17 × 350 nm, isolated from *Peziza ostracoderma*. Bar = 100 nm. [Courtesy of Dieleman-van Zaayen *et al.* (1970), with permission of Academic Press.]

Fig. 9. Electron micrograph of rod-shaped particles, approximately 15 × 700–900 nm, isolated from mushroom, *Lentinus edodes*, stained with 2% uranyl acetate. Bar = 100 nm. (Courtesy of R. Ushiyama and Y. Hashioka, Tottori Mycological Institute, Japan.)

Fig. 10. Electron micrograph of filamentous virus-like particles, approximately 13 × 500 nm, isolated from mushroom, *Boletus edulis*. Bar = 250 nm. [Courtesy of Huttinga *et al.* (1975), Institute of Phytopathological Research, Wageningen, The Netherlands.]

Fig. 11. Ultrathin section of *Thraustochytrium* sp. showing nucleocapsids of herpes-type virus, associated with an electron opaque inclusion within the cytoplasm. Bar = 100 nm. [Courtesy of Kazama and Schornstein (1973) with permission of Academic Press.]

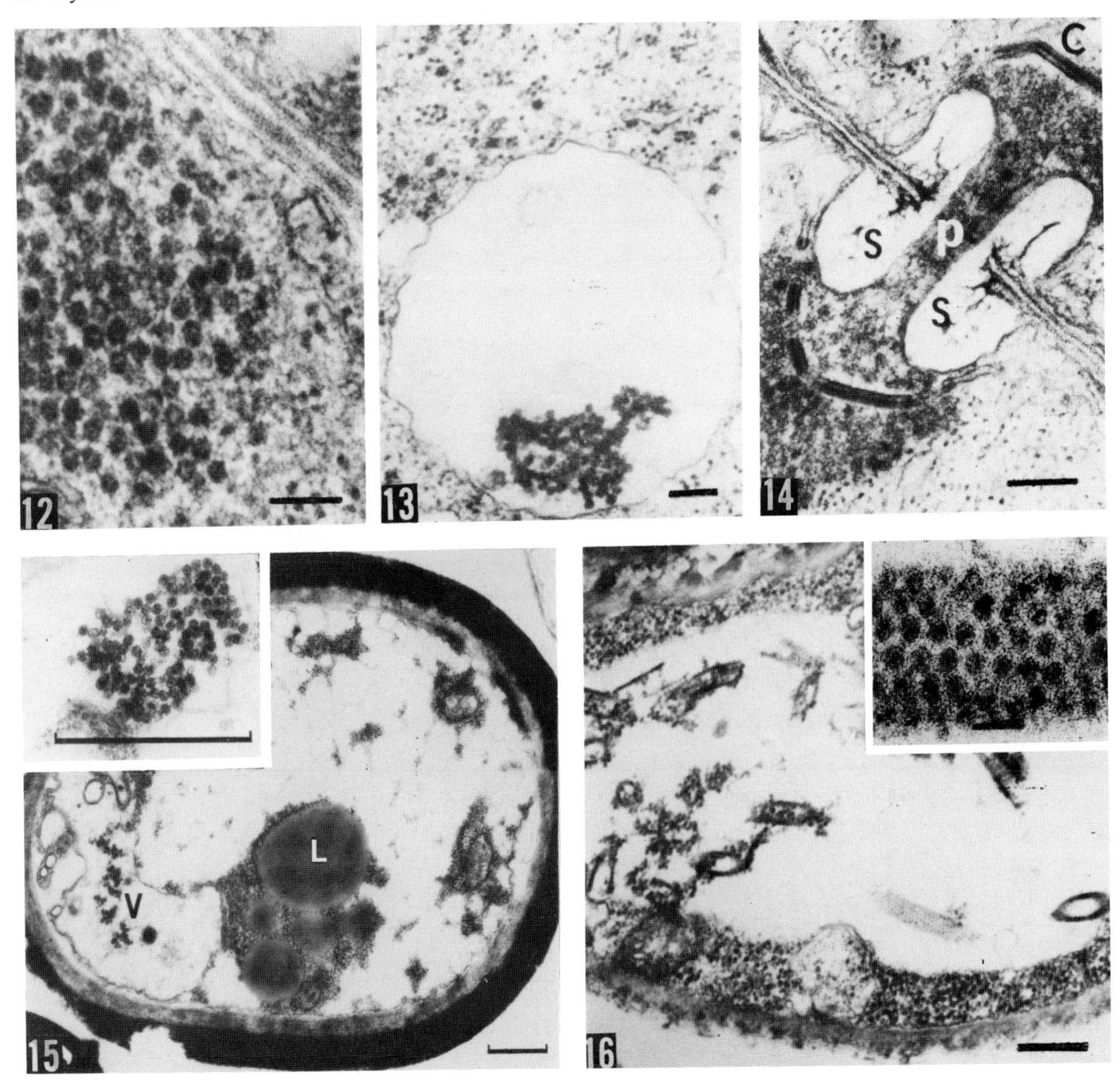

Figs. 12–16. Ultrathin sections of the cultivated mushroom, *Agaricus bisporus*. Figs. 13–15. [Courtesy of Dieleman-van Zaayen (1972) with permission of Academic Press.]

Fig. 12. Longitudinal section of a virus-diseased hyphae with an aggregate of virus particles near the septum. Bar = 100 nm. [Courtesy of Dieleman-van Zaayen and Igesz (1969) with permission of Academic Press.]

Fig. 13. Section of virus-diseased mushroom stipe with AbV-34 particles occurring in a vacuole; some AbV-34 particles are scattered throughout the cytoplasm. Bar = 100 nm.

Fig. 14. Section of virus-diseased mushroom cap showing septum with dolipore and AbV-34 particles on either side of the pore. C, septal pore cap; P, central pore; and S, septal swelling. Bar = 200 nm.

Fig. 15. Section of a spore from virus-diseased mushroom showing virus particles (V) in vacuole, and lipid bodies (L). Inset shows details of a portion of a spore with AbV-34 particles. Bar = 500 nm.

Fig. 16. Ultrathin section of virus-diseased mushroom cap showing crystalline arrays of AbV-34 particles. Bar = 500 nm. Inset shows an enlargement of one such aggregate. Bar = 100 nm. (K. N. Saksena, unpublished.)

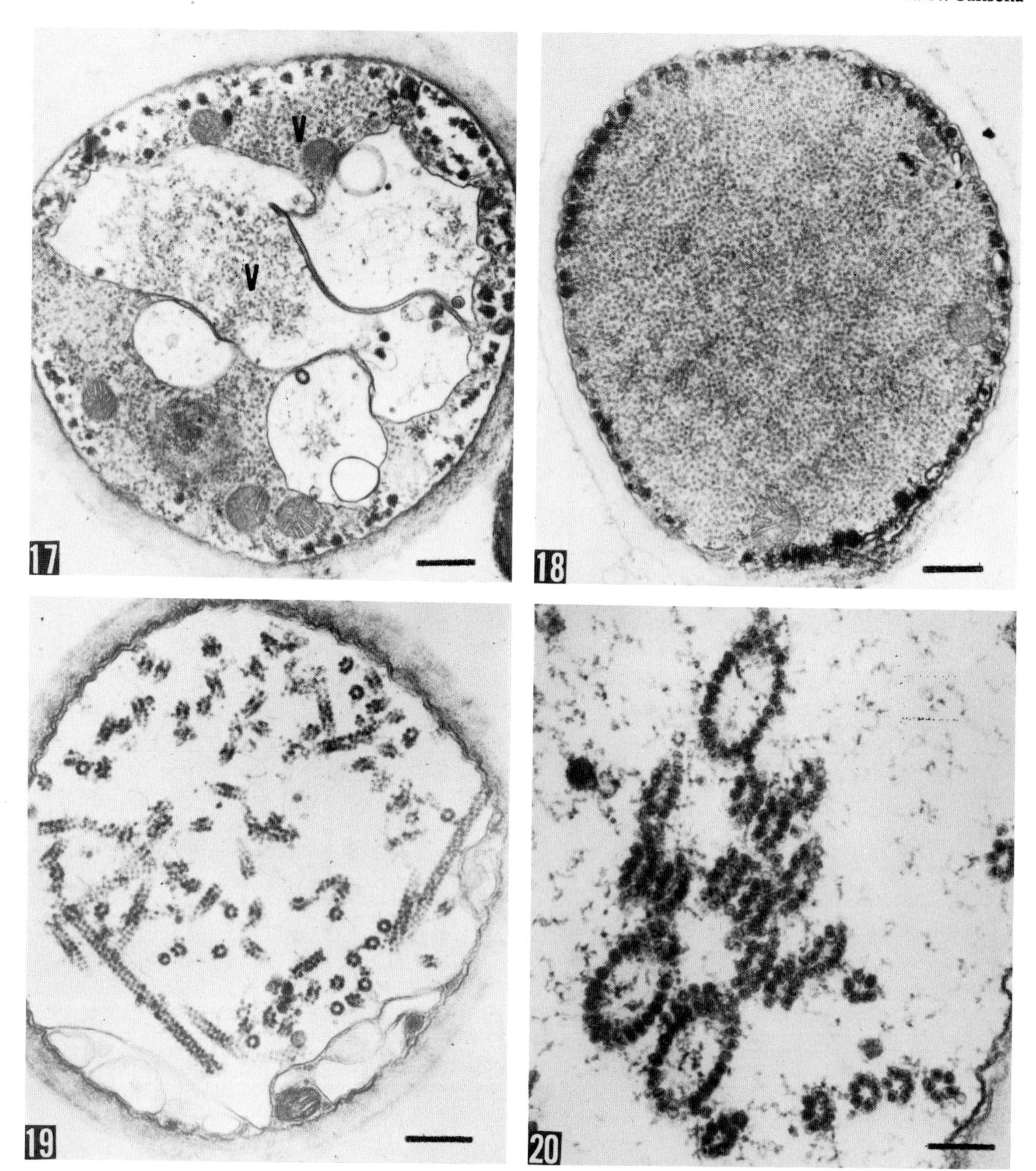

Figs. 17–20. Ultrathin sections of *Penicillium chrysogenum* showing intracellular localization of virus. [Courtesy of Yamashita *et al.* (1973) with permission of Academic Press.]

Fig. 17. Cross section of mycelium with particles (V) in vacuole and cytoplasm. Bar = 400 nm.

Fig. 18. Cross section of a cell filled with virus particles. Cytoplasm has degenerated almost completely. Bar = 400 nm.

Fig. 19. Cylindrical arrays of virus particles in a degenerated cell. Bar = 400 nm.

Fig. 20. Cylindrical arrays of virus particles. Bar = 200 nm.

Chapter 27

Viroids

T. O. DIENER

I. Introduction 431
II. Members of the Group 432
A. Potato Spindle Tuber Viroid (PSTV) 432
B. Citrus Exocortis Viroid (CEV) 432
C. Chrysanthemum Stunt Viroid 432
D. Cucumber Pale Fruit Viroid 432
E. Chrysanthemum Chlorotic Mottle Viroid 433
III. Structure 433
IV. Subcellular Location 434
V. Replication 434
References 435

I. INTRODUCTION

The term viroid has been introduced to denote a newly recognized group of subviral pathogens (Diener, 1971b). Presently known viroids consist solely of a short strand of RNA with a molecular weight in the neighborhood of 75,000–125,000 daltons. Introduction of this low-molecular-weight RNA into susceptible hosts leads to apparent replication of the RNA and, in some hosts, to disease. Viroids are the smallest known agents of infectious disease (Diener, 1974). So far, viroids are only known to occur in higher plants.

The first viroid came to light during attempts to isolate and characterize the agent of the potato spindle tuber disease, a disease which, for many years, had been assumed to be of viral etiology (Diener and Raymer, 1971). Diener and Raymer (1967) reported that the infectious agent of this disease is a free RNA and that virus particles, apparently, are not present in infected tissue. Later, sedimentation and gel electrophoretic analyses conclusively demonstrated that the infectious RNA has a very low molecular weight (Diener, 1971b) and that the agent, therefore, basically differs from conventional viruses.

Four additional plant diseases, citrus exocortis (Semancik and Weathers, 1972a), chrysanthemum stunt (Diener and Lawson, 1973), cucumber pale fruit (Van Dorst and Peters, 1974), and chrysanthemum chlorotic mottle (Romaine and Horst, 1975) are now known also to be caused by low-molecular-weight

RNA's, i.e., by viroids. Recent evidence suggests that a fifth plant disease, coconut cadang-cadang, also may be of viroid causation (Randles, 1975).

II. MEMBERS OF THE GROUP

A. Potato Spindle Tuber Viroid (PSTV)

PSTV has a wide host range, particularly among solanaceous plant species (O'Brien and Raymer, 1964; Singh, 1973), but in most species, no obvious symptoms develop. It is easily transmitted by inoculation of sap (O'Brien and Raymer, 1964). PSTV causes potato spindle tuber disease and tomato bunchy top disease.

Polyacrylamide gel electrophoresis of formaldehyde-denatured PSTV resulted in a molecular weight estimate of 75,000–85,000 daltons (Diener and Smith, 1973). Estimates derived from direct length measurements of molecules visualized by electron microscopy were in the range of 79,000–89,000 daltons (Sogo *et al.*, 1973), which is in excellent agreement with estimates derived from gel electrophoresis.

B. Citrus Exocortis Viroid (CEV)

CEV is the causative agent of citrus exocortis disease. It is mechanically transmissible. Determination of its experimental host range and comparison with that of PSTV showed that in any species or variety so far tested, the two agents elicit similar responses (Semancik and Weathers, 1972b; Singh and Clark, 1973). These results led to the conclusion that PSTV and CEV are closely related, if not identical strains of a single pathogenic agent (Singh and Clark, 1973; Semancik *et al.*, 1973).

Recent results obtained by RNA fingerprinting of ^{125}I-labeled PSTV and CEV (Dickson *et al.*, 1975) indicate, however, that the two RNA's do not have the same primary sequence.

Gel electrophoresis of native CEV led to a molecular weight estimate of 125,000 daltons (Semancik and Weathers, 1972c). So far, CEV has not been visualized by electron microscopy. The following nucleotide composition has been reported: CMP, 28.9%; AMP, 21.4%; GMP, 28.1%; UMP, 21.5% (Semancik *et al.*, 1975).

C. Chrysanthemum Stunt Viroid

The symptoms of stunt disease vary greatly according to the *Chrysanthemum* variety affected. The agent is mechanically transmissible; a large number of composite plant species have been infected and the stunt agent recovered from them (Brierley, 1953).

Diener and Lawson (1973) demonstrated that the causative agent of stunt disease is a low-molecular-weight RNA, i.e., a viroid, and that its electrophoretic mobility in polyacrylamide gels is significantly larger than that of PSTV. No other properties of the stunt viroid are known.

D. Cucumber Pale Fruit Viroid

Cucumber pale fruit disease was first noticed in 1963, in greenhouse-grown cucumber plants in the Netherlands. Van Dorst and Peters (1974) obtained evidence for viroid causation and studied biological characteristics of the disease. The agent is mechanically transmissible and is able to infect a large number of cucurbitaceous species. Further properties of the RNA have not, so far, been determined.

E. Chrysanthemum Chlorotic Mottle Viroid

Romaine and Horst (1975) presented evidence which suggests that the causative agent of chlorotic mottle disease is a viroid. This viroid is not as readily transmitted mechanically as are the other known viroids. Also in contrast to other viroids, the agent of chlorotic mottle disease appears to have a narrow host range (Romaine and Horst, 1975). Biological tests indicate that a latent strain of the chlorotic mottle viroid exists in certain *Chrysanthemum* cultivars which protects against the severe strain of chlorotic mottle viroid (Horst, 1975).

III. STRUCTURE

Although the low molecular weight of viroids has been conclusively demonstrated, a decision as to their exact structure cannot be made unambiguously in light of present knowledge. Uncertainty still exists as to whether viroids are single- or double-stranded molecules.

The total hyperchromicity shift of PSTV in 0.01 × SSC (SSC = 0.15 *M* sodium chloride, 0.015 *M* sodium citrate, pH 7.0) is about 24% and the T_m about 50° (Diener, 1972a). In 0.1 × SSC, T_m is about 54° (Diener, 1974). In either medium, denaturation occurs over a wide temperature range. Semancik *et al.* (1975) reported similar results with CEV and showed that, in the presence of magnesium, the T_m of CEV is conspicuously raised.

The thermal denaturation curves indicate that PSTV and CEV are not regularly base-paired structures since, in this case, denaturation would be expected to occur over a much narrower temperature range and at higher temperatures.

On the other hand, CEV has been shown to be partially resistant to incubation with diethyl pyrocarbonate (Semancik and Weathers, 1972a), a result which suggests a structure that is at least partially double-stranded. Immunological tests, made with antisera that react specifically with double-stranded RNA, indicated that PSTV is not a regularly base-paired double helix (Stollar and Diener, 1971). Similarly, from hydroxyapatite, PSTV elutes mostly at phosphate buffer concentrations lower than those expected with double-stranded RNA (Diener, 1971c; Lewandowski *et al.*, 1971).

As compared with single-stranded viral RNA's, CEV is remarkably resistant to inactivation by heating and by exonucleases (Semancik and Weathers, 1970, 1972a); the latter property, CEV shares with PSTV (Diener, 1971c; Singh and Clark, 1971). Although resistance to inactivation by exonucleases may suggest a circular structure of the RNA's, the degree of resistance of CEV was considered insufficient to postulate circularity (Semancik and Weathers, 1972a).

Two models are compatible with these and other properties of viroids: (a) The RNA's may be single-stranded molecules with some sort of hairpin structure, involving extensive base-pairing, or (b) the RNA's may be double-stranded, but incompletely base-paired, molecules.

In principle, a distinction between these two models should be possible by direct visualization of viroids. So far, electron microscopy has been performed with PSTV only (Sogo *et al.*, 1973).

Purified PSTV was prepared for electron microscopy by the protein monolayer spreading technique (Kleinschmidt and Zahn, 1959).

When PSTV solutions in 8 *M* urea were spread onto distilled water or onto

0.015 *M* ammonium acetate (pH 8), large numbers of individual short strands were seen. The average length of PSTV molecules was about 500 Å. When PSTV was heat denatured in the presence of 1.1 *M* formaldehyde, followed by quenching and spreading onto distilled water, the length of PSTV molecules varied; some molecules, however, were up to 700 Å long (Sogo *et al.*, 1973).

To determine whether PSTV is a single- or a double-stranded molecule, PSTV in 8 *M* urea was mixed with native coliphage T7 DNA (Fig. 1) or with heat-denatured, formylated carnation mottle virus RNA. PSTV and double-stranded T7 DNA had similar widths, but PSTV was clearly thicker than the single-stranded viral RNA. From these data, the authors concluded that PSTV probably is a single-stranded RNA with some kind of hairpin-like structure (Sogo *et al.*, 1973).

IV. SUBCELLULAR LOCATION

Bioassays of subcellular fractions from PSTV-infected tissue disclosed that appreciable infectivity was present only in the original tissue debris and in the fraction containing nuclei (Diener, 1971a). Chloroplasts, mitochondria, ribosomes, and the soluble fraction contained no more than traces of infectivity. Furthermore, when chromatin was isolated from infected tissue, most infectivity was associated with it and could be extracted as free RNA with phosphate buffer (Diener, 1971a). These and other experiments suggest that, *in situ*, PSTV is associated with nuclei and particularly with the chromatin of infected cells.

V. REPLICATION

The molecular weight of PSTV is sufficient only to code for 70 to 80 amino acids; it thus seems unlikely that the RNA contains information for the specification of a replicase in its genome. *In vitro* experiments with several cell-free protein synthesizing systems indicated that neither purified PSTV (Davies *et al.*, 1974) nor CEV (Hall *et al.*, 1974) act as messenger RNA's in these systems.

By what mechanisms an RNA of such small size is replicated in susceptible host cells is, at present, unknown, but, as discussed previously (Diener, 1971b, 1972b, 1973), two schemes are most readily compatible with present views on cellular and viral nucleic acid synthesis. The first scheme postulates that PSTV is replicated on a DNA template. Such a template might be present, in repressed form, in uninoculated plants, or it could be synthesized as a consequence of infection with PSTV. In the latter model, RNA-directed DNA polymerases would have to occur in normal cells. The second scheme presumes that PSTV is replicated independently of DNA; that is to say, that its replication is analogous to that of certain viral RNA's. Implicit in this scheme is the assumption that RNA-directed RNA polymerases occur in normal cells.

A distinction between these two schemes should be possible by use of drugs (such as actinomycin D) which specifically inhibit RNA transcription from DNA templates. Such drugs should inhibit PSTV replication if the first scheme is correct, but not if the second scheme is correct.

Leaf strips from healthy or PSTV-infected tomato plants were incubated in solutions containing [^{3}H]uracil or [^{3}H]UTP. Extraction of nucleic acids and analysis by polyacrylamide gel electrophoresis revealed ^{3}H incorporation into a component with electrophoretic mobility identical with PSTV in extracts from infected, but not in extracts from healthy leaves. No ^{3}H incorporation into PSTV could be detected when leaf strips were pretreated with actinomycin D (Diener and Smith, 1975).

Isolation of low-molecular-weight RNA from *in vitro* reaction mixtures containing nuclei isolated from PSTV-infected leaves revealed ^3H incorporation into a component with the mobility of PSTV, whereas no such component was detectable with mixtures containing nuclei from healthy leaves. Pretreatment of nuclei with actinomycin D inhibited ^3H incorporation into PSTV (Takahashi and Diener, 1975).

These results suggest that PSTV may be transcribed from a DNA template.

REFERENCES

Brierley, P. (1953). *Plant Dis. Rep.* **37,** 343.

Davies, J. W., Kaesberg, P., and Diener, T. O. (1974). *Virology* **61,** 281.

Dickson, E., Prensky, W., and Robertson, H. D. (1975). *Virology* **68,** 309.

Diener, T. O. (1971a). *Virology* **43,** 75.

Diener, T. O. (1971b). *Virology* **45,** 411.

Diener, T. O. (1971c). *In* "Comparative Virology" (K. Maramorosch and E. Kurstak eds.), p. 433. Academic Press, New York.

Diener, T. O. (1972a). *Virology* **50,** 606.

Diener, T. O. (1972b). *Adv. Virus Res.* **17,** 295.

Diener, T. O. (1973). *Perspect. Virol.* **8,** 7.

Diener, T. O., (1974). *Annu. Rev. Microbiol.* **28,** 23.

Diener, T. O., and Lawson, R. H. (1973). *Virology* **51,** 94.

Diener, T. O., and Raymer, W. B. (1967). *Science* **158,** 378.

Diener, T. O., and Raymer, W. B. (1971). *CMI/AAB, Descriptions Plant Viruses* No. 66.

Diener, T. O., and Smith, D. R. (1973). *Virology* **53,** 359.

Diener, T. O., and Smith, D. R. (1975). *Virology* **63,** 421.

Hall, T. C., Wepprich, R. K., Davies, J. W., Weathers, L. G., and Semancik, J. S. (1974). *Virology* **61,** 486.

Horst, R. K. (1975). *Phytopathology* **65,** 1000.

Kleinschmidt, A. K., and Zahn, R. K. (1959). *Z. Naturforsch. Teil B* **14,** 770.

Lewandowski, L. J., Kimball, P. C., and Knight, C. A. (1971). *J. Virol.* **8,** 809.

O'Brien, M. J., and Raymer, W. B. (1964). *Phytopathology* **54,** 1045.

Randles, J. W. (1975). *Phytopathology* **65,** 163.

Romaine, C. P., and Horst, R. K. (1975). *Virology* **64,** 86.

Semancik, J. S., and Weathers, L. G. (1970). *Phytopathology* **60,** 732.

Semancik, J. S., and Weathers, L. G. (1972a). *Virology* **47,** 456.

Semancik, J. S., and Weathers, L. G. (1972b). *Virology* **49,** 622.

Semancik, J. S., and Weathers, L. G. (1972c). *Nature (London), New Biol.* **237,** 242.

Semancik, J. S., Magnuson, D. S., and Weathers, L. G. (1973). *Virology* **52,** 292.

Semancik, J. S., Morris, T. J., Weathers, L. G., Rodorf, B. F., and Kearns, D. R. (1975). *Virology* **63,** 160.

Singh, R. P. (1973). *Am. Potato J.* **50,** 111.

Singh, R. P., and Clark, M. C. (1971). *Biochem. Biophys. Res. Commun.* **44,** 1077.

Singh, R. P., and Clark, M. C. (1973). *FAO Plant Prot. Bull.* **21,** 121.

Sogo, J. M., Koller, T., and Diener, T. O. (1973). *Virology* **55,** 70.

Stollar, B. D., and Diener, T. O. (1971). *Virology* **46,** 168.

Takahashi, T., and Diener, T. O. (1975). *Virology* **64,** 106.

Van Dorst, H. J. M., and Peters, D. (1974). *Neth. J. Plant Pathol.* **80,** 85.

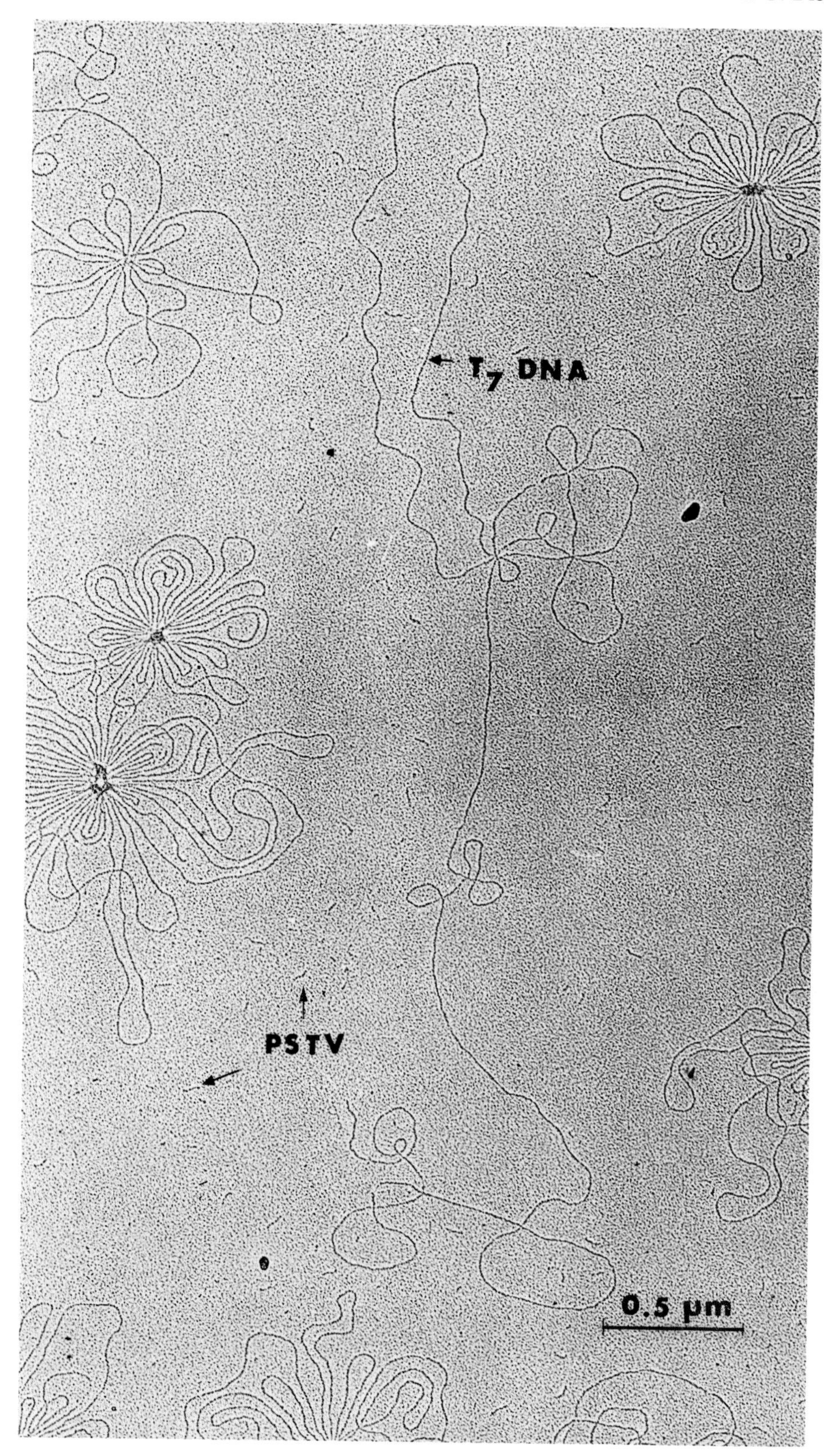

Fig. 1. Electron micrograph of potato spindle tuber viroid (PSTV) mixed with a double-stranded DNA, coliphage T7 DNA. (Courtesy of J. M. Sogo and T. Koller, Swiss Federal Institute of Technology, Zurich.)

PART IV

Mycoplasma- and Spiroplasmaviruses

Chapter 28

Mycoplasmaviruses

JACK MANILOFF, JYOTIRMOY DAS,
AND RESHA M. PUTZRATH

I. Review of the Class Mollicutes 439
A. Properties of the Mycoplasmas 439
B. Isolation and General Properties of Viruses 440
C. Nomenclature 441
II. Group 1 Mycoplasmaviruses 441
A. Occurrence 441
B. Ultrastructure 441
C. Biochemical and Biophysical Properties 442
D. Replication 442
III. Group 2 Mycoplasmaviruses 443
A. Occurrence 443
B. Ultrastructure 443
C. Biochemical and Biophysical Properties 444
D. Replication 444
E. Host Restriction and Modification 444
IV. Group 3 Mycoplasmaviruses 444
A. Occurrence 444
B. Ultrastructure 444
C. Biochemical and Biophysical Properties 445
References 445

I. REVIEW OF THE CLASS MOLLICUTES

A. Properties of the Mycoplasmas

Mycoplasma is a general name for a group of prokaryotes (class Mollicutes; order Mycoplasmatales) which do not have cell walls and are bounded by a single lipoprotein cell membrane. They have small genome sizes and most have DNA with low GC contents. These organisms were previously known as the pleuropneumonia-like organisms or PPLO. Thus far, the various mycoplasma isolates have been classified into six genera: *Mycoplasma, Acholeplasma, Ureaplasma, Spiroplasma, Thermoplasma,* and *Anaeroplasma.* The properties of the genera are summarized in Table I. The mycoplasmas include the etiologic agents of a variety of plant and animal diseases (reviewed by Davis and Whitcomb, 1971; Freundt, 1974).

TABLE I

Genera of the Class Mollicutes[a]

Genus	Genome: %GC	Genome: Molecular weight	Habitat	Sterol required for growth
Mycoplasma	23–41	4.6×10^8	Animals	yes
Acholeplasma	30–36	1.0×10^9	Saprophytic	no
Ureaplasma (T-strains)	28	4.6×10^8	Animals	yes
Spiroplasma	26	1.0×10^9	Plants, insects	yes
Thermoplasma	46	0.9×10^9	Burning coal refuse piles	no
Anaeroplasma	29–33	—	Animals	yes

[a] From review by Maniloff *et al.* (1977).

B. Isolation and General Properties of Viruses

The isolation of a virus that could infect a mycoplasma was first reported by Gourlay (1970). Since then, more than fifty isolates have been reported that can make plaques on *Acholeplasma laidlawii* indicator host strains (reviewed by Maniloff *et al.*, 1977). These isolates form three mycoplasmavirus groups, based on morphology and serology: (1) Group 1, naked bullet shaped particles; (2) Group 2, roughly spherical enveloped particles; and (3) Group 3, polyhedral particles with short tails. All three groups are DNA viruses. The general properties of the three mycoplasmavirus groups are summarized in Table II.

All of the mycoplasmaviruses in the three groups isolated thus far have come from *Mycoplasma* and *Acholeplasma* cultures and have been identified by their ability to form plaques on *A. laidlawii* indicator host strains. Not all *A. laidlawii* strains are indicators (Gourlay and Wyld, 1972; Maniloff and Liss, 1974). Plaques of the nonlytic Group 1 viruses have been shown to be due to infected cells growing slower and making smaller colonies than uninfected cells (Liss and Maniloff, 1973a; Maniloff and Liss, 1973). Group 2 viruses are also nonlytic (Putzrath and Maniloff, 1976), their plaques are similarly due to a differential growth between infected and uninfected cells. Group 3 virus plaques are relatively clear (Gourlay and Wyld, 1973) and these are lytic viruses (Liss, 1977).

TABLE II

Properties of Mycoplasmaviruses[a]

Group	Virion: Morphology	Virion: Size	Density (gm/cm³)	Nucleic acid: Type	Nucleic acid: Molecular weight	Progeny release
1	Naked bullet-shaped particles	14–16 nm × 70–90 nm	1.37	Single-stranded circular DNA	2×10^6	Nonlytic
2	Roughly spherical enveloped particles	80 nm (Range: 52–125 nm)	1.19	Double-stranded DNA	—	Nonlytic
3	Polyhedral particles with short tails	57 × 61 nm head, 25 nm long tail	1.48	Double-stranded DNA	—	Lytic

[a] References cited in Maniloff *et al.* (1977).

In addition to the three groups of mycoplasmaviruses, three different morphological types of "virus-like" particles have been observed in *Spiroplasma* cultures from plants (Cole *et al.*, 1974). It has not been possible to find conditions for propagating these particles. *Spiroplasma* strains associated with the sex-ratio trait in *Drosophila* have been shown to carry viruses (Oishi and Poulson, 1970; Oishi, 1971). The spiroplasmaviruses and "virus-like" particles have been reviewed by Cole (see Chapter 29) and are not discussed further here.

C. Nomenclature

The taxonomic relationships between the mycoplasmas and their viruses are unclear and no standard nomenclature has been adopted. In the interim, the suggestion of Liss and Maniloff (1971) will be used: for a particular isolate, the "MV" designation (for mycoplasmavirus) is followed by a letter or letters indicating the virus origin (e.g., "L" for laidlawii or "Gs" for gallisepticum) and then an isolate number. Maniloff and Liss (1974) proposed the Group 1, 2, and 3 designations for the three morphological types of isolates.

II. GROUP 1 MYCOPLASMAVIRUSES

A. Occurrence

Group 1 virus isolates have been reported from a variety of *Acholeplasma* and *Mycoplasma* species (reviewed by Maniloff and Liss, 1974): *A. laidlawii, A. granularum, A. axanthum, Mycoplasma* sp. strain 14 (goat), *M. hominis, M. pneumoniae, M. gallisepticum, M. arthritidis,* and *M. pulmonis.* All of these isolates appear to be morphologically identical, but they are only serologically related, not identical. In addition to some serological difference, the isolates show differences in biological properties: such as host range, ultraviolet light sensitivity, and one step growth parameters.

B. Ultrastructure

Group 1 virions (Fig. 1) are naked rod-shaped particles 14–16 nm wide and 70–90 nm long with helical symmetry (Gourlay *et al.*, 1971; Bruce *et al.*, 1972; Milne *et al.*, 1972, Gourlay, 1973, 1974; Liss and Maniloff, 1973a; Maniloff and Liss, 1973). The size ranges reflect variations in electron microscopic measurements from different laboratories and differences in negative staining with phosphotungstate, uranyl acetate, or uranyl formate.

Two views of Group 1 virus morphology have been proposed. All investigators agree that one end of the rod-shaped particle is rounded. Liss and Maniloff (1973a) interpret their micrographs as showing a bullet shape for the virion, with the other end of the particle (the basal end that presumably makes contact with the cell membrane) being relatively flat with some sort of short flexible protrusion. The basal structure could not be further resolved but would add about 7 nm to the particle length. However, Bruce *et al.* (1972) reported that the particle is bacilliform and that the basal end is also round but may become degraded to give short protuberances or even a flattened end. It should be noted that the ends of such thin rods have a very small radius of curvature and the drying pattern of a negative stain around any such end structure can give an artifactual rounded appearance.

Optical analysis of electron micrographs of negatively stained Group 1 virions have indicated helical symmetry and, since the optical transforms have a near meridional reflection at 4.8 nm, the distance between helix turns is 4.8 nm (Bruce *et al.*, 1972; Liss and Maniloff, 1973a). From the diffraction data, Bruce *et al.* (1972) have proposed a virion structure based on the models of Hull *et al.* (1969) for helical structures with rounded ends. The Group 1 virion model is a 14–16-nm diameter, $T = 1$ icosahedron, cut across a two-fold axis with structure units added to form a helix. This produces a two-strand helix with 5.6 structure units per turn and a pitch of about 20°.

In addition to infectious virus particles, preparations of Group 1 viruses (Fig. 1) and virus infected cells contain other viral related structures: (1) Unpurified preparations contain 13–14 nm rings (Liss and Maniloff, 1973a; Maniloff and Liss, 1973; Gourlay, 1974), which may be intracellular viral protein subassemblies. (2) A few particles have been found which were penetrated by negative stain, showing a particle wall thickness of 5.2–6.2 nm (Bruce *et al.*, 1972; J. Maniloff, unpublished data); these may be incomplete or damaged virions. (3) Rod-shaped particles have been reported with diameters (and optical transforms) similar to virions and lengths from twice that of virus particles, i.e., about 150 nm (Liss and Maniloff, 1973a), to over 500 nm (Bruce *et al.*, 1972; Milne *et al.*, 1972). No particles with a size between single- and double-length virus particles have been reported, and the double-length particles appear to have both ends rounded. (4) Long hollow tubular structures have been seen. Most widths are 13 to 18 nm, but some as wide as 30 nm have been observed, and lengths vary up to 1000 nm (Bruce *et al.*, 1972; Milne *et al.*, 1972; Liss and Maniloff, 1973a). Bruce *et al.* (1972) reported a greater frequency of these hollow tubes in both older cells (late logarithmic growth phase) infected with virus and infected cells late (over 4 hours) in infection; hence, the hollow tube structures may be some aberrant polymerization product of aged infected cells. (5) Milne *et al.* (1972) found striated masses in infected cells after 72 hours of infection. Since their cell culture was probably no longer viable when the sample was taken and old mycoplasma cells are known to produce many degradative forms, it is not possible to evaluate this report. Other laboratories have been unable to reproduce these results (Maniloff and Liss, 1974; Gourlay, 1974).

C. Biochemical and Biophysical Properties

The Group 1 viral chromosome is single-stranded circular DNA of molecular weight 2×10^6 (Liss and Maniloff, 1973b). This DNA has been shown to be infectious (Liss and Maniloff, 1972, 1974a).

There are four virion proteins with molecular weights of 70,000, 53,000, 30,000, and 19,000 and approximate stoichiometric ratios of 6.9:1.0:1.8:8.8 (Maniloff and Das, 1975). It is not known whether each protein is a unique gene product or whether one or more are cleavage products.

The data on the sensitivity of Group 1 viruses to chemical and physical treatments have been reviewed by Maniloff *et al.* (1977). The viruses are inactivated by specific antiserum and ultraviolet light; however, different isolates show different inactivation constants for these treatments. Group 1 viruses are resistant to detergents and ether but are sensitive to chloroform. The viruses are relatively heat stable. At 60° and pH 8.0, the surviving fraction is 0.37 after 3 minutes and 0.03 after 30 minutes (J. Das and J. Maniloff, unpublished data).

D. Replication

From one step growth, artificial lysis, and single "burst" types of experiments, Liss and Maniloff (1971, and 1973a) showed that viral infection was not lytic and that the viral growth curve has a short latent period followed by a gradual increase in progeny virus titer. After a rise period, the virus titer reaches a plateau. For a Group 1 virus, MVL51, infecting *A. laidlawii* JA1 in tryptose broth, the latent period is about 10 minutes and about 75–100 progeny viruses are then released per cell per hour. Other Group 1 isolates have different latent times and viral production rates.

Adsorption of Group 1 viruses to *A. laidlawii* host cells is ionic, requires mono- or divalent cations, and follows first order kinetics (Fraser and Fleischmann, 1974). The adsorption rate constant is not very temperature-dependent. Fraser and Fleischmann (1974) estimated that each cell has about 10 adsorption sites for Group 1 viruses, but the data of Das and Maniloff (1976a) indicate that there may be only 2 to 3 functional sites for intracellular viral replication.

Data on the intracellular molecular events in Group 1 viral replication have been reported by Das and Maniloff (1975, 1976a,b,c) and are schematically shown in Fig. 2. Upon infection the circular single-stranded parental DNA is converted to membrane-associated double-stranded replicative forms which can be in two configurations: RF I and RF II. RF I is a covalently closed double stranded circular form of the 2×10^6-dalton, single-stranded, circular viral DNA. RF II is a relaxed (nicked) form of RF I. This can take place at 2 to 3 membrane sites per cell. These parental RF molecules undergo symmetric (semiconservative) replication at the membrane to produce progeny RF. This step is blocked in a REP^- host (Nowak *et al.*, 1976); hence, it must require some cellular function. Further replication is asymmetric and takes place in the cytoplasm to produce single-stranded, circular, viral chromosomes (SS I). The SS I viral DNA associates with viral proteins and undergoes assembly and extrusion through the cell membrane without lysing the cell. An intermediate in this pathway, SS II, has been shown to be a complex of SS I with the 70,000- and 53,000-dalton viral proteins. The stoichiometric ratio of these two proteins in SS II is about 1.4:1.0. Chloramphenicol inhibits the SS I to SS II conversion and rifampin blocks assembly somewhere between SS I and the completed virion. The original parental viral DNA strand remains in the cell and is not transferred to progeny viruses (Maniloff and Liss, 1973; Liss and Maniloff, 1974b; Das and Maniloff, 1976a).

Some of the roles of cellular and viral functions in viral replication have been described. The REP^- cells showed a cellular function in RF replication. In addition, using rifampin resistant host cells, it has been shown that rifampin has no effect on viral DNA synthesis. However, the antibiotic inhibits some virus specific transcription necessary for the assembly of mature virus particles.

III. GROUP 2 MYCOPLASMAVIRUSES

A. Occurrence

Three Group 2 isolates have been described; all are from *A. laidlawii* strains (Gourlay, 1971, 1972; Liska, 1972). However, isolation attempts from *Mycoplasma* have not been reported.

B. Ultrastructure

These viruses have a lipid containing envelope. Electron micrographs of negatively stained virions show approximately spherical particles (Fig. 3), with a size range of 50 to 125 nm in diameter and a mean diameter about 80 nm

(Gourlay, 1971; Gourlay *et al.*, 1973). Negative staining does not reveal any internal viral morphology.

Micrographs of thin-sectioned virus (Fig. 3) show the envelope to have a "unit membrane" appearance (Gourlay *et al.*, 1973; Liska and Tkadlecek, 1975; Putzrath and Maniloff, 1977). The internal morphology appears fibrillar, with a densely staining core. These data suggest that Group 2 particles may not have a rigid internal helical or icosahedral capsid structure. Instead the virion could be simply double-stranded DNA complexed with proteins and bounded by a membrane. Freeze etching studies (Fig. 3) show that the virus has fewer particles on the viral membrane fracture faces that are seen on cell membrane fracture faces (Putzrath and Maniloff, 1976).

C. Biochemical and Biophysical Properties

Electron micrographs of Group 2 viral DNA indicate that it is double-stranded (J. Maniloff, R. M. Putzrath, and J. Das, unpublished data). No further data are available on DNA, protein, or lipid compositions.

The viral membrane lipids make the Group 2 viruses sensitive to detergents and organic solvents (Gourlay, 1971) and probably account for their thermal sensitivity (Maniloff *et al.*, 1976).

Group 2 viruses are the most UV resistant of the mycoplasmaviruses (Maniloff *et al.*, 1977). Das *et al.* (1977) have shown that UV damaged Group 2 viral DNA can be repaired by two different host cell DNA repair systems. One involves a host excision repair mechanism and the other appears similar to an inducible error-prone repair mechanism found in *Escherichia coli.*

D. Replication

One-step growth and artificial lysis experiments have shown that Group 2 infection is not lytic (Putzrath and Maniloff, 1977). Thin section and freeze etching preparations show viruses budding from the cell membrane (Fig. 3). The freeze-etch data indicate that infected cells contain a structure which may be a viral replicative intermediate (Putzrath and Maniloff, 1977).

E. Host Restriction and Modification

A Group 2 virus (but not Group 1) has been found to be host restricted and modified (Maniloff and Das, 1975). The virus gives a thousandfold more plaques on its most recent host than on some other (restricting) host. The host-controlled modification is believed to be due to some DNA modification mechanism such as methylation or glucosylation (reviewed by Maniloff *et al.*, 1977).

IV. GROUP 3 MYCOPLASMAVIRUSES

A. Occurrence

Only one Group 3 mycoplasmavirus isolate, from an *A. laidlawii*, has been reported (Gourlay and Wyld, 1973). Isolation attempts from *Mycoplasma* have not been reported.

B. Ultrastructure

Negatively stained preparations of the Group 3 isolate show uniform sized particles with a polyhedral head, about 57 nm × 61 nm, and a short tail, about 9 nm wide and 25 nm long, attached with a collar at one vertex (Fig. 4; Garwes *et al.*, 1975). The collar is more easily seen with phosphotungstate, rather than uranyl acetate. Elongated (perhaps polyhead-like) particles were also observed.

C. Biochemical and Biophysical Properties

The chemical composition of the Group 3 virus was found to be 35% DNA, 63% protein, and 2% fucose (Garwes *et al.*, 1975). The DNA is probably double-stranded, since a thermal hypochromic transition was observed, but conflicting GC values were obtained when different procedures were used. Five proteins were found, with molecular weights 172,000, 81,000, 73,000, 68,000, and 43,000. These account for 1, 7.5, 1, 10.5, and 80%, respectively, of the total virus protein. It is not known which of the proteins are unique or cleavage products. The fucose could be present as glycoprotein or associated with the DNA.

The Group 3 virus isolate is relatively heat stable and is resistant to Nonidet P40 and ether (Gourlay and Wyld, 1973).

ACKNOWLEDGMENTS

We thank the other members of our research group (Dr. A. Liss, Mr. J. Nowak, and Mr. D. Gerling) who have contributed to our information and ideas concerning the mycoplasmaviruses. Throughout these studies, Dr. J. R. Christensen has provided invaluable background information and advice. Support for the studies in this laboratory has come from the United States Public Health Service, National Institute of Allergy and Infectious Diseases, Grant Al-10605.

REFERENCES

Bruce, J., Gourlay, R. N., Hull, R., and Garwes, D. J. (1972). *J. Gen. Virol.* **16,** 215–221.
Cole, R. M., Tully, J. G., and Popkin, T. J. (1974). *Colloq. INSERM* **33,** 125–132.
Das, J., and Maniloff, J. (1975). *Biochem. Biophys. Res. Commun.* **66,** 599–605.
Das, J., and Maniloff, J. (1976a). *Proc. Natl. Acad. Sci. U.S.A.* **73,** 1489–1493.
Das, J., and Maniloff, J. (1976b). *Microbios* **15,** 127–134.
Das, J., and Maniloff, J. (1976c). *J. Virol.* **18,** 969–976.
Das, J., Nowak, J., and Maniloff, J. (1977). *J. Bacteriol.* **129,** 1424–1427.
Davis, R. E., and Whitcomb, R. F. (1971). *Annu. Rev. Phytopathol.* **9,** 119–154.
Fraser, D., and Fleischmann, C. (1974). *J. Virol.* **13,** 1067–1074.
Freundt, F. A. (1974). *Pathol. Microbiol.* **40,** 155–187.
Garwes, D. J., Pike, B. V., Wyld, S. G., Pocock, D. H., and Gourlay, R. N. (1975). *J. Gen. Virol.* **29,** 11–24.
Gourlay, R. N. (1970). *Nature (London)* **225,** 1165.
Gourlay, R. N. (1971). *J. Gen. Virol.* **12,** 65–67.
Gourlay, R. N. (1972). *Pathog. Mycoplasmas, Ciba Found. Symp., 1972* pp. 145–156.
Gourlay, R. N. (1973). *Ann. N. Y. Acad. Sci.* **225,** 144–148.
Gourlay, R. N. (1974). *Crit. Rev. Microbiol.* **3,** 315–331.
Gourlay, R. N., and Wyld, S. G. (1972). *J. Gen. Virol.* **14,** 15–23.
Gourlay, R. N., and Wyld, S. G. (1973). *J. Gen. Virol.* **19,** 279–283.
Gourlay, R. N., Bruce, J., and Garwes, D. J. (1971). *Nature (London) New Biol.* **229,** 118.
Gourlay, R. N., Garwes, D. J., Bruce, J., and Wyld, S. G. (1973). *J. Gen. Virol.* **18,** 127–133.
Hull, R., Hills, G. J., and Markham, R. (1969). *Virology* **37,** 416–428.
Liska, B. (1972). *Stud. Biophys.* **34,** 151–155.
Liska, B., and Tkadlecek, L. (1975). *Folia Microbiol.* (Prague) **20,** 1–7.
Liss, A. (1977). *Virology* **77,** 433–436.
Liss, A., and Maniloff, J. (1971). *Science* **173,** 725–727.
Liss, A., and Maniloff, J. (1972). *Proc. Natl. Acad. Sci. U.S.A.* **69,** 3423–3427.
Liss, A., Maniloff, J. (1973a). *Virology* **55,** 118–126.
Liss, A., and Maniloff, J. (1973b). *Biochem. Biophys. Res. Commun.* **51,** 214–218.
Liss, A., and Maniloff, J. (1974a). *Microbios* **11,** 107–113.

Liss, A., and Maniloff, J. (1974b). *J. Virol.* **13,** 769–774.

Maniloff, J., and Das, J. (1975). *In* "DNA Synthesis and its Regulation" (M. Goulian, P. Hanawalt, and C. F. Fox, eds.), pp. 445–450. Benjamin, New York.

Maniloff, J., and Liss, A. (1973). *Ann. N. Y. Acad. Sci.* **225,** 149–158.

Maniloff, J., and Liss, A. (1974). *In* "Viruses, Evolution and Cancer" (E. Kurstak and K. Maramorosch, eds.). pp. 584–604. Academic Press, New York.

Maniloff, J., Das, J., and Christensen, J. R. (1977). *Adv. Virus Res.* **21,** 343–380.

Milne, R. G., Thompson, G. W., and Taylor-Robinson, D. (1972). *Arch. Gesamte Virusforsch.* **37,** 378–385.

Nowak, J., Das, J., and Maniloff, J. (1976). *J. Bacteriol.* **127,** 832–836.

Oishi, K. (1971). *Genet. Res.* **18,** 45–56.

Oishi, K., and Poulson, D. F. (1970). *Proc. Natl. Acad. Sci. U.S.A.* **67,** 1565–1572.

Putzrath, R. M., and Maniloff, J. (1976). *Biophys. J.* **16,** 48a.

Putzrath, R. M., and Maniloff, J. (1977). *J. Virol.* **22,** 308–314.

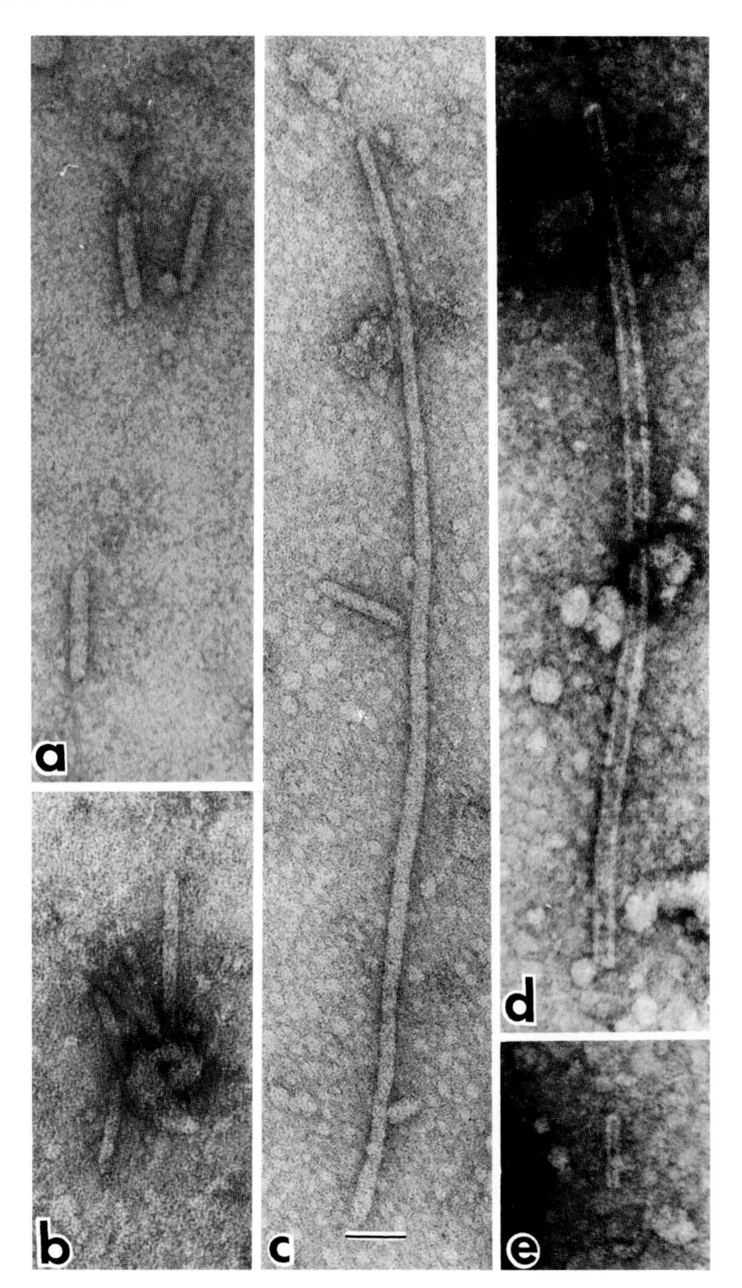

Fig. 1. Electron micrographs of Group 1 mycoplasmaviruses. Purified MVL51 preparations were negatively stained with uranyl formate. (a) Bullet-shaped virions. (b) Double-length particle, near some virions and debris. (c) Long rod-shaped structure and two virions (d) Long hollow tubular structure. (e) Particle penetrated by negative stain. The various morphological forms are discussed in the text. Bar = 50 nm.

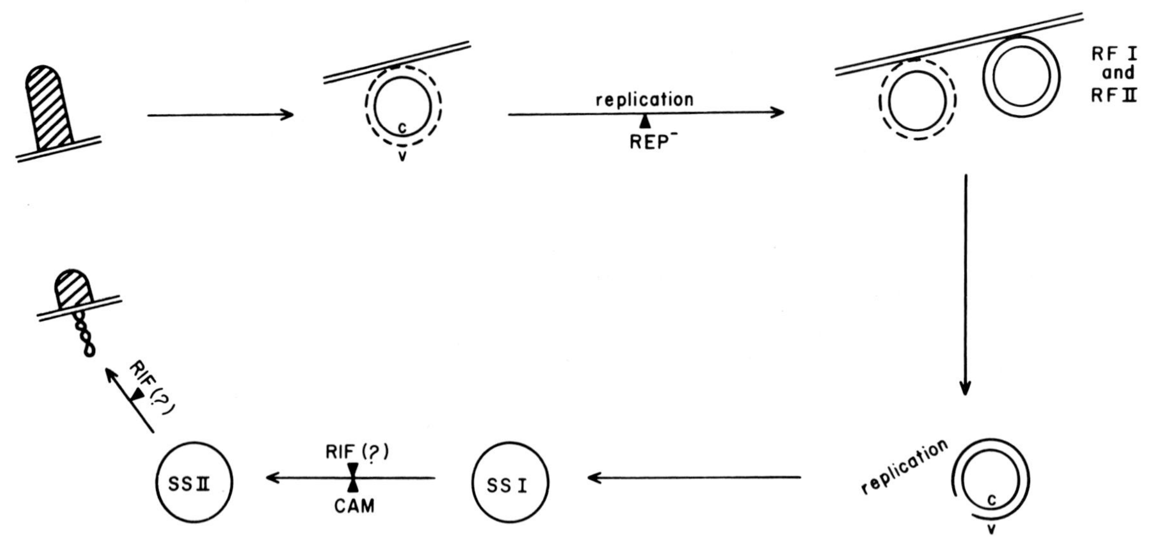

Fig. 2. Schematic of the replication of Group 1 mycoplasmaviruses. Dotted lines show parental viral DNA strands and continuous lines show progeny DNA strands. Steps blocked by rifampin (RIF), chloramphenicol (CAM), and a REP^{-} cell variant are shown. Viral and complementary DNA strands are marked "v" and "c," respectively. The parallel double lines denote cell membranes. From Maniloff *et al.* (1976).

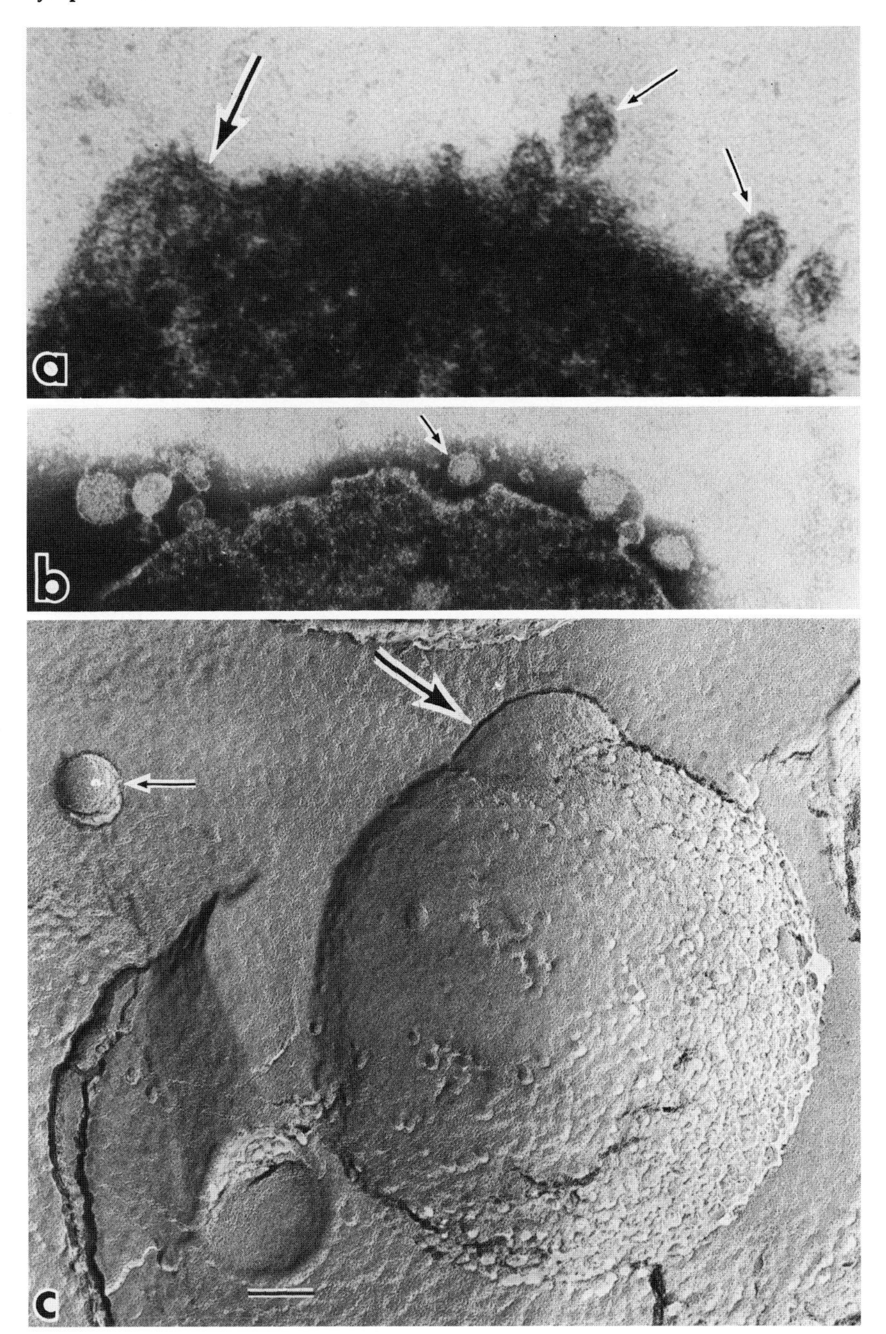

Fig. 3. Electron micrographs of Group 2 mycoplasmaviruses. *A. laidlawii* cells were infected with MVL2 and prepared for electron microscopy after glutaraldehyde fixation. (a) Thin sectioned cells, stained with uranyl acetate and lead citrate. (b) Cells negatively strained with phosphotungstate. (c) Cells freeze etched and platinum shadowed. Large arrows show areas of cell surface where virus budding may be occurring. Small arrows show extracellular progeny viruses. Bar = 100 nm.

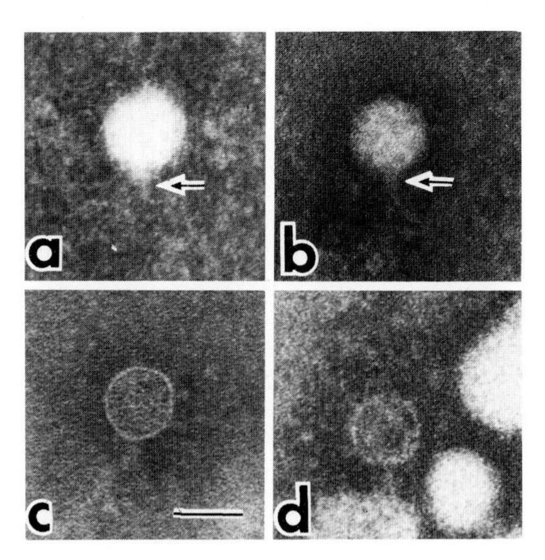

Fig. 4. Electron micrographs of Group 3 mycoplasmaviruses. A wash of MVL3-infected cell lawns was negatively stained with phosphotungstate. (a) and (b) Intact virus particles with tails (arrows). (c) and (d) Empty virion heads. Bar = 50 nm.

Chapter 29

Spiroplasmaviruses

ROGER M. COLE

I. Discovery of Mycoplasmaviruses 451
A. Viruses of *Acholeplasma laidlawii* 451
B. Other Suspected Mycoplasmaviruses 452
C. Viruses of Spiroplasmas 452
II. Spiroplasmas: the Hosts of Spiroplasmaviruses 452
III. The Spiroplasmaviruses 453
A. Morphology and Ultrastructure of SVC1, SVC2, and SVC3 453
B. Detection, Occurrence, Distribution, and Other Properties 455
Addendum 456
References 456

I. DISCOVERY OF MYCOPLASMAVIRUSES

A. Viruses of *Acholeplasma laidlawii*

Until 1970, viruses infecting members of the prokaryotic class Mollicutes (Freundt, 1974) were unknown. At that time Gourlay (1970) isolated the first of three viruses from the organism then known as *Mycoplasma laidlawii*, which was later shown to be a sterol nonrequiring mycoplasma that was consequently renamed *Acholeplasma laidlawii* and placed in a new family of the order Mycoplasmatales (Edward and Freundt, 1970; Freundt, 1974). The virus was subsequently shown to be a DNA-containing rod-shaped particle and was designated MVL1 for Mycoplasmatales virus laidlawii 1 (Gourlay *et al.*, 1971; Gourlay and Wyld, 1972). Isolations of similar rod-shaped particles, infectious for and usually originating from, *A. laidlawii*, were reported by Liss and Maniloff (1971). All these may now be considered, collectively, as group L1 viruses (Maniloff and Liss, 1974). A globular virus designated as MVL2 was then discovered (Gourlay, 1971; Gourlay *et al.*, 1973): a virus (MV-Lg-p32-L172), probably related to MVL2, was reported by Liska and associates (Liska, 1972; Hutková *et al.*, 1973; Liska and Tkadlecek, 1975). MVL3, a polyhedral virion with a short tail, was described later (Gourlay and Wyld, 1973; Gourlay, 1974; Garwes *et al.*, 1975). These *Acholeplasma* viruses, all of which have been propagated in *A. laidlawii* indicator strains, are described in recent articles

(Gourlay, 1974; Maniloff and Liss, 1974; Garwes *et al.*, 1975); and their ultrastructure is discussed in Chapter 28 of this volume.

B. Other Suspected Mycoplasmaviruses

Earlier, there were reports of intracellular virus-like particles detected by electron microscopy in an unspecified human mycoplasma (Swartzendruber *et al.*, 1967) and in *Mycoplasma hominis* (Robertson *et al.*, 1972). In addition, there have been noted several virus-like bodies in, or associated with, presumptive mycoplasmas occurring in some diseased plants (Ploaie, 1971; Gourret *et al.*, 1973; Cadilhac and Gianotti, 1975). None of these have been reported to be propagable, and their natures are unknown.

C. Viruses of Spiroplasmas

A long-tailed phage-like particle in *S. citri* was first reported in 1973 (Cole *et al.*, 1973a,b). Two other particles were later distinguished in the same host (Cole *et al.*, 1974); and as noted below, have also been found in additional microorganisms that appear to be spiroplasmas. Because these hosts represent a new genus and morphologic type of mycoplasma, it is appropriate to discuss them briefly before considering their viruses.

II. SPIROPLASMAS: THE HOSTS OF SPIROPLASMAVIRUSES

The genus *Spiroplasma* was established by Saglio *et al.* (1973). The type species, *S. citri*, was the first cultured in artificial media (Saglio *et al.*, 1971; El-Shafy *et al.*, 1972): it is the microorganism responsible for the yellowing condition of citrus trees known as "stubborn" disease. This sterol-requiring mycoplasma would appear to meet most criteria for inclusion in the family Mycoplasmataceae of the order Mycoplasmatales, but some characteristics (Saglio *et al.*, 1973) have delayed full acceptance in the most recent edition of Bergey's Manual of Determinative Bacteriology (Freundt, 1974). However, the genus is unique among mycoplasmas by virtue of its helical filamentous shape and rapid rotational and flexing motility (Cole *et al.*, 1973a,b). Microorganisms recognized retrospectively to have similar morphology, but considered originally to be spirochetes, were previously found in species of *Drosophilia* exhibiting a sex-ratio imbalance (Poulson and Sakaguchi, 1961; Oishi and Poulson, 1970; Williamson and Whitcomb, 1974) and in a rabbit tick (Pickens *et al.*, 1968). In addition, the mycoplasma-like organism previously associated with corn stunt disease (Chen and Granados, 1970) was later shown to be a helical filament with rotational motility (Davis *et al.*, 1972) for which the trivial name "spiroplasma" was proposed (Davis and Worley, 1973). Now, new evidence (Tully *et al.*, 1976) shows that a pooled isolate of rabbit ticks (*Hemaphysalis leporis-palustris*), passed for many years in embryonated eggs and rodents as the "suckling mouse cataract agent" (SMCA) (Clark, 1964) and shown to be mycoplasma-like ultrastructurally (Zeigel and Clark, 1974), is also a spiroplasma. Only *S. citri* (Saglio *et al.*, 1971; El-Shafy *et al.*, 1972) and, recently the corn stunt organism (CSO) (Chen and Liao, 1975; Williamson and Whitcomb, 1975), have been reported cultivatable in artificial media. While none except *S. citri* have yet been shown to satisfy the criteria (Edward *et al.*, 1972) for designation as new species of mycoplasmas, such descriptions are probably imminent.

Despite the uncertainty of precise taxonomic positions, all the aforementioned spiroplasmas (with the exception of the "spirochete" of Pickens *et al.* (1968)) have been found to contain one or more virus-like particles detectable by electron microscopy.

III. THE SPIROPLASMAVIRUSES

A. Morphology and Ultrastructure of SVC1, SVC2, and SVC3

During the original ultrastructural characterization of *Spiroplasma citri* (Cole *et al.*, 1973a,b), a particle with a polyhedral head and a long, unsheathed, noncontractile tail was found in the cultured mycoplasma. Since then, two additional morphologically distinct particles, appearing also to be viruses, have been discovered infecting *S. citri:* some strains of the host produce two, or all three, of the viruses (Cole *et al.*, 1974). To date, all of 27 strains of *S. citri* examined carry one or more of the viruses. These strains were isolated individually in 4 different laboratories located in France, England, and the United States (R. M. Cole, unpublished data; Cole *et al.*, 1974), and the viruses are therefore believed to represent viruses of *Spiroplasma* and not adventitious agents which could commonly infect mycoplasmas from widely disparate geographic sources.

The three virus-like particles have been designated SVC1, SVC2, and SVC3 for *Spiroplasmavirus-citri* 1, etc. (Cole *et al.*, 1974). This terminology was chosen in analogy with that previously described for the three viruses of *Acholeplasma laidlawii* (Gourlay, 1974; Maniloff and Liss, 1974). Virus detection and identification depends entirely on electron microscopy of morphology and size at the present, because—except for some preliminary findings with SVC3 to be discussed later—no system of indicator strains and detection of formation of infectious centers (plaques) on mycoplasma lawns, has been reported.

1. *Spiroplasmavirus-citri 1.* SVC1 (Cole *et al.*, 1974) is a filamentous or rod-shaped particle that is seen protruding from the host cell surface—frequently in clusters (Figs. 1 and 2). Individual particles measure 10–15 nm in diameter and are most often 230–280 nm in length, although some of double length (Fig. 1) or even much longer filaments are occasionally seen. The rods show one rounded free end and one end with a flat plate (Figs. 3 and 4), which may be either a device for adsorption and/or a remnant of its prior attachment to host surface (Fig. 5), some rods are hollow or partially hollow (Figs. 2 and 5), and the surface is composed of subunits that rarely appear to have a helical arrangement (Figs. 3–5). In section, clusters of rods extending from host cell surface are seen, but details are not clear (Figs. 6 and 7); there is no evidence of rods nor morphological precursors inside the infected cells.

In all these respects, except mean length, SVC1 morphologically resembles the L1 group of *A.laidlawii* viruses. Whether it also contains single-stranded DNA or exhibits other properties of MVL1 (Gourlay, 1974; Maniloff and Liss, 1974), is cross-infective, or is serologically related, are questions that must await further study. The location of the rod-shaped particle on the host cell surface could indicate either adsorption prior to infection, or extrusion of mature particles during virus production; the answer is not known for SVC1, although some evidence for an L1 virus (Liss and Maniloff, 1973) suggests that the latter may be correct.

2. *Spiroplasmavirus-citri 2.* SVC2, which was the first discovered (Cole *et al.*, 1973a,b, 1974), appears to be a classic bacteriophage with a polyhedral head and a relatively long and noncontractile tail (Fig. 8). The head, which usually appears hexagonal in profile, measures 52–58 nm from vertex to vertex, and

48–51 nm between flat sides. The tail is 6–8 nm wide and 75–83 nm in length. No collar nor other differentiation has been defined at the point of attachment of tail to one vertex of the head. At the other end, the tail shows a broadening or base plate that is often poorly defined. Where adsorbed, the tail sometimes shows a suggestion of additional structures between base plate and point of attachment to the outer layer of host cell membrane (Fig. 8d), but spikes or filaments have not been clearly defined.

Particles released from lysing or ruptured cells are usually a mixture of complete or incomplete forms; empty heads are common, and free heads and tails are sometimes seen. In rare instances, bodies appearing to be polyhead structures have also been detected (Fig. 9). In sections, both full and empty heads are seen intracellularly (Fig. 10) and can be distinguished by their larger size from SVC3 (see Section III,A,3); however, intracellular tails are not readily detected.

3. *Spiroplasmavirus-citri 3.* SVC3 (Cole *et al.*, 1974) is a polyhedral particle appearing hexagonal in profile at times, but is more often seen as a sphere (Figs. 11 and 12). Where measurable, the vertex-to-vertex distance of the head is 37–44 nm, and the distance between flat sides is 35–37 nm. A short tail, 13–18 nm in length, projects from one vertex; the details of its slightly wider distal end have not been resolved, but the width of the remainder is 8–10 nm. There are sometimes subunits of the head seen (Figs. 11 and 12), but these do not appear to be arranged in any regular pattern. The particle attaches to the outer layer of the host cell membrane by the distal tip of the tail (Figs. 11–18). No collar nor other tail differentiation have been resolved. In all morphological respects except this latter, SVC3 resembles the short-tailed MVL3—although the dimensions of both head and tail are considerably less than those reported for MVL3 (Garwes *et al.*, 1975). While it is now known that the latter contains double stranded DNA (Garwes *et al.*, 1975), no similar information is yet available for SVC3.

A pecularity of SVC3, which is helpful in electron microscopic identification, is its tendency to emerge from infected cells surrounded by host cell membrane (Cole *et al.*, 1974). Because virus is also seen free or attached to host cells, it is clear that the arrangement is a temporary one and that infectious particles are somehow eventually released from the enveloping host membrane. These attach to other host cells which can be helical or filamentous as well as the round body shown (Fig. 13). The presumptive sequence is nucleic acid injection, followed by appearance of incomplete or empty particles intracellularly which are most often seen in large round bodies (Fig. 14). In other cells, complete particles are also seen (Fig. 15), and in some cultures nearly every cell contains large numbers of dense particles (Fig. 16). The "budding" release of membrane-bounded particles completes the cycle (Figs. 17 and 18).

Particles meeting the morphologic criteria of SVC3 have been observed in 10/27 strains of *S. citri*, 2/4 strains of the CSO, and most abundantly in two strains of the SMCA (R. M. Cole, unpublished observations). They have also been seen in *S. citri* by Calavan and Gumpf (1974) and in the SRO by Williamson *et al.* (Chapter 30). In the latter organism, when it was considered a spirochete, Oishi and Poulson (1970) described a DNA-containing, 50–60 nm spherical virus (spv-1) that emerged encased in surface material of the host cell, that was lytic for the host cells, and that consequently could eliminate the sex-ratio condition (absence of male progeny) in "spirochete"-infected *Drosophila.* It is likely that spv-1 is a member of the SVC3 group, that differences in measured dimensions are not significant, and that a short tail was not detected—the latter recalling the initial difficulty of Gourlay and Wyld (1973) in clearly defining a tail of MVL3, which later became obvious on examination of better preparations (Gourlay, 1974; Garwes *et al.*, 1975). Confirmation of the

SVC3-like morphology and the "budding" release of the particle in SRO is supplied by Williamson *et al.* (Chapter 30).

B. Detection, Occurrence, Distribution, and Other Properties

At present, detection of spiroplasmaviruses depends entirely on the use of electron microscopy (Cole *et al.*, 1974), by which means the characteristic morphologies and sizes of the three particles can be recognized. By this method it was observed that the viruses occurred spontaneously in cultures of *S. citri*, that their detection in each passage of a host strain was sporadic, and that, when present, the numbers of particles differed unaccountably from passage to passage. This information suggested that the virus–host relationship was one of lysogeny, but early attempts at induction were unsuccessful and the nature of the relationship is unproven (Cole *et al.*, 1973a, 1974). Similar sporadicity of virus occurrence has been observed in the other spiroplasma hosts (R. M. Cole, unpublished observations).

As determined by morphologic class of particle, the distribution of the viruses among those spiroplasmas that have been examined is shown in Table I. In *S. citri*, all of 27 strains examined repeatedly contained at least one virus, and some contained two or all three (Cole *et al.*, 1974). An example of multiple infection is shown in Fig. 19. Experience with other spiroplasmas is limited, but it is anticipated that the spectrum of both viruses and hosts will be extended as more spiroplasmas are discovered and cultured and as more strains are examined.

It should be emphasized that current discussions refer only to three morphologic classes of virus-like particles that occur in common among several spiroplasma hosts. The perils of classification based on morphologies alone have been noted by Anderson (1973), and it is expected that differences in proteins, antigenic composition, host range, and nucleic acids will eventually be demonstrated—possibly even within one morphologic class. The reason for the present lack of such information, or for data on biological and physicochemical properties of the particles, is that none have been reproducibly propagated in suitable indicator strains of spiroplasmas. However, such data seems imminent, because Calavan and Gumpf (1974) referred (without details) to the propagation and plaquing of an SVC3-like virus from *S. citri*; and plaques (Fig. 20), from which only SVC3 particles were recovered, have been now produced by SVC3-containing culture fluids on lawns of other strains of *S. citri* in our labora-

TABLE I

Distribution of Viruses among Spiroplasmas

Host	No. strains examined	No. strains positive for[a]		
		SVC1	SVC2	SVC3
Spiroplasma citri	27	11	12	10
Sex-ratio organism	1	1	0	1
Corn stunt organism	4	2	0	2
Suckling mouse agent	2	0	0	2
TOTAL	34	14	12	15

[a] Numbers are not additive to total strains examined because some strains contain more than one virus. In *S. citri*, all 27 strains examined contain at least one virus: 13 strains contain one only, 10 strains contain two, and four strains contain all three. Of the four corn stunt strains examined, two strains contain two viruses as shown, and two contain none.

tory (R. M. Cole, unpublished observations). Once such indicator systems are established for all three spiroplasmaviruses, biological and physicochemical data similar to those for *Acholeplasma* viruses (Gourlay, 1974; Maniloff and Liss, 1974) can be obtained.

ADDENDUM

Since preparation of this manuscript, several pertinent reports have appeared. SMCA has been cultivated in artificial media at temperatures from 30° to 37°C (Tully *et al.*, 1976). The "spirochaete" of M. Pickens *et al.* (1968) has been cultured in artificial media and recognized as a spiroplasma (Brinton and Burgdorfer, 1976). Two new spiroplasmas, also isolated and serially grown in artificial media, have been reported—one apparently responsible for the unique appearance of the ornamental cactus *Opuntia tuna var. monstrosa* (Kondo *et al.*, 1976), and one causing a fatal spring disease of honey bees (Clark, 1977). Except for SMCA, as previously noted, the presence of virus-like particles in these new spiroplasma isolates has not been reported.

REFERENCES

Anderson, T. F. (1973). *In* "Ultrastructure of Animal Viruses and Bacteriophages: An Atlas" (A. J. Dalton and F. Haguenau, eds.), pp. 347–358. Academic Press, New York.

Brinton, L. P., and Burgdorfer, W. (1976). *Int. J. Syst. Bacteriol.* **26,** 554–560.

Cadilhac, B., and Gianotti, J. (1975). *C. R. Hebd. Seances Acad. Sci., Ser. D* **281,** 539–542.

Calavan, E. C., and Gumpf, D. J. (1974). *Colloq. INSERM* **33,** 181–185.

Chen, T. A., and Granados, R. R. (1970). *Science* **167,** 1633–1636.

Chen, T. A., and Liao, C. H. (1975). *Science* **188,** 1015–1017.

Clark, H. F. (1964). *J. Infect. Dis.* **114,** 476–487.

Clark, T. B. (1977). *J. Invertebr. Pathol.* **29,** 112–113.

Cole, R. M., Tully, J. G., Popkin, T. J., and Bové, J. M. (1973a). *J. Bacteriol.* **115,** 367–386.

Cole, R. M., Tully, J. G., Popkin, T. J., and Bové, J. M. (1973b). *Ann. N. Y. Acad. Sci.* **225,** 471–493.

Cole, R. M., Tully, J. G., and Popkin, T. J. (1974). *Colloq. INSERM* **33,** 125–132.

Davis, R. E., and Worley, J. F. (1973). *Phytopathology* **63,** 403–408.

Davis, R. E., Worley, J. F., Whitcomb, R. F., Ishijima, T., and Steere, R. L. (1972). *Science* **176,** 521–523.

Edward, D. G. ff., and Freundt, E. A. (1970). *J. Gen. Microbiol.* **62,** 1–2.

Edward, D. G. ff., Freundt, E. A., Chanock, R. M., Fabricant, J., Hayflick, L., Lemcke, R. M., Razin, S., Somerson, N. L., Tully, J. G., and Wittler, R. G. (1972). *Int. J. Syst. Bacteriol.* **22,** 184–188.

El-Shafy, A., Fudl-Allah, A., Calavan, E. C., and Igwebe, E. C. K. (1972). *Phytopathology* **62,** 729–731.

Freundt, E. A. (1974). *In* "Bergey's Manual of Determinative Bacteriology" (R. E. Buchanan and N. E. Gibbons, eds.), 8th ed., pp. 929–955. Williams & Wilkins, Baltimore, Maryland.

Garwes, D. J., Pike, B. V., Wyld, S. G., Pocock, D. H., and Gourlay, R. N. (1975). *J. Gen. Virol.* **29,** 11–24.

Gourlay, R. N. (1970). *Nature (London)* **225,** 1165.

Gourlay, R. N. (1971). *J. Gen. Virol.* **12,** 65–67.

Gourlay, R. N. (1974). *Crit. Rev. Microbiol.* **3,** 315–331.

Gourlay, R. N., and Wyld, S. G. (1972). *J. Gen. Virol.* **14,** 15–23.

Gourlay, R. N., and Wyld, S. G. (1973). *J. Gen. Virol.* **19,** 279–283.

Gourlay, R. N., Bruce, J., and Garwes, D. J. (1971). *Nature (London), New Biol.* **229,** 118–119.

Gourlay, R. N., Garwes, D. J., Bruce, J. and Wyld, S. G. (1973). *J. Gen. Virol.* **18,** 127–133.

Gourret, J. P., Maillet, P. L., and Gouranton, J. (1973). *J. Gen. Microbiol.* **74,** 241–249.

Hutková, J., Drasil, V., Melichar, A., and Smarda, J. (1973). *Stud. Biophys.* **39,** 217–222.

Kondo, F., McIntosh, A. H., Padhi, S. B., and Maramorosch, K. (1976). *34th Ann. Proc. Electron Microscopy Soc. Amer.* 56–57.

Liska, B. (1972). *Stud. Biophys.* **34,** 151–155.

Liska, B., and Tkadlecek, L. (1975). *Folia Microbiol. (Prague)* **20,** 1–7.

Liss, A., and Maniloff, J. (1971). *Science* **173,** 725–727.

Liss, A., and Maniloff, J. (1973). *Virology* **55,** 118–126.
Maniloff, J., and Liss, A. (1974). *In* "Viruses, Evolution and Cancer" (E. Kurstak and K. Maramorosch, eds.), pp. 583–604. Academic Press, New York.
Oishi, K., and Poulson, D. F. (1970). *Proc. Natl. Acad. Sci. U.S.A.* **76,** 1565–1572.
Pickens, E. G., Gerloff, R. K., and Burgdorfer, W. (1968). *J. Bacteriol.* **95,** 291–299.
Ploaie, P. G. (1971). *Rev. Roum. Biol., Ser. Bot.* **16,** 3–6.
Poulson, D. F., and Sakaguchi, B. (1961). *Science* **133,** 1489–1490.
Robertson, J., Gomersall, M., and Gill, P. (1972). *Can. J. Microbiol.* **18,** 1971–1972.
Saglio, P., Laflèche, D., Bonissol, C., and Bové, J. M. (1971). *C. R. Hebd. Seances Acad. Sci., Ser. D* **272,** 1387–1390.
Saglio, P., L'Hospital, M. Laflèche, D., Dupont, G., Bové, J. M., Tully, J. G., and Freundt, E. A. (1973). *Int. J. Syst. Bacteriol.* **23,** 191–204.
Swartzendruber, D. C., Clark, J., and Murphy, W. H. (1967). *Bacteriol. Proc.* **67,** 151.
Tully, J. G., Whitcomb, R. F., Williamson, D. L., and Clark, H. F. (1976a). *Nature (London)* **259,** 117–120.
Tully, J. G., Whitcomb, R. F., Clark, H. F., and Williamson, D. L. (1976b). *Science* **195,** 892–894.
Williamson, D. L., and Whitcomb, R. F. (1974). *Colloq. INSERM* **33,** 283–290.
Williamson, D. L., and Whitcomb, R. F. (1975). *Science* **188,** 1018–1020.
Zeigel, R. F., and Clark, H. F. (1974). *Infect. Immun.* **9,** 430–433.

Figs. 1–7. *Spiroplasmavirus-citri* 1 (SVCl) particles, from *Spiroplasma citri* strain Arizona 551.

Fig. 1. Cluster of uniform rod-shaped particles, and a few of double length, protruding from surface of fragmenting cell. An adsorbed particle of SVC2 is also present (arrow) (×100,000; bar = 100 nm). In this, and all subsequent electron micrographs of negatively-stained preparations, the stain is 2% ammonium molybdate.

Fig. 2. Clusters and individual particles, including one partially hollow (arrow), protruding from host cell surface. *ol* = outer layer of the spiroplasma. (×125,000; bar = 100 nm.) Negative stain.

Figs. 3 and 4. Free particles showing one rounded end and one with a flat plate or piece (arrows). A particulate, but not obviously helical, substructure can be seen. (×200,000; bar = 50 nm.) Negative stain.

Fig. 5. Particles attached to membrane fragment of disrupted cell. Note one hollow and rounded end (arrow) at membrane attachment, suggesting that the flat piece is a remnant of membrane or its outer layer. (×200,000; bar = 50 nm.) Negative stain. (Micrograph from Cole *et al.* 1974, by permission of the Department des Colloques et des Publications, INSERM, Paris.)

Figs. 6 and 7. Particle clusters protruding from membrane. Ultrathin section: in this and subsequent micrographs of sections, the material was fixed in glutaraldehyde and osmium, embedded in Epon, and stained with uranyl acetate and lead citrate. (×200,000; bar = 50 nm.) (Micrographs from Cole *et al.*, 1974, and by permission INSERM, as cited for Fig. 5.)

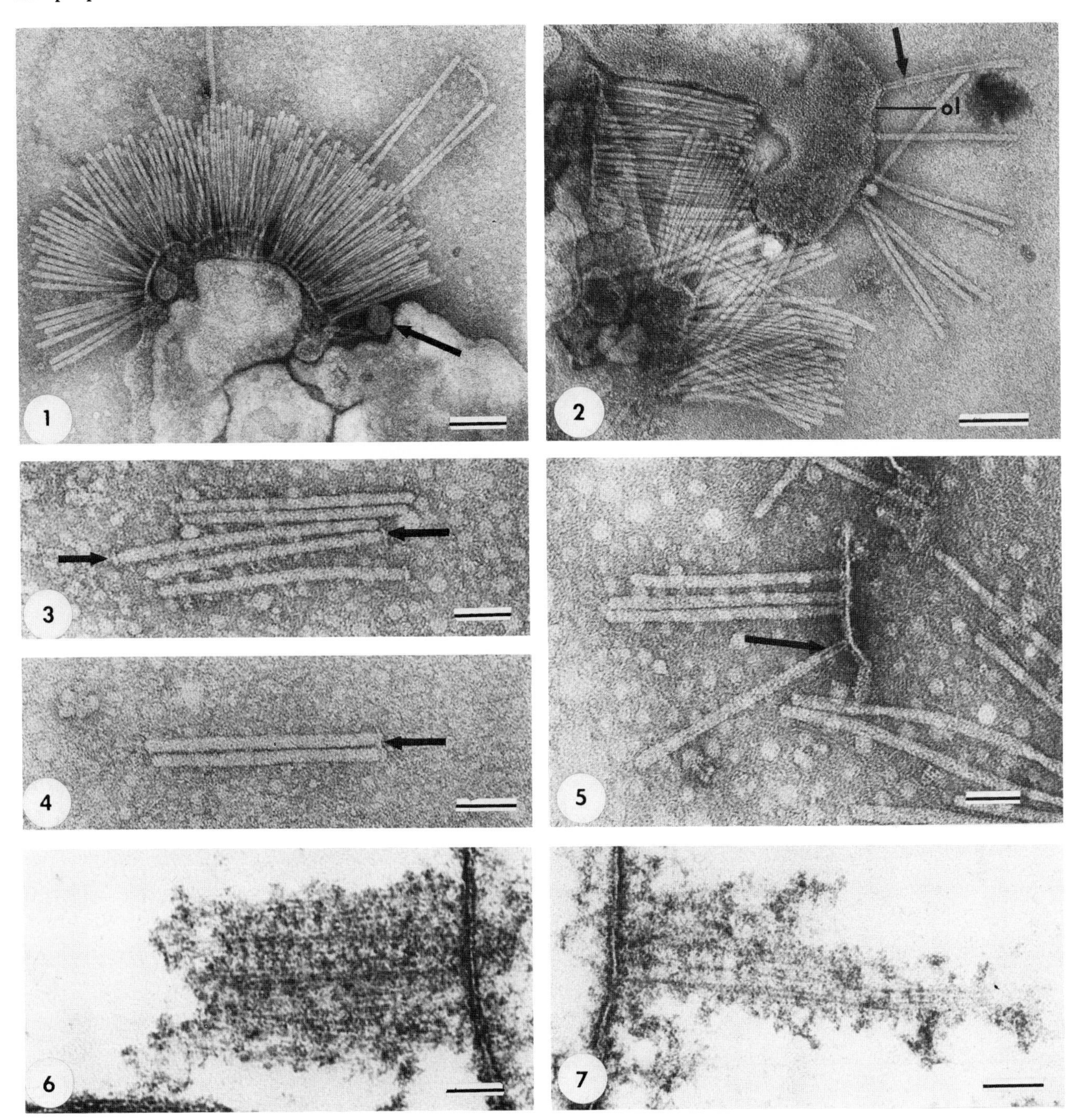

1
2
ol
3
4
5
6
7

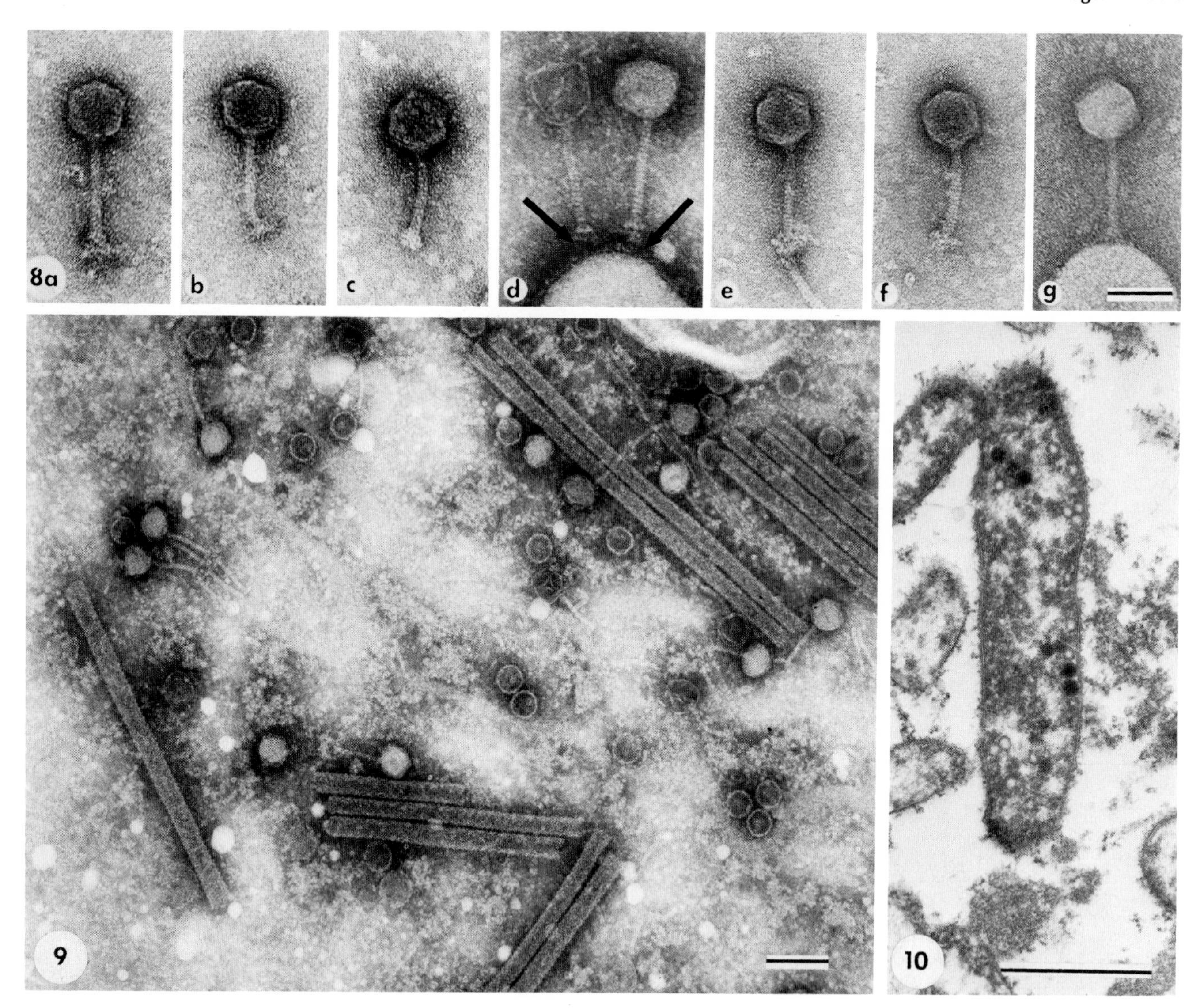

Figs 8–10. *Spiroplasmavirus-citri* 2 (SVC2) from *Spiroplasma citri* strain Morocco (except Fig. 8c, from strain Arizona 539).

Fig. 8. Individual particles free or adsorbed to host surface. Note polyhedral head, long noncontractile tail with particulate substructure, and base plate. Arrows indicate structure between base plate and membrane. (×200,000; bar = 50 nm.) Negative stain. Micrograph 8d from Cole *et al.*, 1973a, by permission American Society for Microbiology; others from Cole *et al.*, 1974, as in Fig. 5.

Fig. 9. Contents of disrupted cells, showing complete and empty tailed SVC2, empty spherical particles of slightly smaller size, free tails, and tubular structures resembling polyheads. (×100,000; bar = 100 nm.) Micrograph, in part, from Cole *et al.*, 1973a, as cited for Fig. 8. Negative stain.

Fig. 10. Thin section of *S. citri* filament containing empty and full heads of SVC2; tails are not discernable in the cytoplasm at any magnification. (×46,000; bar = 500 nm.) Micrograph from Cole *et al.*, 1973a, as above.

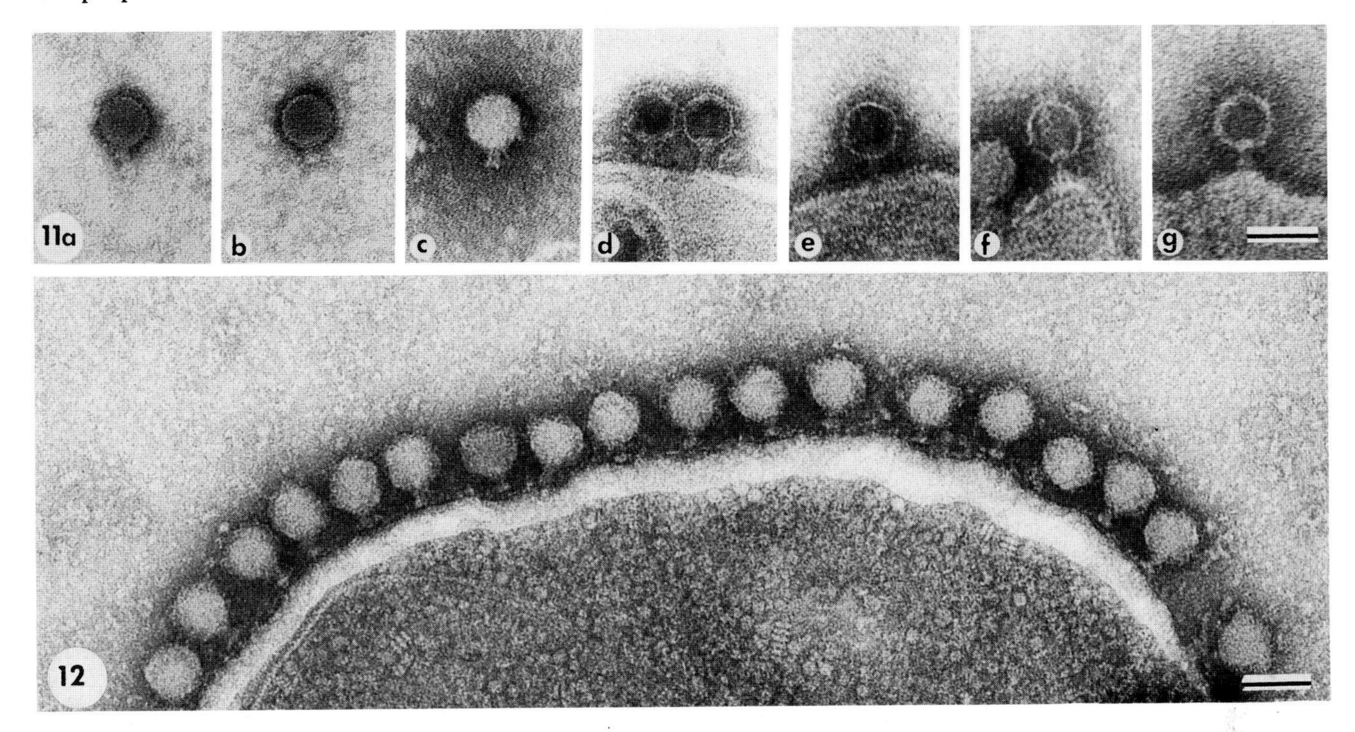

Figs. 11–18. *Spiroplasmavirus-citri* 3 (SVC3) in several host strains.

Fig. 11. Free and adsorbed particles, both complete and empty, showing spherical to polyhedral heads with subunit structure and short tails. All from *S. citri* strain Arizona 608. (×200,000; bar = 50 nm.) Micrographs (except d) from Cole *et al.*, 1974, as cited in Fig. 5. Negative stain.

Fig. 12. Complete particles adherent by tails to the cell of the suckling mouse cataract agent (SMCA), strain GT48. (×200,000; bar = 50 nm.) Negative stain.

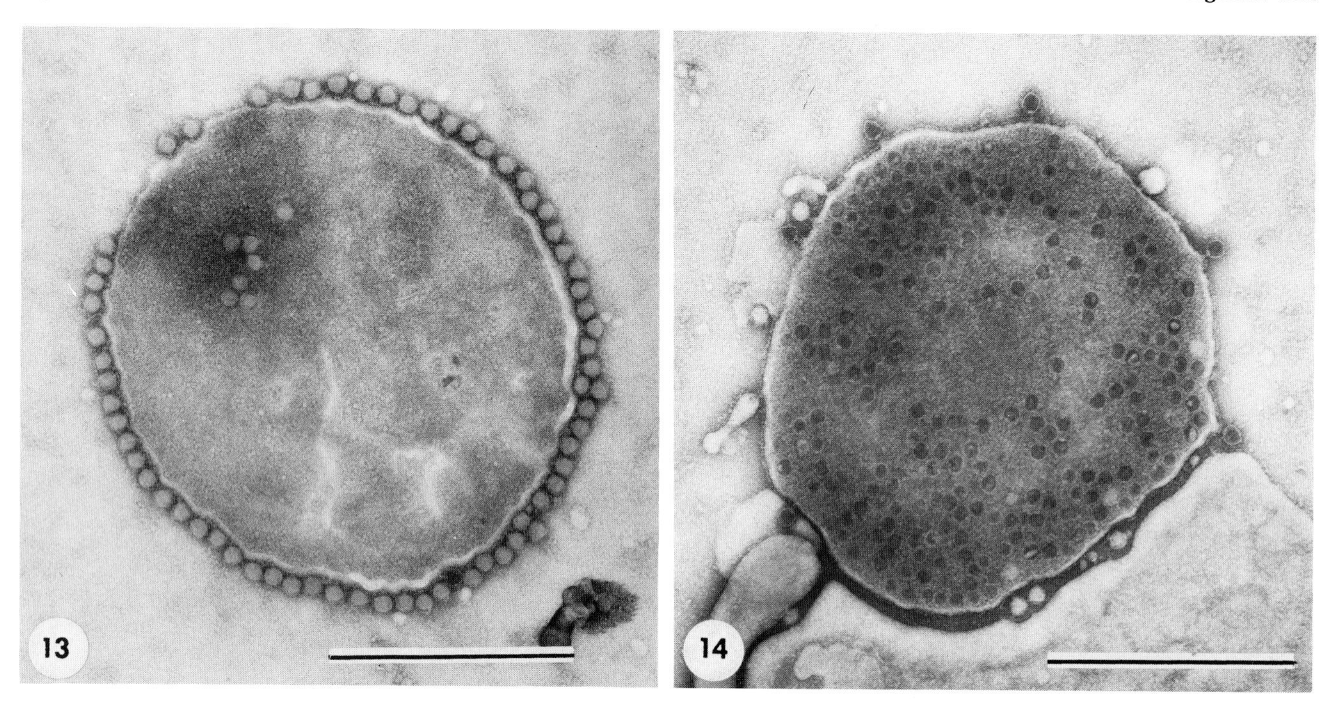

Figs. 13–18. Presumptive stages of SVC3 growth cycle.

Fig. 13. Adsorption of complete particles; SMCA strain GT48. (×68,000; bar = 500 nm.) Negative stain.

Fig. 14. Intracellular development; stage of incomplete (empty) heads in SMCA-GT48. Some other few particles are adsorbed to cell surface. (×68,000; bar = 500 nm.) Negative stain.

Fig. 15. Cell of SMCA, strain 3CA-2EP showing empty and complete intracellular particles as well as adsorption of particles released from other cells. (×68,000; bar = 500 nm.) Negative stain.

Fig. 16. Thin section of pellet of SMCA strain 3CA-2EP, showing heavy production of intracellular particles. (×14,000; bar = 1 μm.)

Fig. 17. Complete, and some empty, particles in characteristic membrane-bounded or "budding" emergence from cell of *S. citri* Arizona 608. Particles free of membrane are also present. The same appearance is seen in SVC3 infections of other spiroplasmas. (×69,000; bar = 500 nm.) Negative stain. Micrograph from Cole *et al.*, 1974, as cited in Fig. 5.

Fig. 18. Thin section of similar cell of *S. citri* Arizona 608 showing intracellular and membrane-bounded particles. (×69,000; bar = 500 nm.) Micrograph from Cole *et al.*, 1974 as cited in Fig. 5.

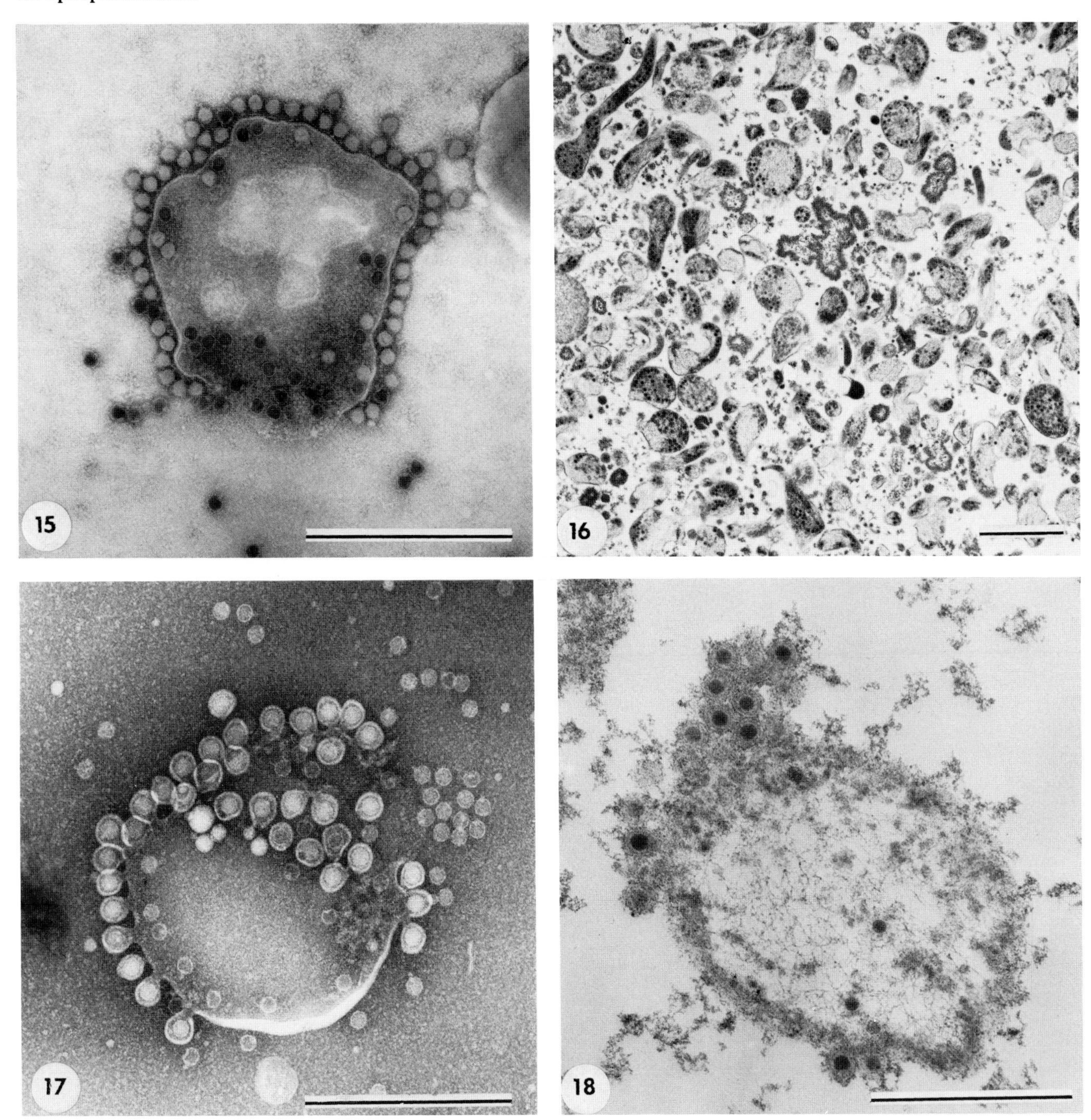
15
16
17
18

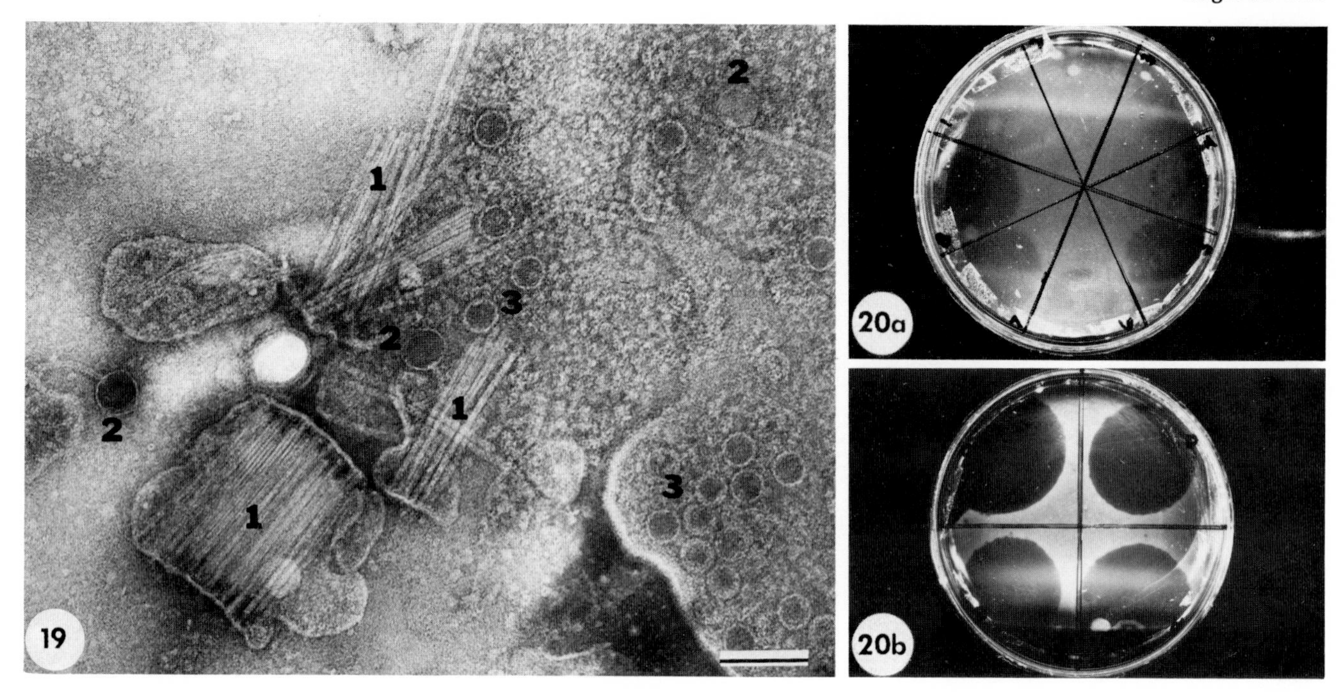

Fig. 19. Example of triple infection with SVC1, SVC2, and SVC3 illustratred by *S. citri* strain Arizona 551. (×120,000; bar = 100 nm.) Negative stain.

Fig. 20. Plaques produced on lawns of *S. citri* strain Arizona 750: (a) by tenfold dilution, counterclockwise from 9 o'clock, of culture supernatant of *S. citri* strain Arizona 608; (b) by undiluted supernates of *S. citri* Arizona 608 (top pair) and *S. citri* Arizona 803 (bottom pair).

Chapter 30

Viruses of *Drosophila* Sex-Ratio Spiroplasma

DAVID L. WILLIAMSON, KUGAO OISHI, AND
DONALD F. POULSON

I. Introduction 465
II. The Sex-Ratio Spiroplasma Viruses 466
III. Structure and Morphogenesis 467
A. Structure 467
B. Morphogenesis 467
References 468

I. INTRODUCTION

The maternally inherited infectious sex-ratio (SR) trait in *Drosophila* that results in the production of unisexual (female) progenies was first described by Malogolowkin (1958). This unusual condition occurs in natural populations of four closely related neotropical species of *Drosophila: D. equinoxialis, D. nebulosa, D. paulistorum,* and *D. willistoni*. Early in the study of the SR trait, Malogolowkin and Poulson (1957) demonstrated that ooplasm from eggs of the SR females could be injected into normal females and, after a short delay, only daughters would appear in their progenies. Those daughters having acquired the SR trait through injection of ooplasm were able to transmit the trait to their daughters, thus making it possible to establish new SR strains artificially through injection.

Poulson and Sakaguchi (1961) studied various adult tissues and organs for their infectivity, and hemolymph was found to be the most efficient in inducing the SR trait in newly infected females. Phase microscopic examination of the hemolymph of SR females revealed the presence of large numbers of spirally shaped organisms not unlike certain spirochetes in their appearance (Poulson and Sakaguchi, 1961). A direct correlation was made between the SR trait and the appearance of the sex-ratio organism (SRO). Hemolymph was then used as a source of the SRO and the SR trait has been artificially transferred to species and strains of *Drosophilia* other than those in which it occurs naturally (Poulson and Sakaguchi, 1961; Sakaguchi and Poulson, 1963; Williamson, 1965, 1966, 1969). In every case the SRO in the new host behaves in basically the same manner as in the donor host, i.e., unisexual (female) progeny and maternal inheritance.

Darkfield microscopy of freshly drawn hemolymph from SR females and electron microscopic studies of fixed and negatively stained preparations of the SRO have shown that it is indeed very similar to spirochetes in its gross morphology (Williamson and Whitcomb, 1974). These helically shaped organisms from each of the four species of *Drosophila* in which they occur naturally are morphologically identical, being 4–8 μm in length and 0.15–0.20 μm in diameter. However, electron microscopic studies of thin sections of pellets of the SRO revealed that two components which characterize spirochetes, namely, the axial filament and the outer wall or envelope, are not present in the SRO. For several years then, the SRO was considered to be some unusual form of spirochete. But with the report of Davis *et al.* (1972) on helical filaments of a mycoplasma-like organism associated with corn stunt disease and the reports of Saglio *et al.* (1973) and Cole *et al.* (1973a) on helical filaments associated with a disease of citrus plants, called citrus stubborn, it seemed more likely that the SRO could also belong to this newly described group of wall-free prokaryotes called "spiroplasma" (Davis and Worley, 1973). The relationship of the SRO to the spiroplasmas is based on their morphological similarities in electron microscopic studies of both negatively stained whole organisms and in sections of pelleted organisms, and on the deformation test in which antibody against one of the spiroplasmas, the corn stunt organism, causes total deformation of the SRO into spherical forms (Williamson and Whitcomb, 1975). However, the reciprocal test, antibody against hemolymph-derived SRO's tested against CSO's, does not result in deformation of the CSO. The reason for this one-way deformation test is not understood and the precise relationship between the SRO and the spiroplasmas must await the development of a suitable medium for *in vitro* culture of the SRO.

Recently, Tully *et al.* (1976) have described still another spiroplasma, the "suckling mouse cataract agent" (SMCA). This agent was originally isolated from a pooled extract of rabbit ticks (*Haemaphysalis leporispalustris* collected near Atlanta, Georgia (Clark, 1964). The agent grows to high titer in the eyes and brain of intracerebrally inoculated newborn mice in which it induces cataract, uveitis, and chronic brain infection. SMCA has also been grown to high titer in embryonated hen's eggs, in which it produces a lethal infection in 4–9 days (Clark, 1964). The citrus organism, the CSO, and the SMCA can all be cultivated in both liquid and solid media. Only the citrus stubborn agent has thus far been fully characterized and given a scientific name, *Spiroplasma citri* (Saglio *et al.*, 1973).

All the spiroplasmas which have been studied thus far have particles associated with them which resemble known viruses. Cole *et al.* (1973a,b, 1974) have described a filamentous or rod-like virus, and two tailed bacteriophages present in cultures of *S. citri*. Tailed bacteriophages have also been observed in cultures of both the corn stunt spiroplasma and the suckling mouse cataract agent (R. M. Cole, personal communication, see Chapter 29).

II. THE SEX-RATIO SPIROPLASMA VIRUSES

In the SRO, virus-like particles were first observed in negatively stained preparations of the SRO derived from hemolymph obtained from flies transmitting the *D. nebulosa* SRO (D. L. Williamson, unpublished). Oishi and Poulson

(1970) showed that extracts of flies carrying the *D. nebulosa* SRO, heated and centrifuged to remove the SRO and injected into *Drosophila* females carrying the *D. willistoni* SRO, could cause the lysis of the latter SRO and a subsequent return of a normal number of males to their progeny, i.e., curing the SR condition. Oishi (1971) demonstrated that a virus is also present in association with the *D. willistoni* SRO. The lysis of the *D. nebulosa* SRO by the *D. willistoni* SR virus does not result in restoration of males to the progeny but does eliminate transmission of the SR condition to daughters. There is evidence that both *D. equinoxialis* and *D. paulistorum* SRO's also have associated viruses (Oishi, 1970; Poulson and Oishi, 1975).

The studies of Oishi (1970) and Oishi and Poulson (1970) showed that the *D. nebulosa* sex-ratio virus is a DNA-containing virus, with a buoyant density of 1.480 in CsCl. This value is very close to the density of *Mycoplasma laidlawii* virus, MVL3 (1.477 in CsCl), as reported by Garwes *et al.* (1975). MVL3 has been shown to contain double-stranded DNA (Garwes *et al.*, 1975), but the strandedness of the SRO virus is still unknown.

The events occurring in cross infection of *D. willistoni* SRO by *D. nebulosa* SR virus leading to the subsequent destruction or lysis of the *D. willistoni* SRO and the release of mature virus particles have not been studied in full detail. The present report will be concerned primarily with the structure of the virus derived directly from the *D. nebulosa* SRO as seen in hemolymph suspensions of infected flies since this virus exists in a manifest form both as free particles in the hemolymph and can also be seen within the body of the SRO itself.

III. STRUCTURE AND MORPHOGENESIS

A. Structure

Hemolymph from *D. nebulosa* containing the SRO and free viruses was placed directly into 2% glutaraldehyde in 0.149 *M* NaCl in 0.05 *M* Tris-maleate buffer (pH 7.2). Droplets of the suspension were picked up on formvar-coated copper grids and negatively stained with 1% phosphotungstic acid, pH 7.2 with 5 *N* potassium hydroxide. Examples of the only free viruses found in such preparations are shown in Fig. 1. This virus, which Oishi and Poulson (1970) called spv-1 is a short-tailed bacteriophage-type virus. It has a polyhedral head of hexagonal outline which measures 35–45 nm from side to side and 35–45 nm from vertex to vertex. The tail, which arises from one vertex, as seen in these preparations, is 10–12 nm in length and 7–9 nm in width. It has a wider tip, suggesting a base plate with spikes and tail fibers, but definitive resolution of these details will require the preparation and examination of highly purified virus.

The spv-1 virus is similar in size and morphology to the SVC3 virus from *S. citri* (Cole *et al.*, 1974, and see Chapter 29) and the Mycoplasma virus (MVL3) first reported by Gourlay and Wyld (1973) and later described as to its morphology by Gourlay (1974). SVC3 is 35–37 nm from side to side and 37–44 nm from vertex to vertex, with a tail length of 13–18 nm (Cole *et al.*, 1974). MVL3 is 57 nm wide and 61 nm long, with a tail 25 nm long (Gourlay, 1974).

B. Morphogenesis

The sequence of events leading to the production of mature virus particles is just beginning to be elucidated, and only for the *D. nebulosa* SRO virus (spv-1). The studies of Oishi (1970) and Oishi and Poulson (1970) have clearly shown that a purified preparation of this virus is capable of infecting, growing, and lysing the *D. willistoni* SRO. The mechanisms of attachment and infection are unknown but presumably involve the use of the tail appendages. Whether spv-1 can, under normal conditions, infect *D. nebulosa* SRO's is not known, but virus particles have not yet been observed attached to the surface of any *D. nebulosa* SRO prepared for negative staining and electron microscopic examination.

Oishi and Poulson (1970) called attention to the fact that spv-1 particles are extruded from infected SRO's in the form of membrane-coated buds. This same type of budding has also been reported to be characteristic of the SVC3 virus of *S. citri* (Cole *et al.*, 1974) and for the viruses in the corn stunt organism and the suckling mouse cataract agent (Cole, Chapter 29). Figures 2 and 3 show negatively stained (phosphotungstic acid) SRO's from *D. nebulosa* displaying buds. Although buds are usually observed to occur at the tips of the SRO, budding may also occur at other points along the SRO as is evidenced by an apparent early-stage bud in a lateral position (see arrow) in the SRO shown in Fig. 2. Viral particles can be seen through the membrane of the bud, and the SRO in Fig. 2 appears to be completely filled with viruses. Multiple buds may also occur, as dramatically shown by the SRO in Fig. 3. The budding stage of viral reproduction is assumed to be followed by a complete loss of the helical morphology of the SRO which is then followed by lysis and release of the viral particles (Fig. 4). It is not known how the membrane-bound particles become freed of host membrane.

Viral particles within the body of the SRO and within buds are also seen in thin sections (Figs. 5 and 6). However, other than the nucleoid, there are no structural details of the nucleocapsid visible in any of the sections.

ACKNOWLEDGMENTS

This work was supported in part by Grant AI-10950 from the National Institutes of Health, U.S. Public Health Service.

REFERENCES

Clark, H. F. (1964). *J. Infect. Dis.* **114,** 476–487.
Cole, R. M., Tully, J. G., and Popkin, T. J. (1973a). *Ann. N. Y. Acad. Sci.* **225,** 471–493.
Cole, R. M., Tully, J. G., Popkin, T. J., and Bové, J. M. (1973b). *J. Bacteriol.* **115,** 367–386.
Cole, R. M., Tully, J. G., and Popkin, T. J. (1974). *INSERM* **33,** 125–132.
Davis, R. E., and Worley, J. F. (1973). *Phytopathology* **63,** 403–408.
Davis, R. E., Worley, J. F., Whitcomb, R. F., Ishijima, T., and Steere, R. L. (1972). *Science* **176,** 521–523.
Garwes, D. J., Pike, B. V., Wyld, S. G., Pocock, D. H., and Gourlay, R. N. (1975). *J. Gen. Virol.* **29,** 11–24.
Gourlay, R. N. (1974). *Crit. Rev. Microbiol.* **3,** 315–331.
Gourlay, R. N., and Wyld, S. G. (1973). *J. Gen. Virol.* **19,** 279–283.
Malogolowkin, C. (1958). *Genetics* **43,** 274–286.
Malogolowkin, C., and Poulson, D. F. (1957). *Science* **126,** 32.
Oishi, K. (1970). Thesis, Yale University, New Haven, Connecticut.
Oishi, K. (1971). *Genet. Res.* **18,** 45–56.
Oishi, K., and Poulson, D. F. (1970). *Proc. Natl. Acad. Sci. U.S.A.* **67,** 1565–1572.
Poulson, D. F., and Oishi, K. (1975). *Int. Virol.* **3,** 145 (abstr.).
Poulson, D. F., and Sakaguchi, B. (1961). *Science* **133,** 1489–1490.
Saglio, P., Lhospital, M., Laflèche, D., Dupont, G., Bové, J. M., Tully, J. G., and Freundt, E. A., (1973). *Int. J. Syst. Bacteriol.* **23,** 191–204.
Sakaguchi, B., and Poulson, D. F. (1963). *Genetics* **48,** 841–861.

Tully, J. G., Whitcomb, R. F., Williamson, D. L., and Clark, H. F. (1976). *Nature (London)* **259,** 117–120.
Williamson, D. L. (1965). *J. Invertebr. Pathol.* **7,** 493–501.
Williamson, D. L. (1966). *J. Exp. Zool.* **161,** 425–430.
Williamson, D. L. (1969). *Jpn. J. Genet.* **44,** Suppl. 1, 36–41.
Williamson, D. L., and Whitcomb, R. F. (1974). *Colloq. INSERM* **33,** 283–290.
Williamson, D. L., and Whitcomb, R. F. (1975). *Science,* **188,** 1018–1020.

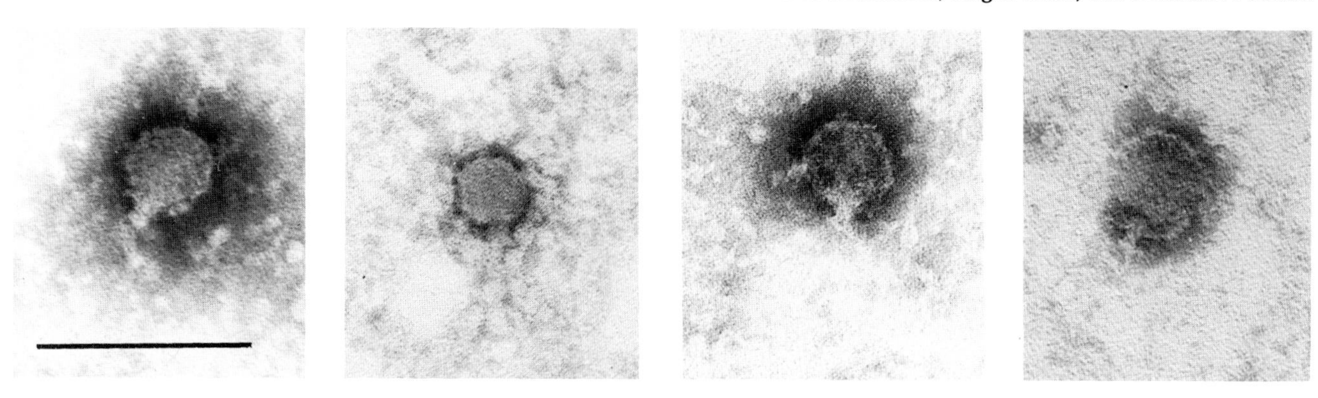

Fig. 1. Free virus particles from *D. nebulosa* SRO. Hemolymph suspension fixed in 2% glutaraldehyde. Preparation stained 1% phosphotungstic acid. Bar = 100 nm.

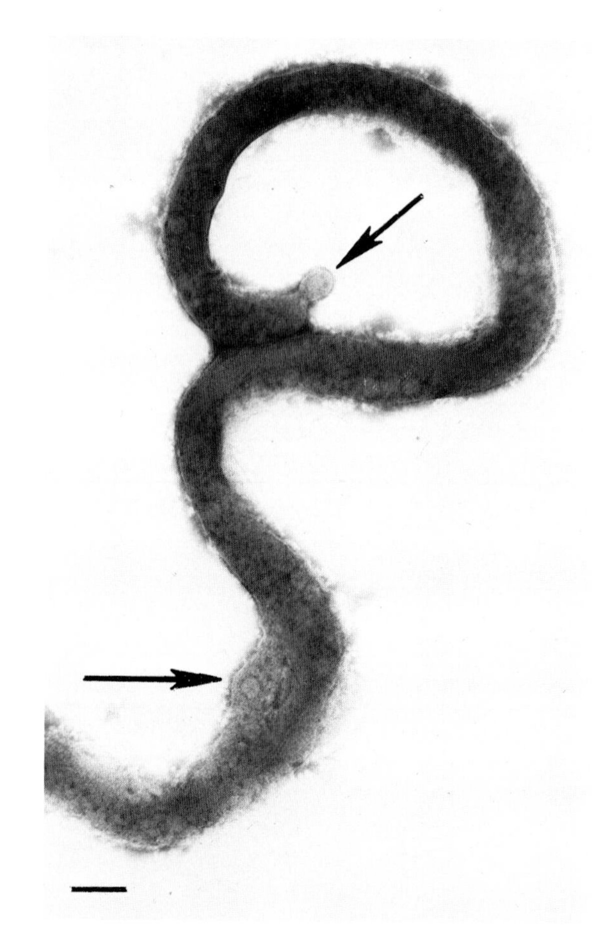

Fig. 2. A portion of an SRO filled with virus particles. Bud formation shown by terminal and lateral buds (see arrows). Viral particles visible through membrane of buds. Same preparation as for Fig. 1. Bar = 100 nm.

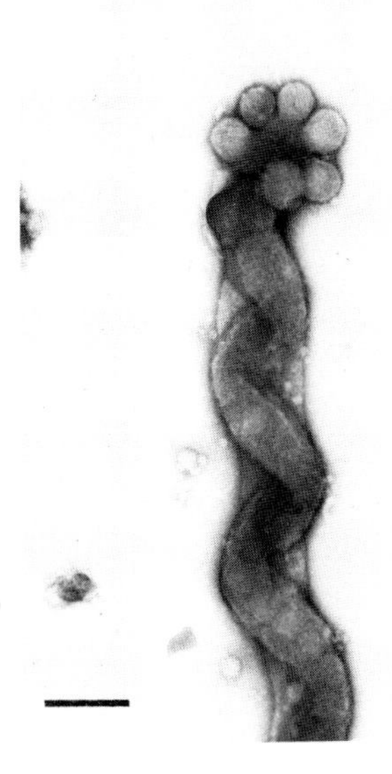

Fig. 3. An SRO showing multiple bud formation. Bar = 100 nm.

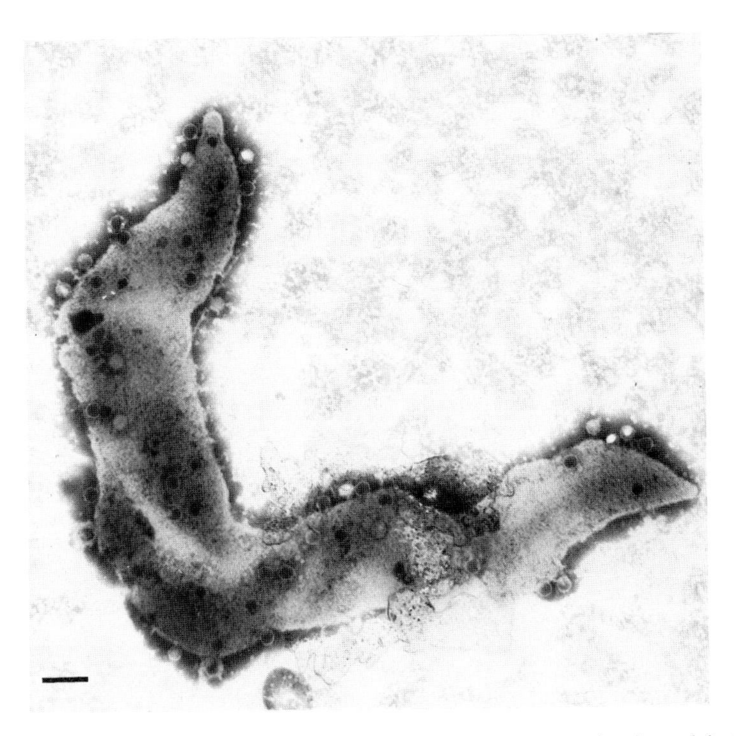

Fig. 4. An SRO in a late stage of viral reproduction. Note the complete loss of the helical nature of the SRO and the presence of particles on its surface. Bar = 100 nm.

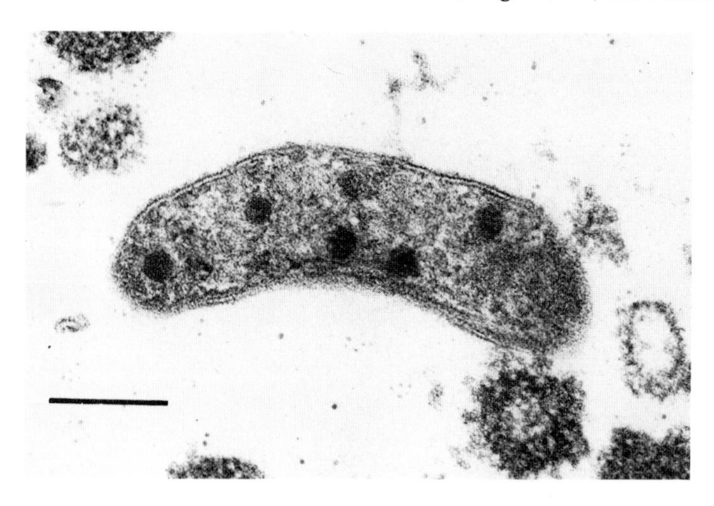

Fig. 5. Thin section of SRO showing intracellular particles. Only the nucleoid is visible; phage tails and other components are not visible in these preparations. Hemolymph suspension prefixed in 2% glutaraldehyde and postfixed in 2% osmium tetroxide. Sections stained in 2% uranyl acetate and Reynold's lead citrate. Bar = 100 nm.

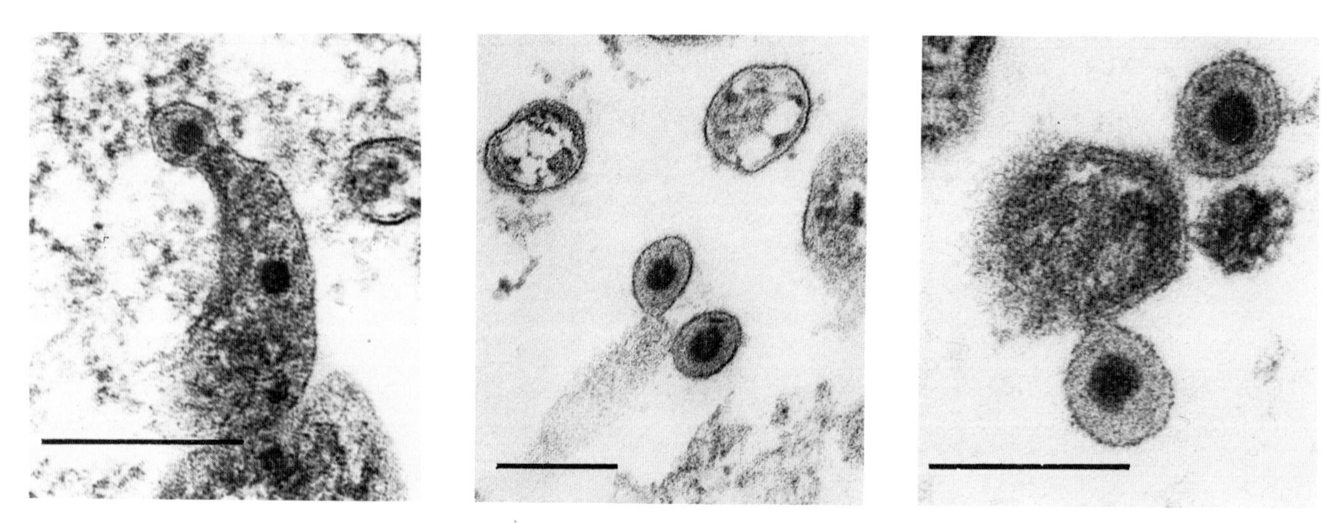

Fig. 6. Thin sections of SRO's showing terminal bud formation. Hemolymph suspension prefixed in 2% glutaraldehyde and postfixed in 2% osmium tetroxide. Sections stained in 2% uranyl acetate and Reynold's lead citrate. Bar = 100 nm.

Index*

A

Acholeplasma laidlawii, viruses of, 451
Acute bee-paralysis virus, 143
Adeno-associated virus, 68, 69, 70, 71, *see also,* Densonucleosis viruses (Parvoviridae)
Aedes densonucleosis virus, 67
American wheat striate mosaic virus, 182
Amsacta moorei, as entomopoxvirus, 32
Andean potato latent virus, as tymovirus group member, 349
Anomala cuprea, as entomopoxvirus, 32
Aphodius tasmaniae, as entomopoxvirus, 32
Apis iridescent virus, 144–145
Apple mosaic virus, as ilarvirus group, subgroup B member, 371
Arabis mosaic virus, as nepovirus group member, 222
Arkansas bee virus, 143
Artichoke curley dwarf virus, as potexvirus group member, 333
Artichoke Italian latent virus, as nepovirus group varient, 222
Artichoke mottled crinkle virus, as tombus virus group member, 258
Atropa belladonna virus, 183

B

Baculovirus (Baculoviridae), 3–27
  chemical properties, 5
  DNA in, 6–7
  genomes, 6–7
  granulosis virus, 3
  host cell interactions, 7–8, 9
  nuclearpolyhedrosis virus, 3
  nucleocapsids, 6
    role in infection, 7
  occlusion of, 4–5
  proteinic crystal, 4–5
  serology of, 9
  structure, 4–7
  surface membrane, 4–5
    viral envelope association, 8
  ultrastructure, *12–27*
Barley yellow dwarf virus, as luteovirus group member, 363
Barley yellow striate mosaic virus, 182
Bean pod mottle virus, as comovirus group member, 167
Bee viruses, 141–156
  ultrastructure, *146–156*
Bee virus X, 143–144
Beet ringspot virus, as nepovirus group varient, 222
Belladonna mottle virus, as tymovirus group member, 349
Black queen-cell virus, 144
Boletus edulis, filamentous particles from as possible potexvirus group member, 332
Broad bean mottle virus, as bromovirus group member, 287
Broad bean stain virus, as comovirus group member, 167
Broad bean true mosaic virus, as comovirus group member, 167
Broccoli necrotic yellows virus, 182
Brome mosaic virus, as bromovirus group member, 287
Bromovirus group (brome mosaic virus), 287–309
  amino acids in, 288
  morphology of, 290–291
  physical properties, 287–289
  replication, 290
  RNA in, 288–289, 290
  transmission of, 289–290
  ultrastructure, *292–302*

C

Cactus virus X, as potexvirus group member, 332

* Page numbers in italic refer to ultrastructure micrographs of virus.

Cacau yellow mosaic virus, as tymovirus group member, 349
Carnation bacilliform virus, 184
Carnation etched ring virus, as caulimovirus group member, 159
Carnation Italian ringspot virus, as tombusvirus group member, 258
Carnation mottle virus, as possible tombusvirus group member, 264
Carnation ringspot virus, as possible tombusvirus group member, 264
Camptochironomus tentans, as entomopoxvirus, 33
Cassava common mosaic virus, as potexvirus group member, 332
Cauliflower mosaic virus (caulimovirus group), 159–166
  biological properties, 159–161
  DNA in, 159, 161–162
  host cell interaction, 162–163
  physical and chemical properties, 161–162
  transmission, 160–161
  ultrastructure, *165–166*
Caulimovirus group, *see* Cauliflower mosaic virus
Celery yellow vein virus, as nepovirus group varient, 222
Cereal tillering disease virus, as plant reovirus group member, 377
Cherry leafroll virus, as nepovirus group member, 221, 222
Chicory chlorotic ringspot virus, as nepovirus group varient, 222
Chironomus attenuatus, as entomopoxvirus, 33
Chironomus iridescent virus, as iridovirus member, 93–94
Chironomus luridus, as entomopoxvirus, 33
Choristoneura biennis, as entomopoxvirus, 32
Choristoneura conflictana, as entomopoxvirus, 32
Choristoneura diversana, as entomopoxvirus, 32
Chronic bee-paralysis virus, 141–142
  associate, 142
Chrysanthemum chlorotic mottle viroid, 433
Chrysanthemum stunt viroid, 432
Citrus exocortis viroid, 432
Clover enation virus, 183
Clover yellow mosaic virus, as potexvirus group member, 332
CO_2 sensitivity, in sigma virus infection, 131–132
Cocoa necrosis virus, as nepovirus group varient, 222
Colocasia esculenta virus, 182
Comoviruses, 167–179
  genetic properties, 169–170
  host cell interaction, 170–171
  morphology, 169–170
  replication, 171–172
  RNA in, 167, 168, 171–172
  ultrastructure, *174–179*
Cowpea chlorotic mottle virus, as bromovirus group member, 287
Cowpea mosaic virus, as comovirus group member, 167–179
Cryptograms of densonucleosis virus (Parvoviridae), 68
  of entomopoxvirus, 30
  of nepovirus group members, 222
  of potato virus X group, 331
  of tobravirus group, 215
  of tombusvirus group, 257
  of watermelon mosaic virus group, 323
Cucumber green mottle mosaic virus, as tabamovirus group member, 237, 238
Cucumber mosaic virus, as cucmovirus group member, 303, 304
Cucumber necrosis virus, as possible tombusvirus group member, 264
Cucumber pale fruit viroid, 432–433
Cucumovirus group (cucumber mosaic virus), 303–309
  RNA in, 304
  structure and morphogenesis, 305–306
  ultrastructure, *308–309*
Cymbidium mosaic virus, as potexvirus group member, 332
Cymbidium ringspot virus, as possible tombusvirus group member, 265
Cynara rhabdovirus, 183
Cytoplasmic polyhedrosis viruses, 105–129
  DNA in, 107–109
    as virus type determinent, 106
  polyhedra of, 106
  replication, 108, 109–111
  ultrastructure, *112–129*
  viral particles
    biochemical properties, 107–109
    morphology of, 107

D

Dahlia mosaic virus, as caulimovirus group member, 159
Demodena boranensis, as entomopoxvirus, 32
Dendrobium and *Phalaenopsis* hybred virus, 182
Densonucleosis viruses (Parvoviridae), 67–91
  biological properties of, 73–74
  cryptogram of, 68
  DNA in, 68, 69
  morphology, 72–73
  proteins in, 69–71
  replication and morphogenesis, 74–75
  ultrastructure, *78–91*
  virions, 71–72
Deoxyribonucleic acid, *see* DNA
Dermolepida alborhirtum, as entomopoxvirus, 32
Desmodium yellow mosaic virus, as tymovirus group member, 349
Dioscorea latent virus, as potexvirus group member, 332
DNA (Deoxyribonucleic acid) in baculovirus, 6–7
  in cauliflower mosaic virus, 159, 161–162
  in densonucleosis virus (parvoviridae), 68–69
  in entomopoxvirus, 30–31
    in infection of, 34
  in iridovirus, 93
  role in viriod replication, 434–435

Drosophila, viruses of sex-ratio spiroplasma, 465–472
Dulcamera mottle virus, as tymovirus group member, 349

E

Eggplant mosaic virus, as tymovirus group member, 349
Eggplant mottled dwarf virus, 183
Elderberry latent virus, as possible tombusvirus group member, 264
Elm mosaic virus, as nepovirus group member, 221, 222
Entomopoxvirus (Poxvirus of Invertebrates), 29–66
  characteristics, 32–33
  cross-transmission, 35
  cryptogram of, 30
  enzymatic activity in, 31
  host cell interaction, 35
  inclusion bodies
    speroids, 31
    spindles, 34
  maturation, 35–37
  morphology of, 35–37
  pathology of, 34–35
  ultrastructure, *40–66*
  virion structure, 30–31
Euonymus fasciations virus, 183
Euxoa auxiliaris, as entomopoxvirus, 32

F

Figulus sublaevis, as entomopoxvirus, 32
Fungus
  pinwheel inclusions, role in control of, 422
  viruses infecting, *see* Mycoviruses

G

Geotrupes silvaticus, as entomopoxvirus, 32
Goeldichironomus holoprasinus, as entomopoxvirus, 32
Golden elderberry virus, as nepovirus group member, 221, 222
Gomphrena virus, 183
Grapevine chrome mosaic virus, as nepovirus group varient, 222
Grapevine fanleaf virus, as nepovirus group varient, 222
Grapevine yellow mosaic virus, as nepovirus group varient, 222
Grape yellow vein virus, as nepovirus group varient, 222

H

H-1 virus, 69, 70, 71
Helper virus in parvovirus replication, 68
Hippeastrum latent virus, as potexvirus group member, 333
Hop line pattern virus, as nepovirus group varient, 222
Hydrangea ringspot virus, as potexvirus group member, 332

I

Ilarvirus group, subgroup B, 371–375
  RNA in, 371–372
  structure, 372
  ultrastructure, *374–375*
Iridovirus (iridoviridae), 93–103
  host cell relationships, 94–95
  ultrastructure, *96–103*
  virions, 93–94

J

Junonia densonucleosis virus, 67

K

Kashmir bee virus, 144
Kilham rat virus, 69, 70, 71

L

Laburnum yellow vein virus, 183
Lettuce necrotic yellows virus, 182, 188
Lettuce ringspot virus, as nepovirus group varient, 222
Lucerne enation virus, 183
Luteovirus group (barley yellow dwarf virus), 363–369
  host cell interactions, 365
  properties, 364–365
  RNA in, 364
  transmission, 363–364
  ultrastructure, *366–369*
  virion, 365

M

Maize mosaic virus, 182
Maize rough dwarf virus, as plant reovirus group member, 377
Malva veinal necrosis virus, as potexvirus group member, 332
Melanoplus sanguinipes, as entomopoxvirus, 33
Melitotus latent virus, 183
Melolontha melolontha, as entomopoxvirus, 32
Melon variegation virus, 183
Mirabilis mosaic virus, as caulimovirus group member, 159
Mollicutes, class, *see* Mycoplasmas
Mycoplasmas, 439–441
  properties of viruses of, 440
  viral relationships, 440–441

Mycoplasmaviruses, 451–452
  group 1
    occurrence, 441
    properties, 442
    replication, 442–443
    ultrastructure, 441–442, *447–448*
  group 2
    host restriction, 444
    occurrence, 443
    properties, 444
    replication, 444
    ultrastructure, 443–444, *449*
  group 3
    occurrence, 444
    properties, 445
    ultrastructure, 444–445, *450*
Mycoviruses, 421–430
  biological implications of, 421–422
  physicochemical properties, 422–423
  RNA in, 422–423
  ultrastructure, 423–424, *426–430*

N

Narcissus mosaic virus, as potexvirus group member, 332
Necrotic ringspot virus, as ilarvirus group, subgroup B member, 371
Negro coffee mosaic virus, as potexvirus group member, 333
Nepovirus group (tobacco ringspot virus), 221–235
  composition and structure, 222–224
  cryptograms of members and varients, 222
  host cell interaction, 224–225
  RNA in, 222–224
  ultrastructure, *226–235*
Nerine virus X, as potexvirus group member, 332
Northern cereal mosaic virus, 182

O

Oat sterile dwarf virus, as plant reovirus group member, 377
Odontoglossum ringspot virus, as tabamovirus group member, 237, 238
Okra mosaic virus, as tymovirus group member, 349
Ononis yellow mosaic virus, as tymovirus group member, 349
Operophtera brumata, as entomopoxvirus, 32
Orchids virus, 182, 192–193
Oreopsyche angustella, as entomopoxvirus, 32
Othnonius batesi, as entomopoxvirus, 32

P

Pangola stunt virus, as plant reovirus group member, 377
Papaya mosaic virus, as potexvirus group member, 332
Parsley rhabdovirus, 183
Parsley virus 5, as potexvirus group member, 332
Parsnip virus 3, as potexvirus group type, 332
Parvoviruses (Parvoviridae), 67–91, *see also* Densonucleosis viruses
Pea early-browning virus, as tobravirus group, 215
Peach rosette mosaic virus, as nepovirus group varient, 222
Peach yellow bud mosaic virus, as nepovirus group varient, 222
Peanut stunt virus, as cucumovirus group member, 305
Pelargonium flower-break virus, as possible tombusvirus group member, 264
Pelargonium leaf curl virus, as tombusvirus group member, 258
Petunia asteroid mosaic virus, as tombusvirus group member, 258
Phyllopertha horticola, as entomopoxvirus, 32
Pigeon pea proliferation virus, 184
Pineapple chlorotic leaf streak virus, 182
Pinwheel inclusions and plant viruses, 405–417
  morphogenesis
    formation, 407
    transformation, 407
  RNA in, 422–423
  serology of, 408
  structure, 406, 407–408, 422–423
  ultrastructure, *410–417*, 423, *426–430*
Pittosporum vein yellowing virus, 184
Plant rhabdovirus group, *see* Rhabdoviruses of plants
Plant viruses
  pinwheel inclusions of, 405–417
Plantain virus, 183
Plant reovirus group, 377–403
  host cell interaction, 379–380
  RNA in, 377–378
  transmission and replication, 377, 380
  ultrastructure, *382–403*
  virion structure, 378–379
Pleuropneumonia-like organisms, *see* Mycoplasma
Potato aucuba mosaic virus, as potexvirus group member, 333
Potato bouquet virus, as nepovirus group varient, 222
Potato pseudo-aucuba virus, as nepovirus group varient, 222
Potato spindle tuber viroid, 432
Potato virus X, as potexvirus group member, 332
Potato virus Y, as potyvirus group type, 311
Potato yellow draft virus, 182
Potato yellow drawf virus, 188
Potexvirus group (potato virus X), 331–345
  host cell interreactions, 336–337
  RNA in, 335
  serological relationships, 339
  structure, 335
  ultrastructure, *340–345*
Potyvirus group (potato virus Y), 311–321

Polyvirus group
inclusions in host cells, 313–314
chemical composition, 312
morphogenesis, 313
morphology, 313
pinwheel inclusions of, 406
RNA in, 312, 313
serology of, 312
transmission of, 312
ultrastructure, 313, *316–321*
watermelon mosaic virus as member, 323
Poxvirus of invertebrates, *see* Entomopoxvirus
Proteinic crystal
of baculovirus, 4–5
Prune dwarf virus, as ilarvirus group, subgroup B member, 371
Prunus necrotic ringspot, as ilarvirus group, subgroup B member, 371

R

Radish mosaic virus, as comovirus group member, 167
Raspberry Scottish leafcurl virus, as nepovirus group varient, 222
Raspberry vein chlorosis, 183
Red clover mosaic virus, 183
Red currant ringspot virus, as nepovirus group varient, 222
Reovirus of plants, *see* Plant reovirus group
Raspberry ringspot virus, as nepovirus group member, 222
Raspberry yellow dwarf virus, as nepovirus group varient, 222
Rhabdoviruses of insects, 131, *see also* Sigma virus of *Drosophila*
Rhabdoviruses of plants, 181–213
host cell interactions
appearance and distribution, 191
morphogenesis, 191–192
hydrodynamic properties, 190
interaction of other viruses, 192–194
nucleocapsid, 188–190
RNA in, 181, 188–189
serological properties, 190
surface projections, 187–188
transmission, 185, 186
ultrastructure, *197*–213
viral envelope, 188
virion, morphology and structure, 186–187
Rhubarb mosaic virus, as nepovirus group varient, 222
Rhubarb virus I and II, as potexvirus group members, 332
Ribgrass mosaic virus, as tabamovirus group member, 237, 238
Ribonucleic acid, *see* RNA
Rice black-streaked dwarf virus, as plant reovirus group member, 377
Rice dwarf virus, as plant reovirus group member, 377
Rice transitory yellowing virus, 182
RNA (ribonucleic acid)
in baculovirus, 5
in bromovirus, 288–289, 290
in comoviruses, 167, 168, 171–172
in cucumovirus group, 304
in entomopoxvirus, 31–32, 34
in ilarvirus group, subgroup B, 371–372
in luteovirus group, 364
in mycoviruses, 422–423
in nepovirus group, 222–224
in pinwheel inclusions, 422–423
in potexvirus group, 335
in potyvirus group, 312, 313
in rhabdoviruses, 181, 188–189
in rhabdoviruses of insects, 131
in tobacco necrosis virus group, 284
in tombusvirus group, 258, 259, 260
in tobravirus group, 215–216
in tymovirus group, 347–348
in viroids, 431, 433–434
Russian winter wheat mosaic virus, 182
Ryegrass bacilliform virus, 182

S

Sacbrood virus, 142–143
Saintpaulia leaf necrosis virus, 183
Sammon's *Opuntia* virus, as possible tabamovirus member, 237, 238
Sarracenia purpurea virus, 183
Satellite virus, activation of, 282
Scrophularia mottle virus, as tymovirus group member, 349
Sericesthis iridescent virus, as iridovirus member, 93–94
Sex-ratio virus of *Drosophila*, 465–472, *see also* Spiroplasmaviruses
Sigma virus of *Drosophila*, 131–139
CO_2 sensitivity from, 131–132
morphogenesis of, 133–134
morphology, 132–133
transmission, 132
ultrastructure, *134–139*
Slow bee-paralysis virus, 144
Sonchus rhabdovirus, 184
Sowbane mosaic virus, as possible tombusvirus group member, 264
Sowthistle yellow net virus, 183
Sowthistle yellow vein virus, 183, 188
Spindle disease, entomopoxvirus in, 29, 34
Spiroplasmas, 440
as virus hosts, 452–453
Spiroplasmaviruses, 451–464
-citri 1, 453
-citri 2, 453–454
-citri 3, 454–455
distribution, 455
properties, 455–456
sex-ratio changes in *Drosophila*, 465–472
structure and morphogenesis, 467–468
ultrastructure, *470–472*
ultrastructure, *458–464*
Squash mosaic virus, as comovirus group member, 167
Strawberry crinkle virus, 183

Strawberry latent ringspot virus, as nepovirus group member, 222
Strawberry vein banding virus, as caulimovirus group member, 159
Stunting of *Chondrilla juncea* virus, 183
Sugarbeet leaf curl virus, 183
Sugarcane Fiji disease virus, as plant reovirus group member, 377
Sunn hemp mosaic virus, as tabamovirus group member, 237, 238

T

Tabamovirus group (tobacco mosaic virus), 237–255
  host cell interaction, 239–241
  RNA in, 237–239
  ultrastructure, *243–255*
  virions, 237–239
Thyme virus, 183
Tipula iridescent virus, as iridovirus member, 93–94
Tobacco mosaic virus, as tabamovirus group member, 237, 238
Tobacco necrosis virus group, 281–285
  inactivation of
    by heat, 282
    by ultraviolet radiation, 282–283
  protein in, 283
  RNA in, 284
  satellite virus, activation of, 281–282
  serology, 281–282
  transmission, 282
  ultrastructure, *285*
  unstable varients of, 283
  viral particles, 283
Tobacco rattle virus, as tobravirus group member, 215–219
Tobacco ringspot virus, as nepovirus group prototype, 221, 222
Tobravirus group (tobacco rattle virus), 215–219
  cryptogram of, 215
  RNA in, 215–216
  structure and composition
    genomes, 216
    particles, 215–216
  ultrastructure, *218–219*
Tomato aspermy virus, as cucumovirus group type, 304–305
Tomato black ring virus, as nepovirus group member, 222
Tomato bushy stunt virus, as tombusvirus group member, 258
Tomato ringspot virus, as nepovirus group member, 222
Tomato top necrosis virus, as nepovirus group varient, 222
Tombusvirus group (tomato bushy stunt virus), 257–279
  host cell interaction
    appearance and distribution, 262
    chloroplasts, 262
    intracellular inclusions, 263–264
    mitochondria, 262–263
    nuclei, 263
  hydrodynamic properties, 260
  protein shell, 259–260
  RNA in, 258, 259, 260
    protein interactions, 260
  serological properties, 261
  ultrastructure, *267–279*
  viral interrelationships, 264–265
  virions, 259
Turnip crinkle virus, as possible tombusvirus group member, 257, 264–265
Turnip yellow mosaic virus, as tymovirus group member, 347, 349
Tymovirus group (turnip yellow mosaic virus), 347–361
  empty protein shells as characteristic of, 351, 352
  members, 348, 349
  replication, 351–352
  RNA in, 347–348
  serological relationships, *354–355*
  structure
    coat protein, 348, 350
    polyamines, 348–349
  transmission, 347–348
  ultrastructure, *354–361*

V

Vesicular stomatitis virus, as rhabdovirus of insects prototype, 131
Viroids, 431–436
  host cell interaction, 434
  replication, 434
  structure, 433–434
  ultrastructure, *436*

W

Watermelon mosaic virus group, 323–329
  biological properties, 324
  cryptogram of, 323
  morphology, 324
  ultrastructure, *326–329*
Wheat chlorotic streak virus, 182
White clover mosaic virus, as potexvirus group member, 332
Wild cucumber mosaic virus, as tymovirus group member, 349
Wound tumor virus, as plant reovirus group member, 377

Z

Zygocactus virus, as potexvirus group member, 333

A 7
B 8
C 9
D 0
E 1
F 2
G 3
H 4
I 5
J 6

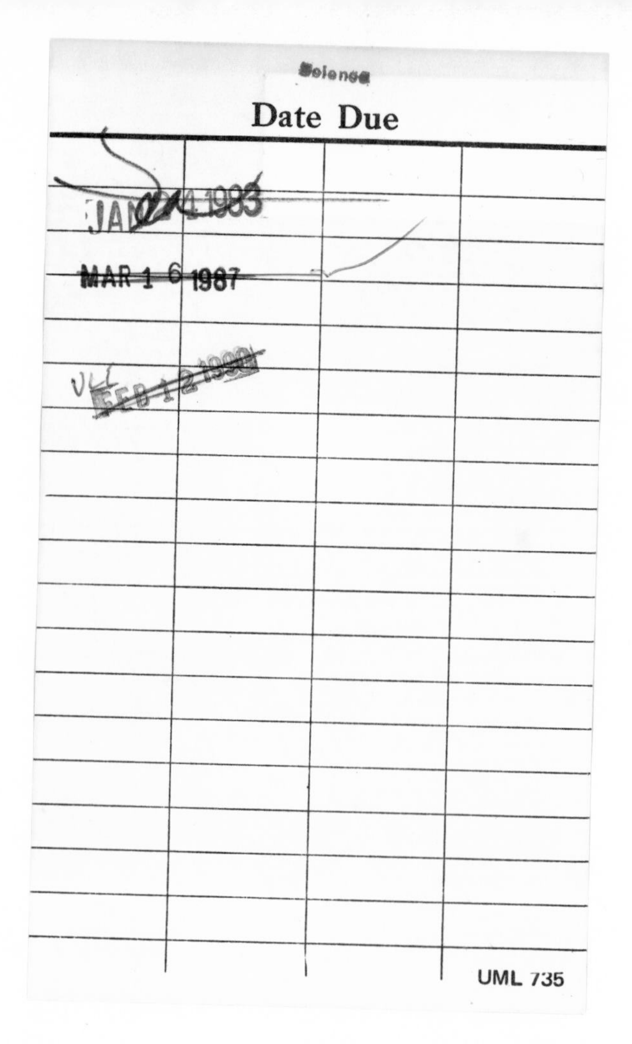

Science
Date Due
JAN 4 1983
MAR 1 6 1987
FEB 12 1990
UML 735